PROCEEDINGS OF THE 2016 INTERNATIONAL CONFERENCE ON ADVANCES IN ENERGY AND ENVIRONMENT RESEARCH (ICAEER 2016), GUANGZHOU, CHINA, 12–14 AUGUST 2016

# Advances in Energy and Environment Research

*Editors*

Bachir Achour
*Department of Hydraulic and Civil Engineering, University of Biskra, Algeria*

Qiyan Wu
*Nanjing Normal University, China*

CRC Press
Taylor & Francis Group
Boca Raton London New York

CRC Press is an imprint of the
Taylor & Francis Group, an **informa** business

A BALKEMA BOOK

CRC Press
Taylor & Francis Group
6000 Broken Sound Parkway NW, Suite 300
Boca Raton, FL 33487-2742

First issued in paperback 2020

ISBN-13: 978-1-138-62682-9 (hbk)
ISBN-13: 978-0-367-73646-0 (pbk)

Typeset by V Publishing Solutions Pvt Ltd., Chennai, India

**Visit the Taylor & Francis Web site at**
**http://www.taylorandfrancis.com**

**and the CRC Press Web site at**
**http://www.crcpress.com**

*Advances in Energy and Environment Research – Achour & Wu (Eds)*
© 2017 Taylor & Francis Group, London, ISBN 978-1-138-62682-9

# Table of contents

Preface     ix

ICAEER 2016 Organizing Committees     xi

*Energy science and energy technology*

A flashover voltage forecasting model of a polluted UHV insulator based on an RBF network     3
*Y. Shi, G. Hu & Y. Chen*

Exploration and discussion on electricity consumption demand forecasting
theories and methods     7
*F.X. Ma*

Load prediction based on support vector regression with genetic algorithms     17
*M. Zou*

Short-term electricity demand forecasting by using a MATLAB-based ANFIS model     23
*A. Balin, M. Sert & I. Besirli*

Use of concentrating solar technology on short solar chimney power plant     27
*O. Motsamai & K. Kebaitse*

LiDFOB in sulfolane-carbonate solvents for high-voltage lithium-ion batteries     33
*L. Dong, L.C. Huang, F.X. Liang & C.H. Li*

The depth load cycling operation characteristic and energy-saving optimization control system
of a power plant     37
*N. Zhao & Y.M. Chen*

Fabrication and mechanical properties of zirconium–yttrium hydrides     43
*J.Q. Peng, Y. Chen, G.Q. Yan, M. Wu, J.D. Zhang, L.J. Wang, J.S. Li & Z.C. Guo*

Carbon paper electrode coated with $TiO_2$ nanorod arrays for electro-assisted photocatalytic
degradation of Cr (VI) and methylene blue     49
*H. Li, Z. Lin & D. Fu*

Inversion of the sound velocity profile in a multibeam survey based on the simulated
annealing algorithm     53
*Z.W. Zhang, J.Y. Bao & F.M. Xiao*

Thermoelectric cooling in downhole measuring tools     61
*R. Weerasinghe*

Measurement and analysis of SH-CCT curves of HD15Ni1MnMoNbCu steel     67
*J.L. Zhang, G.G. Wang, F.Y. Zhang, Z.B. Chen, S.H. Yin & Y.S. Lv*

Research on the interface of a digital measurement device in a smart substation
and its calibration method     71
*H. Chen, H.Y. Chen, L.M. Shen, R. Huang, F.S. Chen, L.M. Bu, X. He, K. Li,
X.Y. Xu & W.N. Wang*

Textured silicon wafer prepared by reactive ion etching for anti-reflection in Si solar cells     75
*X.G. Zhang & M.Z. Shao*

*Environmental science and environmental engineering*

Parametric analysis of a single basin solar still with a point-focus Fresnel lens in Shenzhen 81
*J. Dong, X.H. Xu & S.Y. Zhang*

China's rare earth markets: Value chain and the implications 85
*B.L. Zhou, Z.X. Li, Y.Q. Zhao & C.C. Chen*

Research on the sensitivity of the marine industry to climate change 91
*H. Kong & W. Yang*

Research on the efficiency of regional environmental pollution abatement with consideration of centralized disposal pollution source in China—an analysis based on a model of undesirable outputs 97
*F.B. Kong & Y.J. Yuan*

Preparation of phosphonic acid-functionalized silica magnetic microspheres for uranium(VI) adsorption from aqueous solutions 103
*L.M. Zhou, H.B. Zou, J.Y. Jin, Z.R. Liu & T.A. Luo*

A new scenario for enhancing phosphorus removal in $SBR_s$-AA and $SBR_s$-AO systems during the winter season 109
*S.A.-A. Al-Khalil & A.Z. Hamody*

Chaotic analysis of metallogenic element concentration 115
*L. Wan, X.Y. Xiong & J.J. Lai*

Test study of the degradation ability of the coated sludge microbe-nanometer iron system on perchlorate 119
*Y.J. Tong, B.G. Liu & T.L. Li*

Assessment of the impact of cropland loss on bioenergy potential: The case of Yangtze River Delta, China 123
*F.S. Pei, X.P. Liu, J. Guo, G.R. Xia & C.J. Wu*

The sixth technological revolution in construction industry: Noospheric paths 129
*M.V. Shubenkov, S.D. Mityagin & Z.A. Gaevskaya*

Database design of the Qingyi River basin for multi-objective and multi-department integrated management 135
*Z.H. Tian, J.J. Peng & X.L. Wang*

The design and implementation of the Qingyi River water environment basic information system 141
*H.L. Guo, Y.Y. Wei & X.H. He*

Flotation performance measurement and the DFT study of salicylhydroxamic acid as a collector in niobite flotation 147
*H. Ren, Z.H. Wang, L.F. Hu, F.F. Ji, Y.F. Zhang & T.Z. Huang*

Study of the preparation of a magnetic recyclable emulsifier 151
*H. Ren, Z.H. Wang, F.F. Ji, Z. Li, Y.F. Zhang & T.Z. Huang*

Analysis of the variation of porosity and permeability in different periods of Lamadian Oilfield 155
*L.P. Wang*

Research on greenhouse gas carbon emission reduction in the public bicycle service system 159
*D.-X. Cheng, Y.-N. Huang & J. Chen*

Three mining ideas put forward by the Chinese and their typical applications in the early 21st century 165
*Q.-F. Chen, Z. Yu, Q.-Y. Zhong, C.-H. Feng & W.-J. Niu*

Study on the application of fire-resistant cloth to prevent fire spreading between buildings 169
*X.S. Wu, Z.G. Song & G.Y. Yu*

Study on coal-bed methane isothermal adsorption at different temperatures based on the monolayer adsorption theory 175
*T.T. Cai, D. Zhao & Z.C. Feng*

Evaluation of carbonate reservoir fracture on Shun Nan district in the northern slope
of the central Tarim area                                                                                 183
*B. Zhao, Y.D. Geng, Y. Wang, P.D. Luo, M.H. Duan & X.W. Lu*

Structure and performance of *in situ* $ZrH_{1.8}$ coatings in amounts of urea                            189
*G.Q. Yan, Y. Chen, M. Wu, J.Q. Peng, J.D. Zhang & L.J. Wang*

Glucose biosensor based on glucose oxidase immobilized on branched $TiO_2$ nanorod arrays                 195
*S. Liu & D. Fu*

Hydrothermal fabrication of hierarchically boron- and lanthanum-modified $TiO_2$ nanowire
arrays on FTO glasses                                                                                     199
*W. Zhang & D. Fu*

Spatio-temporal distribution of the heavy metals pollution in the sediments of Minjiang Estuary           203
*Q.Q. Sun, M.X. Luo & H.D. Ji*

Carrying capacity and ecological footprint of Taiwan                                                      207
*Y.-J. Lee, C.-M. Tung & S.-C. Lin*

Water quality index of various Kuwaiti seas, and methods of purification                                  219
*K. Chen*

Identification of bacterial species in Kuwaiti waters through DNA sequencing                              227
*K. Chen*

Topographic feature line extraction from point cloud based on SSV and HC-Laplacian smoothing              235
*W. Zhou, R.C. Peng, L.H. Zhang & W.J. Wang*

Multi-scale representation of a digital depth model based on the positive direction rolling
ball transform algorithm                                                                                  243
*J. Dong, R.C. Peng, L.H. Zhang & J.S. Yuan*

Tectono-stratigraphy evolution of XuanGuang Basin (Anhui, East China) and its implication
for shale gas exploration                                                                                 253
*J.H. Chen*

The main factors analyzing the effect of microstructure cell on the mechanical
properties of L-CFRP                                                                                      257
*Y.Y. Yao, X. Wang & R.S. Dou*

Structural characteristics of genomic islands associated with *RlmH* genes in *Bacillus*                  263
*L.W. Su, X. Peng, Q.Z. Zha, S.D. He, M.J. Yang & L. Song*

Study of the preparation and application of modified luffa sponges in the pollutant
removal process of the eco-pond-ditch system                                                              269
*L.W. Kong, R.W. Mei, S.P. Cheng, Y. Zhang, D. Xu, L. Wang & X. Cui*

Spatio-temporal distribution of $PM_{2.5}$ concentration and its correlation with environmental
factors in different activity spaces of an urban park                                                     275
*H. Xu, J.H. Hao, H. Li & J.J. Zhao*

Effect of leachate concentration on pollutant disposal ratio and electricity production
characteristics by a bio-cathode MFC                                                                      279
*J.F. Hu, L.J. Xu, Q. Jing, D.W. Qing & M. Xie*

Policy orientation and pattern path of sustainable development of modern agriculture
in Suzhou City                                                                                            285
*G.S. Ma & J. Chen*

A study on marketization of land element allocation in Beijing                                            291
*Y.F. Yang & J.Q. Xie*

*Motivation, automation and electrical engineering*

Design and implementation of a $PM_{2.5}$ remote sensing monitoring system based on Hadoop                297
*L. Wang, S. Xu, F.B. Zheng, Y.D. Si & Q. Ge*

A novel design of the MANFIS-PSO PID controller for quadrotors 305
C.-C. Ko, C.-N. Huang, D.T.W. Lin & J.-K. Kuo

A mechanism study of power grid ancillary service cost change by intermittent power supply
incorporated into the power grid 309
K.J. Wang, G.Y. Wu, Z.D. Du, D.N. Liu & Q. Li

A probabilistic evaluation method of energy-saving benefit for daily generation based
on the Monte Carlo stochastic simulation method and wind power and load uncertainties 315
L. Guo, J.M. Wang, C. Ma, C.C. Gao, D.N. Liu & M. Yang

Research and design of excavation face remote reproduction and remote control systems 323
C. Liu, J.G. Jiang, W.H. Qu, Z.Z. Zhou, J.J. Li, H. Liu, S. Ye & S.Y. Yao

An automatic electricity sales quota allocation algorithm applied in national companies
with branches 329
S.Q. He & S.Y. Wu

Cooling of oil-filled power equipment 335
O.S. Dmitrieva & A.V. Dmitriev

Integration of SolidWorks and Simulink models of an open-closed cycle air-fueled Stirling engine 339
A. Rehman, W.Q. Xu, F. Zakir & M.L. Cai

3D surface MRS method based on variable geometry inversion 345
Y. Zhao, J. Lin & C.D. Jiang

Hierarchy and development framework of military requirement system 351
J. Ren & X.L. Zheng

Research on the WRSPM reference model for the development of system requirements 357
X.L. Zheng & J. Ren

Study and simulation of titanium pipe deformation in the multi-pass backward
spinning process 361
Z.Y. Xue, Y.J. Ren, Y. Ren & W.B. Luo

Nonlinear mechanical analysis for contact between the main shaft and the supporting wheel
in a large mine hoist 365
J.-C. Li, J.-C. Cao, J. Hu & X. He

Dynamic analysis and optimization of the extracting cartridge case system of Breechblock 371
Y.M. Zhu & L.Y. Ni

Anvil configuration optimization of an underwater pile driving hammer 375
L.Q. Wang & D.H. Chen

Study on thermo-mechanical coupling characteristics for linear rolling guideway 381
Y.M. Zhang, S.X. Jia & X.Z. Zhu

Effect of superfine grinding on the physicochemical properties of Moringa leaf powder 389
Y.P. Cao, N. Liu, W. Zhou, J.H. Li & L.J. Lin

Analysis and discussion of a household heat pump dryer 393
Z. Yan & L.X. Hou

Fano resonance based on the coupling effect in a metal–insulator–metal nanostructure 397
X.G. Zhang & M.Z. Shao

Design of a smart grid dispatching operation analysis system based on data mining 401
J. Li, K.P. Wang, Y.L. Yu, H.D. Tan & Y. Tang

Author index 407

*Advances in Energy and Environment Research – Achour & Wu (Eds)*
*© 2017 Taylor & Francis Group, London, ISBN 978-1-138-62682-9*

# Preface

The 2016 International Conference on Advances in Energy and Environment Research (ICAEER 2016) was held on August 12–14, 2016 in Guangzhou, China. We have invited scholars and researchers from all over the world to present the conference.

ICAEER 2016 attaches importance to presenting and publishing novel and fundamental advances in energy and environment research fields. Scholars and specialists on ICAEER 2016 will share their knowledge and amazing research. During the conference, an international stage is prepared for the participants to present their study theories. They will also find practical applications for these theories in related fields.

In order to organize ICAEER 2016, we have sent our invitation to scholars and researchers from all around the world. Eventually, over 200 submissions were submitted to us. These papers have gone through a strict reviewing process performed by our international reviewers. All the submissions were double-blind reviewed, both the reviewers and the authors remaining anonymous. First, all the submissions were divided into several chapters according to the topics, and the information of the authors, including name, affiliation, email and so on, removed. Then the editors assigned the submissions to reviewers according to their research interests. Each submission was reviewed by two reviewers. The review results should be sent to chairs on time. If two reviewers had conflicting opinions, the paper would be transmitted to the third reviewer assigned by the chairs. Only papers which were approved by all reviewers were accepted for publication.

With the hard work from the reviewers, only 71 papers were finally accepted for publication. These papers were divided into three chapters:

- Energy Science and Energy Technology
- Environmental Science and Environmental Engineering
- Motivation, Automation and Electrical Engineering

To prepare ICAEER 2016, we have received a lot help from a lot of people.

We thank all the contributors for their interest and support to ICAEER 2016. We also feel honored to have the support from our international reviewers and committee members. Moreover, the support from CRC Press/Balkema (Taylor & Francis Group) is also deeply appreciated; without their effort, this book will not be able to come into being.

# ICAEER 2016 Organizing Committees

## CONFERENCE CHAIRS

Dr. Arefieva Olga, *Far Eastern Federal University, Russian Federation*
Prof. Chung-Neng Huang, *National University of Tainan, Taiwan*
Prof. Maria Moshkevich, *Southwest State University (Kursk), Russia*
Prof. Vladimir Badenko, *Peter the Great Saint-Petersburg Polytechnic University, Russia*

## TECHNICAL PROGRAM COMMITTEES

A. Prof. Md. Didarul Islam, *The Petroleum Institute, United Arab Emirates*
Prof. Zhanfeng Dong, *Chinese Academy for Environmental Planning, China*
Prof. Chih-Huang Weng, *I-Shou University, Taiwan*
Assistant Prof. Cheung-Chieh Ku, *National Taiwan Ocean University, Taiwan*
Prof. Hongwei Wang, *Hebei University, China*
Prof. Cheng-Fu Yang, *National University of Kaohsiung, Taiwan*
Prof. Fengwu Bai, *The Institute of Electrical Engineering Chinese Academy of Sciences, China*
Dr. Amimul Ahsan, *Universiti Putra Malaysia (UPM), Malaysia*
Dr. Naichuk Anatoli, *Branch of RUE "Institute BelNIIS"—Scientific and technical center, Belarus*
Prof. Xiangbin Zeng, *Huazhong University of Science and Technology, China*
Dr. Tongzhen Wei, *Institute of Electrical Engineering, Chinese Academy of Sciences, China*
Prof. Jianhua Chen, *Chinese Research Acaemy of Environmental Sciences, China*
Prof. Bing Chen, *Xi'an Shiyou University, China*
Prof. Jingpei Cao, *China University of Mining & Technology, China*
Prof. Chen Chen, *Jilin University, China*
Dr. Baixin Chen, *Heriot Watt University, UK*
Prof. Jianhua Chen, *Chinese Research Academy of Environmental Sciences, China*
Prof. Petros Gikas, *Technical University of Crete, Greece*
Prof. Dingsheng Chen, *South China Institute of Environmental Science (SCIES), Ministry of Environmental Protection (MEP), China*
Prof. Dr.-Ing. Joachim Burghartz, *Institute for Microelektronics Stuttgart (IMS), Germany*
Prof. Hsin-ping Chen, *National Chengchi University, Taiwan*
Dr. J.P. Keshri, *Delhi College of Engineering, Delhi*
Prof. A. M. El-Tamimi, *King Saud University, Saudi Arabia*
Prof. T.R. Abdurashidov, *Tashkent Chemical Technological Institute, Russia*
Prof. Adel Sharif, *University of Surrey, UK*
Prof. Ahmad Fauzi Mohd Noor, *Universiti Sains Malaysia, Malaysia*
Prof. Ali Hussain Reshak, *South Bohemia University, Republic Czech*
Dr. Qing Fu, *Sun Yat-sen University, China*
Prof. Ying Fu, *University of Jinan, China*
Dr. Pankaj Sharma, *Chungnam National University, Republic of Korea*
Dr. Rongbing Fu, *Shanghai Academy of Environmental Sciences, China*
Dr. Vikneswaran Munikanan, *National Defense University of Malaysia, Malaysia*
Prof. Seokcheon Lee, *Purdue University, USA*
Prof. Sebastian Saniuk, *University of Zielona Góra, Poland*
Prof. Saad Abu-Alhail, *University of Basrah, IRAQ*
Ph.D. Jozef Martinka, *Slovak University of Technology in Bratislava, Slovak Republic*

*Energy science and energy technology*

*Advances in Energy and Environment Research – Achour & Wu (Eds)*
*© 2017 Taylor & Francis Group, London, ISBN 978-1-138-62682-9*

# A flashover voltage forecasting model of a polluted UHV insulator based on an RBF network

Yan Shi & Gang Hu
*Chongqing University of Science and Technology, Chongqing, China*

Yong Chen
*State Grid Electric Power Research Institute, Wuhan City, Hubei Province, China*

ABSTRACT:   According to the test result of a class insulator string of 48 units, a method of calculating 50% AC flashover voltage $U_{50}$ based on an RBF network is put forward in this paper. Inputs of the model are equivalent to the salt deposit density $\rho_{ESDD}$ and non-soluble deposit density $\rho_{NSDD}$ and the output of the model is $U_{50}$ of polluted insulator strings; the orthogonal least squares algorithm was used to select right hide layer neurons and center vectors and the weight from hide layer neurons to output layer neurons is determined by using the pseud-inverse method. The calculating result is in concordance with the test result. It is indicated that the model can give some references to assess the pollution conditions of external insulation.

## 1   INTRODUCTION

The technical feasibility of UHV systems has been demonstrated by many investigations during the last 40 years, but there are many problems that remain unsolved. Some countries need not develop a UHV power system because of their geographical feature, but there are some countries that still have a strong interest in UHV. Many high voltage laboratories and companies in the world have been studying on UHV's technical aspects, such as SGE-PRI of China, IREQ of Canada, CESI of Italy, STRI of Sweden, EPRI of U.S., etc.

The distribution of resource and energy load centers in the geographical region is very uneven in China, the conventional energy sources used in power generation, such as hydroelectricity and coal resources, are mainly distributed in the western region. More than 90% of the coal and hydro power resources are concentrated in the western region, but the center of power load is located in the developed southeast coastal area. In order to satisfy the requirements for the development of national economy, strategies of the Development of West China and the West-to-East Power Transmission have been implemented in China; several 1000 kV AC transmission lines are to be constructed from the west to east of China, so as to guide the effective development of large scale coal electricity, hydroelectricity and nuclear power resources, and then realize resources optimization disposition nationwide. The external insulation of these lines will be polluted by industrial and natural contaminants and there are different pollution features of every line because of special geographical conditions in there areas. Contamination flashover performance is a key factor in the external insulation design of UHV transmission lines. Although this issue has been widely studied in China and other countries, there are many problems that require investigation.

This is the first time that AC UHV transmission lines are constructed in the world and there are few references for selection and design of external insulation and the operational experience of test lines is only about a few years in China. In order to provide a technology reference for construction of UHV transmission lines, in this paper, a series of AC artificial pollution tests were conducted on an EHV porcelain insulator string of 48 units. The relationships between flashover voltages of polluted insulators and $\rho_{ESDD}$ and $\rho_{NSDD}$ were studied; the radial basis function network is selected to build the calculation model of the insulator's flashover voltage. By means of the Orthogonal Least Squares (OLS) algorithm, right hide layer neurons and center vectors were chosen and the weight from hide layer neurons to output layer neurons are determined by using the pseud-inverse method. The results could be used as references to exterminate the contamination conditions of insulators based on flashover voltage calculation.

## 2 TEST METHOD

The specimen insulators used are normal porcelain insulators. The technical parameters and profiles of the samples are shown in Figure 1 and Table 1, in which H is the configuration height, L is the leakage distance, and D is the shed diameter.

Before tests, all samples were carefully cleaned so that all traces of dirt and grease were removed and dried naturally in a laboratory. After preparation, the solid layer method was used to pollute the insulator surface, in which sodium chloride and diatomite were conductive and inert materials, respectively. Sodium chloride and diatomite were mixed with a small amount of water uniformly. And then, the sample was polluted and suspended vertically on the hook of an artificial climate chamber; polluted insulators would dry naturally in about 24 h.

After the polluted insulators were dried, the insulator string of 48 units was hanged and wetted by steam fog in an artificial fog chamber. And then, the flashover test began as soon as the pollution layers on insulators surfaces had been completely wetted for 10–15 min. AC voltage was applied to the insulators and it rose until the occurrence of flashover. Three strings of each type of insulator and 4–5 times of flashover tests per string were carried out at one pollution degree, in which 4–5 flashover voltages, deviating from their average lower than 10% were valid values. The flashover voltage $U_f$ of polluted insulators under a given pollution degree was the average of all measured flashover voltages.

Figure 1. Picture showing configuration of insulators.

Table 1. Parameters of tested insulators.

| Type | Materials | Roted failing Load/kN | H/mm | D/mm | L/mm |
|------|-----------|------------------------|------|------|------|
| A596EZ | Porcelain | 400 | 205 | 340 | 550 |

At least 10 effective tests should be conducted on the insulator string in a certain contamination degree and the 50% AC flashover voltage $U_{50}$ of the class insulator string can be calculated as follow:

$$U_{50} = \frac{\sum (n_i \times U_i)}{N} \quad (1)$$

where $U_i$ is a certain value of applied voltage; $U_i$ is the number of tests carried out under the same level of $U_i$; and $N$ is the number of tests.

## 3 THE MODEL OF CALCULATION POLLUTED INSULATOR'S FLASHOVER VOLTAGE

A typical RBF neural network with multi-input and single-output is shown in Figure 2. The network consists of three layers, which are as follows: the input layer, hidden layer and output layer. The information of input layers is not treated after the input vectors are assigned to the hidden layer. The transformation from the input layer to the hidden layer is nonlinear, because the form of radial basis functions has little effect on the performance of the RBF network; and so, the Gaussian kernel function was selected as the form of RBF. The hidden layer to output layer mapping is performed by using the linear superposition technique. The data set was obtained after performing many experiments under different values of $\rho_{ESDD}$ and $\rho_{NSDD}$ and it consisted of 50 input–output pairs. The input variables should be normalized before initiation of training of the RBF, so that they fall in the interval [−1, 1]. If the input value is negatively correlated to the flashover voltage of insulators, then the normalized input parameters $f_n(x)$ can be determined as follows:

$$f_n(x) = \frac{f_{max} - f(x)}{f_{max} - f_{min}} \quad (2)$$

The input parameters of RBF, such as $\rho_{ESDD}$ and $\rho_{NSDD}$, are negatively correlated to the 50% AC flashover voltage $U_{50}$ of the class insulator string, so that the normalized input parameters $f_n(x)$ is in accordance with equation (2).

Input parameters of the model are $\rho_{ESDD}$ and $\rho_{NSDD}$ and the output is $U_{50}$ of the polluted insulator string. By means of the orthogonal least squares algorithm, right hide layer neurons and center vectors are chosen and the weight from hide layer neurons to output layer neurons are determined by using the pseud-inverse method, an error goal or performance goal of 0.001 was introduced into the network. From 100 pairs of the data set, 8 pairs were employed for testing samples and the rest of the data are used to train the RBF network. The 8 group testing samples were calculated by using the BRF network and compared with test results, as shown in Figure 2. Compared with the experimental results, the $U_{50}$ value has good consistency and the maximum relative error is only 3.4%, as shown in Figure 3; basically consistent with the experimental results, it meets the engineering need to estimate the insulator contamination degree.

For example, with regard to the 1000 kV UHV system, the phase voltage is about 577 kV. By measuring the monitoring insulator located near the power line, if $\rho_{ESDD}$ is 0.175 mg/cm$^2$ and $\rho_{NSDD}$ is 0.523 mg/cm$^2$, the 50% flashover voltage calculated by using the RBF network is 545 kV, which is less than 577 kV; it suggests that the possibility of flashover is very large under pollution conditions.

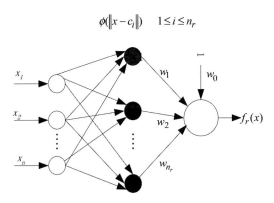

$$\phi(\|x - c_i\|) \quad 1 \le i \le n_r$$

Figure 2. Schematic of the radial basis function network.

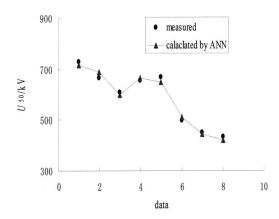

Figure 3. Graph showing test results and calculated results.

If $\rho_{ESDD}$ is 0.067 mg/cm$^2$ and $\rho_{NSDD}$ is 0.437 mg/cm$^2$, the 50% flashover voltage calculated by using the RBF network is 694 kV, which is larger than 577 kV; it suggests that the possibility of flashover is very small under pollution conditions. This provides a reference for the UHV line pre-pollution flashover.

## 4 CONCLUSION

The 50% AC flashover voltage $U_{50}$ of the UHV porcelain string of 48 units is reduced with the increase of $\rho_{ESDD}$ and $\rho_{NSDD}$ values. According to the test results of insulator, a calculation model of $U_{50}$ is put forward based on the radial basis function network in this paper. Inputs of the model are $\rho_{ESDD}$ and $\rho_{NSDD}$, the output of the model is $U_{50}$ of polluted insulator strings, the hide layer neurons and center vectors are selected by means of the Orthogonal Least Squares (OLS) algorithm, and the weight from hide layer neurons to output layer neurons are determined by using the pseud-inverse method. The calculated result is in accordance with the test result. This method would give some reference for the state assessment system on polluted insulators.

## ACKNOWLEDGMENTS

This research work was sponsored by the Chongqing Research Program of Foundation and Advanced Technology (Project No. cstc2013 jcyjA90006) and Foundation of Chongqing University of Science and Technology (Project No. CK2014B17), and Foundation Chongqing Municipal Education Committee (Project No. KJ1501338).

## REFERENCES

Ahmad S. & Ghosh P.S. Modeling of various met eorological effects on contamination level for suspension type of high voltage insulators using ANN (2002). *Transmission and distribution conference and exhibition, Asia Pacific. IEEE/PES*, pp. 465–473.
Artificial pollution tests on high-voltage insulators to be used on A.C. systems (1991). *IEC60507*.
Chen, S., C.F.N. Cowan, and P.M. Grant (1991). Orthogonal Least Squares Learning Algorithm for Radial Basis Function Networks. *IEEE Transactions on neural networks*, 2, pp. 302–309.
Enemenlis, C., Harbec, G., Hould, A., Shperling, B.R., Pokorny, W., and Zelingher, S. (1981). Air insulation design of UHV station based on switching surges. *IEEE Transaction on Power Apparatus and Systems*, PAS-100, pp. 891–898.
GB/T 4585–2004. Artificial pollution test method for AC 1000V and above (2004). in Chinese.

Hu, J., C. Sun, X. Jiang, Z. Zhang and L. Shu (2007). Flashover performance of pre-contaminated and iced-covered insulators to be used in 1000 kV ac transmission lines. *IEEE Transaction on Dielectrics and electrical and insulation*, 14, pp. 1347–1353.

IEEE committee report (1983). UHV ac substation design criteria summary of world trends. *IEEE Transaction on Power Apparatus and Systems*, PAS-102, pp. 513–520.

Jiang, X., J. Yuan, Z. Zhang, J. Hu and C. Sun (2007). Study on ac artificial-contaminated flashover performance of various types of insulators. *IEEE Transaction on Power Delivery*. 22, pp. 2567–2574.

Jiang, X., J. Yuan, Z. Zhang, J. Hu and L. Shu (2007). Study on pollution flashover performance of short samples of composite insulator intended for ±800 kV UHV dc. *IEEE Transaction on Dielectrics and electrical and insulation*. 14, pp. 1192–1200.

Lloyd K.J. (1981). Testing contaminated insulators at project UHV for voltage of the future. *IEEE Transaction on Electrical Insulation*, EI-16, pp. 220–223.

Sforzini, M., R. Cortina, G. Marrone and P.D. Bernardelli (1980). A laboratory investigation of the pollution performance of insulator strings for UHV transmission systems. *IEEE Transaction on Power Apparatus and Systems*, V PAS-99, pp. 678–682.

Takasu, K., N. Arai, Y. Imano, T. Shindo and T. Seta (1981). AC flashover characteristics of long air gaps and insulator string under fog conditions. *IEEE Transaction on Power Apparatus and Systems*, PAS-100, pp. 639–645.

*Advances in Energy and Environment Research – Achour & Wu (Eds)*
© *2017 Taylor & Francis Group, London, ISBN 978-1-138-62682-9*

# Exploration and discussion on electricity consumption demand forecasting theories and methods

Faxuan Ma
*Dongguan Power Supply Bureau, Guangdong Power Grid Co. Ltd., Dongguan, China*

ABSTRACT: It is of great significance to accurately forecast electricity demand for supply and demand interaction of multi-agents under the electricity market's competitive circumstances, such as the grids, the users, the electric energy providers, and the load integrators. Firstly, the home and abroad electricity demand forecasting theories are summarized briefly, including grey theory and artificial neural network theory, among which, the forecasting principle of the Grey Model GM(1,1) is elaborated, several main electricity demand forecasting methods of the power market were introduced, including the econometric model method, the comprehensive analysis method, the analysis prediction method, other methods, etc. which could be applied in short-term, middle-term, and long-term electricity demand forecasting, and then, based on the historical electricity consumption data of Guangzhou city from the year 2000 to 2008, an example using the electricity elasticity coefficient method was analyzed, which forecasted the energy consumption and the electricity consumption rising rate of Guangzhou city in the year 2009 and it is shown that this forecasting method was accurate and effective. Finally, several measures which could improve the electricity forecasting accuracy are provided and considered that the forecasting accuracy could be improved via application of the newly developed machine learning smart algorithms, such as the deep learning, Q-learning, extreme learning, etc.

## 1 INTRODUCTION

The load characteristics of grid corporations, the user structures, and their electricity consumption have been changed sharply, which put forward more rigorous requirements for the deep electricity analysis, forecasting, marketing management, and service work. Currently, the load and electricity forecasting objects in China are extensive when compared with those abroad; moreover, we cannot accurately grasp the electricity characteristics of concrete industries or large users; it is hard to achieve fine forecasting and these are limited to play a role in fine management and deep development of marketing work. In trend of large data analysis, for improvement of the marketing management and service work, a comprehensive analysis on the features and relationships of the electricity consumption information and multi-source information is required. The above-mentioned work is lacking, although the industrial condition indexes and its analysis can completely grasp the economic situations of the industry and can also analyze the industry alteration, while it cannot be applied in the electric power industry.

Forecasting of the electricity demand helps in understanding the trends of power consumption, and for the enterprises, it may assist in determining its energy conservation methods for implementation of energy conservation and emission reduction and improvement of the energy consumption per unit output value; for the power corporations, forecasting has great significance for implementation of reasonable generation dispatching. The electricity demand forecasting, according to power customer classification, can be classified into residents, industrial, and commercial users' electricity demand forecasting; according to forecasting methods, can be divided into the artificial intelligence method, the statistical analysis method, the econometric analysis method, the terminal energy demand analysis method, the software hourly dynamic simulation method, etc.

In this paper, firstly, a brief summation of home and abroad electricity demand forecasting theories is made, including the grey theory and artificial neural network theory and especially the GM(1,1) model is elaborated; secondly, several main electricity demand forecasting methods are introduced, including the economic model method, comprehensive analysis method, classification forecasting method, and other methods, and one of the forecasting methods, that is the electricity elasticity coefficient method, is focused. An example of electricity demand forecasting is given according to the historical electricity data of Guangzhou city from the year 2000 to 2008; and finally, some measures that improve the forecasting accuracy are provided.

## 2 BRIEF SUMMARIZATION ON ELECTRICITY DEMAND FORECASTING THEORIES

As aforementioned, accurate forecasting of the electricity demand of residents, the industry, and commerce users has great significance for the power enterprises to implement energy-saving generation dispatch, while if the electricity utilization trend of one certain area can be forecasted accurately, then it both has importance for the power corporations and power users in this area. The mathematical theory and artificial intelligence method have been developed and applied in power system electricity demand forecasting since 1980s and especially, the grey theory method and Artificial Neural Network (ANN) theory are widely applied.

### 2.1 *Electricity demand forecast by using the grey theory*

The key of the grey theory in forecasting the electricity demand is to build the Grey Model (GM), and the building approaches include the sequence forecast, calamity forecast, topological forecast, and comprehensive forecast [1]. Generally, we adopt the sequence forecast method, that is, make data transformation for the electricity consumption data of residents, industry, and commerce users, then the sequence is generated, and we utilize the sequence to generate the generation function with strong regularity. And then, a differential equation model of this function is built to forecast the future electricity demand. The GM model based on the sequence forecasting method is not limited by the sample that whether it has statistical regularity and whether the statistical regularity is a typical process, and this model is simple, has advantages such as small data size, high prediction accuracy, strong adaptability, etc. and which can be applied for short-term prediction, as well as middle and long-term prediction [2,3]. The grey theory-based GM forecasts of the electricity demand is carried out by using a grey variable or a grey process along with the time variation, instead of treating the electricity consumption monitoring data as a stochastic process [4]. According to the accumulation or subtraction generation method, the grey variable is gradually whitened, making the original sequence with no or little regularity into one with strong regularity. Thus, the corresponding differential equation model is built and applied for future electricity demand prediction [5].

The GM is the core, while the GM(1,1) is the most widely used and simple one, which is composed of only one-covariant and first-order differential equations. The GM has a certain accuracy, as well as a simple calculation process, while it only considers the sold electricity, instead of other factors that may have an impact on the sold electricity, such as the GDP. While utilizing the GM(1,1) not only the common influence of multi-certainty and uncertainty factors of the sold electricity is took into consideration, but also the impact of each main factor on the sold electricity is understood; hence, the influence degree of model prediction accuracy is lowered and the prediction accuracy and validity of sold electricity are effectively improved and ensured. From previously reported literature [6], the electricity demand prediction modeling procedures of GM(1,1) include the following steps:

a. Make a first accumulation of the users' original electricity consumption data and obtain the generation sequence; suppose the original sequence is $X^{(0)} = [X^{(0)}(1), X^{(0)}(2), \cdots, X^{(0)}(n)]$ and the first accumulation sequence is obtained as 1-AGO (Accumulated Generating Operation, AGO), that is $X^{(1)} = [X^{(1)}(1), X^{(1)}(2), \cdots, X^{(1)}(n)]$, among which, $X^{(1)}(k) = X^{(0)}(1) + X^{(0)}(2) + \cdots + X^{(0)}(k)$, $k = 1, 2, \cdots, n$. After this cumulative transformation, the impacts of bad data in original electricity data are weakened, thus making it become a generation sequence with stronger regularity and then we can build a model. We know that $X^{(1)}(k)$ follows an exponential increasing law; what is more, the first-order differential equation also has a form of exponential rising, and then we can consider that the newly formed sequence meets the following form of first-order differential equation model, that is the first-order whitening differential equation:

$$\frac{dX^{(1)}}{dt} + \alpha X^{(1)} = u \tag{1}$$

where $\alpha$ is the development parameter, which reflects the developing trend of original electricity sequence $X^{(0)}$ and first-accumulation sequence $X^{(1)}$; $u$ is the cooperation coefficient, which reflects the transformation relationship between the data.

b. Solve the coefficients of $\alpha$ and $u$; we can utilize the least square method to obtain them and its solving equation is as follows:

$$\begin{bmatrix} \hat{\alpha} \\ \hat{v} \end{bmatrix} = (\boldsymbol{B}^T B)^{-1} \boldsymbol{B}^T \boldsymbol{Y} \tag{2}$$

In equation [2], the matrixes $B$ and $Y$ are given as follows:

$$B = \begin{bmatrix} -\dfrac{1}{2}\left(X^{(1)}(2)+X^{(1)}(1)\right) & 1 \\[2ex] -\dfrac{1}{2}\left(X^{(1)}(3)+X^{(1)}(2)\right) & 1 \\[1ex] \vdots & \vdots \\[1ex] -\dfrac{1}{2}\left(X^{(1)}(n)+X^{(1)}(n-1)\right) & 1 \end{bmatrix} \quad (3)$$

$$Y = \begin{bmatrix} X^{(0)}(2) \\ X^{(0)}(3) \\ \vdots \\ X^{(0)}(n) \end{bmatrix} \quad (4)$$

According to equation [2], the solved parameters of $\hat{\alpha}$ and $\hat{u}$ are substituted in equation [1], and then equation [1] can be modified into the following form:

$$\frac{dX^{(1)}}{dt} + \hat{\alpha}X^{(1)} = \hat{u} \quad (5)$$

Then, we solve equation [5] and we can obtain the grey prediction model, which is shown as follows:

$$\hat{X}^{(1)}(k+1) = \left(X^{(1)}(1) - \frac{\hat{u}}{\hat{\alpha}}\right)e^{-\hat{\alpha}k} + \frac{\hat{u}}{\hat{\alpha}} \, (k=0,1,2,\cdots) \quad (6)$$

c. Implement Inverse Accumulated Generating Operation (IAGO), that is

$$\hat{X}^{(0)}(k+1) = \hat{X}^{(1)}(k+1) - \hat{X}^{(1)}(k)$$
$$= (1-e^{\alpha})[X^{(0)}(1) - \frac{\hat{u}}{\hat{\alpha}}]e^{-\hat{\alpha}k}\,(k=0,1,2,\cdots) \quad (7)$$

when $k \geq n$, $\hat{X}^{(0)}(k+1)$ is the forecasting value.

### 2.2 Electricity demand prediction by using the Artificial Neural Network (ANN) theory

The ANN theory is applied for future electricity demand prediction because of its excellent learning features and nonlinear relationship of input and output variables; hence, the theory is widely used to build the electricity demand forecasting model [7]. The principle of electricity demand prediction by using the ANN is based on simulation of information processing and storage of search mechanisms

of the human brain' meanwhile, the electricity consumption prediction of residents, industry, and commerce users is performed by using the ANN model via its advantages, for example, the nonlinearity, self-organization, and Massively Parallel Processing (MPP), etc. The differences of ANN models are determined by differences between the artificial neurons' structure and its inter-connecting modes. The commonly used models include the Multi-layer Feed-forward Neural Network (MFNN), which is also called the BP model, which also includes the self-learning Kohonen self-organization neural network model, etc.

We can use the BP model to fit any complex nonlinear relationship quickly. When the historical data sample of user electricity consumption has an appropriate size, we can use the BP model to finish specialized artificial training, and then a proper functional relationship can be generated automatically, through which the neural network is formed to meet the requirements of electricity demand prediction. The method of electricity demand prediction based on the BP model utilizes the errors after outputting to estimate the direct leading layer errors of the output layer. This mode is transferred one layer by one layer and finally we can get error statistics of other layers [7].

While the electricity demand prediction method based on the Kohonen self-organization neural network model adopts the self-organization feature of the Kohonen network and finishes the neural network training via inputting the proper load sample, and then the formed network can realize the classification of compound models automatically.

Among the relative studies, ref. [8] proposed a combinational optimization grey neural network model, which, based on better increasing features of GM(1,1), can build the model of the increasing trend of seasonal time sequences. Meanwhile, based on the ability of better description of the complex nonlinear function of ANN, a model of the seasonal load can be built and finally, according to the optimal combination forecasting theory, an optimal combination forecasting model, which is called GANN and combining the features of GM(1,1) and ANN, was built. Combined with each single model prediction, GANN has a better improvement in prediction accuracy. In ref. [9], a structure adaptive clustering neural network model was put forward, based on actual marketing data, the time feature analysis of user electricity consumption was assessed and its conclusion had an important reference value for targeted adjustment of electricity price and reasonable implementation of electric power production. In ref. [10], while proposing a whole society electricity demand prediction model based on the utilization of additional momentum and self-adaptive learning rate of the Back Propagation (BP) neural network, the input variables of the model were GDP, electricity consumption of primary, secondary, and tertiary industry and urban and rural residents; the output variable was the whole society's electricity consumption of the subsequent year, and in addition, it simulated the whole society's electricity consumption data from the year 1986 to 2006, and the validity of the proposed model was verified by the simulation results.

# 3 COMMONLY USED METHODS OF ELECTRICITY DEMAND FORECASTING IN THE POWER MARKET

## 3.1 Economic model method

The economic model methods are applied for electricity demand prediction, which can be classified as the econometric forecasting method, the regression analysis method, and the aforementioned grey prediction method.

The econometric forecasting method has placed an emphasis on finding main factors that influence electricity demand and its cause-and-effect relationship, and solved the equation to forecast future electricity demand. The variables of the forecasting function built by this method are selected as follows: the primary, secondary, and tertiary industry added values; the urban and rural resident's incomes; the population; the GDP per person; and the living space. The function itself is a first-order or multiple order function.

The regression analysis method implements regression analysis by adopting the historical electricity consumption data of residents, industry, and commerce users, and then a group of model equations of the electricity demand variable and exogenous variable is obtained. Generally, the region's GDP is selected as the exogenous variable and the commonly used single-element regression models have the following variables: a) linear regression, $Y = a + bx$; b) S curve, $Y = 1/(c + a*e^{bx})$; c) exponential function, $Y = a + e^{0.001*x}$; d) power function, $Y = a*x^b$; e) logarithmic function, $Y = a + b*\log x$; and f) exponential function, $Y = a*b^x$. The regression analysis method is usually applied for the mid-and-long term predictions and we can make a multivariate regression analysis according to the requirements and conditions.

The grey forecasting method, aforementioned, which uses the grey theory to forecast the electricity demand of the power market and we always use the sequence prediction method, which transfers original data of user electricity into the function with strong regularity, and then builds the differen-

tial equation model to forecast the future electricity demand and is usually applied for short-term and mid-and-long term predictions with advantages of simple, small size of data, higher forecasting accuracy, and strong adaptability.

## 3.2 Comprehensive analysis method

This method is also commonly used for electricity demand prediction and includes the electricity elasticity coefficient method, the analogy method, and the average growth method.

The electricity elasticity coefficient method includes the power production and power consumption elasticity coefficients, among which, the latter is referred to the ratio of power consumption growth rate and the national economy growth rate. The electricity elasticity coefficient, which is selected as a macro-index of the relevance between the economy development and power demand growing speed, which can be used for load forecasting of long-term rough line style planning [11] and can comprehensively reflect the promoting function of economy growth to power demand. We can apply the electricity elasticity coefficient to forecast the power demand and its accuracy depends on the statistical accuracy of the national economy historical materials and correct appraisal of the future economic structure and scientific–technical progress to the electricity demand [12]. The value of electricity elasticity coefficient should be smaller than 1 generally and larger than 1 means the marginal cost of GDP exceeds its average electric power cost, and then the sustainable development of economy may be influenced. For the electricity demand prediction in the power market, aimed at the electricity elasticity coefficient in a period, its value is equal to the ratio of the whole society's electricity consumption average growth rate and the GDP in the calculation period, while the electricity growing speed in the prediction period is equal to the product of electricity consumption elasticity coefficient and GDP growing speed in the prediction period. Under normal circumstances, we can use the industry electricity consumption elasticity coefficients that has been removed from the household demand to forecast or after forecasting of which the industry uses the accumulation value to forecast. Aimed at electricity demand forecasting of one certain region in China, because the electricity elasticity coefficients of each year lack regularity, it is only suitable for one period forecasting when we use the electricity elasticity coefficient method.

The analogy method includes the load–density method, per capita domestic electricity consumption method, etc. By using this method, the planning regions and the regions that have experienced a similar development process are compared, so as to determine the electricity demand. We can draw on the development situations in domestic developed areas for analysis and comparison, when allowing for the differences between the areas in aspects of the economic system, economic construction, resource, production, consumption, living habit, electricity utilization constitution, etc. Thus, we can forecast the electricity demand in the local region. We find the common features via analogy and forecast the electricity demand in the planning period combined with the actual situations in the forecasting region. This method is not a suitable main approach for electricity demand prediction, but its prediction results can be selected as corrections for other electricity demand forecasting methods and the method is commonly used in long-term rough-and-ready predictions.

The average growth method is usually essential to obtain the yearly growing rate of year electricity consumption via comprehensive analysis and calculation to compute the annual electricity consumption demand value, when it is used for electricity demand forecasting, that is, $A_n = A \times (1+k)^n$. This method has a strong applicability in those areas where the collection age of historical data is long, the economy develops regularly, the electricity consumption grows evenly, and the proportion of the large customer electricity consumption in the whole regional electricity consumption is smaller.

### 3.3 Analytical prediction method

The analytical prediction method includes several parameters when it is used for electricity demand forecasting, such as the prediction made by using the department method, the large customer comprehensive analytical method, the prediction made by using the region method, the conversion analytical prediction method, the progress marching analytical prediction method, the equivalent hour analytical prediction method, etc.

The prediction made by using the department method lies in accurate industry electro-sort, which is combined with the actual development situation and different features of the industry in the local region and performs electricity demand forecasting for different industries with different methods, and then gather the prediction values for the final prediction.

The large customer comprehensive analytical method divides the users into large users and general users and speculates the electricity demand according to the respective prediction of large user electricity consumption and natural growing electricity of general users. Concretely, first process the historical electricity data, remove the large user electricity data among them, and then calculate the natural growing speed of the general user electricity demand, which is selected as the foundation of forecasting general user electricity demand's growing speed. This method is suitable for those regions where the amount of large users is larger and the proportion of large user electricity consumption in the whole region's electricity consumption is larger as well.

The prediction by using the region method is used to forecast via the above-mentioned prediction methods by region, gather the prediction values, and conclude the whole region's prediction value in the planning period.

The conversion analytical prediction method [13], based on the analysis of the business expand capacity and electricity consumption's growing trend of power enterprises, analyzes and forecasts the demand situation of the power market in the subsequent year from the data. This method is one of most commonly used methods for electricity analysis and prediction of power supply enterprises under electricity market circumstances. Concretely, the electricity demand to be forecasted of the subsequent year deducts that of last year and then divides the business expand capacity of power enterprise two years ago, that is the equivalent number of hours of the subsequent year. The growing analysis data of electricity consumption of the subsequent year is equal to the equivalent house of the same year when it multiplies its business expand capacity. This method is mainly used for analysis of quarter, half-year, and the whole year of electricity consumption and the future electricity demand prediction.

The progress marching analytical prediction method is mainly used for electricity demand prediction in the local region and in each electricity consumption area, according to the trend of daily electricity consumption data, to judge the electricity situations under weather conditions, seasonal variation and holidays, to implement electricity habit analysis of power enterprises and electricity residents in local area, to judge the impacts of electricity situations on demand trend, and finally obtain the law of electricity consumption, so as to better forecast the electricity demand in this local area [14,15].

The equivalent hour analytical prediction method is used mainly depending on electricity generation mode per hour and applies this variable in the electricity dependent prediction work, to improve the quality of power market electricity demand forecasting effectively.

### 3.4 Other forecasting methods

The other methods include the trend analysis method, the exponential smoothing method, the per capita domestic electricity consumption method, the load–density method, etc.

The key of the trend analysis method is to build an accurate trend analysis model, which is used for trend curve analysis, curve fitting, or curve regression analysis, and it is a commonly used prediction method. Aimed at the historical electricity consumption data of residents, industry, and commerce users, the first step is to form the fitting curve, which should reflect the growing trend of electricity or load, then according to this growing trend curve, to estimate the electricity demand value or load prediction value in this moment of one future point, this process is a determined extrapolation, and it is unnecessary to consider the random errors. The basic mode of several typical trend models applied in this method is as follows:

$$Y_t = f(t, \theta) + \varepsilon_t \qquad (8)$$

where $Y_t$ is the prediction object, $\varepsilon_t$ is prediction error, and $\theta$ is the undetermined parameter.

Based on equation [8], the several representative trend analysis models included are the following: a) polynomial trend model, $Y_t = a_0 + a_1 t + \ldots + a_n t^n$; b) linear trend model, $Y_t = a + bt$; c) logarithmic trend model, $Y_t = a + b\ln t$; d) power function trend model, $Y_t = at^b$; e) exponential trend model, $Y_t = ae^{bt}$; f) logistic model, $Y_t = L/(1 + \mu e^{-bt})$; and g) Gompertz model, $Y_t = \exp(-\beta e^{-\theta t})$, $\beta > 0$, $\theta > 0$.

The exponential smoothing method is different from the trend analysis method and its prediction depends on modeling of actual values of the time sequence, and then utilizes the exponential weighted array of the historical electricity consumption data to directly forecast the future values of time sequence. The basic mode of the exponential smoothing method-based forecasting model is given by equation [16]:

$$X'(t+1/t) = \sum_{i=0}^{\infty} \alpha(1-\alpha)^i X(t-i) \qquad (9)$$

where $\alpha$ is the attenuation factor, which highlights the effect of past historical data when it is smaller, or otherwise, the effect of recent data. For the power demand prediction, the importance of this method is that when the curve closes to the current moment, the prediction should be more accurate, while for the long past data, it is unnecessary to make accurate fittings, which is just similar to the inertia effect.

The per capita domestic electricity consumption method utilizes the growing rate method and regression method to finish numerical computation or comprehensive analysis of the per person's electricity consumption value in the planning period and which multiples with the annual population,

generating the annual electricity consumption in the planning year.

The load–density method uses the average load value per square kilometer, firstly to calculate the current and historical sectionalized load density, and then according to the features of region development and planning and load development of each subregion, to reckon the target year's load density prediction value of each subregion [17]. This method provides a strong guidance for stationing of future 110 kV substations, and a distribution network construction of 10 kV in the urban center, while the amount of fundamental information investigation work is larger.

## 4  EXAMPLE ANALYSIS OF ELECTRICITY DEMAND FORECASTING BASED ON THE ELECTRICITY ELASTICITY COEFFICIENT METHOD

### 4.1  *Electricity consumption elasticity coefficient*

Electricity consumption elasticity analysis is a commonly used prediction method in the current industry. The electricity consumption elasticity coefficient can reflect the corresponding relationship between the power consumption's growing speed and national economic growing speed. The coefficient is calculated as follows:

$$\vartheta = \frac{(E_2 - E_1)/E_1}{(G_2 - G_1)/G_1} \qquad (10)$$

where $E_1$ is the electricity consumption of the previous year, $E_2$ is the electricity consumption of the subsequent year, $G_1$ is the GDP of the previous year, and $G_2$ is the GDP of the subsequent year.

The economy, environment, technology, and other aspects will influence the elasticity coefficient, but the total trend is gradually reduced. The influence factors include the following: a) constant adjustment of the economic structure, such as the changes of proportion for secondary and tertiary industries; b) strong impacts of world finance forms on the industrial structure; c) adjustment of the energy policy; d) scientific and technological progress and factors of bringing in power demand side management; e) price relationship between the electric power and current energy; f) variable factors of climate and population; g) authenticity of GDP statistical data; etc.

### 4.2  *Relations between elasticity coefficient and unit power consumption variation*

When the GDP growth occurs under certain conditions, there is a one-to-one corresponding relation between the elasticity coefficient and unit power consumption variation. We suppose that the GDP gross of the first year and second year are $G_1$ and $G_2$ (ten thousand Yuan), the GDP growth rate of the second year is $g$ with respect to that of the first year, that is, $G_2 = G_1*(1 + g)$. The per 10,000 Yuan GDP of the first and second year are $D_1$ and $D_2$ (ten thousand Yuan), and for the per 10,000 Yuan GDP energy consumption, the second year is $\sigma$ times the first year, that is, $D_2 = D_1*\sigma$, where $\sigma$ is the power consumption variation factor.

The power consumptions of the first and second years are $E_1$ and $E_2$ (kilowatt-hour), respectively, that is, $E_1 = D_1* G_1$, $E_2 = D_2* G_2$. And then, the elasticity coefficient $\vartheta$ is calculated as follows:

$$\vartheta = \frac{\text{The growth rate of electricity consumption}}{\text{The growth rate of GDP}}$$
$$= \frac{(E_2 - E_1)/ E_1}{(G_2 - G_1)/ G_1}$$

$$(11)$$

The $G_1$, $G_2$, $E_1$, and $E_2$ values are substituted in equation [11] above, and then, the following equation can be obtained:

$$\vartheta = \frac{\sigma(1+g)-1}{g}$$

$$(12)$$

According to equation [12], we can deduce the variation of the electricity elasticity coefficient via the growth rate of the GDP and power consumption coefficient.

### 4.3 Example analysis

Based on the historical electricity consumption data of Guangzhou city from the year 2000 to 2008, the growth rate of electricity consumption of the year 2009 is forecasted and its analysis flow is shown in Figure 1.

The electricity consumption of Guangzhou city from the year 2000 to 2008 is shown in Table 1, among which, the actual electricity consumption is equal to the accumulation of per ten thousand kWh primitive electricity consumption and the output at the end of year and negative input at the beginning of year.

The computational values of GDP based on the constant-price of the year 2000 are shown in Table 2, among which, the GDP growth is $V_j$, the GDP index value is $\mu_j$, the GDP value based on constant-price of the year 2000 is $\varepsilon_j$, where $j$ denotes the years from 2000 to 2009, for example, $V_{2000}$, $\mu_{2000}$ and $\varepsilon_{2000}$ are respectively the corresponding values in year of 2000, $j = 2000, 2001, \cdots, 2008$, for the three variables, $V_j = (\mu_j/100)-1$; for the year 2000, $\varepsilon_j$ is equal to the

| Step 1: search the GDP growth rate of year 2007 according to Table 1, calculate the constant-price per ten thousand Yuan GDP power consumption of year 2008 (Table 2) |

↓

| Step 2: calculate the power consumption of year 2008 according to Table 3, analyze the variation of power consumption coefficient of year 2009 |

↓

| Step 3: based on analysis of power consumption coefficient of year 2008, forecast that of 2009 with the following considerations: a) continuous energy conservation and emission reduction policy (about 3%~4%); b) reduction of line loss rate (about 0.2%); c) impacts of economic crisis (judge whether the impacts exist and its size according to the actual circumstances) |

↓

| Step 4: according to the power consumption coefficient and elasticity coefficient formulas, which are put into the GDP forecasting growth rate and power consumption coefficient solving formulas of year 2009, to get the elasticity coefficient |

↓

| Step 5: forecast the electricity consumption growth rate of year 2009, namely, the electricity consumption growth rate is equal to elasticity coefficient and GDP growth rate (Table 4) |

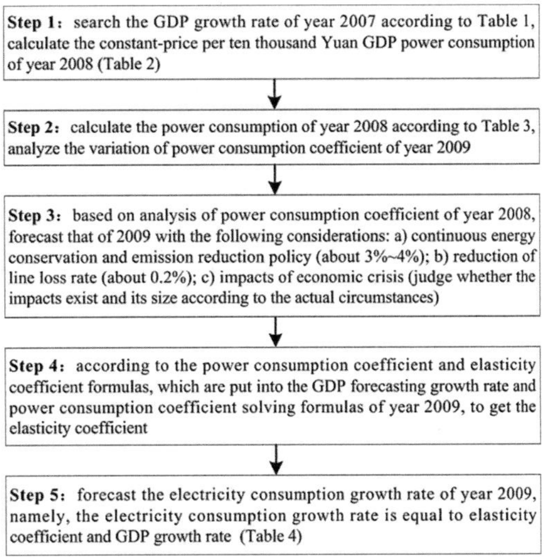

Figure 1. Flowchart showing forecasting and analysis flow of the electricity consumption growth rate in 2009.

Table 1. Electricity consumption of Guangzhou city from the year 2000 to 2008.

| Year | Per ten thousand kWh primitive electricity consumption | Output at end of the year/kWh | Input at the beginning of the year/kWh | Actual electricity consumption/ kWh |
|------|------|------|------|------|
| 2000 | 2,387,780 | 0 | 0 | 2,387,780 |
| 2001 | 2,556,452 | 0 | 0 | 2,556,452 |
| 2002 | 2,849,015 | 0 | 0 | 2,849,015 |
| 2003 | 3,349,711 | 0 | 0 | 3,349,711 |
| 2004 | 3,846,374 | 0 | 0 | 3,846,374 |
| 2005 | 4,256,676 | 1,000 | 0 | 4,257,676 |
| 2006 | 4,694,234 | 25,280 | 1,000 | 4,718,514 |
| 2007 | 5,271,257 | 40,708 | 25,280 | 5,286,685 |
| 2008 | 5,459,185 | 90,486 | 40,708 | 5,508,963 |

actual GDP value of this year and for the other years, $\varepsilon_j = \varepsilon_{j-1}*\mu_j/100, j = 2001, 2002, \cdots, 2008$.

The computational values of energy consumption and elasticity coefficient based on the constant-price GDP of year 2000 are shown in Table 3, among which, the energy consumption is $\eta_k$, the energy consumption coefficient is $\xi_k$, the elasticity coefficient is $\kappa_k$, where $k = 2000, 2001, \cdots, 2008$, for example, $k = 2000$, $\eta_{2000}$, $\xi_{2000}$, and $\kappa_{2000}$ are respectively the corresponding values in the year 2000. Set the actual electricity consumption of each year as $y_k$ (see Table 1), for the three variables, $\eta_k = (y_k*10000)/\varepsilon_j$, $\xi_k = \eta_k/\eta_{k-1}$, $\kappa_k = (\xi_k*(1 + V_j)-1)/ V_j$,

13

Table 2. The computational values of GDP based on the constant-price of the year 2000.

| Year | Actual value of GDP/ten thousand Yuan | GDP index value | GDP growth rate/% | The GDP based on the constant-price of year 2000/ ten thousand Yuan |
|------|------|------|------|------|
| 2000 | 24,927,434 | 113.35 | 13.4% | 24,927,434 |
| 2001 | 28,416,511 | 112.74 | 12.7% | 28,103,189 |
| 2002 | 32,039,616 | 113.24 | 13.2% | 31,824,051 |
| 2003 | 37,586,166 | 115.21 | 15.2% | 36,664,490 |
| 2004 | 44,505,503 | 115.04 | 15.0% | 42,178,829 |
| 2005 | 51,542,283 | 112.92 | 12.9% | 47,628,333 |
| 2006 | 60,738,277 | 114.82 | 14.8% | 54,686,852 |
| 2007 | 71,091,814 | 114.90 | 14.9% | 62,835,193 |
| 2008 | 82,158,200 | 112.30 | 12.3% | 70,563,922 |

Table 3. The computational values of the energy consumption and elasticity coefficient based on the constant-price GDP of the year 2000.

| Year | Energy consumption | Energy consumption coefficient | Elasticity coefficient |
|------|------|------|------|
| 2000 | 957.89 | 0.000 | 0.000 |
| 2001 | 909.67 | 0.950 | 0.554 |
| 2002 | 895.24 | 0.984 | 0.864 |
| 2003 | 913.61 | 1.021 | 1.155 |
| 2004 | 911.92 | 0.998 | 0.986 |
| 2005 | 893.94 | 0.980 | 0.828 |
| 2006 | 862.82 | 0.965 | 0.730 |
| 2007 | 841.36 | 0.975 | 0.808 |
| 2008 | 780.71 | 0.928 | 0.342 |

$k, j = 2001, 2002, \cdots, 2008$. For the year 2000, the energy consumption coefficient and elasticity coefficient are both zero.

Based on this truth, the elasticity coefficient of the year 2008 has reduced sharply and is aimed at achieving the prediction of the energy consumption coefficient in the year 2009; three prediction schemes in low, middle and high schemes are given and the energy consumption coefficient prediction in 2009 is determined by forecasting the variation of the overall financial environment, so that the GDP growth in 2009 can be divided into the low-, middle- and high-scheme, and the predicted growth rate values with respect to each scheme are 10%, 9%, and 8%.

The corresponding energy consumption coefficient is also predicted as low-, middle- and high-scheme and the corresponding predicted values are 0.959, 0.950 and 0.941. The prediction of the energy consumption coefficient in 2009 is based

on the sharp reduction of that of last year and the factors that cause sharp reduction of the elasticity coefficient in 2008 include the following factors: a) 2.98% contribution based on energy conservation and emission reduction in past years; b) 0.76% contribution based on the reduction of the line loss rate; c) 0.73% contribution based on the reduction of temperature-lowering load; d) 0.00% contribution based on 16 high energy-consuming enterprises; d) 0.03% contribution based on the off-peak power consumption in the first quarter of the year; e) 2.70% contribution based on the industrial adjustment shocked by financial crisis; and f) 0.90% contribution based on an average quarterly impact. Hence, aimed at prediction of energy consumption in 2009, the factors that cause reduction of its value include the following: a) continuous energy conservation and emission reduction contributes 2.98%; b) line loss rate contributes 0.23%; c) contribution of forced adjustment by financial crisis, which is divided into three schemes, the high-scheme (influence one quarter), the medium-scheme (influence two quarters), and the low-scheme (influence three quarters). Finally, summarize the two aforementioned factors, the energy consumption coefficients are respectively reduced to 0.959, 0.950, and 0.941 under the three schemes.

Finally, the elasticity coefficient and electricity consumption growth rate of Guangzhou city in 2009 are shown in Table 4, among which, under the high-, medium- and low-scheme, suppose the elasticity coefficients are respectively $\kappa_i$, the energy consumption coefficient prediction values are $\xi_i$, the GDP growth rates are $V_i$, the electricity consumption growth rates are $z_i$, the electricity consumption prediction values are $w_i$, the face values are $F_i$, the face growth rates are $H_i$, $i = 1, 2, 3$. The electricity demand forecasting values of Guangzhou city in 2009 are shown as follows:

1. The elasticity coefficient is $\kappa_i$. $\kappa_i = (\xi_i*(1 + V_i)-1)/V_i$, $i = 1, 2, 3$, with respect to the high-, medium- and low-schemes;
2. The electricity consumption growth rate is $z_i$. $z_i = \kappa_i * V_i$, $i = 1, 2, 3$, with respect to the high-, medium- and low-schemes;
3. The electricity consumption prediction value is $w_i$. $w_i =$ (actual electricity consumption value in 2008)*$(1+ z_i)$, $i = 1, 2, 3$, with respect to the high-, medium- and low-schemes;
4. The face value is $F_i$. $F_i = w_i +$ (output value at end of the year 2008), $i = 1, 2, 3$, with respect to the high-, medium- and low-schemes;
5. The face growth rate is $H_i$. $H_i = F_i/$(per ten thousand kWh primitive electricity consumption value in 2008)$-1$, $i = 1, 2, 3$, with respect to the high-, medium- and low-schemes.

Table 4. Elasticity coefficient and electricity consumption growth trend forecasting values of Guangzhou city in the year 2009.

| Schemes | Elasticity coefficient | Electricity consumption growth rate | Electricity consumption value in 2009 | Face value in 2009 | Face value growth rate in 2009 |
|---|---|---|---|---|---|
| High-scheme | 0.55 | 5.47% | 5,810,505 | 5,900,991 | 8.09% |
| Medium-scheme | 0.39 | 3.53% | 5,703,602 | 5,794,088 | 6.13% |
| Low-scheme | 0.20 | 1.61% | 5,597,692 | 5,688,178 | 4.19% |

## 5 MEASURES OF IMPROVING ELECTRICITY DEMAND FORECASTING ACCURACY

As aforementioned, it is of great significance to improve the accuracy of electricity demand forecasting and the measures included are as follows [18–21]:

a. Accomplish relation material accumulation between the climate and climate-load, grasp the weather changes in time, build an effective communication mechanism with the meteorological station, and record some relations between climate and climate-load in different periods of each year. Based on the collection and analysis of materials, gradually grasp the law and make dynamic corrections on the forecasting results, to improve the short-term electricity demand forecasting accuracy;

b. Adopt suitable electricity forecasting methods, for example, build the seasonal scalar model, the multivariate fuzzy linear regression model, etc. Based on the multivariate linear regression analysis method, build an easily acceptable analysis model to provide scientific evidence for power forecasting;

c. Adopt the neural network method and constantly improve neural network training modes in neural network methods to increase the size of training sample, so as to reduce errors of electricity prediction and improve the prediction accuracy;

d. Adopt newly developed machine learning methods, such as the Q-learning algorithm [22–24], reinforcement learning, deep learning, and other artificial intelligent algorithms. Use these algorithms to constantly train the historical electricity consumption sample, to obtain significant "experience", so as to improve forecasting accuracy of future electricity demand;

e. Adopt the quantitative method with qualitative method. Firstly, read the historical electricity basic data from the database, forecast based on relative prediction model theories, such as the seasonal scale model theory, improved BP network model, multivariate fuzzy linear regression model, etc. to respectively solve the next phase electricity sale prediction values in each market segment; the prediction values of above three models are set as inputs of the assorted forecasting model to solve the forecasting values of electricity sales of the subsequent month in each market segment, and then, based on these prediction values and errors existing in the analysis process, analyze whether the prediction values are reliable, and if they are, then use these values to improve supply ability of grid or reduce electricity sales in each market segment; otherwise, continuously make corrections with the working personnel;

f. Design a better power consumption scheme and focus on the national financial economy development trend in time and accurately grasp the economic pulse and energy demand growth and the relations between national and social development; meanwhile, maintain a fine train of thought;

g. Build a forecasting database. Based on building such a detailed fundamental database, analyze the historical data, find objective laws, and improve electricity prediction accuracy;

h. Investigate the big users and deeply understand their production and business operation and completely implement market investigation work, and know customer products sale, overstock status, fund flow status, etc., and strengthen the management of big industrial customers and local power factories and completely eradicate disorder production.

## 6 CONCLUSIONS

Electricity demand forecasting is an important part of national power system planning. It is of great significance to accurately forecast the electricity demand under power market competitive conditions, which is conducive to index evaluation in the same business and provide basics for grasping the electricity sale market by using power supply enterprises, directions for grid development, references for ensuring tariff in full recovery, etc. In this paper, the electricity demand forecasting theories and commonly used methods, such as the econometric model method, comprehensive analysis method, etc. are explored and then, an example analysis based on the electricity elasticity coefficient method and some measures to improve the electricity demand forecasting accuracy, especially combined with newly developed artificial intelligent algorithms, such as the machine learning algorithm, deep perception theory, reinforcement learning, deep learning, extreme learning, emotion learning, etc. are provided. Apply these smart methods in electricity prediction and gain reliable experience according to large amount training on the historical electricity data sample and hence, form a forecasting model with high accuracy, which will provide significant guidance and reference for multi-agents supply and demand interaction, such as the grids, users, power sellers, and load integrators, in the future power market competitive environment.

ACKNOWLEDGMENT

This research work was supported by the Science and Technology Projects of China Southern Power Grid (No. GDKJ00000052).

REFERENCES

[1] Hu N. 2012. The electric power consumption forecasting method. *Jiangxi Electric Power* 36(4): 77–79.
[2] Chen M.C., Mu G., Sun Y. et al. 2005. Application of DFT-based grey model in daily load forecast. *Electric Power Automation Equipment* 25(9): 29–32.
[3] Zhang Y.Q. 1999. The middle & long term power consumption forecasting based on the grey system theory. *Power System Technology* 23(8): 47–50.
[4] Niu D.X., Cao S.H., Zhao L., et al. 1998. *Power system load forecasting technique and its application*. Beijing: China Electric Power Press.
[5] Cao G.J., Huang C., Long H., et al. 2004. Load forecasting based on improved GM (1,1) model. *Power System Technology Technology* 28(13): 50–53.
[6] Pang Y.Y., Wang S.Y. & Han L. 2015. Study of power system load electricity forecasting. *Electric Age* 35(7): 104–105.
[7] Niu D.X., Chen Z.Y., Xing M. et al. 2002. Combined optimum grey neural network model of the seasonal power load forecasting with the double trends. *Proceedings of the CSEE* 22(1): 29–32.
[8] Zhong J. & Chen G. 2007. Clustering analysis for time feature of user power consumption based on structural self-adaptation ANN. *Jourrnal of Chong-Qing University: Natural Science Edition* 30(8): 44.

[9] Tan X.D., Hu Z.G., Li C.B., et al. 2007. Study on predictive model to social power consumption based on the improved BP neural network. *Journal of North China Electric Power University* 34(3): 85.
[10] Qin H.T., Liu Y., Xiao H., et al. 2015. Forecast of medium and long-term power demand and distribution in Chongqing. *Electric Power Construction* 36(4): 115–122.
[11] Fang L. 2005. *China electricity demand forecasting and analysis*. Master's Thesis, Dongbei University of Finance & Economics, Shenyang, China.
[12] Qian W. 2001. Load forecasting using the electricity elasticity coefficient method. *Yunnan Electric Power* 29(1): 41–44.
[13] Liu D. 2014. Power market electricity analysis and forecasting analysis. *Technology Innovation and Application* 3(35): 162.
[14] Xia R.C., Xie K.G., Cao K., et al. 2011. Error correction model for short-term electricity price forecasting. *East China Electric Power* 39(2): 161–167.
[15] Wu J. 2010. Discussion on reactive power service pricing under power market circumstance. *Science & Technology Information* 8(23): 225.
[16] Pan G.M. 2000. Discussion on several electricity utilization forecasting methods. *Shanghai Electric Power* 13(2): 25–28.
[17] Zheng G. 2012. Discussion on electric power consumption forecasting method. *Science Mosaic* 19(6): 94–96.
[18] Ji H.Q. 2015. Brief discussion on measures of improving electricity forecasting accuracy. *Scientific Time* 19(10): 84–86.
[19] Mai Q. 2015. Brief analysis on necessity and measures of improving power electricity forecasting accuracy. *Technology and Market* 22(12): 105–107.
[20] Jiang Z.Y. 2013. *Design and implementation of power grid system based on electricity demand forecast and customer behavior analysis*. Master's Thesis, Sun Yat-sen University, Guangzhou, China.
[21] Guo Y.J., Huang C., Zhu H.B., et al. 2015. Distribution automation technology development and demonstration projects. *The Journal of New Industrialization* 5(11): 20–26.
[22] Wu W.M., Lu J., Tan M., et al. 2015. Multi-regional reactive power optimization based on correlated equilibrium Q-learning collaborative algorithm. *The Journal of New Industrialization* 5(6): 33–40.
[23] Tang J., Zhang X.S., Cheng L.F., et al. 2015. Intelligent Materials and Mechatronics in China: An international conference. *Hierarchical correlated Q-Learning for multi-layer optimal generation command dispatch; Hong Kong, 29–30 October 2015*. Pennsylvania: DEStech Publications, Inc.
[24] Tang J., Zhang X.S., Cheng L.F., et al. 2015. Intelligent Materials and Mechatronics in China: An international conference. *TOPSIS-Q($\lambda$) learning for multi-objective optimal carbon emission flow of power grid; Hong Kong, 29–30 October 2015*. Pennsylvania: DEStech Publications, Inc.

*Advances in Energy and Environment Research – Achour & Wu (Eds)*
*© 2017 Taylor & Francis Group, London, ISBN 978-1-138-62682-9*

# Load prediction based on support vector regression with genetic algorithms

Min Zou

*School of Electronics and Electricity Engineering, Wuhan Textile University, Wuhan, China*

ABSTRACT: Load prediction is an important activity in either real-time or day-ahead electricity markets. In this paper, a combination approach based on Least Square Support Vector Regression (LSSVR) and Genetic Algorithm (GA) parameters' optimization is proposed for load prediction. An effective prediction model can only be built by using optimal parameters. The genetic algorithm is applied to search for optimal parameters of the above-mentioned load prediction model. The experimental results based on the above-mentioned model for a sample load series show that the model proposed in this paper outperforms the BP neural network approaches and the simple LSSVR methods on the mean absolute percent error criterion.

## 1 INTRODUCTION

Load prediction plays an important role in all aspects of reliable, economic, and secure strategies for power systems. The power load series are influenced by many factors such as climate, economic policy, price, etc. Thus, it is difficult to obtain an accurate prediction. Many load prediction models are constructed (Nima & Farshid 2008, Nima & Ali 2011), of which the neural network is often used. The neural network method is generally used in nonlinear time series predictions (Huang et al. 2007). But the over-fitting problems in the neural network method retard the performance improvement to achieve higher levels of accuracy. With the development of the Statistic Learning Theory (SLT) and Support Vector Machines (SVM) algorithm, SVM and its extended versions, such as Least Squares Support Vector Machines (LS-SVM) are gradually applied to nonlinear time series prediction research with good performance (Catalão et al. 2007, Xu et al. 2004). However, parameters' determination in the time series prediction based on SVM is generally a complicated and important problem to be solved carefully, which is directly related to the model prediction performance.

Genetic algorithms provide the general framework to solve optimization problems of complicated systems, which is independent of the specific areas. After the kernel function of SVM methods is determined, the kernel function parameter and penalty parameter need to be set properly. In the regression prediction research based on SVM, the model parameters are generally determined by the trial-and-error method, which determines the

final optimal set of parameters with best performance from sets of selected parameters through analyzing the influences of these parameters on the model. With the development of SVM research, the research of optimal parameters' determination for SVM is gradually becoming a common problem in SVM research areas (Yang et al. 2006, Liu et al. 2006). With genetic algorithms global optimization features, a novel prediction model based on Least Squares Support Vector Regression (LSSVR) with parameters optimized through genetic algorithms, also named as the GA–LSSVR model, is proposed in this paper, which searches for the optimal parameters of the LSSVR model using real-value genetic algorithms and adopts the optimal parameters to construct the LSSVR model.

This paper is organized as follows: In section 2, LSSVR is introduced and the GA–LSSVR forecast model based on genetic algorithms and LSSVR is discussed. Experimental data applied in the GA–LSSVR model are proposed in section 3. Finally, the conclusions are given in section 4.

## 2 GA–LSSVR

### 2.1 *GA–LSSVR optimization procedure*

The Support Vector Regression (SVR) algorithm is a machine learning algorithm, which is proposed by Vapnik (1999) based on the Statistical Learning Theory (SLT). The LS-SVM algorithm is proposed by Suykens (2001) based on the standard SVM theory, which is a least squares version of the standard SVM algorithm and involves equality instead of inequality constraints and works with a least squares object function. Owing to the excellent performance in nonlinear forecasting problems, support vector machine prediction models, including models based on its extension, have been generally applied for prediction models.

Given data $\{(x_1, y_1), \cdots, (x_N, y_N)\}$, where $x_i \in R^n$ denotes the inputs of sample data and has a corresponding target value $y_i \in R$ for $i = 1 \cdots N$. The SVR algorithm maps the input space into high dimension feature space through nonlinear functions and constructs a linear regressive function as follows in this feature space:

$$y(x) = \omega^T \phi(x) + b \qquad (1)$$

where $\phi(x)$ is a nonlinear function, $\omega^T$ is an m-dimensional vector, and $b$ is a bias.

To solve the above-mentioned regressive function, LS-SVR defines the optimization problem as follows (Suykens 2001):

$$\begin{cases} \min_{\omega, b, e} J_2(\omega, b, \xi) = \frac{1}{2}(\omega^T \omega) + \gamma \frac{1}{2}\sum_{i=1}^{n}\xi_k^2 \\ y_k\left[\omega^T \phi(x_k) + b\right] = 1 - \xi_k, \quad k = 1, \ldots, n \end{cases} \qquad (2)$$

where $\xi$ is a slack variable, which is necessary to allow regression error and $\gamma$ is the penalty parameter, which compromises the effect between maximum margin and the minimum regression error.

To build an efficient LSSVR prediction model, LSSVR's parameters must be determined carefully. These parameters include the following factors:

1. *Kernel function*: the kernel function is important in the LSSVR model and is used to construct a nonlinear decision hyper-surface in the LSSVR input space. The RBF kernel is the most typical of all kernel functions and exhibits the same performance under some conditions with other kernel functions, such as linear polynomial and sigmoid kernel functions.
2. *Regularization parameter $\gamma$*: $\gamma$ determines the trade-off cost between minimizing the training error and minimizing the model's complexity.
3. *Bandwidth of the kernel function ($\sigma^2$)*: $\sigma^2$ represents the variance of the Gaussian kernel function.

The GA–LSSVR uses genetic algorithms to seek the optimal value of LSSVR's parameters and improve the model prediction performance. The proposed GA–LSSVR prediction model dynamically optimizes the optimal value of LSSVR's parameters during the genetic algorithm evolution process to construct the optimized LSSVR model.

Figure 1 illustrates the process of the GA–LSSVR model. Details of the proposed GA–LSSVR model are presented as follows:

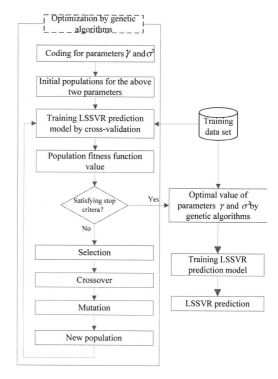

Figure 1. Flowchart showing the functioning of the GA–LSSVR model.

1. *Representation*: when genetic algorithms are used for optimization, different kinds of representations of chromosomes are concerned, in which the real-value representation is a direct one. In real-value genetic algorithms, the real-value parameters or variables are directly used to form chromosomes without being translated into the binary code and the chromosome representation is straightforward. The two LSSVR parameters $\gamma$ and $\sigma^2$ are directly coded to generate the chromosome in the proposed model. The chromosome $X$ can be represented as $X = [p1, p2]$, in which $p1, p2$ represent the above parameters $\gamma$ and $\sigma^2$, respectively.
2. *Population initialization*: in this model, initial population is composed of 30 random chromosomes. The population size of 30 is the trade-off between model complexity and population diversity.
3. *Fitness function*: Fitness function is directly related to the performance of genetic algorithms. If the fitness value of training data is calculated directly, genetic algorithms may lead to over-fitting. In order to solve this problem, $k$-fold cross-validation (Duan et al. 2003) is concerned. In $k$-fold cross-validation, the training data are randomly split into $k$ mutually

exclusive subsets (the folds) of approximately equal size. The regression model is obtained with a given set of parameters $\{\gamma, \sigma^2\}$, using $k-1$ subsets as the training set and the performance of the parameter set is measured by the Mean Absolute Percent Error (MAPE) on the subset left out. The above procedure is repeated $k$ times and in this fashion, each subset is used for testing once. Averaging the MAPE over the $k$ trials (MAPE$_{CV}$) gives an estimate of the expected generalization error. Conventionally, the training error of $k$-fold cross-validation is applied to estimate the generalization error ($k = 5$ is suggested in Duan et al. (2003)). The fitness function of the proposed model is defined as the MAPE$_{CV}$ of the 5-fold cross-validation method on the training data set, which is as follows:

$$\min f = \text{MAPE}_{CV} \qquad (3)$$

$$\text{MAPE}_{CV} = \frac{1}{k}\sum_{i=1}^{k}\text{MAPE}_i \qquad (4)$$

$$\text{MAPE} = \frac{1}{n}\sum_{i=1}^{n}\left|\frac{x_i - \hat{x}_i}{x_i}\right| \qquad (5)$$

where n is the number of training data samples, $x_i$ is the actual value, and $\hat{x}_i$ is the predicted value.

4. *Genetic operator*: the processes carried out by using a genetic operator include selection, crossover, and mutation. A standard tournament selection method is employed to select 8 chromosomes from the current population. The standard arithmetic crossover method is applied to randomly pair chromosomes with a crossover probability of 0.8. The standard uniform mutation method is used in this model and the mutation probability is 0.05.
5. *Stopping criteria*: The process is repeated as shown in Figure 1 until the generation number is equal to 50 or the fitness function value variation is not beyond 0.005.

## 2.2 *Embedded dimension*

In time series prediction problems, the time series are generally expanded into three dimensions or higher dimensions space to explicit the implicit information of the series, which is called as the state reconstruction (Kennel et al. 1992). Embedding dimension $m$ is an important factor in this method. The given series $\bar{x} = \{x_1, x_2, \ldots, x_N\}$ are transformed as follows by the embedding dimension, where $X$ is the input matrix and $Y$ is the corresponding output matrix.

$$X = \begin{bmatrix} \bar{x}_1 \\ \bar{x}_2 \\ \vdots \\ \bar{x}_{N-m} \end{bmatrix} = \begin{bmatrix} x_1 & x_2 & \cdots & x_m \\ x_2 & x_3 & \cdots & x_{m+1} \\ \vdots & \vdots & \ddots & \vdots \\ x_{N-m} & x_{N-m+1} & \cdots & x_{N+1} \end{bmatrix},$$

$$Y = \begin{bmatrix} \bar{y}_1 \\ \bar{y}_2 \\ \vdots \\ \bar{y}_N \end{bmatrix} = \begin{bmatrix} x_{m+1} \\ x_{m+2} \\ \vdots \\ x_N \end{bmatrix} \qquad (6)$$

According to the nonlinear and nonstationary characteristics of vibration of HGU, the False Nearest Neighbors (FNN) method is adopted in this paper to estimate the embedding dimension of the given series.

## 2.3 *Error criterion*

There are many criterions that are used to evaluate the performance of prediction models; here the MAPE criterion shown in equation [5] is adopted in this paper to evaluate the performance of the hybrid model for prediction.

## 3 GA–LSSVR MODEL APPLICATION

Week max load series of some areas are used in this paper to testify to the proposed forecasting model performance. The week max load series are shown in Figure 2.

## 3.1 *Estimation of embedding dimension*

Embedding dimension is very important in time series prediction with state reconstruction. Figure 3 shows the embedding dimension of load series as can be seen from the figure; the embedding

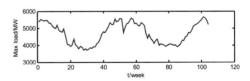

Figure 2. Graph showing the load series of some areas.

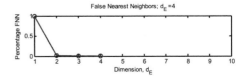

Figure 3. The FNNP value of load series.

19

dimension is estimated while the FNN percentage (FNNP) reduces to 0 and the dimension is 4.

### 3.2 *Result analysis*

In order to contrast the performance of PSO-LSSVR with other existing prediction methods, a BP neural network prediction model and a simple LSSVR prediction model are concerned, whose outputs are shown in Figures 4 and 5, in which, the solid line indicates original data and the dashed-line indicates regression and prediction data. In the BP neural network prediction model, the sample data are normalized before being used in the model. In the LSSVR prediction model, a grid search and cross-validation method is adopted to find the optimal value of the model parameters.

In the GA–LSSVR model, the search domains of parameters $\gamma$ and $\sigma^2$ should be given at the beginning. After optimal searching in the above domains, the optimal value of parameters $\gamma$ and $\sigma^2$ are found, which are used to construct the LSSVR forecasting model. The performance of the GA–LSSVR is shown in Figure 6.

As can be seen from Figures 4–6, the BP neural network model exhibits better performance in the regression curve, but worse performance in the prediction curve, which indicates the over-fitting problems in the BP neural network model, while

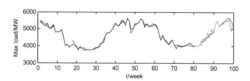

Figure 4. Graph showing the prediction result of the BP neural network.

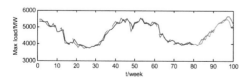

Figure 5. Graph showing the prediction results of LSSVR.

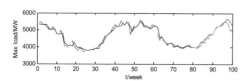

Figure 6. Graph showing the prediction results of GA–LSSVR.

Table 1. Regression and prediction error of load series.

| Forecasting model | Regression error MAPE | Prediction error MAPE |
|---|---|---|
| BP NNs | 0.1623 | 0.1815 |
| LSSVR | 0.0219 | 0.0288 |
| GA–LSSVR | 0.0208 | 0.0198 |

the GA–LSSVR model has an approximate performance to the simple LSSVR model with parameters optimized by grid search and cross-validation. But the performances of both the simple LSSVR model and the GA–LSSVR model are better than that of the BP neural network model. The over-fitting problems are resolved better in the last two models. From error indices shown in Table 1, we can see that the GA–LSSVR performance is slightly better than that of the simple LSSVR model.

Upon achieving desirable regression and prediction performance, the GA–LSSVR model exhibits better fitting performance to the original load series curve.

## 4 CONCLUSION

Load prediction is one of the important processes in the power dispatch automation and electricity market. By combining LSSVR with GA, a model named as GA–LSSVR is proposed in this paper. The K-fold cross-validation fitness function is used to search for optimal parameters to construct the LSSVR prediction model. When compared to the BP neural network model and simple LSSVR with grid search and cross-validation, the GA–LSSVR exhibits great performance in regression and prediction. The GA–LSSVR model is an adoptable technique for load prediction.

ACKNOWLEDGMENT

This work was supported by the Educational Commission of Hubei Province (Grant No. B20111606).

REFERENCES

Catalão J. P. S., Mariano S. J. P. S., Mendes V. M. F. & Ferreira L. A. F. M. 2007. Short-term electricity prices forecasting in a competitive market: A neural network approach. Elect. Power Syst. Res. 77(10): 1297–1304.

Duan K., Keerthi S. S. & Poo A. N. 2003. Evaluation of simple performance measures for tuning SVM hyper-parameters. Neurocomputing 51: 41–59.

Huang R., Xi L., Li X., Liu C. R., Qiu H. & Lee J. 2007. Residual life predictions for ball bearings based on self-organizing map and back propagation neural network methods. Mechanical Systems and Signal Processing 21(1): 193–207.

Kennel M. B., Brown R. & Abarbanel H. D. I. 1992. Determining embedding dimension for phase-space reconstruction using a geometrical construction. Physical Review A (General Physics) 46: 3404–3411.

Liu J. & Zhang T. 2006. A method determining parameters of SVR model based on Probability and Statistics. Proceedings of the World Congress on Intelligent Control and Automation (WCICA) 1: 1553–1557.

Nima A. & Ali D. 2011. Midterm Demand Prediction of Electrical Power Systems Using a New Hybrid Forecast Technique. IEEE TRANSACTIONS ON POWER SYSTEMS 26(2): 755–765.

Nima A. & Farshid K. 2008. Mid-term load forecasting of power systems by a new prediction method. Energy Convers. Manage 49(10): 2678–2687.

Suykens J. A. K. 2001. Nonlinear modelling and support vector machines. Conference Record—IEEE Instrumentation and Measurement Technology Conference 1: 287–294.

Vapnik V. N. 1999. An overview of statistical learning theory. Neural Networks, IEEE Transactions on 10: 988–999.

Xu, T., He, R., Wang, P. & Xu, D. 2004. Input dimension reduction for load forecasting based on support vector machine. in Proc. IEEE Int. Conf. Electric Utility Deregulation, Restructuring and Power Technologies (DRPT) 2: 510–514.

Yang H.Z., Jiao X.N., Zhang L.Q., & Li F.C. 2006. Parameter Optimization for SVM using Sequential Number Theoretic for Optimization. Proceedings of the 2006 International Conference on Machine Learning and Cybernetics 2006: 3461–3464.

*Advances in Energy and Environment Research – Achour & Wu (Eds)*
*© 2017 Taylor & Francis Group, London, ISBN 978-1-138-62682-9*

# Short-term electricity demand forecasting by using a MATLAB-based ANFIS model

Abit Balin, Merve Sert & Ilknur Besirli
*Department of Industrial Engineering, Faculty of Engineering, University of Yalova, Yalova, Turkey*

ABSTRACT: Demand for electricity is a very important issue in the development policies of countries. Due to industrialization, growth of cities, and population growth, the demand for electricity has increased and with that increase in electricity, the electric energy demand forecast becomes important. Because of this reason, demand forecasting models are requested in order to meet customer expectations. In this study, an ANFIS model was developed for electricity demand by dint of MATLAB. In this model, the daily total electrical energy consumption for the last week of April 2016 are obtained and compared with actual electrical energy consumption and it best fits with the 0.982133R2 value.

## 1 INTRODUCTION

Energy consumption is rapidly increasing around the world due to population growth and industrial development. The great need for energy thus raised the importance of electricity generation methods in terms of efficient use of existing sources. Within this context, variety of sources, economical operation of systems in production and consumption, more efficient use of resources, realistic decisions, and energy management are becoming increasingly important.

In Turkey, the generation of electricity is fueled by coal, natural gas, and renewable energy sources. These renewable energy sources are as follows: hydropower, biomass, wind, solar, and geothermal. The annual average electricity demand in Turkey is constantly increasing by 6–7%. The power capacity of 69516.4 MW was installed by the end of 2014. Of this installed power, 40.3% has been obtained from renewable energy sources and the rest is obtained from other sources (Ministry of Energy and Natural Resources Activity Report, 2014).

## 2 LITERATURE REVIEW

Artificial string networks are widely used in many fields, particularly in classification, modeling, and forecasting studies. By using ANN, general studies forecasting electricity load, energy consumption, electric energy demand, forecasting activities related to electricity generation, and consumption have been conducted. An ANFIS method is used in various fields such as the evaluation of bank loan application, risk assessment of structures like bridges, and diagnosis of diabetes.

Yıldırım and et al. (2006) estimated the level of daily air pollution by using the ANFIS method. Hamzaçebi (2007), taking electricity demand on a sectoral basis by using the ANN method, made a prediction on the net electricity consumption of Turkey till 2020. Sözen (2007) developed three different models by dint of the ANN, so as to demonstrate the relation between Turkey's net energy consumption and economic indicators. Fırat and Güngör (2007) established a model of predicting river flows, thanks to the ANFIS method. Wang and Elhag (2008) applied the ANFIS method in risk assessment of bridges. In another data, Sözen (2009) created a model by using YSA to estimate the future position of Turkey's energy dependency, by using basic energy indicators and sectoral energy consumption data. Hocaoğlu et al. (2009) designed a model by the ANFIS method to estimate monthly wind speeds despite deficient data in the case of temporal failure or maintenance at wind observation stations.

## 3 MATERIALS AND METHODS

In this work, we used the ANFIS method, which stands for Adaptive Network Base Blurred Extraction System. It is also known as blurred artificial string network. This method aims to achieve true determination of electricity demand in advance, which is required for electric energy. We will bring out the fact that the demand for electric energy could be estimated by using the ANFIS method. ANFIS is a hybrid artificial intelligence method. ANFIS is a hybrid blurred logic extraction method, which applies learning skills by parallel estimation to artificial string networks. One of the basic features of ANFIS is that it uses data in blurred form. It also consists of a combination of blurred learning and artificial string networks. ANFIS could self determine the rules to solve the problems as well as an expert might create the rules (OnatandErsoz, 2011; Jang, 1993; Nayak, et al., 2004). The graph xy shown in Figure 2 illustrates the flow schema for the ANFIS method (see Balin and Baracli 2014 for further details about the model). The performance of the model was measured by using the Mean-Squared Error (MSE) statistical parameter.

$$MSE = \frac{1}{n}\sum_{i=1}^{n}\varepsilon_i^2 = \frac{1}{n}\sum_{i=1}^{n}(x_i - y_i)^2 \qquad (1)$$

In equation (1), i refers to data index, n is the number of output data, which are employed in error measurement, $x_i$ and $y_i$ are data of the system's real output and system's output, respectively and $\varepsilon_i$ denotes error.

## 4 A REAL FORECASTING FOR SHORT-TERM DEMAND ESTIMATION

Our model, examined in the light of the literature, estimated with the data, spanning between 2014 and 2016 (the most convenient time for electric energy consumption) for the import-export, was obtained from Turkish Electric Flow Inc. (TEIAS).

A frame, containing 6 inputs and an output, is shown in Figure 1 for the model ANFIS within the scope of this problem. The import and export values for the electric energy was used for input values for the period 2014–2016. Each of the input values was blurred with the activation function.

For model 6, the exact estimation values' output was given by entering blurring and 100 epoch values into the 2 inputs. A trapez shaped membership function (trapmf) was employed in our model, which gives the best results. In work 6, while the energies of electricity export and import values, as training data, were examining for the test data, 2014–2016 daily total consumption values were employed for the target data by entering 100 epoch values. As a result, it best fits with 0.982133R$^2$ value.

Each work was examined separately with 2, 3, 4, 5, 6 inputs and the best result was obtained with 6 input model, whereas the blurring method worked best with the 2 inputs method and also it was tried with each of the 2, 3, 4, 5, 7 inputs separately.

A 7 membership function was tried for the ANFIS' linear and hybrid models and it was observed that the 3 angle shaped membership method (trimf) gives the best result for the linear model.

The epoch values were appointed differently between 50 and 30000 in the works. The model gives the most accurate results with the 50-epoch value in the last sample. In Figure 2, the data for training in response to the mean squared error values were illustrated.

10% of the data is used as test data and 90% as training data. In Figure 3, the level of accuracy of the results provided by the training and test data is shown. It can be observed that the estimates gave the best approximation to the reality.

The possibly estimated values of electricity on a weekly basis and real values are demonstrated in Figure 4. The data set represented by points is our estimated value. As seen in Table 1, *mean-square-error* is the least value.

The model is feedforward. The total electricity consumption per day for the last week of April 2016 was determined. The daily outcome data are

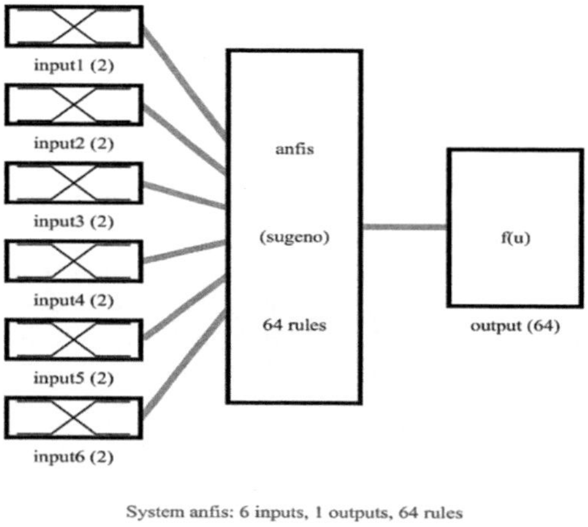

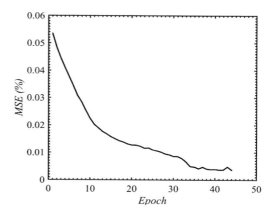

System anfis: 6 inputs, 1 outputs, 64 rules

Figure 1. Schematic of the structure of the model.

Figure 2. The mean squared error values for the training data.

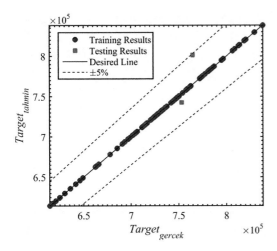

Figure 3. Compatibility of training and test data with estimation.

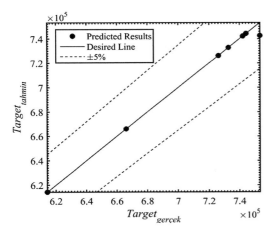

Figure 4. Estimated results and real values.

Table 1. Forecast values error values.

| No | TTCO April last week | ANFIS April last week | R2 | MSE |
|----|------|-------|-----|-----|
| 1 | 732773.8 | 732773.7304 | 0.982133 | 0.000519 |
| 2 | 666055.55 | 666057.5108 | 0.791546 | 0.006051 |
| 3 | 614276.17 | 614279.9659 | 0.794936 | 0.005953 |
| 4 | 726124.77 | 726123.5118 | 0.968693 | 0.000909 |
| 5 | 742096.24 | 742094.8934 | 0.827703 | 0.005002 |
| 6 | 744215.94 | 744211.8504 | 0.97959 | 0.000593 |
| 7 | 753254.55 | 742600.7517 | −0.09918 | 0.03191 |

acquired, thanks to the ANFIS model and data drawn from the official website of TETCO (Turkey Electricity Transmission Company) are presented in Table 1.

## 5 CONCLUSION

Because of the impossibility of storing electricity, the supply–demand stability is the primary objective of the states.

As a result of this study, we have suggested that the ANFIS model is possible to be used in the estimation of electricity demand and therefore, the model will be benefited when the actors in the energy industry face a competing demand and consumers may be offered a constant energy flow in almost-accurate values.

This study can also be carried out by using the ANN and genetic algorithms. Besides, adding data to the ones used in the study, the same demand estimation is possible, thanks to the ANFIS model.

## REFERENCES

Balin, A. and Baracli, H., (2014). "Modeling Potential Future Energy Demand for Turkey in 2034 by Using an Integrated Fuzzy Methodology," Journal of Testing and Evaluation, Vol. 42, No. 6, pp. 1–13.

Fırat M., Güngör M., (2007). "River Flow Estimation Using Adaptive Neuro Fuzzy Inference System", Mathematics and Computers in Simulation, 75, 87–96.

Hamzaçebi, C., (2007). Forecasting of Turkey's net electricity energy consumption on sectoral bases, Energy Policy, 35: 2009–2016.

Hocaoğlu, F., Oysal, Y., Kurban, M., (2009). Missing wind data forecasting with adaptive neuro fuzzy inference system, Neural Computation & Application, 18(3): 207–212.

Jang, J. S., (1993). "ANFIS: Adaptive-Network-Based Fuzzy Inference System," IEEE Trans. Syst. Man Cybernet., Vol. 23, No. 3, pp. 665–685.

Nayak, P. C., Sudheer, K. P., Rangan, D. M., and Ramasastri, K. S., (2004). "A Neuro-Fuzzy Computing Technique for Modeling Hydrological Time Series," J. Hydrol., Vol. 291, No. 1, pp. 52–66.

Onat, N. and Ersoz, S., (2011). "Analysis of Wind Climate and Wind Energy Potential of Regions in Turkey," Energy, Vol. 36, No. 1, pp. 148–156.

Republic of Turkey Ministry of Energy and Natural Resources (T.C. EnerjiveTabiiKaynaklarBakanlığı) 2014 Annual Report (Web page: http://www.enerji.gov.tr/File/?path = ROOT%2F1%2FDocuments%2FFaaliy et+Raporu%2F2014.pdf), (Erişimtarihi: Aralık 2015).

Sözen, A., (2009). Future projection of the energy dependency of Turkey using artificial neural network, Energy Policy, 37(11): 4827–4833.

Sözen, A., Arcaklioğlu E., (2007). Prediction of net energy consumption based on economic indicators (GNP and GDP) in Turkey, Energy Policy, 35: 4981–4992.

Wang Y. M., Elhag T. M. S., (2008). "An Adaptive Neuro-Fuzzy Inference System for Bridge Risk Assessment", Expert Systems with Applications, 34, 3099–3106.

Yıldırım, Y., Bayramoglu, M., (2006). Adaptive neuro-fuzzy based modelling for prediction of air pollution daily levels in city of Zonguldak, Chemosphere, 63(9): 1575–1582.

*Advances in Energy and Environment Research – Achour & Wu (Eds)*
*© 2017 Taylor & Francis Group, London, ISBN 978-1-138-62682-9*

# Use of concentrating solar technology on short solar chimney power plant

Oboetswe Motsamai & Kealeboga Kebaitse
*Department of Mechanical Engineering, University of Botswana, Gaborone, Botswana*

ABSTRACT: This paper investigated the influence of solar concentration on the performance of solar chimney power plant with the intension to reduce the solar chimney height. Theoretical analysis of the solar collector and the solar concentrator was presented. Simulation was carried out using Matlab. The parameters investigated were the temperature difference between the ambient air and the collector outlet, the air mass flow rate and the heat output of the solar collector. A reflective area of 527 m² was designed for purposes of concentrating solar energy into the solar collector. The area was segmented into 33 solar reflector of 16 m² focusing solar radiation to a focal area 25 m away from the center of each reflector. The maximum air temperature, mass flow rate and heat output achieved for a collector without solar concentrator are 325.51 K, 6.55 kg/s and 147.03 kW respectively whereas the maximum air temperature, air mass flow rate and heat output obtained for a collector with solar concentrator are 335.54 K, 9.48 kg/s and 308.6 kW respectively.

## 1 INTRODUCTION

Solar chimney power plant is one of the exciting innovations that have for some years attracted a lot of researchers throughout the world. It is a simple technology which is not yet well established. A prototype solar chimney power plant with a chimney of 194.5 *m*, chimney radius of 5.08 *m*, collector mean radius of 122.0 *m* and design power of 50 *kW* was built in Manzanares, Spain (S. J, 1988). The plant managed to produce electricity for seven years proving the viability and reliability of the system (Pasumarthi, 1998). Since then a lot of research have been carried out in many areas of the world to investigate the performance of the technology and also to test if the technology is also viable in those respective areas. Despite that, there is still need to carry out research that is directed to improving the power output, efficiency and reducing the startup costs for the solar chimney power plant (Motsamai, 2013). The efficiency and power output of the solar chimney power plant are proportional to the chimney height and the collector area (S. J, 1988; Dai, 2003; Gannon, 2000; Haaf, 1983; Hamdan, 2011; Schiel, 2005). As a result a lot of researchers are tempted to design solar chimney power plants with very tall chimneys and large collector areas. The disadvantage of constructing solar chimney power plants with large collector area and tall chimneys is that construction costs are high and there is limited height because of the technological constraints and restrictions on the construction material (Ngala, 2013). Additionally Li *et al.*

(2012) indicated that, up to date, no humans have acquired the experience of constructing large scale chimney of more than 1000 *m*. In order to overcome the challenges of constructing solar chimney power plants with very high solar chimney, Bilgen and Rheault (2005), Koonsrisuk (2013), Cao *et al.* (2011) and Zhou *et al.* (2013) investigated the performance of sloped chimney power plants. The system is constructed in such a way that the solar collector is built along the slope of a suitable hill and a short chimney is installed at the top of the hill. Indeed this approach will reduce the construction costs of the solar chimney but again the construction challenges will be shifted to the solar collector. Koonsrisuk (2013) suggested construction of small plants with high efficiency as this will encourage investments. No known literature has been reported on the application of concentrating solar energy technology on the solar chimney power plant. The objective of this paper is to develop a concept solar chimney using concentrating solar power concept in order to increase the power output of the system.

## 2 DESCRIPTION OF THE CONCEPT

The concept under study is a combination of four distinct components namely; solar collector, solar concentrator, solar chimney and a turbine coupled to a generator. The concentrator is made up of an array of solar reflectors arranged systematically on a defined area. The purpose of the reflectors is to

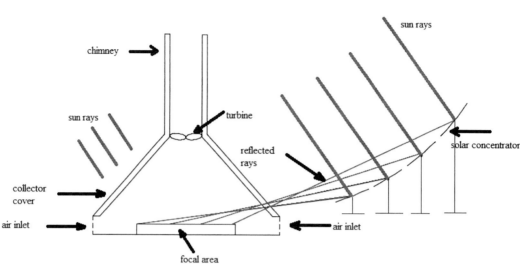

Figure 1.   A 2 dimensional schematic diagram of a solar chimney concept incorporating solar concentrators.

focus solar radiation into a focal area as shown in Figure 1. A combination of solar radiation coming directly from the sun through the collector cover and the one reflected from solar concentrators is absorbed by the thermal storage medium under the collector cover. As the thermal storage medium loses heat to the surrounding through the air inside the collector, some of the heat will be gained by the air. This reduces the density of air resulting in an upward movement of air. The kinetic energy of air will then be converted to mechanical energy as the air hits the turbine blades. Hence electricity will be produced because the turbine is coupled to a generator.

## 3   THEORETICAL MODEL

### 3.1   *Combined energy balance for the solar collector*

As stated by the law of conservation of energy, the rate of energy input $(E_{in})$ is equal to the sum of the rates of useful heat obtained from the solar collector $(q_u)$ and energy lost from the collector to the ambient $(E_{out})$ as indicated in equation (1).

$$E_{in} = q_u + E_{out} \tag{1}$$

The rate of energy input $(E_{in})$ refers to the amount of thermal energy absorbed by thermal storage medium $(G_{abs})$ and it is calculated as in equation (2).

$$G_{abs} = \tau_c \alpha_{ground} G_{coll} \tag{2}$$

where; $(\tau_c)$ is the transmittance of the collector cover, $(\alpha_{ground})$ is absorptivity of the thermal storage medium and $(G_{coll})$ is the rate of solar energy received by the collector cover.

The rate of solar energy received by the collector cover is expressed as the sum of the rates of reflected radiation from the concentrator to the collector $(G_{ref})$ and the solar radiation that falls directly into the solar collector from the sky $(G_{dir})$ as shown in equation (3).

$$G_{coll} = G_{ref} + G_{dir} \tag{3}$$

The rate of useful heat gain from the collector $(q_u)$ is obtained through the relation:

$$q_u = Mc_p \frac{dT_{air}}{dt} \tag{4}$$

where; $(M)$ is the mass of air, $c_p$ is specific heat capacity of air and $\left(\frac{dT_{air}}{dt}\right)$ is the rate of change of air temperature with time.

The rate of energy lost from the collector to the ambient $(E_{out})$ is calculated as in equation (5).

$$E_{out} = U_L \left(T_{surf} - T_{amb}\right) \tag{5}$$

where; $(U_L)$ is the overall heat transfer coefficient, $(T_{surf})$ is the temperature at the surface of the thermal storage medium and $(T_{amb})$ is the ambient temperature.

Re-writing equation (1) we get:

$$G_{coll}\tau_c\alpha_{soil} = U_L\left(T_{surf} - T_{amb}\right) + mc_p\frac{dT_{air}}{dt} \quad (6)$$

Introducing the heat removal factor $(F_R)$ into equation (6), Khartchenko and Kharchenko (2013) expresses the rate of useful heat output of a flat plate solar collector as:

$$mc_p\frac{dT_{air}}{dt} = AF_R\left[G_{coll}\left(\tau\alpha\right)_e - U_L\left(T_{air-in} - T_{amb}\right)\right] \quad (7)$$

where A is the absorber surface area (m²), $F_R$ is the heat removal factor of the solar collector, $\left(\tau\alpha\right)_e$ is the effective product of the transparent cover transmissivity $\tau$ and absorber absorptivity $\alpha$, and $T_{in}$ and $T_a$ are the collector fluid inlet temperature and ambient temperatures respectively, in °C.

Duffie and Beckman (1980) expressed the heat removal factor as:

$$F_R = \frac{\dot{m}C_p}{AU_L}\left[1 - exp\left(-\frac{AU_LF'}{\dot{m}C_p}\right)\right] \quad (8)$$

where; $F'$ is the collector efficiency factor.

The useful heat output of a solar concentrator is shown by equation (9).

$$Q_{useful} = \dot{m}C_p\Delta T = A_{ref}G_{dir}\rho_{A1}\tau_c\alpha_{soil} \\ -A_rU_L\left(T_{surf} - T_{amb}\right) \quad (9)$$

where; $\left(A_{ref}\right)$ is reflector aperture area, $(A_r)$ is receiver/absorber/focal area, $\left(\rho_{A1}\right)$ is the reflectivity of the concentrator and $\Delta T$ is the temperature difference between the collector outlet and collector inlet.

Making reflector area $\left(A_{ref}\right)$ in equation (9) subject of the formula, the equation becomes;

$$A_{ref} = \frac{\dot{m}C_p\Delta T + A_rU_L\left(T_{surf} - T_{amb}\right)}{G_{dir}\rho_{A1}\tau_c\alpha_{soil}} \quad (10)$$

The dimensions of the solar chimney collector used in this study are the same as the those from a study by Motsamai et al. (2013). The only modification performed on the above mentioned solar collector is the introduction of solar concentrators to increase the amount of solar radiation received by the solar collector. Simulation

Table 1. Dimensions of the solar collector.

| Description | Symbol | Unit | Value |
|---|---|---|---|
| Air passage | | mm | 250 |
| Collector thickness | $t_c$ | mm | 6 |
| Collector outlet from ground | $h_{coll}$ | m | 2 |
| Solar collector length | $L_c$ | m | 15 |
| Solar collector Width | $W_c$ | m | 15 |
| Solar collector Area | $A_c$ | m² | 225 |
| Reflective area | $A_{ref}$ | m² | 527 |
| Focal area | $A_r$ | m² | 154 |
| Focal length | | m | 25 |

was performed based on the dimensions shown on Table 1. The equations which were programmed into the Matlab code were equations (3), (6) and (9).

## 4 RESULTS AND DISCUSSIONS

Figure 2(a) and (b) shows how air mass flow rate changes with change in collector outlet temperature and the variation of collector heat output with change in air mass flow rate respectively. It can be observed that the collector outlet air temperature is directly proportional to the air mass flow rate. The reason behind this is that mass flow rate is a function of solar radiation incident on the solar collector, therefore if solar radiation increases, the collector outlet air temperature will also increase and hence this causes an increase in air mass flow rate. The maximum air temperature and mass flow rate reached are $325.51\,K$ and $6.55\,kg/s$ respectively. Figure 2(b) indicates that as mass flow rate increases, the heat output also increase. The maximum heat output attained without is $147.03\,kW$.

The design of the reflector area was based on the maximum heat output achieved by the solar collector without the solar concentrator. The assumption put across was that; the introduction of the solar concentrator will result in doubling the useful heat output under the collector. This means, the concentrator alone will produce a useful heat output of $147.03\,kW$. For a focal area of $154\,m^2$, concentrator reflectivity of $0.86$ and absorber maximum temperature of 370 K; the reflector area was discovered to be $527\,m^2$ using equation (10). The area of $527\,m^2$ presents a big structure that will give huge challenges during construction. Due to these challenges, it is suggested that smaller concentrators covering a space of $527\,m^2$ shall be considered. In this regard, it means if reflectors with an aperture area of $16\,m^2$ are to be planted on that area, therefore, 33 of them will be desired.

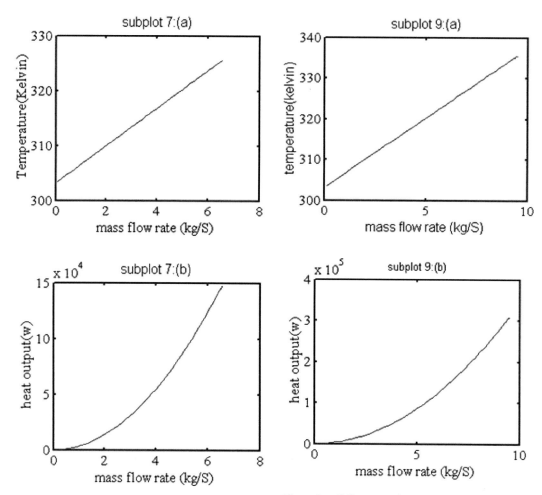

Figure 2. (a) Collector outlet temperature Vs. Air mass flow rate (b) Collector Heat output Vs. Air mass flow rate (without booster mirror).

Figure 3. Collector outlet temperature Vs. Air mass flow rate (b) Collector Heat output Vs. Air mass flow rate (with newly designed booster mirror).

Figure 3(a) and (b) presents the air temperature as a function of air mass flow rate and heat out as a function of air mass flow rate respectively. The results were obtained through simulations involving a separately designed booster mirror. The aperture area and the focal area of the booster mirror are $527 \, \text{m}^2$ and $154 \, \text{m}^2$ respectively. It is observed that there is a considerable increase of values as compared to values obtained in Figure 2. The maximum air temperature, air mass flow rate and heat output obtained are $335.54 \, K$, $9.48 \, kg/s$ and $308.6 \, kW$ respectively. The heat output obtained is almost double that of a collector without booster mirror and the air temperature increased by $10°C$ whereas the air mass flow rate increased by $3 \, kg/s$ as compared to the one without booster mirror.

## 5 CONCLUSIONS

The objective of this paper was to develop a concept solar chimney using concentrating solar power concept in order to increase the power output of the system. Mathematical models for the collector without and with booster mirror were developed and simulations were carried out using Matlab 6.1. An analysis of the simulated results was also carried out. The results indicate that indeed it is possible to increase the power output of a solar chimney power plant by applying the concept of solar concentration technology. Solar concentrators that can focus solar radiation $25 \, m$ away from the focal point were designed.

# REFERENCES

Alberti, S., "Analysis and Optimozation of the Scheffler Solar Concentrator," 2014.

Bilgen E., and J. Rheault, "Solar chimney power plants for high latitudes," *Solar Energy,* vol. 79, pp. 449–458, 2005.

Cao, F., L. Zhao, and L. Guo, "Simulation of a sloped solar chimney power plant in Lanzhou," *Energy Conversion and Management,* vol. 52, pp. 2360–2366, 2011.

Dai, Y. J., H. B. Huang, and R. Z. Wang, "Case study of solar chimney power plants in Northwestern regions of China," *Renewable Energy,* vol. 28, pp. 1295–1304, 2003.

Duffie J. A., and W. A. Beckman, *Solar engineering of thermal processes* vol. 3: Wiley New York etc., 1980.

Gannon A. J., and T. W. v. Backstrom, "solar chimney cycle analysis with system loss and solar collector performance," *solar energy engineering,* vol. 122, pp. 133–137, 2000.

Haaf, W., K. Friedrich, G. Mayr, and J. Schlaich, "Solar chimney Part I: Principle and Construction of the Pilot Plant in Manazanares," *Solar Energy,* vol. 2, pp. 2–20, 1983.

Hamdan, M. O., "Analysis of a solar chimney power plant in the Arabian Gulf region," *Renewable Energy,* vol. 36, pp. 2593–2598, 2011.

Jason Rapp, D. P. S., "Construction and improvement of a Scheffler reflector and thermal storage device," 2010.

Khartchenko N. V., and V. M. Kharchenko, *Advanced energy systems*: CRC Press, 2013.

Koonsrisuk A., and T. Chitsomboon, "Mathematical modeling of solar chimney power plants," *Energy,* vol. 51, pp. 314–322, 2013.

Koonsrisuk, A., "Comparison of conventional solar chimney power plants and sloped solar chimney power plants using second law analysis," *Solar Energy,* vol. 98, pp. 78–84, 2013.

Li, P.-h. J.-y., Guo, and Y. Wang, "Effects of Collector Radius and Chimney height on power output of a solar chimney power plant with turbines," *Renewable Energy,* vol. 47, pp. 21–28, 2012.

Motsamai, O., L. Bafetanye, K. Mashaba, and O. Kgaswane, "Experimental Investigation of Solar Chimney Power Plant," *energy and Power Engineering,* vol. 7, pp. 1980–1984, 2013.

Munir, A., O. Hensel, and W. Scheffler, "Design principle and calculations of a Scheffler fixed focus concentrator for medium temperature applications," *Solar Energy,* vol. 84, pp. 1490–1502, 2010.

Ngala, G. M., A. T. Sulaiman, and I. Garba, "Review of Solar Chimney Power Technology and Its Potentials in Semi-Arid Region of Nigeria," *International Journal of Modern Engineering Research,* vol. 3, pp. 1283–1289, 2013.

Pasumarthi N., and S. A. Sherif, "Experimental and theoretical performance of a demonstration Solar Chimney Model-Part I: Mathematical Model Development," *energy research,* vol. 22, pp. 277–288, 1998.

Schiel, W., J. Schlaich, R. Bergermann, and G. Weinrebe, "Design of Commercial Solar Updraft Tower Systems—Utilization of Solar Induced Convective Flows for Power," *Solar Energy,* vol. 127, pp. 117–124, 2005.

"Solar S. J, chimneys solar electricity from solar radiation," in *Advances in solar energy technology*, ed, 1988, pp. 1838–1842.

Zhou, X., S. Yuan, and M. A. d. S. Bernardes, "Sloped-collector solar updraft tower power plant performance," *International Journal of Heat and Mass Transfer,* vol. 66, pp. 798–807, 2013.

*Advances in Energy and Environment Research – Achour & Wu (Eds)*
*© 2017 Taylor & Francis Group, London, ISBN 978-1-138-62682-9*

# LiDFOB in sulfolane-carbonate solvents for high-voltage lithium-ion batteries

Liang Dong, Licong Huang, Fuxiao Liang & Cuihua Li
*College of Chemistry and Enviromental Engineering, Shenzhen University, Shenzhen, China*

ABSTRACT: In this work, we report a mixed electrolyte based on Sulfolane (SL) and lithium difluoro (oxalato) borate (LiDFOB) to improve electrochemical performance of high-voltage $LiNi_{0.5}Mn_{1.5}O_4$ cathode. In presence of sulfone-based electrolyte, the discharge capacity retention of cell achieves 88.2% after 100 cycles comparing to 78.6% of the cell with carbonate-based electrolyte at elevated temperature (55°C). It is verified by Inductively Coupled Plasma Atomic Emission Spectroscopy (ICP-AES) that the dissolution of transition metal in the sulfone-based electrolyte is heavily alleviated. Additionally, Scanning Electron Microscopy (SEM) shows that the decomposition of LiDFOB creates a uniform SEI film on the cathode surface and so suppresses the electrolyte decomposition at high voltage, especially at high temperature.

## 1 INTRODUCTION

In recent decades, driven by rapid development of plug-in hybrid electric vehicles, exploiting of Lithium-Ion Batteries (LIB) with increased energy density is waiting to be solved. Increasing its working potential is an effective way to enhance the energy density of the LIB. Spinel $LiNi_{0.5}Mn_{1.5}O_4$ has high operating potential (~4.7 V vs. $Li/Li^+$) and energy density (650 W h $kg^{-1}$), and three-dimensional $Li^+$ transmission channels (Liu et al., 2014). However, traditional carbonate electrolytes are easily decomposed under high voltage due to low oxidation potential (Kim et al., 2014). Moreover, traditional lithium salt $LiPF_6$ is susceptible to heat and trace water (Lux et al., 2012). Therefore, high voltage electrolyte is critical for LIB.

Lithium difluoro (oxalato) borate (LiDFOB) has high thermal stability, good solubility in organic carbonates, and can form a very stable and protecting Solid Electrolyte Interphase (SEI) on cathode and anode surface. Furthermore, LiDFOB-based electrolyte shows a better cycling performance at high temperature and low temperature. To improve the electrochemical stability under high voltage, researchers have developed several high oxidation potential solvents such as ionic liquid (Borgel et al., 2009, Salem and Abu-Lebdeh, 2014, Wongittharom et al., 2014), fluorinated carbonates (Kalhoff et al., 2014, Bolloli et al., 2015), alkyl sulfone (Xue et al., 2014, Wu et al., 2015, Xue et al., 2015). Sulfolane (SL) possesses characteristics with high dielectric constant (43.4), high flash point (166°C), high oxidation potential (6.35 V), SL-based electrolyte has high chemical stability and safety.

In the present study, a mixed electrolyte based on sulfolane and lithium salt LiDFOB was designed to improve the electrolyte stability and capacity retention of $LiNi_{0.5}Mn_{1.5}O_4$ cathode especially at high temperature. Electrochemical measurements, including chronoamperometry test and cycling test, were carried out to investigate the effect of sulfone-based electrolyte on $Li/LiNi_{0.5}Mn_{1.5}O_4$ cell.

## 2 EXPERIMENTAL

The sulfone-based electrolyte was prepared by 1 mol/L LiDFOB (Suzhou Fosai Co. Ltd) in a mixture of SL (Acros, 99%), PC, EMC in 2:2:1 volume ratio. The carbonate-based electrolyte was prepared by 1 mol/L $LiPF_6$ in a mixture of EC, EMC, DEC in 3:5:2 volume ratio. All these steps were operating in an argon-filled glove box (MBruan $O_2$ < 0.5 ppm, $H_2O$ < 0.5 ppm). The cathode slurry was prepared with $LiNi_{0.5}Mn_{1.5}O_4$ (Shenzhen Tianjiao Technology Co. Ltd), acetylene carbon black and polyvinylidene fluoride binder at a weight ratio of 8:1:1 in N-methyl pyrrolidone. Then the slurry was pressed on an Al foil and dried for 12 h at 80°C, the obtained electrode was punched into discs of 14 mm diameter, and dried again under vacuum at 100°C for 12 h. The CR2032 type coin cells were assembled in an argon-filled glove box (MBruan $O_2$ < 0.5 ppm, $H_2O$ < 0.5 ppm).

The flammability test of the electrolytes was examined by observing the flame on the surface of the sample after the electrolyte was ignited, the Self-Extinguish Time (SET) in s $g^{-1}$ was recorded (Yang et al., 2014). The thermal stability test of the electrolytes was operating on a thermogravimetric analysis instrument (NETZSCH STA 409PC), the sample was heated to 550°C at a heating rate of 5 K min⁻¹. Chronoamperometry test was carried on a cell system (Solartron 1470E). The Li/LiNi$_{0.5}$Mn$_{1.5}$O$_4$ cells with carbonate-based and sulfone-based electrolytes were charged to 5.2 V and 5.5 V individually at 0.5 C, and then holding constant potential at 5.2 V and 5.5 V for 24 h. The cyclic performances of Li/LiNi$_{0.5}$Mn$_{1.5}$O$_4$ cells were evaluated on LAND test system (Wuhan Land Electronics Co. Ltd., China) at room temperature (25°C) and elevated temperature (55°C) at 0.5 C (Chen et al., 2016). Surface morphology of the cathodes was observed by a Hitachi S3400 N Field Emission Scanning Electron Microscopy (FE-SEM). The dissolution of the transition metal of the LiNi$_{0.5}$Mn$_{1.5}$O$_4$ electrode in the electrolyte was measured by means of Inductively Coupled Plasma Atomic Emission Spectroscopy (ICP-AES), two pristine LiNi$_{0.5}$Mn$_{1.5}$O$_4$ electrodes were placed in two glass bottles filled with 500 µL carbonate and sulfone mixtures individually, the bottles were stored after sealed in a glovebox with Ar atmosphere at 60°C for 5 days.

## 3 RESULTS AND DISCUSSION

To clarify the effect of sulfone-based electrolytes on battery safety, flammability tests on electrolytes with the same volume were carried out, as shown in Figure 1. When exposed to a free flame, the carbonate-based electrolyte ignited instantaneously and kept the flame alive for more than 35 s. In comparison, the flame on the sulfone-based electrolyte was extinguished within 6 s. The extinction time was 91 s g⁻¹ for the carbonate-based electrolyte and 20 s g⁻¹ for the sulfone-based electrolyte. A shorter extinction time indicates stronger flame retardancy. The addition of 40 vol% SL shortened the extinction time by 78.0%, compared to the organic solvent. The non-flammability character-

istic of the sulfone electrolyte plays an important role in battery safety.

Figure 2 shows the TGA curves of two electrolytes. As shown by the curve for carbonate-based electrolyte, the volatilization was 89.5% when it was heated to 170°C, and there is essentially no weight loss after 170°C, which indicates solvents and lithium salts in the electrolyte has basically volatilized and decomposed completely, leaving a small amount of inorganic substance. However, the sulfone-based electrolyte decomposed completely at 215°C. At 81.4°C, the volatilization of organic addictive was 39.7% for carbonate-based electrolyte, 10.3% for sulfone-based electrolyte. This result indicates that the sulfone-based electrolyte possess a better thermal stability. The sulfone-based electrolyte using SL and PC with high boiling point as the main solvent and LiDFOB with higher thermal stability as lithium salt, reduces the evaporation kinetics of the organic solvent, which is advantageous to improve batteries safety.

In order to verify the anodic stability of sulfone-based electrolyte, the leakage current of Li/LiNi$_{0.5}$Mn$_{1.5}$O$_4$ cell was monitored at constant charging voltages of 5.2 and 5.5 V for 24 h. As shown in Figure 3a, the cell with carbonate-based electrolyte shows larger leakage current at both 5.2 V and 5.5 V, suggesting that severe decomposition of the carbonate-based electrolyte. However, incorporation of sulfone-based electrolyte reduces the leakage current, especially at 5.5 V, which is thought to be more stable and thus helps the electrolyte tolerate high voltages.

Figure 3b shows the discharge capacity retention of Li/LiNi$_{0.5}$Mn$_{1.5}$O$_4$ half cells during 200 cycles at 0.5C under 25°C. The cell cycled in carbonate-based electrolyte delivers a capacity of 129.8 mAh g⁻¹ initially and maintains 117.8 mAh

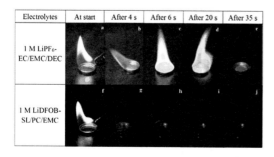

Figure 1.   Images of flammability tests of the sulfone-based and carbonate-based electrolyte.

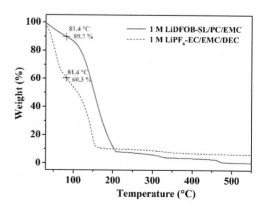

Figure 2.   TGA curves of the sulfone-based and carbonate-based electrolyte.

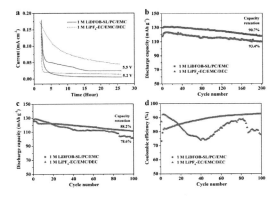

Figure 3. Potentiostatic profiles of Li/LiNi$_{0.5}$Mn$_{1.5}$O$_4$ cells with sulfone-based and carbonate-based electrolyte maintained at charging voltage of 5.2 V and 5.5 V respectively (a). Cyclic performances of LiNi$_{0.5}$Mn$_{1.5}$O$_4$ electrode in sulfone-based and carbonate-based electrolyte at 25°C (b). Cyclic performances of Li/LiNi$_{0.5}$Mn$_{1.5}$O$_4$ cells with different electrolytes at 55°C: discharge capacity (c) and its corresponding coulombic efficiency (d).

Figure 4. SEM images of fresh LiNi$_{0.5}$Mn$_{1.5}$O$_4$ cathode (a) and the cathode after 100 cycles at 55°C in carbonate-based electrolyte (b) and sulfone-based electrolyte (c).

g$^{-1}$ after 200 cycles, and the capacity retention is 90.7%. The initial discharge capacity of sulfone-based electrolyte is 117.6 mAh g$^{-1}$, the discharge capacity increases first and then decreases, after 200 cycles the discharge capacity is 109.8 mAh g$^{-1}$, capacity retention is 93.4%. The initial discharge capacity of sulfone-based electrolyte is lower than that of carbonate-based electrolyte. However, the cell cycled in sulfone-based electrolyte has better cycle stability, which is due to that sulfone-based electrolyte has higher oxidation potential. What's more, the oxidation reaction of LiDFOB on the LiNi$_{0.5}$Mn$_{1.5}$O$_4$ electrode surface helps the formation of SEI film, which increases the stability of interface between electrode and electrolyte.

To investigate the positive effect of sulfone-based electrolyte at elevated temperature, galvanostatic charge/discharge cycles of Li/LiNi$_{0.5}$Mn$_{1.5}$O$_4$ cells were carried out at a high temperature of 55°C. Comparison of the cycling performance between the cathode cycled in carbonate-based electrolyte and sulfone-based electrolyte is shown in Figure 3c. The discharge capacity retention of half cells with sulfone-based electrolyte is dramatically improved compared to the carbonate-based electrolyte from 78.6% to 88.2% after 100 cycles at 55°C. The improved cycling performance is because of the suppression of the electrode polarization and transition metal ions dissolution. The comparison of the coulombic efficiency between these two different electrolytes is shown in the Figure 3d. The coulombic efficiency of the cell with carbonate-based electrolyte is low and unstable during cycling, whereas that of the cell with

sulfone-based electrolyte shows much more stable and reaches above 93%. This result is due to the high quality SEI film formation, which also can be confirmed in the Figure 4.

The surface morphologies of the fresh and cycled LiNi$_{0.5}$Mn$_{1.5}$O$_4$ cathode were observed by SEM. As is seen in the Figure 4a, the surface of cathode before cycling is very clean and smooth. After 100 cycles at high temperature in carbonate-based electrolyte, the surface of LiNi$_{0.5}$Mn$_{1.5}$O$_4$ particle is covered with a rough film, which is due to that the decomposition products of the electrolyte continuously deposit at the unprotected cathode surface (Figure 4b). However, the cathode cycled in sulfone-based electrolyte exhibits a uniform surface in Figure 4c. The stable SEI film derived from the LiDFOB protects the cathode structure and thus ensures better cycling performance.

In order to investigate stability of LiNi$_{0.5}$Mn$_{1.5}$O$_4$ cathode in the carbonate-based and sulfone-based electrolyte, the fresh cathodes were placed in 500 μL two different electrolytes at 60°C for 5 days individually, after this we measured the dissolution amount of Mn and Ni in the electrolytes by means of ICP-AES. As is shown in the Figure 5, the dissolution amount of Mn and Ni in the carbonate-based electrolyte is 9.645 ppm and 2.525 ppm, much larger than 0.511 ppm and 0.145 ppm in the sulfone-based electrolyte. This result indicates that LiNi$_{0.5}$Mn$_{1.5}$O$_4$ cathode is more stable in sulfone-based electrolyte, in which the transition metal ions dissolved very little, so that the structure of LiNi$_{0.5}$Mn$_{1.5}$O$_4$ particle is retained. Differently, HF produced by side reactions of LiPF$_6$ in the carbonate-based electrolyte, can dissolve the Mn and Ni from LiNi$_{0.5}$Mn$_{1.5}$O$_4$, which results in structure damage.

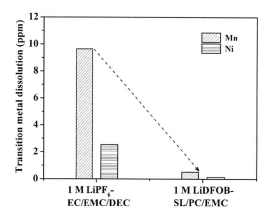

Figure 5. The amount of Ni and Mn ions in two different electrolytes after storage with $LiNi_{0.5}Mn_{1.5}O_4$ at 60°C.

## 4 CONCLUSIONS

A novel sulfolane-carbonate electrolyte containing LiDFOB as the lithium salt is prepared to investigate its physical and electrochemical properties and compatibility with $LiNi_{0.5}Mn_{1.5}O_4$ cathode. The chronoamperometry results revealed that sulfone-based electrolyte possesses better electrochemical stability and alleviates electrolyte decomposition. Due to the SEI film formed by LiDFOB, the sulfone-based electrolyte presents a better cycling stability than carbonate-based electrolyte in $Li/LiNi_{0.5}Mn_{1.5}O_4$ half cells. Furthermore, the sulfone-based electrolyte has better safety due to its superior thermal stability. We believe that sulfone-based electrolyte is a promising application for lithium-ion battery.

## ACKNOWLEDGMENT

This work was financially supported by National Natural Science Foundation (51574166) and Scientific and Technological Research and Development Foundation of Shenzhen City (JCYJ20140418193546111).

## REFERENCES

Bolloli, M., Kalhoff, J., Alloin, F., Bresser, D., Phung Le, M. L., Langlois, B., Passerini, S. & SANCHEZ, J.-Y. 2015. Fluorinated Carbamates as Suitable Solvents for LiTFSI-Based Lithium-Ion Electrolytes: Physicochemical Properties and Electrochemical Characterization. *The Journal of Physical Chemistry C,* 119, 22404–22414.

Borgel, V., Markevich, E., Aurbach, D., Semrau, G. & Schmidt, M. 2009. On the application of ionic liquids for rechargeable Li batteries: High voltage systems. *Journal of Power Sources,* 189, 331–336.

Chen, J., Zhang, H., Wang, M., Liu, J., Li, C. & Zhang, P. 2016. Improving the electrochemical performance of high voltage spinel cathode at elevated temperature by a novel electrolyte additive. *Journal of Power Sources,* 303, 41–48.

Kalhoff, J., Bresser, D., Bolloli, M., Alloin, F., Sanchez, J. Y. & Passerini, S. 2014. Enabling LiTFSI-based electrolytes for safer lithium-ion batteries by using linear fluorinated carbonates as (Co) solvent. *ChemSusChem,* 7, 2939–2946.

Kim, J.H., Pieczonka, N.P. & Yang, L. 2014. Challenges and approaches for high-voltage spinel lithium-ion batteries. *Chemphyschem,* 15, 1940–1954.

Liu, D., Zhu, W., Trottier, J., Gagnon, C., Barray, F., Guerfi, A., Mauger, A., Groult, H., Julien, C.M., Goodenough, J.B. & Zaghib, K. 2014. Spinel materials for high-voltage cathodes in Li-ion batteries. *RSC Adv.,* 4, 154–167.

Lux, S.F., Lucas, I.T., Pollak, E., Passerini, S., Winter, M. & Kostecki, R. 2012. The mechanism of HF formation in LiPF6 based organic carbonate electrolytes. *Electrochemistry Communications,* 14, 47–50.

Salem, N. & Abu-Lebdeh, Y. 2014. Non-Flammable Electrolyte Mixtures of Ringed Ammonium-Based Ionic Liquids and Ethylene Carbonate for High Voltage Li-Ion Batteries. *Journal of the Electrochemical Society,* 161, A1593–A1601.

Wongittharom, N., Lee, T.-C., Hung, I. M., Lee, S.-W., Wang, Y.-C. & Chang, J.-K. 2014. Ionic liquid electrolytes for high-voltage rechargeable $Li/LiNi_{0.5}Mn_{1.5}O_4$ cells. *Journal of Materials Chemistry A,* 2, 3613.

Wu, F., Zhu, Q., Chen, R., Chen, N., Chen, Y. & Li, L. 2015. Ionic liquid electrolytes with protective lithium difluoro (oxalate) borate for high voltage lithium-ion batteries. *Nano Energy,* 13, 546–553.

Xue, L., Lee, S.-Y., Zhao, Z. & Angell, C.A. 2015. Sulfone-carbonate ternary electrolyte with further increased capacity retention and burn resistance for high voltage lithium ion batteries. *Journal of Power Sources,* 295, 190–196.

Xue, L., Ueno, K., Lee, S.-Y. & Angell, C.A. 2014. Enhanced performance of sulfone-based electrolytes at lithium ion battery electrodes, including the $LiNi_{0.5}Mn_{1.5}O_4$ high voltage cathode. *Journal of Power Sources,* 262, 123–128.

Yang, B., Li, C., Zhou, J., Liu, J. & Zhang, Q. 2014. Pyrrolidinium-based ionic liquid electrolyte with organic additive and LiTFSI for high-safety lithium-ion batteries. *Electrochimica Acta,* 148, 39–45.

*Advances in Energy and Environment Research – Achour & Wu (Eds)*
© *2017 Taylor & Francis Group, London, ISBN 978-1-138-62682-9*

# The depth load cycling operation characteristic and energy-saving optimization control system of a power plant

Ning Zhao
*North China Electric Power Research Institute Co. Ltd., Beijing, China*

Yami Chen
*North China Institute of Science and Technology, Beijing, China*

ABSTRACT: At present, the variable working condition calculation of thermodynamic system and the variable load operation of a large capacity power plant are important subjects for research in thermodynamic engineering field all the time. When the current operation point of the power plant is far away from the rated mode, it is very important to analyze the thermo-economy of thermodynamic system and offer appropriate operation guidance in engineering practices. Based on the steam turbine's variable working condition theory, this paper reveals the inherent characteristic of the power plant under off-design condition. From the viewpoint of systematic engineering, according to the thermodynamic system topological structure and system thermodynamic properties, the state model of thermo-economy has been studied and set up. Combining the optimization theory and the analytic equation of the heat consumption rate, the optimum operational main steam pressure set point has been offered to advanced control strategy. Thus, the further intellectual optimum control can be put into practice, and the unit can meet the demand for safe and economical operation within the wider range of load.

## 1 INTRODUCTION

With the development of power industry and the adjustment of industrial structure, there is great change on the operational demand and running mode for a large capacity power plant. When the unit is running on low load condition, its running variables are far away from the value of manufacturer's desiring, and its economical and controllable characteristics cannot always be satisfied. At present, the traditional control strategy is mostly based on safety and stability. The research on control method and thermal system analysis is developing parallelingly. Nowadays, with the development of computer technology and measuring technology, great progress has been made on ability to acquire data and process information. As a large and complicated energy transform system, there is practical meaning to analyze and optimize system in order to increase its efficiency for thermal power plant. Through exact deduction, a steamswater distribution standard equation of thermo-system for coal-fired power plant is obtained.

The equation reveals the inner law between the topological structure of thermo-system for a power plant and its corresponding mathematic configuration. They are constructed on the measurable properties of system entirely and the heat consumption rate can be obtained through the thermodynamic laws and principles among them, which is different from the traditional method based on flow rate parameters measured. Therefore, it can eliminate these errors resulting in flow measurement and increase the analytic accuracy greatly. Based on the analytic solution of the heat consumption rate and the inner law of large turbine's sliding running mode, an optimized pressure set point can be obtained through genetic algorithms or other optimizing methods. Then, the result pressure set point can be used in coordinated control strategy of DCS to realize the safe and economical running indeed. By practical engineering projects, its validity has been proven.

## 2 THE STEAM-WATER DISTRIBUTION EQUATION

In the papers [1] [2] referenced, we have discussed the typical thermal system. Through exact deduction, the steam-water distribution equation (Formula 1) was obtained.

The terms for the equation (Formula 1) are as follows:

$D$: flow quantity; note that the measurement of extracted steam's flow quantity comes from the extraction outlet of turbine;

$h$: specific enthalpy of the corresponding $D$;

$q$: heat transferred by extracted steam per unit mass in heater. For the closed feedwater heater unit, it is equal to the specific enthalpy difference of inlet steam and outlet drain water. For the open feedwater heater unit, it is equal to the specific enthalpy difference of inlet steam and the heater unit's inlet feedwater.

$\gamma$: heat transferred by drain water per unit mass in heater. For the closed feedwater heater unit, it is equal to the specific enthalpy difference of the upper drain water and its own drain water. For the open feedwater heater unit, it is equal to the specific enthalpy difference of the upper drain water and the heater unit's inlet feedwater.

$\tau$: specific enthalpy increase in feedwater through the heater unit.

$\tau_p$: specific enthalpy increase in feedwater through pump.

$$
\begin{bmatrix} q_1 \\ \gamma_2 & q_2 \\ \gamma_3 & \gamma_3 & q_3 \\ \tau_4 & \tau_4 & \tau_4 & q_4 \\ \tau_5 & \tau_5 & \tau_5 & \gamma_5 & q_5 \\ \tau_6 & \tau_6 & \tau_6 & \gamma_6 & \gamma_6 & q_6 \end{bmatrix} \begin{bmatrix} D_1 - D_{df1} \\ D_2 \\ D_3 + D_{sg3} + D_{df3} \\ D_4 + D_{df4} \\ D_5 - D_{df5} \\ D_6 + D_{sg6} \end{bmatrix}
$$

$$
+ \begin{bmatrix} -D_{df1}q_{df1} \\ \left(D_0 + D_{bl} + D_l - D_{wf2}\right)\tau_p \\ D_{df3}q_{df3} + D_{sg3}q_{sg3} \\ D_{wf4}q_{wf4} + D_{f4}q_{f4} \\ 0 \\ D_{sg6}q_{sg6} + D_{sg6'}q_{sg6'} \end{bmatrix} = \begin{bmatrix} \tau_1 \cdot \left(D_0 + D_{bl} + D_l - D_{wf2}\right) \\ \tau_2 \cdot \left(D_0 + D_{bl} + D_l - D_{wf2}\right) \\ \tau_3 \cdot \left(D_0 + D_{bl} + D_l\right) \\ \tau_4 \cdot \left(D_0 + D_{bl} + D_l\right) \\ \tau_5 \cdot \left(D_0 + D_{bl} + D_l - D_{wf4}\right) \\ \tau_6 \cdot \left(D_0 + D_{bl} + D_l - D_{wf4}\right) \end{bmatrix}
$$

$$(1)$$

where different subscripts are used to distinguish the steam or water stream, such as f stands for auxiliary steam from or to extraction pipe, wf for water stream into or out of feedwater pipeline, df for water stream from or to drain pipe, sg for shaft gland steam, fw for feedwater, and wc for condensate water. The number in the subscripts stands for the relative heater.

From Formula 1, the steam-water distribution equation for a practical system can be simply written as:

$$[A][D_i] + \left[Q_{fi}^0\right] = \left[D_{fw}\tau_i\right] \tag{2}$$

It is a common practice to express the equation based on main steam flow quantity $D_0$. From boiler side, the mass balance equation is:

$$D_{fw} = D_0 + \sum_{m=1}^{t} D_{bm} - D_{ss},$$

where the item $\Sigma_{m=1}^{t} D_{bm}$ denotes the total flow quantity of the working substance leaving system from the boiler side (such as continuous blow down, periodical blow down, and the steam for soot blower system, that is, usually t = 3), and $D_{ss}$ denotes the desuperheating spray flow rate. Substituting $D_{fw}$ into Formula 2, the steam-water distribution equation can be transformed to the form based on $D_0$ and then mathematic transformation to the equation can give the following equation (Formula 3):

$$[A][D_i] + \left[Q_f\right] = D_0[\tau_i] \tag{3}$$

where the item $[Q_f]$ consists of all the auxiliary steam, auxiliary water stream, etc. We call it as heat disturbance item attached to the MS (the system excluding any auxiliary steam or auxiliary water stream is called Main System (MS), that is, only the extracted steam is considered in MS).

## 3 ANALYTIC EQUATION OF THE HEAT CONSUMPTION RATE

It is a requisite to obtain the current heat consumption rate of a thermal system for further detailed analysis. In contrasted to the traditional method, an analytic formula of the heat consumption rate can be obtained by taking advantage of the steam-water distribution equation.

The various terms appearing in the below equations are as follows:

$w$ = work done per unit working substance, and defined as $w = W/D_0$, where $W$ denotes power and $D_0$ denotes main steam mass flow rate;

$h$ = enthalpy per unit mass of the working substance;

$\sigma$ = heat transferred by reheating per unit working substance;

$\alpha$ = fraction of steam extracted or fraction of auxiliary steam or auxiliary water stream, and defined as $\alpha_x = D_x/D_0$,

where the subscript x is used to distinguish the kinds of steam or water stream.

$[q_f]$ = heat disturbance item arising from auxiliary steam and water stream per unit working substance, defined as $[q_f] = \frac{[Q_f]}{D_0}$, (also see equation Formula 3).

$\dot{q}$ = heat transferred by boiler per unit mass of the working substance;

$\hat{q}$ = heat consumption rate, $kJ/kWh$.

### 3.1 *The equation for power output of system*

1. The power output equation for the ideal Rankine cycle is:

$$w = h_0 - h_c$$

where the subscript 0 stands for main steam at turbine inlet, and c for the last-stage exhaust steam.

2. For the ideal reheat Rankine cycle:

$$w = h_0 - h_{eh} + h_r - h_c$$

where the subscript eh stands for exhaust steam of HP and r for reheated steam.

3. For an actual cycle of power plant:
Defining $\sigma = h_r - h_{eH}$ gives $w = h_0 + \sigma - h_c$. When certain steam $x\,(\alpha_x, h_x)$ leaves from HP, the power output decrease is $w_x = \alpha_x(h_x - h_c)$, and when it leaves from IP or LP (that is, after reheating), the power output decrease is $w_x = \alpha_x(h_x + \sigma - h_c)$.

For simplicity, the uniform term $h_x^\sigma$ is defined as follows:
When the steam leaves from HP:

$$h_x^\sigma = h_x + \sigma - h_c;$$

When the steam leaves from IP or LP:

$$h_x^\sigma = h_x - h_c;$$

For an actual thermal system, the steam leaving from turbine consists of all kinds of extraction steam for heaters and leaking steam from shaft gland. Thus, a complete power output equation for an actual system is obtained according to mass and energy balance.

$$w = (h_0 + \sigma - h_c) - \sum_{i=1}^{p}\alpha_i h_i^\sigma - \sum_{n=1}^{u}\alpha_n h_n^\sigma + \alpha_{rs} h_{rs}^\sigma$$

Written in matrix form:

$$w = (h_0 + \sigma - h_c) - [\alpha_i]^T[h_i^\sigma] - [\alpha_n]^T[h_n^\sigma] + \alpha_{rs} h_{rs}^\sigma \quad (4)$$

where the subscript i stands for the steam extracted for heater as well as their serial number, n stands for shaft gland steam as well as their index number, and rs stands for reheating spray water, and $h_{rs}^\sigma = h_r - h_c$.

### 3.2 The equation for the heat transferred by boiler

The equation for the heat transferred by boiler for the ideal Rankine cycle is

$$\dot{q} = h_0 - h_w$$

For the ideal reheat Rankine cycle:

$$\dot{q} = h_0 + \sigma - h_w$$

where the subscript w stands for feedwater (boiler's inlet). For an actual system, considering all kinds of steam that are not reheated and all kinds of working substance flowing out of and into system from boiler side such as continuous blow down, periodical blow down, the steam for soot blower system, desuperheating spray flow, and reheating spray flow, then the complete equation for the heat transferred by boiler becomes

$$\dot{q} = (h_0 + \sigma - h_w) - \sum_{i=1}^{c}\alpha_i\sigma - \sum_{n=1}^{d}\alpha_n\sigma + \alpha_{ss}(h_w - h_{ss})$$
$$+ \alpha_{rs}(h_r - h_{rs}) + \sum_{m=1}^{t}\alpha_{bm}(h_{bm} - h_w)$$

where c stands for the total number of the extracted steam leaving before reheating, d for the total number of shaft gland steam of HP, and the subscript ss for desuperheating spray water.

The above equation can be written in matrix form:

$$\dot{q} = (h_0 + \sigma - h_w) - [\alpha_i]_c^T[\sigma]_c - [\alpha_n]_d^T[\sigma]_d$$
$$+ \alpha_{rs}(h_r - h_{rs}) + \alpha_{ss}(h_0 - h_{ss}) + [\alpha_{bm}]_t^T[h_{bm} - h_w]_t \quad (5)$$

### 3.3 The analytic formula of heat consumption rate

$[\alpha_i]$ can be obtained.

$$[\alpha_i] = [A]^{-1}\big[[\tau_i] - [q_f]\big] \quad (6)$$

Doing matrix transform, thus

$$[\alpha_i]^T = \big[[\tau_i] - [q_f]\big]^T[A^T]^{-1} \quad (7)$$

and

$$[\alpha_i]_c^T = \big[[\tau_i]_c - [q_f]_c\big]^T[A_c^T]^{-1} \quad (8)$$

The equation of heat consumption rate is defined as

$$\hat{q} = 3600\dot{q}/w \quad (9)$$

Substituting $[\alpha_i]^T$ and $[\alpha_i]_c^T$ into equation. (5) and (5), the equation of heat consumption rate $\hat{q}$ can be expressed as equation (10):

$$A = \big[[\tau_i]_c - [q_f]_c\big]^T[A_c^T]^{-1}[A_c^T][\sigma]_c - [\alpha_n]_d^T[\sigma]_d$$
$$B = \alpha_{rs}(h_r - h_{rs}) + \alpha_{ss}(h_0 - h_{ss}) + [\alpha_{bm}]_t[h_{bm} - h_w]_t^T$$
$$C = \big[[\tau_i] - [q_f]\big]^T[A^T][h_i^\sigma] - [\alpha_n]^T[h_n^\sigma] + \alpha_{rs}h_{rs}^\sigma$$
$$\hat{q} = 3600\frac{(h_0 + \sigma - h_w) - A + B}{(h_0 + \sigma - h_c) - C} \quad (10)$$

Equation (10) is constructed completely on properties and the auxiliary steam-water flow rate of the thermal system. Under the assumption that all the measuring instruments work well, the only important thing left is the ascertainment of the steam properties of the final several stages of the turbine, like turbine exhaust steam enthalpy. To fix their states, our other research achievements [3] can be referenced. In addition, the auxiliary steam-water flow rate can be obtained through turbine's varying working condition calculation. Thus, the equations can also be called the equation of state.

## 4 ASCERTAINMENT OF OPTIMUM PRESSURE SET POINT

Figure 1 illustrates the efficiency of velocity stage under constant pressure (according to some manufacturer's data). In the figure, the points I, II, III, and IV stand for one, two, three, and four governing valves opening wide, respectively. Supposing that the point E is an actual working load condition, the crossing point F is the working point with the pressure value sliding from PE to P0d (rated pressure value). It indicates that the efficiency of the velocity stage at NgH in the sliding mode is equal to that at NgF in the constant pressure mode. The improvement of the inner efficiency of velocity stage is obvious. However, the change in main steam pressure can result in the whole system's changing, such as steam extraction pressure, exhaust pressure, and steam consumption of feedwater pump turbine. In nature, selecting an appropriate pressure is an optimism problem with multi-constraints. Based on the analytic solution of the heat consumption rate, it can be got along.

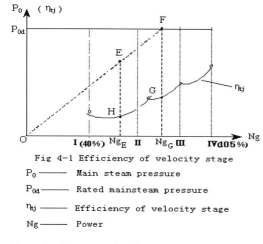

Fig 4-1 Efficiency of velocity stage

$P_0$ ———— Main steam pressure

$P_{0d}$ ———— Rated mainsteam pressure

$\eta_{hj}$ ———— Efficiency of velocity stage

$Ng$ ———— Power

Figure 1. The scope of optimum pressure set point.

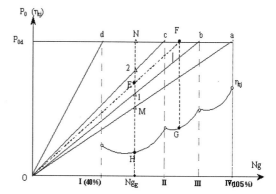

Figure 2. Ideal sliding pressure diagram.

Figure 2 is a more detailed P-Ng curve. Theoretically, the line Oa stands for the sliding pressure curve from zero load to maximum load with four valves wide open, Ob for three valves wide open, Oc for two valves wide open, and Od for one valve wide open. Obviously, the pressure scope available is from PM to PN at the load point NgE.

The analytic solution of the heat consumption rate $\hat{q}$ (equation 3–7) consists of properties variables, but as well known, any actual system is a non-linear system, and properties variables are determined by load. Therefore, for an actual thermal system, the heat consumption rate can be expressed as

$$q = f(Ng_E, P_0, X) \qquad (11)$$

where NgE stands for the working load point, $P_0$ for the pressure needed to be optimized, and X for the current working state and environments condition. Obviously, the function (5-1) can be used as the object function for optimizing along with turbine's varying working condition calculation. Substituting the constraint conditions, such as $N_{gmin} \leq N_g \leq N_{gmax}$,

$$\frac{P_{0d} \cdot N_g}{1.05 N_{gd}} \leq P_0 \leq P_{0d}.$$

The function is very flexible for different object optimization. Therefore, we can also select constraint conditions according to other factors such as safety and equipment characteristic.

## 5 THE PRACTICING CASE ON SOME 300 MW UNIT

Figure 3 is the system structure diagram for a whole optimizing control system. In order to abide by the power council policy, the system works on unilat-

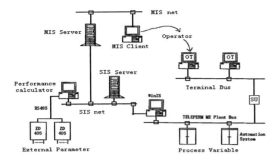

Figure 3.  System structure.

eral mode at present, that is, data can transmit from DCS to performance calculation sever through interface WinIs automatically. The resulting data optimized to DCS must be set through the operator. The system works well, and the user has benefits from it. With the development of SIS (Supervising Information System), the system will realize closed-loop running to make the economical and automatic level of power plant to a new stage before long.

## 6   CONCLUSIONS

The analytic formula of the heat consumption rate for the thermal power plant is presented for the first time. It indicates that the current heat consumption rate of the system is determined by the system structure and its thermodynamic properties as well as the fraction of all kinds of auxiliary steam-water flow.

The analytic formula of heat consumption rate is the base of online optimizing and control. As a multi-variable system, many constraint conditions are hidden in turbine's varying working condition calculation.

Only the energy consumption index is concerned, can optimizing control possess of more practical meaning for a complicated energy transform system and lead its automatic level up to a new stage.

## ACKNOWLEDGMENTS

This paper was supported by North China Electric Power Research Institute Co., Ltd.

## REFERENCES

[1] Zhang Chunfa, Shunlin Yan, Hansong Fan, etc. A steam-water distribution matrix equation of the whole thermal system for coal-fired power plant and its general construction regulation [C]. PWR-Vol. 34, 1999 Joint Power Generation Conference Volume 2.

[2] Chunfa Zhang, Liping Li, Mingzhi Zhang, etc. Steam-Water distribution standard equation of themo system for coal-fired power plant Procceding of International conference on Power Engineering (ICOPE-2003), Japan.

[3] Zhang Chunfa, Cui Yinghong, Yang Wenbing, etc. Proof for part of the Sttla flow experimental conclusion and the improvement of Flügel formula. Science in China, series E, Vol. 45, No. 1, 2002, pp. 35–46.

[4] Yunus A. Çengel, Michael A Boles. Thermodynamics: An engineering Approach (4th Edition). New York: McGraw-Hill 2002.

[5] Wang Shuqi, A new parameter of combustion control, Power Engineering, 2000, 20(1): 535–538.

[6] Ricardo Dobson, Plant's Computerized Maintenance System Improves Operations, Power Engineering, 1989, No. 5, p. 30–32.

[7] Alan Reeve, Energy Management Maximizes Efficiency, Control Instrument, 1998, No. 7, p. 51–53.

[8] Fl. Muhlhauser. Bader, The Influence of Present Economic Constrains on Design of Conventional Steam Power Plants, Brown Boveri Review, 1984, No. 3–4, p. 120–128.

[9] Edgarc Hutchins. OG&E's Seminole Plant Operates Under Free Pressure, Power Engineering, 1997, No. 12, p. 34–39.

*Advances in Energy and Environment Research – Achour & Wu (Eds)*
© 2017 Taylor & Francis Group, London, ISBN 978-1-138-62682-9

# Fabrication and mechanical properties of zirconium–yttrium hydrides

Jiaqing Peng, Yang Chen, Guoqing Yan, Ming Wu, Jiandong Zhang & Lijun Wang
*Division of Rare Metals Metallurgy and Materials, General Research Institute for Nonferrous Metals, Beijing, China*

Jingshe Li
*School of Metallurgical and Ecological Engineering, University of Science and Technology Beijing, Beijing, China*

Zhancheng Guo
*State Key Laboratory of Advanced Technology, University of Science and Technology Beijing, Beijing, China*

ABSTRACT: The crack-free bulk hydrides of Zr–Y alloys have been fabricated by regulating the cooling rate of the furnace temperature and the flow rate of the hydrogen gas. The phase structure and lattice parameters were studied by XRD. The relationship between hydrogen content and lattice parameters is derived based on literature data and employed to determine the H/Zr and H/Y ratios of the hydrides, which were consistent with the H/(Zr+Y) ratios measured by mass change method. The mechanical properties of the Zr–Y alloy hydrides were characterized by nanoindentation and Vickers hardness test. The nanohardness, elastic modulus, Vickers hardness and fracture toughness of the hydrides of Zr–Y alloys increased with the Y content. The results were in the intermediate range of pure zirconium hydride and yttrium hydride, and were in good agreement with the literature. It was concluded that the Zr–Y alloy hydrides were superior to inhibit cracks formation during hydriding and working at the nuclear reactor, thus retaining the physical integrity of the hydride.

## 1 INTRODUCTION

Zirconium alloys have been widely used as structural materials in nuclear industry, such as fuel claddings, due to their superior mechanical properties and low neutron absorption cross-section. However, zirconium can absorb hydrogen to form zirconium hydride, leading to hydrogen embrittlement (Northwood & Kosasih 1983). Yttrium addition is proposed to mitigate hydrogen embrittlement of zirconium (Batra et al. 2009, Li et al. 2015). Besides, zirconium hydride is also used as neutron moderators in nuclear reactors due to its high hydrogen atomic density and negative prompt temperature coefficient of reactivity (Olander et al. 2009). Nevertheless, it is difficult to fabricate crack-free zirconium hydride in the bulk form because of the brittleness of the hydride and the large volume expansion accompanied by the phase transformation from metal to hydride (Barraclough & Beevers 1970). It is possible that yttrium is able to inhibit cracks formation because of its strong grain refining effect (Zhao et al. 2013). Therefore, it is of importance to gain knowledge on the mechanical properties of the hydrides of Zr–Y alloys.

In the present study, the crack-free bulk hydrides of Zr–Y alloys have been fabricated by regulating the cooling rate of the furnace temperature and the flow rate of the hydrogen gas. The phase structure and lattice parameters of the as-received hydrides of Zr–Y alloys are studied. The relationship between hydrogen content and lattice parameters is derived based on literature data and employed to determine the H/Zr and H/Y ratios of the hydrides. The mechanical properties of the Zr–Y alloy hydrides, such as nanohardness, elastic modulus, Vickers hardness and fracture toughness are studied by nanoindentation and Vickers hardness test both in the microscopic and macroscopic scale.

## 2 MATERIAL AND METHODS

### 2.1 Sample fabrication

The Zr–Y alloys containing 1 at%, 4 at%, 10 at%, 20 at%, 30 at% and 57 at% Y were fabricated by arc-melting the high-purity Zr and Y metals (>99%) under an argon atmosphere. The alloys were turned over and re-melted at least four times to achieve good homogeneity. The details of alloy preparation were described elsewhere

(Peng et al. 2016). The arc-melted alloys were cut into 2-mm-thick disks, and then abraded and polished to obtain a flat and smooth surface.

The alloy disks were hydrided in a stainless steel tube furnace. Firstly, Zr–Y alloys were annealed at 900°C in vacuum ($10^{-5}$ Pa) for an hour for degassing. Then the high-purity hydrogen gas (5 N) was slowly introduced into the reaction chamber to make sure that the hydrogen absorption of the alloys was kept at a very low speed. The furnace temperature was held at 900°C until the equilibrium was reached under a hydrogen pressure of 1 atm. Subsequently, the temperature was gradually lowered to 700°C at a rate of 1°C/h. In order to avoid cracks formation during hydrogenation, both the cooling rate of the furnace temperature and the flow rate of the hydrogen gas should be kept quite low. Finally, after the hydrogen pressure reached the equilibrium, the samples were furnace-cooled to room temperature. The bulk Zr–Y alloy hydrides without any micro-cracks were fabricated by this method and confirmed by Scanning Electron Microscopy (SEM). The pure zirconium hydride was prepared by the same method for comparison. The H/(Zr + Y) ratios were determined by measuring the weight before and after hydrogenation.

## 2.2 *Characterization*

The phase structures and lattice parameters of the Zr–Y alloy hydrides were studied at room temperature by X-ray diffraction (Rigaku D/max 2550/HB) using Cu-Kα radiation with a current of 200 mA and a voltage of 40 kV.

The nanoindentation tests were performed at room temperature using MTS Nano Indenter XP with the maximum indentation depth of 1 μm. The maximum load was held for 10 seconds. The average nanohardness was obtained at the same maximum indentation depth from at least eight tests. The elastic modulus can be calculated using the data obtained by nanoindentation tests. The Vickers hardness was measured using a hardness tester (Wilson Tukon 2500) with the applied load of 200 gf. The Vickers hardness was also performed at a larger applied load of 500 gf for at least three times to produce cracks. The average length of radial cracks was measured in situ by optical microscopy of the hardness tester to evaluate the fracture toughness of the alloy hydrides.

## 3 RESULTS AND DISCUSSION

### 3.1 *Phase structure and hydrogen content*

Figure 1 shows the XRD patterns of the hydrides the Zr–Y alloys. It can be seen that the alloy hydrides mainly consisted of ε-ZrH$_{1.950}$ and δ-YH$_2$

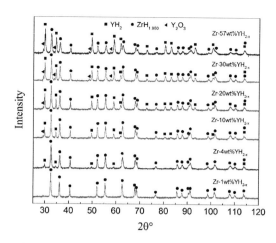

Figure 1. XRD patterns of Zr–Y alloy hydrides.

phase. The intensity of peaks associated with δ-YH$_2$ tended to increase with the increase of the Y content, whereas that of peaks associated with ε-ZrH$_{1.950}$ tended to decrease. When the Y content increased up to 10 wt%, Y$_2$O$_3$ phase started to appear because yttrium has a strong affinity with oxygen. The relationship between hydrogen content and lattice parameters of zirconium hydride and yttrium hydride are summarized in Figures 2–4. The H/Zr and H/Y ratios were calculated using the lattice parameters examined by XRD according to the following equations which were fitted from the literature data (Moore & Young 1968, Barraclough & Beevers 1970, Cantrell et al. 1984, Bowman et al. 1985, Daou & Vajda 1992, Setoyama et al. 2005):

$$H/Zr = 61.51779 - 124.98858 \times \frac{c}{a}$$
$$+ 65.25545 \times \left(\frac{c}{a}\right)^2, \ (H/Zr = 1.7 - 2.0)$$
(1)

$$H/Y = 762.34564 - 1460.36486 \times a(nm) \quad (2)$$

The hydrogen content of the Zr–Y alloy hydrides are listed in Table 1. The calculated H/Zr and H/Y ratios confirmed the existence of ε-ZrH$_{1.950}$ and δ-YH$_2$ phase. The H/Zr ratios were almost the same irrespective of Y content because all the samples were hydrided under the same equilibrium temperature and pressure. The H/Y ratios were determined to be roughly around 2.0, corresponding to δ-YH$_2$ phase. However, the H/Y ratios of the samples with 1 wt% and 4 wt% Y cannot be accurately determined due to the low Y content. The H/(Zr + Y) ratios determined by mass change were in agreement with the H/Zr and H/Y ratios

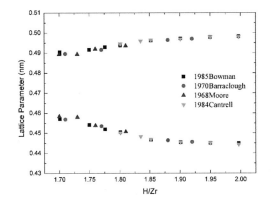

Figure 2. Change in the lattice parameters with the hydrogen content.

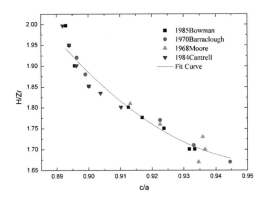

Figure 3. Hydrogen content as a function of c/a of zirconium hydride.

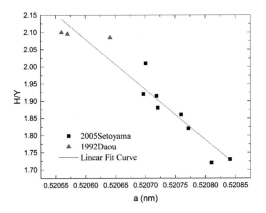

Figure 4. Hydrogen content as a function of lattice parameter of yttrium hydride.

calculated by lattice parameters considering different Y content. This implied that the Zr–Y alloy hydrides existed as zirconium hydride and yttrium hydride mixtures but not their solid solutions.

Table 1. Hydrogen content of the Zr-Y alloy hydrides determined by mass change and lattice parameters.

| Y content/ wt% | H/ (Zr+Y) | Lattice parameters of ZrH$_x$/Å | | | Lattice parameters of YH$_x$/Å | |
|---|---|---|---|---|---|---|
| | | a/nms | c/nm | H/Zr | a/nm | H/Y |
| 1 | 1.95 | 4.9756 | 4.4515 | 1.93 | – | – |
| 4 | 1.94 | 4.9754 | 4.4530 | 1.92 | – | – |
| 10 | 1.93 | 4.9717 | 4.4528 | 1.92 | 5.2057 | 2.12 |
| 20 | 1.95 | 4.9765 | 4.4503 | 1.93 | 5.2044 | 2.31 |
| 30 | 1.94 | 4.9786 | 4.4500 | 1.93 | 5.2064 | 2.02 |
| 57 | 1.95 | 4.9786 | 4.4500 | 1.93 | 5.2069 | 1.95 |

3.2 *Nanohardness and elastic modulus*

The nanohardness was calculated using the indentation load-displacement data based on the method of Oliver and Pharr (Oliver & Pharr 1992), and can be formulated as follows:

$$H_n(GPa) = \frac{P_{max}}{A_c} \qquad (3)$$

where $P_{max}$ is the maximum indentation load, and $A_c$ is the projected contact area at the peak load. The nanoindentation technique is able to measure the mechanical properties in microscopic scale to avoid the defected zone, thus representing the intrinsic property of the materials. Figure 5 shows the obtained nanohardness of the Zr–Y alloy hydrides. The nanohardness increased with the increase of the Y content, which can be described as following equation:

$$H_n(GPa) = 1.51605 + 0.0155 \times C_Y(wt\%) \qquad (4)$$

The elastic modulus of the specimen can be measured using the equation below (Oliver & Pharr 2004):

$$\frac{1}{E_{eff}} = \frac{1-\nu^2}{E} + \frac{1-\nu_i^2}{E_i} \qquad (5)$$

where $E$ and $\nu$ are the elastic modulus and Poisson's ratio of the specimen, respectively; and $E_i$ and $\nu_i$ are the elastic modulus and Poisson's ratio of the indenter, respectively. And the effective elastic modulus $E_{eff}$ can be calculated by relation between the contact area $A_c$ and the unloading stiffness $S$:

$$S = \frac{dP}{dh} = \frac{2}{\sqrt{\pi}} E_{eff} \sqrt{A_c} \qquad (6)$$

where $P$ is the indentation load and $h$ is the contact depth. Figure 6 shows the elastic modulus of

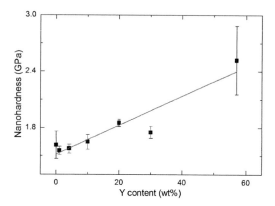

Figure 5.  Nanohardness of the Zr–Y alloy hydrides.

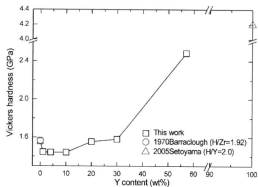

Figure 7.  Vikers hardness of the Zr–Y alloy hydrides.

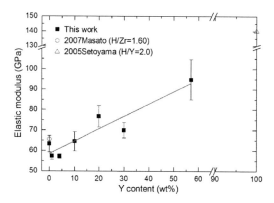

Figure 6.  Elastic modulus of the Zr–Y alloy hydrides.

the Zr–Y alloy hydrides, together with the literature data of pure zirconium hydride and yttrium hydride (Setoyama et al. 2005, Ito et al. 2007). The elastic modulus of pure zirconium hydride was determined to be 63.2 GPa, which is consistent with literature result, demonstrating the reliability of the method. By contrast, the elastic modulus of pure yttrium hydride was much higher than that of pure zirconium hydride and the Zr–Y alloy hydrides. It was found that the elastic modulus increased with the Y content, as expressed as following equation:

$$E(GPa) = 58.53458 + 0.60563 \times C_Y (wt\%) \qquad (7)$$

3.3  *Vickers hardness and fracture toughness*

Figure 7 shows the Vickers hardness of the Zr–Y alloy hydrides, together with the literature data of pure zirconium hydride and yttrium hydride (Barraclough & Beevers 1970, Setoyama et al. 2005). The obtained Vickers hardness of the pure

zirconium hydride agreed well with the literature (Barraclough & Beevers 1970). The Vickers hardness of the Zr–Y alloy hydrides was in the intermediate range of the pure zirconium hydride and yttrium hydride, and was comparable with the nanohardness as shown in Figure 5. It can be seen that the Vickers hardness initially increased slightly with the Y content, and surged when the Y content increased from 30 wt% to 57 wt%.

The fracture toughness can be measured by the Vickers indentation crack length method according to the following equation (Quinn 2008):

$$K_{IC}(MPa \cdot m^{1/2}) = 0.016 \left( \frac{E}{H} \right)^{1/2} \frac{P}{c^{3/2}} \qquad (8)$$

where $K_{IC}$ is the fracture toughness, $E$ is the elastic modulus, $H$ is the hardness, $P$ is the applied indentation load, c is the length of the radial cracks. The values of $E$ and $H$ were directly obtained from the nanoindentation test assuming that they are independent of the indentation load. The average crack radius was measured by the optical microscopy, and the indentation photographs were presented in Figure 8. It is obvious that the cracks became fewer in number and smaller in size as the Y content increased. This suggests that zirconium hydride containing higher Y content may be more resistant to crack formation and propagation. Therefore, it is superior to produce crack-free hydride and retain its physical integrity during service.

Figure 9 shows the calculated fracture toughness of the Zr–Y alloy hydrides. It can be seen that the fracture toughness of the Zr–Y alloy hydrides was in the intermediate range of pure zirconium hydride and yttrium hydride. And the fracture toughness of the pure zirconium hydride is in good agreement with that of literature (Xu & Shi 2004). The fracture toughness increased sharply when the Y content increased up to 4 wt%, and then

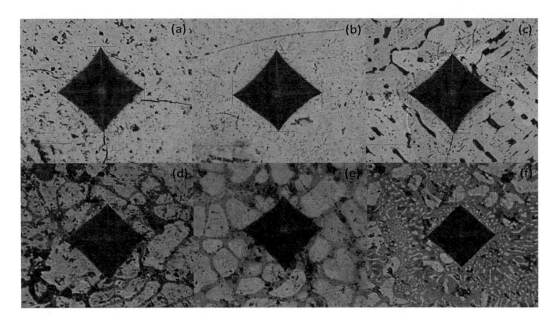

Figure 8. Indentation photograph for fracture toughness measurement: (a) 1 wt% Y, (b) 4 wt% Y, (c) 10 wt% Y, (d) 20 wt% Y, (c) 30 wt% Y, (c) 57 wt% Y.

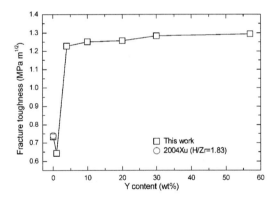

Figure 9. Fracture toughness of the Zr–Y alloy hydrides.

increased slightly when the Y content ranged from 10 wt% to 57 wt%. This is in line with indentation cracks observed by optical microscopy.

## 4 CONCLUSIONS

The crack-free bulk hydrides of Zr–Y alloys have been fabricated by regulating the cooling rate of the furnace temperature and the flow rate of the hydrogen gas. The phase structure and lattice parameters were studied by XRD. The Zr–Y alloy hydrides were composed of $\varepsilon$-ZrH$_{1.950}$, $\delta$-YH$_2$ and small amount of Y$_2$O$_3$ phase. The H/Zr and H/Y ratios were calculated by lattice parameters, and were consistent with the H/(Zr+Y) ratios determined by mass change method. The H/Zr ratios were almost the same irrespective of Y content, and the H/Y ratios were determined to be roughly around 2.0.

The mechanical properties of the Zr–Y alloy hydrides were characterized by nanoindentation and Vickers hardness test. The results were in the intermediate range of pure zirconium hydride and yttrium hydride, and were in good agreement with the literature. It was found that nanohardness, elastic modulus, Vickers hardness and fracture toughness of the hydrides of Zr–Y alloys increased with the increase of the Y content. This inhibits the cracks formation of the Zr–Y alloy hydrides during hydriding and working at the nuclear reactor, thus retaining the physical integrity of the hydride.

## ACKNOWLEDGEMENTS

This work was financially supported by the National Natural Science Foundation of China (No. 5140041034).

## REFERENCES

Barraclough, K. & Beevers, C. (1970). Some observations on the phase transformations in zirconium hydrides. *Journal of Nuclear Materials*. 34, 125–134.

Batra, I. S., Singh, R. N., Sengupta, P., Maji, B. C., Madangopal, K., Manikrishna, K. V., Tewari, R. & Dey, G. K. (2009). Mitigation of hydride embrittlement of zirconium by yttrium. *Journal of Nuclear Materials.* 389, 500–503.

Bowman, R. C., Craft, B. D., Cantrell, J. S. & Venturini, E. L. (1985). Effects of thermal treatments on the lattice properties and electronic structure of $ZrH_x$. *Physical Review B.* 31, 5604–5615.

Cantrell, J. S., Bowman, R. J. & Sullenger, D. (1984). X-ray diffraction and nuclear magnetic resonance studies of the relationship between electronic structure and the tetragonal distortion in zirconium hydride ($ZrH_x$). *The Journal of Physical Chemistry.* 88, 918–923.

Daou, J. N. & Vajda, P. (1992). Hydrogen ordering and metal-semiconductor transitions in the system $YH_{2+x}$. *Physical Review B.* 45, 10907–10913.

Ito, M., Nishioka, S., Muta, H., Kurosaki, K., Uno, M. & Yamanaka, S. Indentation study of titanium, zirconium, and hafnium hydrides. MRS Proceedings, 2007. Cambridge Univ Press, 1043-T09–11.

Li, C., Xiong, L., Wu, E. & Liu, S. (2015). Effect of yttrium on nucleation and growth of zirconium hydrides. *Journal of Nuclear Materials.* 457, 142–147.

Moore, K. & Young, W. (1968). Phase studies of the Zr-H system at high hydrogen concentrations. *Journal of Nuclear Materials.* 27, 316–324.

Northwood, D. & Kosasih, U. (1983). Hydrides and delayed hydrogen cracking in zirconium and its alloys. *International Metals Reviews.* 28, 92–121.

Olander, D., Greenspan, E., Garkisch, H. D. & Petrovic, B. (2009). Uranium–zirconium hydride fuel properties. *Nuclear Engineering and Design.* 239, 1406–1424.

Oliver, W. C. & Pharr, G. M. (1992). An improved technique for determining hardness and elastic modulus using load and displacement sensing indentation experiments. *Journal of materials research.* 7, 1564–1583.

Oliver, W. C. & Pharr, G. M. (2004). Measurement of hardness and elastic modulus by instrumented indentation: Advances in understanding and refinements to methodology. *Journal of materials research.* 19, 3–20.

Peng, J. Q., Chen, Y., Yan, G. Q., Wu, M., Wang, L. J. & Li, J. S. (2016). Solid solubility extension and microstructure evolution of cast zirconium yttrium alloy. *Rare Metals.* 35, 325–330.

Quinn, G. D. (2008). Fracture toughness of ceramics by the Vickers indentation crack length method: A critical review. *Mechanical Properties and Performance of Engineering Ceramics II: Ceramic Engineering and Science Proceedings.* 27, 45–62.

Setoyama, D., Ito, M., Matsunaga, J., Muta, H., Uno, M. & Yamanaka, S. (2005). Mechanical properties of yttrium hydride. *Journal of Alloys and Compounds.* 394, 207–210.

Xu, J. & Shi, S.-Q. (2004). Investigation of mechanical properties of ε-zirconium hydride using micro- and nano-indentation techniques. *Journal of Nuclear Materials.* 327, 165–170.

Zhao, C., Song, X., Yang, Y. & Zhang, B. (2013). Hydrogen absorption cracking of zirconium alloy in the application of nuclear industry. *International Journal of Hydrogen Energy.* 38, 10903–10911.

*Advances in Energy and Environment Research – Achour & Wu (Eds)*
*© 2017 Taylor & Francis Group, London, ISBN 978-1-138-62682-9*

# Carbon paper electrode coated with TiO₂ nanorod arrays for electro-assisted photocatalytic degradation of Cr (VI) and methylene blue

H. Li, Z. Lin & D. Fu
*State Key Laboratory of Bioelectronics, Southeast University, Nanjing, China*
*School of Biological Science and Medical Engineering, Southeast University, Nanjing, China*
*Suzhou Key Laboratory of Environment and Biosafety, Suzhou, China*

ABSTRACT: The TiO₂ nanorod arrays decorated Carbon Paper (CP) were successfully synthesized by a simple hydrothermal method. The structure and morphology of the prepared samples were characterized by Scanning Electron Microscopy (SEM), X-Ray Diffraction (XRD), Raman spectra. The electro-assisted photocatalytic degradation of Cr (VI) and Methylene Blue (MB) was performed using as-prepared TiO₂/CP. The effect of initial pH and applied potential are investigated. A synergistic enhancement effect for the degradation of Cr (VI) and MB was found in their mixed solution. This work demonstrates the promising photocatalytic activity of the samples, and may provide better understanding on the treatment of real wastewater.

## 1 INTRODUCTION

Recently, the application of semiconductor TiO₂ in polluted drinking water and industrial wastewater treatment gets more and more attention because of its simple operation, mild reaction conditions, and no secondary pollution (Wang, Y 2012). In UV irradiated TiO₂ system, varieties of organic and inorganic substance can be either photo-oxidized or photo-reduced effectively. Cr (VI) is in the list of priority heavy metals of most countries due to its serious toxicity, carcinogenic action and high solubility in water, while Cr (III) is less noxious relatively and can be removed easily through precipitation. Therefore, converting Cr (VI) into Cr (III) is an excellent method to remove Cr (VI) pollution. During past decades, much works has been done to investigate degradation of dyes such as Methyl Orange (Yao, W 2012) or Cr (VI) (Xu, S C 2015) using TiO₂. Cr (VI) with high toxicity is widely applied in several industrial processes such as electroplating, leather tanning and paint making (Cappelletti G 2008). During some of these processes, Cr (VI) and dyes often coexist in most wastewater. Therefore, the photo-catalytic of such coexistence system is more significant from the view of practical application.

The redox reaction processes can be induced by reactive electrons and holes generated at the surface of TiO₂, which is illuminated by light with energy larger than its band gap (Yang, Y 2014, Papadam T 2007). However, the inevitable recombination of electrons and holes is quite adverse for photocatalysis. Applying a bias potential can prompt the separation of electrons and holes effectively, and significantly improve photocatalytic activity (Asahi R 2001, Jung, D 2012, Ojani, R 2012, Yang, Y 2014).

In this work, we fabricated TiO₂ nanorod arrays at Carbon Paper (CP) electrode to perform electro-assisted photocatalytic degradation of Cr (VI) and Methylene Blue (MB). An anodic bias voltage was applied at the electrode during photocatalysis to prevent the recombination of photo-generated carriers. The photo-generated hole moved to electrode surface reacted with water molecules and produced highly oxidizing hydroxyl radical (·OH) as shown in reaction (1), while the photo-generated electrons transferred to counter electrode platinum wire and was helpful for reaction (2).

$$H_2O + h^+ \rightarrow \cdot OH + H^+ \tag{1}$$

$$CrO4^{2-} + H^+ + e^- \rightarrow Cr^{3+} + H_2O \tag{2}$$

## 2 EXPERIMENTAL SECTION

### 2.1 Preparation of TiO₂ nanorod arrays covered CP electrode

The TiO₂ nanowires were synthesized by a modified hydrothermal route (Yang, Y 2014). In a typical process, carbon paper was cut into 2 cm × 7 cm and cleaned under sonication in acetone, ethanol and deionized water for 20 min respectively, then dried under 80°C. Colloidal sol consisted of 6 ml alcohol and 1 ml tetrabutyl titanate was stirred at 25°C. The cleaned CP was put into the sol for

30 s and then dried under 80°C for 30 min. Such dipping and drying process was repeated 3 times to cover the CP with a TiO₂ seed layer. After that, the seeded CP was put into sealed Teflon-lined stainless steel autoclave consisting of 15 ml hydrochloric acid (36.5%), 15 ml water, 0.6 ml tetrabutyl titanate and allowed to react at 180°C for 10h. After the reaction, the substrate was rinsed with deionized water, and dried with N₂ blow. Then the substrate was placed in a furnace and maintained at 450°C for 2h in N₂ to get TiO₂ nanorod array covered CP electrodes.

## 2.2 Characterizations

The morphology and microstructures of as-prepared electrodes were examined using emission Scanning Electron Microscopy (SEM, FEI Sirion200). The phase structure was identified by X-Ray Diffraction (XRD, XD-3A) and Raman spectra (Renishaw in Via). The concentration was determined from absorption spectra obtained by UV–vis spectrophotometer (U3900H).

## 2.3 Photo-catalytic activity measurement

The evaluation of electro-assisted photo-catalytic activity of the samples for removal of Cr (VI) and MB in aqueous solution was performed at ambient temperature. $K_2Cr_2O_7$ was used as the sources to obtain Cr (VI) containing wastewater. The initial concentrations of Cr (VI) and MB in either single-component solution or mixture solution are 2 mg/L and 5 mg/L respectively. The photocatalysis was performed in a cuvette containing 3 ml model wastewater and two electrodes. The pH value of the wastewater was adjusted using HCl or NaOH. The apparent catalytic area of TiO₂/CP electrode was 0.8 cm*2 cm and Pt wire was used as counter electrode. Before irradiation, the solution was placed in dark for 30 min to establish adsorption/desorption equilibrium. The TiO₂/CP electrode was then irradiated by UV light (MVL-210, 200 mW·cm⁻²). The concentration of MB was determined by the absorption at 664 nm at the interval of 10 min. The concentration of Cr (VI) was measured by the spectrophotometric method using chromogenic agent diphenylcarbazide at 544 nm.

## 3 RESULT AND DISCUSSION

### 3.1 Morphology characteristics

Fig. 1a shows the low magnification SEM image of the TiO₂/CP electrode. Flower-like branches of TiO₂ grow on the TiO₂ nanorods, which can increase active sites and boost sorption for the pollutant. Fig. 1b shows TiO₂ rods grow on the CP

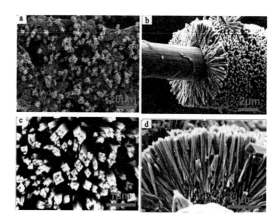

Figure 1. SEM images of the TiO₂/CP electrode (a); TiO₂ rods grow on the CP fiber (b); top of TiO₂ nanorod arrays (c); longitudinal cross section of TiO₂ nanorod arrays (d).

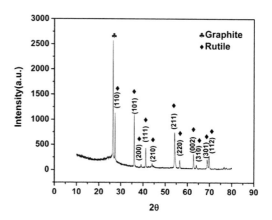

Figure 2. XRD pattern of TiO₂/CP.

fiber vertically. Fig. 1c and 1d were the high magnification SEM images that further reveal the features of nanorods.

XRD pattern of as prepared sample is shown in Fig. 2. It can be easily seen that the XRD data for the TiO₂/CP match well with the standard rutile pattern (PDF#21–1276) (Papadam T 2007) and there is no detection of the anatase and brookite phases of TiO₂.

The peaks at 2θ of about 26.4° is assigned to CP (Huang H 2013). The Raman spectra is shown in Fig. 3. The Raman lines at 145, 240, 445 and 608 cm⁻¹ conform well to $E_g$, $E_g$, $B_{1g}$, and $A_{1g}/B_{1g}$ modes of typical rutile (Wang W 2016). The peak at ~1597 cm⁻¹ were the G bands of carbon paper (Eunwoo L 2011) and shows in inset image. Raman result is consistent with that of XRD.

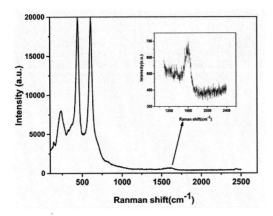

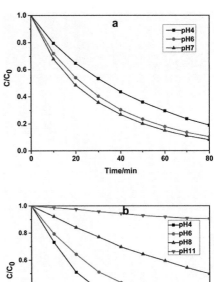

Figure 3. Raman of TiO$_2$/CP, the inset is peak of graphitic CP.

## 3.2 Photocatalytic activity measurement

The pH effect on photo-catalysis was examined firstly. Fig. 4a reveals that the lower pH is adverse for MB degradation because the surface of TiO$_2$ arrays possess positive charge in acidic solution that prevent the sorption of cationic MB. Fig. 4b indicates the lower pH is beneficial for Cr (VI) degradation because Cr (VI) ions have stronger oxidizing ability in more acidic solution as seen from reaction (1). Although pH has opposite effect to photo-catalytic degradation of Cr (VI) and MB. The following experiments were performed at pH 4 considering the pH of most real wastewater was in acidic range. The anodic bias voltage was then applied at TiO$_2$/CP electrode to improve photo-catalytic efficiency. Fig. 5a and 5b show appropriate bias potential is beneficial for the photo-catalytic process. Within the bias used in this research which is no more than 0.6V, Cr (VI) degradation is increased with applied bias voltage but has a small difference between 0.4V and 0.6V bias. As aforementioned, applied bias can prevent the recombination of electrons and holes and prompt their separation effectively, so more photo-generated electrons could reach counter electrode with higher bias which is helpful for the reduction of Cr (VI). However, the MB degradation is maximum at bias 0.4V. Because the separation of photo-generated carriers gets almost saturated at 0.4V, but larger electrostatic repulsion between MB and TiO$_2$/CP surface is induced with increasing bias which prevent the adsorption of MB at active sites. Thus the photo-catalytic efficiency of MB shows best with 0.4V bias.

For real application, it's very important to know the mutual impact of coexist pollutants. Fig. 6a gives the comparison of MB degradation in its single-component solution and in mixed solution

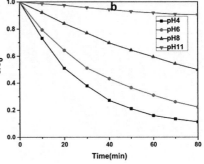

Figure 4. Photodegradation curves of MB (a) and Cr (VI) (b) at different initial pH.

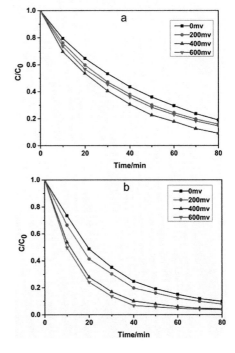

Figure 5. Photodegradation curves of MB (a) and Cr (VI) (b) under different bias potential (initial pH = 4).

51

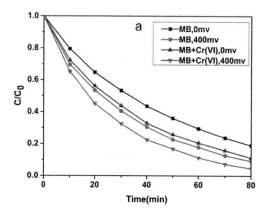

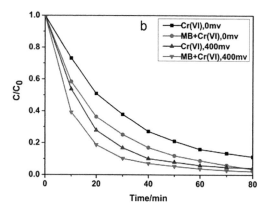

Figure 6. Photodegradation of MB, MB+Cr (VI) (a) and Cr (VI), MB+Cr (VI) (b) under different bias potential. Initial pH = 4.

with Cr(VI). It's clearly found that more MB are degraded in mixed solution containing Cr(VI) without bias or with 0.4V bias. In the mixture system, the degradation rates of MB are 89.2% and 95.5% after 80 min when the bias is 0V and 0.4V respectively, which have the increases of 8.3% and 4.5% relative to in pure MB solution. The similar phenomena are also found for Cr (VI) degradation as seen in Fig. 6b. The results show that there is a synergistic enhancement effect between MB and Cr (VI) at acidic solution during photocatalysis.

## 4 CONCLUSIONS

CP electrode had large specific surface area and good stability and conductivity which is a suitable substrate for loading $TiO_2$ to realize elec-tro-assisted photo-catalysis. We obtained $TiO_2$ nanorod arrays covered CP electrode by a simple way. The height of the $TiO_2$ nanorods ranges from 4 to 6 μm. Anodic bias was applied to prevent the recombination of electron and holes and prompt the photo-catalytic efficiency for degradation of Cr (VI) or MB containing wastewater. The waste-water containing both Cr (VI) and MB was also examined. A synergistic enhancement effect was found in mixed solution using as-prepared photo-catalyst, which showed its promising application in industrial wastewater treatment.

## REFERENCES

Asahi, R., Morikawa, T., Ohwaki, T., et al. (2001), Visible-light photocatalysis in nitrogen-doped titanium oxides. Science, 293(5528): 269–271.

Cappelletti, G., Bianchi, C.L., Ardizzone, S. (2008), Nano-titania assisted photoreduction of Cr (VI): the role of the different $TiO_2$ polymorphs. Appl Catal B-environ, 78(3): 193–201.

Eunwoo, L., et al. (2011). Synthesis of $TiO_2$ nanorod-decorated graphene sheets and their highly efficient photocatalytic activities under visible-light irradiation. J Hazard Mater. 219–220: 13–18.

Huang, H., et.al. (2013). Construction of sheet–belt hybrid nanostructures from one-dimensional mesoporous $TiO_2$ (B) nanobelts and graphene sheets for advanced lithium-ion batteriest. J. Mater. Chem. A, 2013 (1) 2495–2500.

Jung, D., et al. (2012). Evaluation of photocatalytic activity of carbon-doped $TiO_2$ films under solar irradiation. Korean J Chem Eng 29(6): 703–706.

Ojani, R., et al. (2012). Photoelectrocatalytic degradation of 3-nitrophenol at surface of $Ti/TiO_2$ electrode. J Solid State Electrochem 17(1): 63–68.

Papadam, T., Xekoukoulotakis, N.P., Poulios, I., et al. (2007), Photocatalytic transformation of acid orange 20 and Cr (VI) in aqueous $TiO_2$ suspensions. J Photoch Photobio A, 186(2): 308–315.

Wang, Y., et al. (2012). Design of a novel $Cu_2O/TiO_2/$ carbon aerogel electrode and its efficient electrosorption-assisted visible light photocatalytic degradation of 2,4,6-trichlorophenol. ACS Appl Mater Interfaces 4(8): 3965–3972.

Wang, W., Chen, J., et al. (2016), Photocatalytic degradation of atrazine by boron-doped $TiO_2$ with a tunable rutile/anatase ratio. Appl Catal B-environ, 195:69–76.

Xu, S.C., Pan, S.S., Xu, Y., et al. (2015), Efficient removal of Cr (VI) from wastewater under sunlight by Fe (II)-doped $TiO_2$ spherical shell. J Hazard Mater, 283: 7–13.

Yang, Y., et al. (2014). Microwave-assisted fabrication of nanoparticulate $TiO_2$ microspheres for synergistic photocatalytic removal of Cr (VI) and methyl orange. ACS Appl. Mater. Interfaces 6(4): 3008–3015.

*Advances in Energy and Environment Research – Achour & Wu (Eds)*
*© 2017 Taylor & Francis Group, London, ISBN 978-1-138-62682-9*

# Inversion of the sound velocity profile in a multibeam survey based on the simulated annealing algorithm

Z.W. Zhang, J.Y. Bao & F.M. Xiao
*Department of Hydrography and Cartography, Dalian Naval Academy, Dalian, Liaoning, China*

ABSTRACT: In this paper, a method of inversion of sound velocity profile in multibeam survey is presented. First, the sound velocity profile samples are analyzed with EOF (Empirical Orthogonal Function) and the reconstruction coefficients are calculated by sound velocity perturbation matrix and previous orders of EOF. Second, the reconstruction coefficients of EOF are estimated by simulated annealing algorithm, the target function is constructed with the seafloor distortion, which is obtained by multibeam sounding, and the annealing control parameters are set reasonably. Finally, the inversion of sound velocity profile is reconstructed. The analysis of examples shows that the inversion of sound velocity profile is closer to the real sound velocity profile and the seafloor corrected with inversion of sound velocity profile is closer to the real seafloor. The method presented in this paper can effectively reduce the negative influence of typical errors of sound velocity profile and significantly improve the accuracy of bottom topographic survey, which will be of great importance in detecting and exploiting the ocean energy resources.

## 1 INTRODUCTION

In general, the ocean is very rich in mineral resources and food resources. In order to exploit the oil and the nature gas resources under the sea, the top priority should be to obtain the high-accuracy bottom topographic map. Multibeam echo sounding system is one of the important tools for bottom topographic survey. However, in the process of multibeam survey, some areas will lack sufficient sound velocity profiles because of the incomplete coverage of the sound velocity profile stations. Usually, we choose the alternative sound velocity profile according to the time-nearby principle. However, because of the spatial and temporal variation characteristics of sound velocity profile (Li 1999), the alternative sound velocity profile can only approximately describe the distribution of sound velocity structure within a certain space-time. There will be representative errors of sound velocity profile if the distribution is beyond a certain range. Moreover, using the alternative sound profile to correct sound velocity will cause position reduction errors and finally trigger the distortion of seafloor topography. Because of the timeliness nature of the sound velocity profile, it is impossible to replace the sound velocity profile with errors by re-survey. Therefore, combining the receiving acoustic signal with existing hydrological data to inverse the velocity profile can yet be regarded as a feasible method (He et al. 2006).

Scholars worldwide have proposed some methods of sound velocity profile inversion. For example, Tolstoy et al. (1991) proposed an inversion method of sound velocity profile by adopting matched field. Skarsoulis et al. (1996) used the concept of time to peak and adopted the peak matching method of the inversion of sound velocity profile in deep sea. Zhang et al. (2002) presented a method of inversion of sound velocity profile in shallow sea by calculating the arriving time difference between the rays and a single hydrophone. Tang et al. (2006) used the acoustic signal propagation time to inverse velocity profile with the real data of the experiments in the sea. More recently, He & Li (2011) have used the vertical receiving array to record the broadband explosion of the sound source signal, and then used the method of inversion of sound velocity profile to achieve good results. It is well known that the inversion of sound velocity profile is a multi-dimensional optimization problem, while the number of inversion parameters should be as little as possible to reduce computational complexity. Davis (1976) & LeBlanc (1980) showed that EOF with previous orders can accurately describe sound velocity profile. Shen et al. (1999) also analyzed the feasibility of sound velocity profile, which is described via EOF in shallow sea. Peng et al. (2003) tried to describe sound velocity profile with EOF in deep sea and inverse sound velocity profile simultaneously. Therefore, sound velocity profile with EOF representing method can significantly reduce the inversion parameters, and only previous orders of inversion coefficients of EOF are needed. In addition, we need to search the EOF coefficients in a certain range, so it is necessary to choose fast and efficient search algorithms. As a type of general stochastic search algorithm (Ingber 1993), Simulated Annealing (SA) algorithm can be used to quickly search the global optimal solution of the target function or approximate global optimal solution in the process of parameter optimization, thus providing an effective way for inversion of sound velocity profile.

On the basis of the above analysis, a method combined with EOF analysis of inversion of sound velocity profile in multibeam survey is presented. That is, according to the analysis of sound velocity profile with EOF in surveyed area, the variation range of the coefficients is obtained. Meanwhile, the target function is built with submarine topography distortion amount and the optimal coefficients of reconstruction sound velocity profile are searched by using the simulated annealing algorithm. Finally, the inversion sound velocity profile is obtained. The goal of this paper is to achieve less representative error of sound velocity profile and to improve the quality of multibeam data.

## 2 EXPRESSION OF SOUND VELOCITY PROFILE WITH EOF

The representing method of sound velocity profile with EOF is to decompose sound velocity profile samples into orthogonal vectors and space vectors (Zhang 2010). Normally the previous orders of EOF can be used to reconstruct any profile accurately in the sequence. Assuming that the sampling sequences are composed of $N$ sound velocity profiles, we insert $M$ standard vertical layers and obtain the velocity matrix $C_{M \times N}$. By taking average of every row in the sound velocity matrix, we can obtain the average velocity $\bar{C}_{M \times N}$. The perturbation matrix of sound velocity profile $\Delta C_{M \times N}$ is obtained by subtracting the column of sound velocity matrix from the velocity matrix. Finally, the covariance matrix is obtained:

$$R_{M \times M} = \Delta C_{M \times N} \Delta C_{N \times M}^T / N \quad (1)$$

By eigenvalue decomposition, we can get:

$$R_{M \times M} F_{M \times M} = D_{M \times M} F_{M \times M} \quad (2)$$

where $D_{M \times N}$ is the eigenvalue matrix, whereas $F_{M \times N}$ is eigenvector matrix that is corresponding to eigenvalues, i.e. the EOF space function. Projecting EOF onto $\Delta C_{M \times N}$, we can get all the space eigenvector corresponding to the time factor:

$$A_{M \times N} = F_{M \times M}^T \Delta C_{M \times N} \quad (3)$$

where each row in $A_{M \times N}$ is the time coefficient of each eigenvector. Finally, we can obtain the representation of sound velocity profile with EOF:

$$C_{M \times N} = \bar{C}_{M \times N} + F_{M \times M} A_{M \times N} \quad (4)$$

By arranging eigenvalues in descending order, we can represent a certain sound velocity profile with $k$ orders of EOF as:

$$c(z) = c_0(z) + \sum_{i=1}^{k} \alpha_i f_i(z) \quad (5)$$

where $z$ indicates as the depth value, $c_0(z)$ denotes the average sound velocity profile, and $c(z)$ is the sound velocity profile represented with EOF. The subscript $k$ is orders of EOF, $\alpha_i$ is the EOF coefficient, and $f_i(z)$ is the empirical orthogonal function.

It is important to note that in (5), $c(z)$ and $\alpha_i$ are correspondent to each other. Correcting the sound velocity profile will only fine-tune the previous orders of EOF coefficients, and thus the number of search parameters will significantly reduce. Therefore, the process of solving the optimal sound velocity profile is converted to search a set of optimal processes of EOF coefficients.

## 3 INVERSION OF SOUND VELOCITY PROFILE

### 3.1 Basic ideas of inversion for sound velocity profile

The multibeam sounding is satisfied with the following basic mathematical relationship:

$$H = f(C, \theta, T) \quad (6)$$

where $C$ indicates the sound velocity profile, $\theta$ is the beam direction angle, and $T$ is echo time. The subscript $H$ is the depth of each beam point, which is calculated according to the constant gradient sound ray tracking model (Lu 2012). The depth values of each beam can be determined using the initial velocity profile, echo time, and beam angle of a certain ping. If deviation exists between the measured sound velocity profile and the initial velocity profile, the correction seafloor topography is prone to be distorted. Moreover, if the distortion quantity meets the requirements of a certain size, we regard $C_0$ as the demanding sound velocity profile. Otherwise, we can revise $C_0$ in accordance with certain principle until the distortion quantity size meets the requirements. According to (5), we can

54

correct the sound velocity profile by fine-tuning the previous orders of EOF coefficient. The process of solving the optimal sound velocity profile is converted to search a set of optimal processes of EOF coefficient, i.e. to search the global optimal solution of problems in the solution space.

### 3.2 Sound velocity profile inversion process

Searching the optimal EOF coefficient using simulated annealing algorithm mainly includes the following steps:

Step 1: the construction of target function. Target function can be represented with topographic distortion quantity size. Because the beam direction angle is not sensitive to the sound velocity profile error near ±45°, there exists "the depth of the zero difference point phenomenon" (Zhu 2005 & Capell 1999). Thus, the target function can be defined as follows:

$$E = \sum_{i=1}^{n} \left( H_i - H_\theta \right)^2 \qquad (7)$$

where $n$ is the beam number of each ping, $H_\theta$ is the depth value, and beam direction angle is 45°. We can find the best solution of the EOF coefficient until the target function value is minimum.

Step 2: the coefficient perturbation and initial solution are set. In the process of modified EOF coefficient, we can write:

$$\alpha = \alpha_0 + \Delta\alpha \qquad (8)$$

where $\Delta\alpha$ is a random number, which is uniformly distributed in $[-1,1]$. The value of $\alpha$ lies in the range of EOF coefficient values, and $\alpha_0$ is the initial solution, which can be a random value within the scope of the coefficient values.

Step 3: the new solution acceptance mechanism. Metropolis criterion is generally used as the state function of simulated annealing algorithm. The target function difference is calculated as:

$$\Delta E = E\left(\alpha\right) - E\left(\alpha_0\right) \qquad (9)$$

where, if $\Delta E < 0$, $\alpha$ is accepted as a new current solution. Otherwise, a new current with probability solution $P = exp(-\Delta E/T) > \delta$ is accepted. The parameter $\delta$ is a uniformly distributed random number in $[0, 1]$. For the new solution to be accepted,

we should replace the current solution with the new solution. Meanwhile, target function values should be modified.

Step 4: the initial temperature and the temperature updating function. In order to ensure that the acceptance probability of algorithm solution is 1 at the beginning of the algorithm running, a sufficiently high initial temperature is required. Then, the initial temperature can be estimated:

$$T_0 = K\varphi \qquad (10)$$

where $K$ is a sufficiently large number, whereas $\varphi = \max\left(|\Delta E_i|\right)$ and the value $\varphi$ can be simply estimated through several iterations.

Temperature updating function is used to modify the temperature outer loop. In this paper, the update function of temperature is set as:

$$t_{k+1} = \lambda t_k \qquad (11)$$

where $\lambda$ is a constant that is close to 1. Usually, $\lambda \in [0.85, 0.98]$.

Step 5: iteration length control of each temperature (inner loop). When the temperature is high, the probability of each state is accepted, and almost all values are accepted. We can reduce the iteration steps as far as possible for the same temperature. As the temperature gradually decreases, more and more states are rejected. At this moment, we can increase the number of iteration steps correspondingly. Then, maximum iterative steps and the accepted number limits can be given. The iteration continues until the actual accept times reach the limits at a certain temperature. Otherwise, the iterative steps should reach the upper limit.

Step 6: algorithm termination criterion (outer loop). First, a small positive number $T_{min}$ is given. When the temperature is less than $T_{min}$, the algorithm stops indicating that the temperature has reached the minimum value.

When the searching process is stopped, the EOF coefficient can be regarded as optimal. The sound velocity profile obtained by substituting the EOF coefficient into (5) is close to the actual sound velocity profile, and the seafloor corrected with inversion of sound velocity profile is close to the real seafloor. The flow chart showing the above process is depicted in Figure 1.

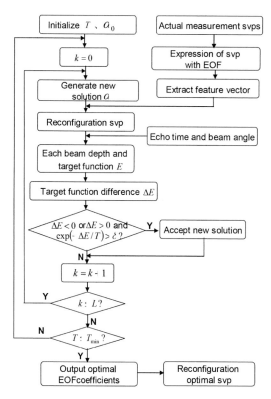

Figure 1. Flowchart for inversion of sound velocity profile.

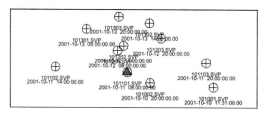

Figure 2. Spatial and temporal distribution of sound velocity profile station.

# 4 EXAMPLE ANALYSIS

## 4.1 *Experimental data selection and EOF analysis*

We select 11 sound velocity profiles for analyzing in a certain area of northern China on 10 October 2001. For convenience, we numbered all the sound velocity profiles consecutively. The distribution of sound profile stations and measuring time is shown in Figure 2, where sound velocity profile marked with triangle symbol indicates inversion (the sound velocity profiles have been measured, which are only used to authenticate inversion algorithm).

It is a minimum mean square error fitting method that the sound velocity profile is represented with EOF. As only previous orders for fitting are used, there exists deviation between actual sound velocity profile and fitting sound velocity profile. We can analyze the precision of fitting results with different orders by calculating the maximum error and the mean square error along the depth direction. Finally, the fitting results are shown in Figure 3.

We suppose that the research depth is the sound velocity profile maximum detection depth. Then, we preset beam angle and the depth of the water and use the original sound profile to obtain echo time by the constant gradient layered sound ray tracing method. Furthermore, we combined echo time with reconstructed sound profile to calculate the depth by the constant gradient sound ray tracing method. Finally, we compared the depth value with the original depth value and obtained a set of depth deviation. Figure 4 shows the maximum deviation of depth as a function of EOF orders (the dotted line indicates 0.25% water tolerance).

Combining Figures 3 and 4, we know that with the increase of EOF orders, the maximum error and the mean square error of sound velocity profile decrease along the depth direction, and the maximum depth deviation of each beam decreases.

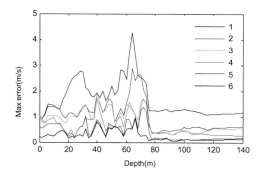
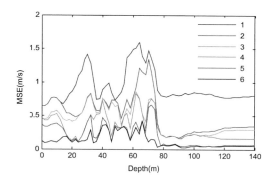

Figure 3. Maximum deviation and mean square error of sound velocity profile with different orders as a function of depth.

In this case, assuming coefficient of six orders or above to represent the sound velocity profile with EOF can meet 0.25% depth of the water tolerance requirements.

### 4.2 Inversion of sound velocity profile

We can take the sound velocity profile, which is marked in five-star symbol in Figure 2, as the inversion profile. The acquisition time of sound velocity profile is on 10 October 2001. The sound profile is regarded as reference profile, which is only used to test the accuracy of the inversion results. We suppose that the seafloor is flat, and then combine the constant gradient ray tracing method of mathematical simulation with the corresponding multibeam echo time. Finally, the remaining 10 sound velocity profiles are considered for EOF analysis. Table 1 shows the six lowest orders of EOF coefficients and their searching ranges.

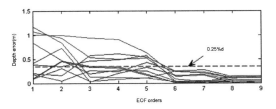

Figure 4. Maximum deviation of depth as a function of EOF orders.

We can search the optimal coefficients according to the process of simulated annealing algorithm for the inversion of sound velocity profile shown in Figure 1. Because the SA algorithm is less dependent on initial values, we can set the fitting coefficients of Svp1 as the initial solution of the algorithm, that is, $\alpha_0 = (19.091\ -9.737\ -2.032\ -4.736\ -1.887\ -1.032)^T$. The largest annealing temperature (i.e. the initial temperature) can be obtained from (9), and we set the minimum temperature $T_{min} = 0.01°$. Figure 5 shows the convergence of SA algorithm, which uses double ordinate form to describe the target function and the target function difference as a function of temperature in the iteration process. The figure shows that when the outside loop iteration reaches 80, i.e. the temperature $T = 1.42°$, both the target function and the target function difference attain a stable state. Finally, the searching EOF coefficients are optimal, that is, $\alpha_0 = (2.400\ 8.322\ 5.376\ 3.979\ 3.457\ -1.1.595)^T$.

When the optimal EOF coefficient $\alpha$ is searched, by substituting $\alpha$ into (5), the inversion of sound velocity profile can be obtained. Suppose the five-star symbol of sound velocity profile in Figure 2 is unknown, we usually choose the alternative sound velocity profile according to the time-nearby principle for sound velocity correction. Figure 6 shows the distribution of original sound velocity profile, the alternative sound velocity profile, and the inversion sound velocity profile. The precision of the two sound profiles and the depth error caused

Table 1. EOF coefficients and search range with six lowest orders.

|  | Svp1 | Svp2 | Svp3 | Svp4 | Svp5 | Svp6 | Svp7 | Svp8 | Svp9 | Svp10 | Max | Min |
|---|---|---|---|---|---|---|---|---|---|---|---|---|
| $a_1$ | 19.091 | 0.191 | −21.538 | 28.281 | 18.194 | −21.083 | −3.589 | −24.769 | −22.802 | 28.024 | 28.281 | −22.802 |
| $a_2$ | −9.737 | −8.260 | 8.289 | 8.680 | −5.387 | −1.194 | −3.524 | 3.513 | −1.065 | 8.687 | 8.687 | −9.737 |
| $a_3$ | −2.032 | 5.484 | 4.728 | −0.666 | 2.130 | −5.143 | −2.885 | −1.526 | 0.677 | −0.767 | 5.484 | −5.143 |
| $a_4$ | −4.736 | −0.058 | 0.012 | −0.362 | 4.241 | 2.912 | 2.312 | −2.563 | −1.560 | −0.199 | 4.241 | −4.736 |
| $a_5$ | −1.887 | 3.736 | −2.502 | 0.656 | −1.289 | 0.868 | −0.704 | 3.032 | −2.571 | 0.662 | 3.736 | −2.571 |
| $a_6$ | −1.032 | 1.137 | −0.846 | 0.686 | −2.111 | −0.791 | 2.345 | −1.610 | 1.773 | 0.448 | 2.345 | −1.610 |

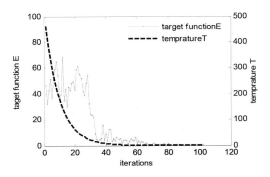
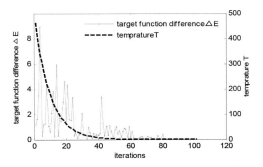

Figure 5. Convergence of the iterative algorithm based on SA.

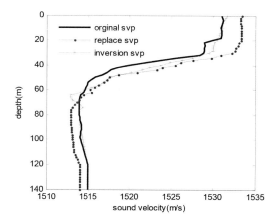

Figure 6. Distribution of the three sound velocity profiles.

Table 2. Statistics of sound velocity profile error and depth error.

| Error | Svp | Max error | Mean error | MSE |
|---|---|---|---|---|
| SVP error (m/s) | Alternative | 5.60 | 1.79 | 2.08 |
| | Inversion | 3.83 | 0.72 | 1.05 |
| Depth error (m) | Alternative | 1.04 | 0.25 | 0.23 |
| | Inversion | 0.07 | 0.05 | 0.01 |

by the two sound profiles after sound velocity correction are presented in Table 2. The depth error caused by the two sound profiles is a U-shaped distribution with the maximum value at the edge of the beam.

Figure 6 and Table 2 show that the inversion error of sound velocity profile is superior to the alternative sound velocity profile, and the depth error with sound velocity correction is superior to the alternative sound velocity profile. Meanwhile, the depth error can meet the requirements of 0.25% depth tolerance with the inversion of sound velocity profile. Therefore, the inversion of sound velocity profile with the method is closer to the real sound velocity profile and the seafloor corrected with the inversion of sound velocity profile is closer to the real seafloor.

## 5 CONCLUSIONS

By combining the simulated annealing algorithm with EOF representing method, this paper presents a new method of inversion of the sound velocity profile, which regards the seafloor topography distortion amount as the target function. Then, the inversion process of sound velocity profile is given. On the basis of the analysis of examples, the following conclusions are drawn:

1. With the increase of EOF orders, the maximum error and mean square error of the sound velocity profile decrease along the depth direction, and the maximum depth deviation of each beam decreases. In this case, we can take six orders or above to represent the sound velocity profile. After the sound velocity correction, each beam depth can meet 0.25% of the depth tolerance requirements.

2. In the execution process of inversing sound velocity profile based on simulated annealing algorithm, the algorithm's effect depends on the selection of a set of control parameters including initial temperature, temperature updating function, each iteration length, and termination criteria. In the process of practical application, the parameters must be set properly. Otherwise, it will affect the efficiency and running performance of the search algorithm.

3. When using the inversion of sound velocity profile in the way presented in this paper, the search of EOF coefficients is optimal as the target function achieves stable state. Furthermore, the inversion of sound velocity profile is obtained, which is closer to the real sound velocity profile, and the seafloor corrected with inversion of sound velocity profile is closer to the real seafloor. The new method can effectively reduce the effect of typical sound velocity profile errors and significantly improve the accuracy of multibeam survey data.

## REFERENCES

Capell, W.J. 1999. Determination of sound velocity profile errors using multibeam data. *Oceans'99 MTS/IEEE*, Seattle, WA, 1144–1148.

Davis, R.G. 1976. Predictability of sea surface temperature and sea level pressure anomalies over the north pacific ocean. *Phys Ocean*, 6: 249–266.

He, L., Li, Z.L. & Peng, Z.H. 2011. Inversion for sound speed profiles in the northern of South China Sea. *Scientia Sinica Phys, Mech & Astron*, 41(1): 49–57.

He, L., Li, Z.L. & Zhang, R.H. 2006. Expression of sounding sound velocity profile with EOF and matched filed inversion in East China Sea. *Progress in Natural Science*, 16(3): 351–355.

Ingber, L. 1993. Simulated annealing: practice versus theory. mathematical and computer modeling, 18(11): 29–57.

LeBlanc, L.R. & Middleton, F.H. 1980. An underwater acoustic sound velocity data model. *Acoust Soc Am*, 67(6): 2055–2062.

Li, J.B. 1999. Principles multibeam survey techniques and methods. Beijing: Ocean Press.

Lu, X.P, Bian, S.F. & Huang, M.T. 2012. An improved method of calculating average sound speed in constant gradient sound ray tracing technology. *Geomatics and Information Science of Wuhan University*, 37(5): 590–593.

Peng, L., Wang, L. & Qiu, X.F. 2003. Modal wave number tomography for South China Sea front. *China Ocean Engineering*, 17(2): 289–294.

Shen, Y.H., Ma, Y.L. & Tu, Q.P. 1999. Feasibility of description of the sound speed profile in shallow water via empirical orthogonal function (EOF). *Applied Acoustics*, 18(2): 21–25.

Skarsoulis, E.K., Athanassoulis, G.A. & Send, U. 1996. Ocean acoustic tomography based on peak arrivals. *J Acoust Soc Am*, 100(2): 797–813.

Tang, J.F. & Yang, S.E. 2006. Sound speed profile in ocean inverted by using travel time. *Journal of Harbin Engineering University,* 27(5): 733–736.

Tolstoy, A., Diachok, O. & Frazer, L.N. 1991. Acoustic tomography via matched field processing. *J. Acoust. Soc. Am*, 89(3): 1119–1127.

Zhang, X., Zhang, Y.G. & Zhang, J.X. 2010. EOF analysis of sound speed profile in east water of Taiwan. *Advances in Marine Science*, 28(4): 498–506.

Zhang, Z.B., Ma, Y.L. & Ni, J.P. 2002. A new and practical method of inverting sound speed profile in shallow water. *Journal of Northwestern Polytechnical University*, 20(1): 36–39.

Zhu, X.C., Xiao, F.M. & Liu, Y.C. 2005. Analysis of the simulative images of mutibeam echo sounding in different sound velocity gradients. *The 4th International Symposium on Acoustic Engineering and Technology*. Harbin, China, 29–30.

*Advances in Energy and Environment Research – Achour & Wu (Eds)*
*© 2017 Taylor & Francis Group, London, ISBN 978-1-138-62682-9*

# Thermoelectric cooling in downhole measuring tools

Rohitha Weerasinghe

*University of the West of England, Bristol, UK*

ABSTRACT: Thermoelectric cooling is a highly effective and novel technique used for cooling solid-state circuitry inside downhole measuring equipment. A Peltier device in a deep well measuring tool works as a heat pump. The understanding of the thermal behavior inside is important in designing the tools. For this study, thermal modeling is used, but accurate modeling of the Peltier device is crucial in predicting the thermal behavior and subsequent thermal management of the tool. This paper is based on numerical simulations using computation fluid dynamics and experimental analysis of the thermal behavior inside an industry-standard deep well measuring tool using a model equipped with Peltier cooling device. The present modeling results are based on two mathematical models to predict the thermal behavior of the Peltier device. Results show a technique that can be used in designing cooling performance of downhole equipment and generally in Peltier cooling at elevated temperatures.

## 1 INTRODUCTION

### 1.1 Downhole tooling

Vertical Seismic Profiling (VSP) is a common geophysical technology used in oil and gas investigation in deep wells. VSP provides high resolution and dependable results, but is subjected to high temperatures and harsh well environments. As these wells become deeper and hotter, the need for effective thermal management in the instrumentation becomes paramount in keeping the electronics of the measuring tool within allowable working temperatures. Seismic surveys are a critical element of the search for oil and gas exploration. In a seismic survey, the propagation of elastic waves through the rock gives an indication of the sub-surface distribution of different rock types and thus the probability of finding a viable source of hydrocarbons. The seismic source varies depending on the application, but may be dynamite, an air gun, or a vibrating plate to produce waves at a range of frequencies. In a surface seismic study, usually both the source and the receivers are located at the surface and the reflected waves are analyzed. These sensitive ground velocity sensors are called geophones. In a VSP borehole investigation, the receivers are located within the borehole and the source is usually located on the surface or, less frequently, downhole. Another important and fast growing application for borehole seismic logging tools is the monitoring of hydraulic fracturing (fracking) sites.

### 1.2 The cooling problem

Instrumentation of borehole seismic investigations needs to be extremely sensitive in order to capture the micro-seismic waves at the receiver and large volumes of data must be sent back to the surface; in the case of continuous monitoring surveys, this must occur in near real time. This process requires sensitive and sophisticated electronics to be exposed to extremely hostile environments; depending on the well being surveyed, this may be pressures of up to 30,000 psi (2000 bar or 206 MPa), temperatures of higher than 200°C, and often in environments with high concentrations of $H_2S$. The electronics manufacturers are continually striving to increase the service temperatures of their products. Various strategies have been deployed to achieve this over the last 30 years, including vacuum (Dewar) insulation, eutectic alloys, which extend the time available downhole, and various active cooling technologies as evidenced by Bennett et al. (1986). Unlike in conventional electronics cooling applications, the need for the system to be hermetically sealed and withstand high pressures prevents the use of forced air cooling. In addition to shielding the electronics from the heat of the borehole fluid, a further challenge is in dissipating the heat produced by the electronics themselves. Various active cooling techniques have been evaluated in the past.

## 2 PROBLEM DEFINITION

### 2.1 Peltier cooling for downhole tools

Thermoelectric cooling has been around since the early 1950 s and extensive research has taken place to aid cooling at or around room temperature. The phenomenon that a voltage is generated in a conductor or semiconductor subject to a temperature gradient was discovered around the 1800 s, which is known as the Seebeck effect. The inverse process where heat is pumped across a conductor or a semiconductor due to a voltage difference is known as the Peltier effect. The current uses of the Seebeck effect are mainly for thermoelectric heat recovery. Thermoelectric cooling is used in applications where space is limited and where conventional refrigerants cannot be used, for example, in computers and small machinery. Most these applications take place around ambient temperature. Manufacturers of Peltier conductors have tested their devices at room temperature and within a band of ±30 K. Seismic downhole measuring devices operate under harsh environmental conditions, a few kilometers below ground at temperatures that are 200 K above standard room temperature. These tools contain sensitive electronic circuitry and the performance of the devices depends on whether they can be kept within the operating temperatures of the electronics. Several studies have been conducted in thermoelectric heat recovery at elevated temperatures and representative data are available. However, representative data in cooling at these temperatures is rare. As a representative technique according to Dirker et al. (2005), the properties are being extrapolated from room-temperature results. Thermoelectric cooling in seismic measuring tools is a novel technology. The performance of such devices has been hampered by the relatively low overall Coefficient of Performance (COP). However, this low COP resulted mostly from the losses in the thermal passage of such devices. Thermal analysis showed that the performance can be enhanced by better insulating the device and carefully arranging the thermal paths. Heat transfer modeling helps improve this aspect of the measuring tools. The overall device performance can be analyzed by thermal analysis of Peltier performance through identifying the heat transfer patterns, paths, and the actual cooling performance of the Peltier device. This paper focuses on the performance of the Peltier devices in cooling at elevated temperatures and the validity of heat transfer models used to predict the properties.

### 2.2 Thermoelectric devices

The amount of heating or cooling obtained using a thermoelectric device depends on its thermal efficiency. The maximum efficiency of a thermoelectric device for both power generation and cooling is determined by:

$$ZT = \frac{S2\sigma}{\kappa}T \quad (1)$$

Thermoelectric materials have been studied recently in the forms of bulk thermoelectric materials, individual nanostructures, bulk nanostructures, and interfaces in bulk thermoelectric materials. Peltier heating and cooling occur when two materials with different Peltier coefficients are joined by the imbalance of the Peltier heat flowing in and out of the junction. The cooling or heating (Q) occurring at the junction is given by:

$$Q = \left(\Pi_2 - \Pi_1\right)I \quad (2)$$

where I is the electric current, $\Pi$ is the Peltier coefficient, and subscripts 1 and 2 represent the two materials.

The total heat pumped by the device can be found from the number (N) of junctions (pairs of N-type and P-type semiconductors), the ratio of the length to area of these junctions (G), and the thermal conductance of the unit ($\kappa$) for a given temperature distribution of the hot ($T_H$) and cold ($T_C$) faces, as a function of the supplied current [4]:

$$\dot{Q}_C = 2N\left[\alpha I T_C - \left(\frac{I^2}{2G}\right) - \kappa(T_H - T_C)G\right] \quad (3)$$

However, these equations can be simplified to obtain the linear resistance Seebeck coefficient and conductance. These equations can be written after Luo and Bons (2008) as:

$$Q_C = -\kappa T_H + (\alpha I + \kappa)T_C - \frac{I^2 R}{2} \quad (4)$$

$$\dot{Q}_H = \kappa T_C - (\alpha I - \kappa)T_H + \frac{I^2 R}{2} \quad (5)$$

$$V = \alpha(T_H - T_C) + IR \quad (6)$$

The resistance of the unit is available from manufacturer data (validated by manufacturer measurements), which helps derive the Seebeck coefficient and the conductance [10]:

$$\alpha = \sqrt{\frac{2RQ_{max}}{T_H^2}} \quad (7)$$

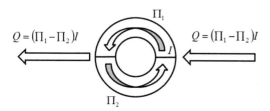

$$Q = \left(\Pi_1 - \Pi_2\right)I \qquad Q = \left(\Pi_1 - \Pi_2\right)I$$

Figure 1. Mechanism of Peltier effect at the junction of two materials.

$$\kappa = \frac{\alpha^2 \left(T_H - \Delta T_{max}\right)^2}{2R\Delta T_{max}} \tag{8}$$

Coefficient of performance ($Z$) and figure of merit ($zT$) are also very useful in determining the performance of thermoelectric devices according to Scherrer et al. (2006).

## 3 MODELING AND SIMULATION

### 3.1 Mathematical modeling

The mathematical model above can be used in a simulation tool if the temperature and heat flux are to be solved for the region. The values defined above can be used to obtain the Seebeck coefficient, conductance, and resistance as a function of temperature. The values of $\alpha$, $\kappa$ (derived using equations 5 and 6 above), and $R$ (obtained from the manufacturer data) were used in a commercially available CFD package Star CCM+ to provide heat flux boundary conditions for the thermoelectric cooling (TEC) model. Temperature boundary conditions were set for the hot side of the tool. The above system of six equations would be solved to obtain the heat flux and the temperature field. A simplified version of the boundaries defined is given in Figure 2.

In the simulation model in Figure 2, the well temperature is set to 160, 180, and 225°C for two simulations.

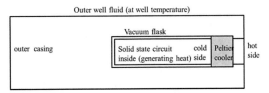

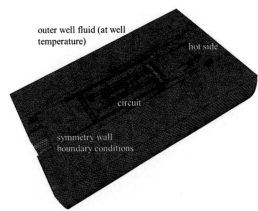

Figure 2. Boundaries and computational mesh.

Initial validation of the TEC was done using a linear interpolation and a curve fitting technique. Two test modules were considered in the analysis: Bismuth Telluride (HT2) and custom hybrid bi-Te doped with lead (TESH127). In order to derive the linear equations for resistance and Seebeck coefficients, data from the manufacturers were used. These values were verified with in-house testing of the data. The so developed TEC model was valid in the tested temperature range. At elevated temperatures, however, the performance could vary as the thermal conductance, electrical resistivity, and Seebeck coefficient for the thermoelectric modules vary with temperature, each material having differing characteristics. In order to derive the linear equations that define the performance of the TECs at elevated temperature, it is necessary to re-evaluate the manufacturer data values for these temperatures. In the case of the Laird module, it is possible to use the manufacturer analytical design tool, Aztec (Scillasoft, 2014), to find these values at the system temperature. When such data are not available, an alternative approach has to be followed. The resistance of the unit at a range of mean temperatures was measured in a laboratory oven and a digital multi meter to record voltage drop across the unit at a fixed current. From these data, a linear equation for the resistance could be derived. In the absence of direct experimental data and extrapolation, a technique based on measurements and curve fitting was used to find values of $zT$ for a range of temperatures for materials. If the composition of the module was known, values could be estimated from these curves. In the absence of these values, the data from the experimental oven testing were used to approximate $zT$ with temperature. A copper heat sink was used to dissipate heat from the hot side, and the cold side was fixed to an insulated mass. Thermocouples were used to measure the hot and cold side temperatures.

The temperature performance has been predicted using derived linear equations (6). Table 1 shows the values available from manufacturer data. These values are used in the Computational Fluid Dynamics (CFD) numerical tool in order to simulate the thermal performance.

Table 1. Summary of mean temperatures and cooling observed in experimental testing and simulation.

| | Fluid Temp. (°C) | Flask Temp. (°C) | Cooling (°C) | Voltage (V) |
|---|---|---|---|---|
| Experiment HT2 | 161.06 | 134.82 | 26.20 | 12.5 |
| CFD HT2 | 160.00 | 133.77 | 26.20 | 11.6 |

# 4 MEASUREMENTS

## 4.1 *CFD and measurements*

The thermoelectric cooling performance was initially tested with a CFD cooling model of the Peltier device alone. Tests were carried out to characterize the TEC performance. The heat flux and cold side temperatures were measured at different hot side temperatures: 160, 180, and 225°C. The modules are fitted into an oven test rig comprising a copper heat sink and vacuum flask. The system was placed in the oven at 160, 180, and 225°C and the time to heat the slug was observed. Figure 3 shows the temperature results of the TEC module using a linear interpolation technique. It shows that the predicted COP values are coherent with the measured values and are able to predict the TEC performance satisfactorily. The tool modeled used for the purposes of this study was supplied by Avalon Sciences Ltd. It comprises a steel pressure barrel, which houses the geophones, a mechanism to operate an arm, which clamps the tool to the wall of the borehole and a module containing the digital electronics, which performs signal processing. These electronics are housed within a vacuum insulated vessel. Active cooling is provided by a TEC module similar to the one modeled above.

The numerical model is a three-dimensional representation of the downhole tool similar to one that is commercially manufactured. To expedite the simulation, the regions adjacent to the digital electronics module are excluded from the model, as there is no active component in this region and thus have no impact on the cooling of the electronics. Fully resolving all of the electronic components housed within the module would incur a high computational cost to accurately solve the geometry and thus a simplified representation of the printed circuit board is used. This simplification does not hinder the performance analysis of heat transfer. The model takes advantage of the symmetry of the tool; only one-half of the system is modeled, cut down the central axis of symmetry (Figure 2). Planar symmetry conditions are applied to the cut faces. The external region of the model, representing the well fluid, has a fixed temperature boundary condition on the far face, representing the large thermal capacity of the borehole fluid.

The fluid is modeled in the laminar regime, with convection driven by gravity in the direction of orientation of the tool to the well. The well fluid is modeled as water. The solid regions of the tool were modeled with appropriate material properties, sourced from the manufacturers' data sheet. The vacuum region of the flask is modeled as a gas with a conductivity of $1 \times 10^{-6} \, \mathrm{Wm^{-1}K^{-1}}$.

# 5 RESULTS

## 5.1 *Text and indenting*

Figure 4 shows the temperature profiles obtained with the computational simulation. The temperature profiles show that the cooling effect of the standard TEC is based on the cooler. It is evident from the bright red colors in the heat sink side that the TEC performs well during cooling.

The main aim of the tool thermally is to isolate the hot fluid from the electronic circuitry and

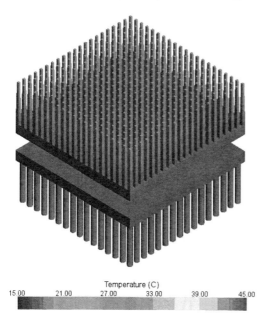

Temperature (C)
15.00   21.00   27.00   33.00   39.00   45.00

Figure 3. Simulation results for a typical thermoelectric module.

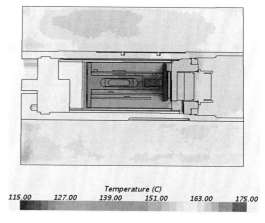

Temperature (C)
115.00   127.00   139.00   151.00   163.00   175.00

Figure 4. Temperature profile of the tool with the hybrid TEC module (TESH 127).

pump the heat generated out of the flask. Table 2 shows the fluid temperatures and flask temperatures with the cooling effect at a temperature difference.

A pumped hot oil bath was used to emulate the well fluid at elevated temperature and pressure. It is intended to take measurements in a well, but the controlled conditions in the laboratory represent the conditions down the hole. The limit of the fluid in the open system was 160°C. The tool was submersed in the fluid, which was then heated. Once the system reached steady state, the temperature, as reported by a sensor embedded within the onboard electronics, was recorded. The model was validated using the data from the experimental results at external temperature of 160°C. The temperatures observed in the model are equivalent to those in the experimental testing.

The results are in line with the results that were obtained with the simulation of the thermoelectric cooling device. The cooling effect of the TEC module is represented by the dark blue region near the TEC in Figure 4.

The area adjacent to the TEC unit shows a good cooling effect with external temperature of 190°C of TEC in the model. This represents the heat pumping capacity of the TEC to the well fluid. This reduction is significant in the performance, as it defines whether it leads to exceeding the threshold or not. The experimental results and simulations prove the data shown in Table 1.

## 6 CONCLUSIONS

The main aim of the simulation work was to evaluate the performance of the downhole tool using TEC cooling module and then use the model as a base model in evaluating the performance of similar tools. The numerical model was able to show the difference in performance within the tool geometry. The temperatures recorded in the electronic region of the module show good agreement with the experimental values. However, the agreement on the voltage predictions is less impressive. This is most likely due to the estimation of $\alpha$ value.

The first-order linear approximation method gives realistic values of the TEC properties at elevated temperatures. This is more true for $Q$ and $T$. It is not yet confirmed how accurate it is for $\alpha$. A nonlinear approximation method based on experimental values is under development at the moment, which is expected to produce more accurate results at higher temperatures. Initial investigations have shown that the relationship is better explained with the second- and third-order terms. This should give better predictions at elevated temperatures. However, with current capabilities and available knowledge base, the numerical model is able to predict the thermal behavior of the downhole measuring tool accurately. This has lead to optimal designing of the tool geometry and TEC sizing. Accurate model prediction has also helped in making the customers understand the tool geometry and cooling process. With more advanced prediction models and further measurements, numerical models are expected to produce accurate models that will eliminate the use of testing.

## ACKNOWLEDGMENTS

The authors would like to thank Dr Tom Hughes for his contribution to this work and Avalon Sciences Ltd. for the continued support for the project given as the industry partner.

The research work was funded by the knowledge transfer partnership (Grant No. KTP009448) of Innovate UK.

## REFERENCES

Bennett, G., Sherman, G. (1983), Analysis and thermal-design improvements of downhole tools for use in hot-dry wells http://www.osti.gov/pages/190 servlets/purl/6491119, doi:10.2172/6491119.

Dirker, J., Liu, W., Van Wyk, J.D., Meyer, J.P., and Malan, A.G., (2005) Embedded solid state heat extraction in integrated power electronic modules, Accepted for publication in the *I.E.E.E. Transactions on Power Electronics*, vol. 20, No. 3.

Kim, W. et.al. (2006), Thermal conductivity reduction and thermoelectric figure of merit increase by embedding nanoparticles in crystalline semiconductors. Phys. Lett. Rev., 96.

Kutasov, V. A., Lukyanova, L. N. and Vedernikov M. (2006), Thermoelectrics Handbook Macro to Nano. CRC, Boca Raton. Ch. 37.

Luo, Z., (2008) A simple method to estimate the physical characteristics of a thermoelectric cooler. Electronics Cooling.

Scherrer, H., and Scherrer, S. Thermoelectrics Handbook Macro to Nano. CRC, Boca Raton, 2006. Ch. 27.

Shengqiang Bai, Hongliang Lu, Ting Wu, Xianglin Yin, Xun Shi, and Lidong Chen. Numerical and experimental analysis for exhaust heat exchangers in automobile thermoelectric generators. Case Studies in Thermal Engineering, 4: 2014, 99–112.

Sootsman, D.Y., Chung J.R. and Kanatzidis. M.G. (2009). New and old concepts in thermoelectric materials. Angewandte Chemie (International ed. in English), 48: 2009, 8616–8639.

Zhao, D., and Tan, G.A. (2014) Review of thermoelectric cooling: Materials, modeling and applications. Applied Thermal Engineering, 66: No. 1–2. 2014, 15–24.

Zhiting Tian, Sangyeop Lee, and Gang Chen. A comprehensive review of heat transfer in thermoelectric materials and devices. Ann. Rev. Heat Transfer, 17: 2014, 425–483.

*Advances in Energy and Environment Research – Achour & Wu (Eds)*
© *2017 Taylor & Francis Group, London, ISBN 978-1-138-62682-9*

# Measurement and analysis of SH-CCT curves of HD15Ni1MnMoNbCu steel

Jianlin Zhang & Gangang Wang
*Suzhou Nuclear Power Research Institute Co. Ltd., Suzhou, China*

Fayun Zhang
*Nuclear and Radiation Safety Center, Beijing, China*

Zhongbing Chen, Shaohua Yin & Yishi Lv
*Suzhou Nuclear Power Research Institute Co. Ltd., Suzhou, China*

ABSTRACT: The microstructural evolution and Vickers hardness measurement in the welding Heat-Affected Zone (HAZ) of HD15Ni1MnMoNbCu steel in nuclear power station were investigated by Gleeble-3180 thermal mechanical simulator, and the Simulated HAZ Continuous Cooling Transformation (SH-CCT) curves were measured simultaneously. With $t_{8/5}$ increasing from 3.75 to 15000 s, products martensite, bainite, ferrite, and pearlite were obtained successively.

## 1 INTRODUCTION

With the rapid development of domestic nuclear power plant construction, it is necessary to realize the supply of home-made seamless steel pipes for nuclear power generation industry. HD15Ni1MnMoNbCu steel based on the Germanic 15NiCuMoNb5 steel has emerged recently. HD15Ni1MnMoNbCu steel is also called WB36CN1, representing Chinese first batch steel for nuclear power, which is a Ni–Cu–Mo-alloyed structural steel with good weldability and heat-resistant properties. Because of its excellent properties, it has been widely used in the main steam pipe and main water supply pipe systems of conventional nuclear power station (Wu, 2010). Compared with 15NiCuMoNb5 steel, HD15 Ni1MnMoNbCu steel retards the flow-accelerated corrosion phenomenon by controlling chromium content to extend the service life of the pipe (Bindi, 2008). Considerable number of studies have been carried out on steel (Guo, 2010; Zhang, 2013; Gan, 2013; Zhang, 2015), while the research on the basis theory of the steel, such as the microstructure and mechanical properties during welding and heat treatment, is inadequate.

SH-CCT diagram plays a significant role in welding. It makes it possible to predict the microstructures and properties of coarse-grained HAZ (CGHAZ) and serves an incredibly useful purpose to evaluate the weldability of steel (Li, 2005; Spanos, 1995). In this paper, the SH-CCT diagram of HD15Ni1MnMoNbCu steel was obtained by using thermal expansion measurement combined with microscopic analysis and Vickers hardness test. The research of HD15Ni1MnMoNbCu steel can provide the correct theory formulation for its welding technology design to produce an excellent weld quality.

## 2 EXPERIMENTAL MATERIAL AND METHOD

The experimental material was obtained from the quenched and tempered HD15Ni1MnMoNbCu steel pipe of $\Phi$ 813 mm × 64 mm for nuclear power station, whose main chemical composition is listed in Table 1, and the original microstructure in Figure 1 demonstrates the mixture of ferrite and bainite. Specimens were cut from the steel pipe and machined to the size of $\Phi$6 mm × 90 mm.

Table 1. Chemical composition of experimental HD15 Ni1MnMoNbCu steel (wt%).

| C | Si | Mn | P | S | Cr | Ni | Mo | Cu | N | Nb | Al | Fe |
|---|----|----|---|---|----|----|----|----|---|----|----|----|
| 0.16 | 0.38 | 0.92 | 0.005 | 0.001 | 0.21 | 1.11 | 0.32 | 0.6 | 0.007 | 0.02 | 0.006 | Bal. |

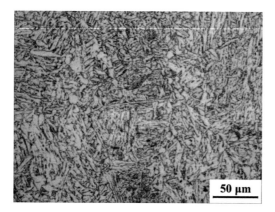

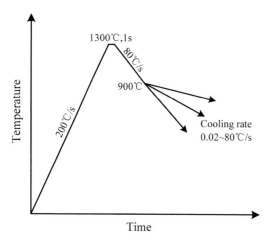

Figure 1. Original microstructure of HD15Ni1MnMoNbCu steel.

Figure 2. Thermal simulation process of HD15Ni1MnMoNbCu steel.

In this study, the thermal cycle simulation was conducted with a Gleeble-3180 thermal mechanical simulator. The phase transformation temperatures of HD15Ni1MnMoNbCu steel under various cooling rates were determined according to the principle of tangent. First, the points of Ac1 and Ac3 were determined from room temperature to 1000°C at a heating rate of 0.5°C/s, and after holding the temperature for 10 min, the point of Ms was measured from 1000°C to room temperature at a cooling rate of 100°C/s. Second, specimens were heated at 200°C/s from room temperature to the peak temperature of 1300°C, and cooled to 900°C at 80°C/s after holding for 1 s, and finally cooled to room temperature at different linear rates ranging from 0.02 to 80°C/s, in which $t_{8/5}$ is between 3.75 and 15000 s accordingly. The schematic of the schedule for the thermal simulation process is shown in Figure 2. Moreover,

SH-CCT diagram of HD15Ni1MnMoNbCu steel was drawn with the help of microstructure analysis and hardness measurement.

## 3 SH-CCT RESEARCH OF HD15NI1MNMONBCU STEEL

### 3.1 *SH-CCT diagram*

On the basis of thermal expansion curve shown in Figure 3, the critical temperatures of Ac1, Ac3, and Ms were calculated to be 753, 860, and 412°C, respectively. Analogously, the temperatures at the beginning and end of phase transformation under various cooling rates can also be determined. Combining the metallographic analysis and hardness measurement, the SH-CCT diagram of HD15 Ni1MnMoNbCu steel was developed as shown in Figure 4, which exhibits consecutive characteristic of

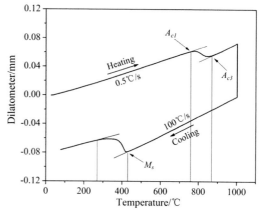

Figure 3. Curve of phase transformation critical temperatures.

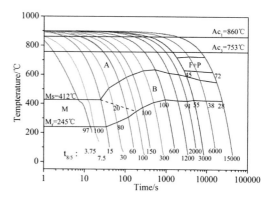

Figure 4. SH-CCT diagram of HD15Ni1MnMoNbCu steel.

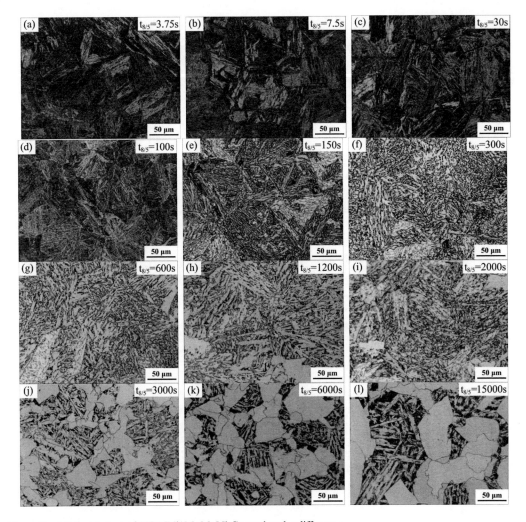

Figure 5. Microstructure of HD15Ni1MnMoNbCu steel under different $t_{8/5}$.

the phase transformation and multi-transformation curves, revealing complex microstructures with $t_{8/5}$ increasing from 3.75 to 15000 s. In the figure, A is austenite, F is ferrite, P is pearlite, B is bainite, and M is martensite. It is clear that martensite is observed at $t_{8/5}$ less than 100 s, bainite is produced at $t_{8/5}$ ranging from 15 to 15000 s, and ferrite and pearlite form throughout at $t_{8/5}$ higher than 3000 s. The phase transformation into whole bainite can occur in a wide range of $t_{8/5}$ between 100 and 3000 s.

### 3.2 Microstructure in CGHAZ

Figure 5 shows the micrographs of the simulated CGHAZ under different $t_{8/5}$ conditions. It is obvious that the microstructure characteristic of steel changes with the increase of $t_{8/5}$ distinctly. The

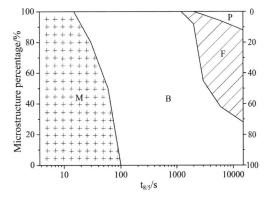

Figure 6. Relationship between microstructure percentage and $t_{8/5}$.

micro-structure acquired at high cooling rate is martensite as shown in Figure 5 (a, b, c), when $t_{8/5}$ is less than 30 s. As $t_{8/5}$ increases, the structure of bainite is formed at intermediate cooling rate according to Figure 5 (d, e, f, g, h, i). With cooling rate decreasing, acicular bainite transforms to granular bainite. The granular bainite is defined as the aggregation of M-A constituent distributed on polygonal ferrite, which is mainly observed when $t_{8/5}$ is higher than 100 s. Figure 5 (j, k, l) shows that the mixture structure of polygonal ferrite and lath-like bainite forms when $t_{8/5}$ exceeds 3000 s, the amount of ferrite increases with the increase of $t_{8/5}$, and the grain size of CGHAZ is increased with the increase of $t_{8/5}$. Furthermore, it can also be found that a small quantity of pearlite occurs in the grain boundary of ferrite. According to the metallographic structure analysis of specimens at different cooling rates using image analyzer, the relationship between the various microstructure percentage and $t_{8/5}$ is drawn in Figure 6.

### 3.3 Vickers hardness of CGHAZ

Figure 7 shows the Vickers hardness of the simulated CGHAZ under different $t_{8/5}$ conditions. It is clear that the hardness decreases gradually with the increase of $t_{8/5}$, because the volume fraction of soft structure (ferrite and pearlite) increases and that of hard structure (bainite and martensite) decreases with the above metallographic analysis. The microstructure of CGHAZ mainly consists of martensite at high cooling rate and the maximum is 405 HV10, which usually correlates with the increased brittleness and reduced ductility. The microstructure is almost whole bainite when $t_{8/5}$ is in the range of 100–3000 s, and the hardness value is between 225 HV10 and 255 HV10. With $t_{8/5}$ increasing, the hardness becomes less than 225

HV10 because of the appearance of the ferrite phase. The above results indicate that the hardness in the CGHAZ can be used as a criterion for the organization evaluation standard under appropriate welding conditions.

The SH-CCT diagram of HD15Ni1MnMo NbCu steel presented above shows that the whole bainite is archived at CGHAZ and $t_{8/5}$ should be chosen from 100 to 3000 s.

## 4 CONCLUSIONS

1. The SH-CCT of HD15Ni1MnMoNbCu steel was obtained by Gleeble-3180 thermal mechanical simulator, which can provide a theoretical basis for the optimization of the steel welding technology.
2. As $t_{8/5}$ increases from 3.75 to 15000 s, the following microstructures are obtained in succession: martensite, bainite, ferrite, and pearlite. This also leads to the decrease of hardness.

## REFERENCES

Bindi Chexal. Flow-Accelerated Corrosion in Power Plants [M]. Office of Secretary Rulemakings and Adjudications Staff. 2008.

Gan Huan-chun, Peng Xiang-yang, Cao Hui, et al. Welding Technology of WB36CN1 Steel Composite Piping of Nuclear Power [J]. Welding & Joining, 2013, 12: 29–32. (In Chinese).

Guo Yuan-rong, Wu Hong, Hu Maohui, et al. R&D of WB36CN1 Seamless Steel Pipe Used for Steam/Water Supply Piping Circuit of Conventional Island of Nuclear Power Plant [J]. New Product Development, 2010, 39(4): 31–35. (In Chinese).

Li Ya-jiang. Microstructures, Properties and Quality Control for Welding [M]. Chemistry Industry Press, 2005. (In Chinese).

Spanos G, Fonda R.W., Vandermeer R.A., et al. Microstructural Changes in HSLA-100 steel Thermally Cycled to Simulate the Heat-Affected Zone during Welding [J]. Metallurgical and Materials Transactions A, 1995, 32: 77–93.

Wu Jia-kai, Chen Juan. Selection of Main Thermodynamic Piping Materials of AP1000 Nuclear Power Station Conventional Island [J]. Guangdong Electric Power, 2010, 23(2): 75–79. (In Chinese).

Zhang Jian-lin, Zhu Ping, Wang Gan-gang, et al. Measurement and Analysis on Continuous Cooling Transformation Curves of HD15NiMnMoNbCu steel [J]. Transactions of Materials and Heat Treatment, 2015, 36(9): 114–118. (In Chinese).

Zhang Jin-ping, Qi Yan-li, Tian Tian, et al. Material Selection and Localization of Main Steam and Main Feedwater Pipeline for Conventional Island in Nuclear Power Plants [J]. Thermal Power Generation, 2013, 42(10): 9–12, 20. (In Chinese).

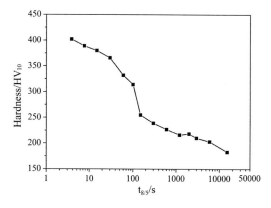

Figure 7. Curve of Vickers hardness versus $t_{8/5}$.

*Advances in Energy and Environment Research – Achour & Wu (Eds)*
*© 2017 Taylor & Francis Group, London, ISBN 978-1-138-62682-9*

# Research on the interface of a digital measurement device in a smart substation and its calibration method

Hao Chen, Huiyu Chen, Liman Shen, Rui Huang, Fusheng Chen, Longmin Bu,
Xing He, Kai Li, Xianyong Xu & Weineng Wang
*State Grid Hunan Electirc Power Company, Changsha City, Hunan Province, China*

ABSTRACT: In order to meet the demand of intelligent substation metering system, a research on IEC 61850 and other relevant standards was done, the digital interface features of metering device describing the function of model merging unit and the design based on merging unit ARM & FPGA embedded processor platform solution were given, and the combined unit hardware and software implementations were described. The test results indicated that the digital interfaces of this study could meet the requirements of IEC 60044 standard fully with a good application value.

## 1 INTRODUCTION

With the development of smart grid, intelligent substations have become more and more advanced. The intelligent substation digital metering system is based on IEC 61850, and according to the distributed structure, the system is composed of electronic transformers, protection devices, intelligent control terminals, digital measuring instruments, optical fiber communication networks, and substation model, in order to achieve information sharing and interoperability between electrical equipment within the entire intelligent substation (RICHARD (2008), Wang Lu (2008), L. Andersson (2003)).

## 2 INTERFACE CHARACTERISTICS OF INTELLIGENT DIGITAL SUBSTATION METERING DEVICE

IEC 60044-7/8 and IEC 61850-9-1/2 define the basic functionality of the interface digital measuring device as follows (IEC 61850-9-1(2003)):

1. The device should be compatible with high-pressure side of the interface, and can receive real-time information including not only digital samples using Manchester encoding format sent by the electronic transformer output, but also the analog signal by the amount of conventional electromagnetic transformer and switch output state. The secondary equipment interface uses frame format with IEC standards for Ethernet transmission over data packets.
2. The device should be configured with a synchronization and it sends a synchronization signal based on the sampling synchronization signal

to the electronic transformer sampler (A/D converter module) to ensure that each transformer sample value is consistent over time, both measured on the secondary side to protect the output device synchronization signal abnormality flag function (IEC60044-7(2002)).
3. The device should have a laser energy supply module, in order to supply energy to the high-pressure side of the electronic transformer through the optical fiber (IEC 60044-8(2002)).

Figure 1 shows the block diagram of the basic functional model for the merging unit, which mainly contains the following sections: synchronous sampling control module; multi-channel digital, analog, and switch collection module; and sample data processing and communication modules.

The synchronous sampling control module can capture and identify the external sync pulse in seconds input (GPS), in accordance with the need for secondary metering and protection devices, determine the sampling rate, and transmit the synchronization signal to the electronic transformer

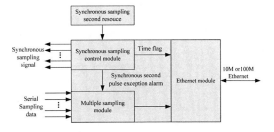

Figure 1. Digital interface functional block diagram of a metering device.

sampling sampler to ensure the consistency of each transformer sampling time.

The multi-channel digital, analog, and switch collection module receives real-time electronic transformer multi-channel transmission over the sampling frame made after the Manchester decoding Cyclic Redundancy Check (CRC), and sends the processed data to the data processing module for subsequent processing. When the input is an analog signal, the synchronous sampling module, according to a given sampling rate of the signal, is sampled and the samples are sent to the data processing module. The input signal is a multi-way switch, switching state recognition, and in the agreement status bit.

# 3 MERGING UNIT BASED ON ARM AND FPGA

## 3.1 Implementation of merging units

In this paper, the merging unit is developed based on ARM & FPGA. Its principle block diagram is shown in Figure 2. In this scheme, ARM and FPGA constitute the core processor platforms. The FPGA makes sampling synchronization control and serial data reception. The ARM is also used to control and coordinate the implementation of various functional modules, because of its high-speed operating frequency. It enables complex data processing, and its rich interface controller can implement Ethernet communication, man–machine interface, and other functions.

## 3.2 Digital interface processing platform based on ARM & FPGA architecture

The LPC2214 ARM7 processor chip produced by NXP is used, which exhibits excellent performance meeting the design requirement. The chip ARM7TDMI-S core, which supports 32-bit RISC processor ARM/Thumb instruction set, has a processing speed of up to 60 MHz, containing 16 KB on-chip SRAM and 256 KB on-chip Flash memory. The integrated on-chip, with 32-bit data bus width, and 24-bit address bus width, dedicated external memory control interface (EMIF). External bus, LPC2214 processor can control a variety of peripherals, such as external expansion FLASH, network chips, and its scalability is very strong. In addition, the LPC2214 internal interrupt controller has 16 interrupts priority control, and multiple support interrupt sources can be set up to meet the needs of embedded real-time operating system (PHILIPS (2004)).

When choosing FPGA, the following aspects are mainly considered: operating speed, number of logical units, available I/O ports and high-speed I/O port number, the size of on-chip memory resources, clock management, PLL, hardware multiplier, a hard IP core processor soft IP core, power consumption, and prices. Performance of programmable logic device (FPGA) depends on its production technology, capacity, and speed. According to the design, Altera's programmable logic array EP2C8 CycloneII series chip is selected. The chip is made of copper, with 1.2V power supply, 90 nm production process, on-chip resources including 8256 logic cells, M4 K RAM blocks of 36 KB, and 20741B RAM modules, embedded within the 18-chip hardware multiplier, and two on-chip phase-locked rings. EP2C8 is widely used in industrial applications, has superior performance, and stability can meet the functional requirements of the merged cell (CIRRUS (1999)).

## 3.3 Hardware design of Ethernet control interface

The merging unit receives electronic transformer sampling data, which are defined with IEC 61850-9-1 and sent to the second device via Ethernet. The merging unit is designed with IEC 61850-9-1standard, using point-to-point communication protocol. According to IEC 61850-9-1, in a 12-way channel sampling, each electronic transformer week of sampling points is 80, the total data transfer rate is $(80 \times 50) \times 984$ bps = 3.936 Mbps < 10 Mbps, and 10 Mbps Ethernet can meet the requirements of the selection of communication protocol. Therefore, we selected the Ethernet controller CS8900 network communication functions.

CIRRUS LOGIC CS8900 is a 16-bit network controller chip with low power consumption. Therefore, the user can dynamically configure the data transfer mode depending on needs, working methods, and application of physical layer interfaces, to accommodate a variety of different networks in the application environment. Its interior contains industrial communication standard in line with IEEE802.3 MAC block (MAC), which has 4 KB of RAM unit with ISA interface that supports full-duplex communication to achieve sending and

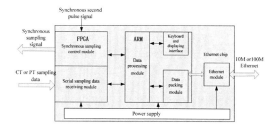

Figure 2. Block diagram of merging unit based on ARM & FPGA.

receiving of Ethernet data frames. Then, collision detection, generation, and checking of header can be performed simultaneously to generate the CRC checksum. When the transmission control register (TxCMD) works after configuration, MAC frames can automatically detect conflict by conflict, which can be automatically re-transmitted if the data portion of the frame is less than 46 bytes; the data are automatically populated segments, which cause the frame length 802.3 standard to specify the minimum length requirement (Z.L. Yin (2004)).

## 4 TASK SCHEDULING OF MERGING UNIT

The sampling electronic transformer is under the control of the synchronization signal, so the correct identification and transmission with the synchronization control signal is the key to sample value accurately. The synchronization signal interface module is implemented by the FPGA. When the FPGA receives the correct second pulse synchronization signal, it sends an interrupt request signal to notify ARM synchronous control to the ARM processor. Therefore, ARM synchronous sampling control program should be processed by interrupt execution with the highest priority interrupt. After synchronization control, the ARM sends electronic transformer synchronous sampling pulse and receives transmitted electronic transformer data, since the serial transmission method is used to ensure real-time data reception, which should be completed by interrupt execution.

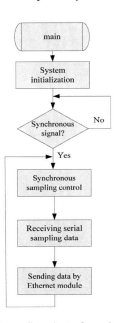

Figure 3. Software flow chart of merging unit.

After receiving 12 channels, with ARM data frame format according to IEC 61850-9-1 standards for the data framing and packing process, the control chip then sends data to the Ethernet device (Tor Skeie (2002), Li Meng (2004)). For ARM processors, these tasks work sequentially, whose order of priority is sampling synchronization control, receiving serial sample value, and Ethernet communication control. The HMI program is set to the lowest priority. According to these principles, a modular unit design incorporates software programming, whose flow chart is shown in Figure 3.

## 5 TEST OF DIGITAL METERING DEVICE INTERFACE

As shown in Figure 4, the system is composed of standard voltage transformer (PT), Agilent 3458 type of digital multimeter, electronic voltage transformers, synchronous second pulse generator, merging unit, and IPC. The verification process is as follows: the output voltage of the autotransformer is grid signal and the signal is sent to the standard PT and electronic voltage transformer. A digital multimeter is used to measure the standard PT output signals via bus GBIP, and the sampling results will be sent to the IPC. The measured value of the sample from electronic transformer is sent to the merging unit, which can transmit through the Ethernet interface to the IPC. The calibration software on the IPC processes is used to calculate the ratio errors and angle errors.

The test results are shown in Table 1:

Data transmission time per thousand frames is about 1.25 s and offset is only a few milliseconds, indicating that the packet transmission rate is uniform, while the received data packet and the sent packet are equal, so the merging unit does not appear as the last package phenomenon.

The synchronous second pulse generator provides a sync signal for eight semi-digital multimeter

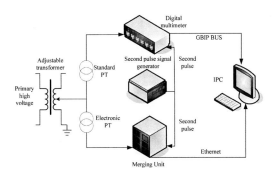

Figure 4. Merging unit test system connection diagram.

Table 1. Data reception time of the merging unit.

| No. | Receiving time (μs) | Data frame | Time interval (μs) |
|---|---|---|---|
| 0001 | 0 | – | – |
| 1001 | 1250071 | 1000 | 1250071 |
| 2001 | 2500614 | 1000 | 1250543 |
| 3001 | 3750050 | 1000 | 1249436 |
| 4001 | 5000447 | 1000 | 1250397 |
| 5001 | 6250669 | 1000 | 1250222 |
| 6001 | 7500918 | 1000 | 1250249 |
| 7001 | 8751328 | 1000 | 1250410 |
| 8001 | 10000871 | 1000 | 1249543 |

Table 2. Ratio error and angle error of the merging unit.

| Input voltage percentage | Ratio error (%) | Angle error (') |
|---|---|---|
| 120% | −0.014~−0.017 | 0~0 |
| 100% | −0.006~−0.007 | 0~1 |
| 20% | 0.028~0.041 | 5~7 |
| 5% | 0.131~0.149 | 9~12 |
| 1% | 0.065~0.12 | 13~17 |

and electronic transformers, and the IPC transformer calibration software is used to calculate the ratio error and angle error. Table 2 shows the primary voltage ratings of 120%, 100%, 20%, 5%, and 1%, ratio error, and angle error. The results show that the measurement system can achieve 0.2% accuracy.

## 6 CONCLUSION

In this paper, a merging unit based on IEC 61850-9-1 is developed. However, in order to achieve two-way seamless communication between the process layer and the spacer layer, it should meet the IEC 61850-9-2 standards. Considering the network communication program, only achieving some basic data transmission and processing functions for the development of TCP/IP protocol is not deep enough, and the communication software is weak and less portable. It is important to investigate how to use an embedded operating system for optimal task scheduling and improve communication software structure to meet the requirements of IEC 61850-9-2.

## REFERENCES

Andersson, L., C. Brunner, F. Engler (2003). Substation automation based on IEC 61850 with new process-close technologies. Power Tech Conference Proceedings. IEEE Bologna, 2: 6.
CIRRUS (1999). CS8900 A Crystal LAN TM ISA Ethernet Controller Handbook.
IEC 60044-7 (2002). Instrument Transformers-part 7: Electronic Voltage Transformers.
IEC 60044-8 (2002). Instrument Transformers-part 8: Electronic Current Transformers.
IEC 61850-9-1 (2003). Communication Networks and Systems in Substations-Part 9-1: Specific Communication Service Mapping (SCSM)-Sampled analogue values over serial unidirectional multidrop point to point link.
Meng Li, Li Hongbin, Feng Kai (2004). Ethernet-based power system automation electronic transducer embedded digital output port, 28(13): 93–96.
PHILIPS (2004). LPC2214 16/32-bit ARM Micro—controllers Datasheet.
Richard D, Cherry T (2008). Standards for the smart grid. Proceedings of IEEE Energy Conference. November 17–18: 1–7.
Tor Skeie, Svein Johannessen, Christoph Brunner (2002). Ethernet in Substation Automation. IEEE control system Magazine, 22(3): 43–51.
Wang Lu, Wang Buhua, Song Lijun, Guo Zhihong (2008). Research and application of IEC 61850 Digital Substation [J]. Power System Protection and Control, 36: 90.
Yin, Z.L., W.S. Liu (2004). A Novel FPGA-Based Method to Design the Merging Unit Following IEC61850. Power System Technology. Vol. 1, 260–263.

*Advances in Energy and Environment Research – Achour & Wu (Eds)*
© *2017 Taylor & Francis Group, London, ISBN 978-1-138-62682-9*

# Textured silicon wafer prepared by reactive ion etching for anti-reflection in Si solar cells

X.G. Zhang & M.Z. Shao
*Department of Physics, South University of Science and Technology of China, Shenzhen, China*

ABSTRACT: Because of the refractive index mismatch of air and Silicon (Si), the light collection efficiency of Si usually suffers from a relatively high reflection at the air–silicon interface. Silicon surface texturing by wet or dry etching increases the efficiencies of solar cells. Reactive Ion Etching (RIE) techniques are very effective for application in low-cost, large-area crystalline Si solar cells. In this paper, we studied $SF_6/O_2$ RIE process texturing of Si. After the RIE process, the measured reflection of textured Si wafer was reduced significantly compared with polished Si wafer. This demonstrated that the texturing technique of RIE could be applied in Si solar cells.

## 1 INTRODUCTION

Solar cells, as one type of environmentally friendly energy sources, can efficiently harness the energy of sunlight and have many appealing advantages such as inexhaustible, pollution-free, and ability to supply electrical power to rural areas and satellites (Lewis 2007). At present, the current production of solar cells is dominated by crystalline Silicon (Si) modules, because of their mature fabrication techniques and relatively high efficiency (Tamang et al. 2016).

Two types of energy loss occur in solar cells: optical loss and electricity loss. Because of the refractive index mismatch of air and Si, the light collection efficiency of Si usually suffers from a relatively high reflection at the air–Si interface. The reflection loss of light at polished Si surface can reach more than 33% in the spectral range of 300–1100 nm (Wan et al. 2010). As a result, the conversion efficiency of photovoltaic devices is significantly reduced.

Therefore, new techniques such as anti-reflection coating and surface modifications have been developed to enhance the absorption and reduce the reflection loss in Si solar cells (Dewan et al. 2012, Jovanov et al. 2013, Haase et al. 2007). An ideal anti-reflection structure should have minimum reflection loss on solar cell surfaces over an extended solar spectral range for all angles of incidence (Sai et al. 2012). As for the technique of anti-reflection coating, a single layer of Silicon Nitride (SiNx) with the thickness of quarter-wavelength was generally used because of the compatibility with the silicon fabrication process (Yang et al. 2015). However, such a single-layer anti-reflection coating reduces reflectivity only at a certain wavelength range under normal incident conditions. Multilayer quarter-wavelength film stacks have been developed to overcome the limitation of single-layer anti-reflection. Relative studies have proved the ability of multilayer film stacks to reduce the reflection over a certain spectral range for wide ranges of incident angles (Hsu et al. 2012, Ferry et al. 2012). The major technical challenges for this approach are precise controlling of the refractive index and thickness of each individual layer in the multilayer.

For surface modification, sub-wavelength texturing was accomplished by femtosecond laser pulses (Shen et al. 2004) or plasma etching (Hitoshi et al. 2006). Wet chemical approaches are most commonly used to minimize reflection losses. However, the wet chemical etching in aqueous alkaline solution is carried out by an anisotropic method, which is limited to (100) oriented silicon only. Therefore, this approach does not apply to polycrystalline Si solar cells.

In recent years, with the dramatic development of microfabrication tools and techniques, different textured Si surfaces for photovoltaic application can be realized (Tan et al. 2015). Nevertheless, implementing most of these existing texturing techniques is too expensive and/or of too low reproducibility to apply in the solar cells. Reactive Ion Etching (RIE) is a popular choice for etching silicon. In this paper, ordered Si needle-like pyramid arrays were fabricated to obtain the textured Si wafer. The isotropic etching profile and surface of pyramid arrays were improved by appropriately adding $O_2$ to $SF_6$ etching gas. The reflection of textured Si wafer was measured and the results demonstrated that the textured Si wafer can be applied in solar cells.

## 2  EXPERIMENTS

Our experiments were performed on polished silicon substrate. First, a 200-nm-thick photolithography resist was spin-coated on Si substrate and pre-baked. Then, the patterns consisting of two-dimensional square arrays with the side/space sizes of 0.5 μ/0.5 μ were defined by photolithography. After exposure, the patterns were developed with a standard developer at room temperature. Subsequently, Al film with thickness of about 50 nm was thermally evaporated and lifted-off by dipping the samples into acetone with ultrasonic agitation. Finally, under the patterned Al mask, the samples were further processed by RIE using $SF_6$ and $O_2$ plasma under a pressure of 4.0 Pa. After etching, the Al mask was removed. The obtained textured Si wafer was inspected by a Scanning Electron Microscope (SEM). The reflection of textured Si wafer was also measured.

## 3  RESULTS AND DISCUSSION

In order to investigate the etching effect of $SF_6$ plasma for etching silicon, SEM was used to characterize the surface of etched samples. Figure 1(a) shows the SEM images of Si nanostructure etched by $SF_6$ plasma. Furthermore, $SF_6$ is the source of F* radicals, which are responsible for the chemical etching of the Si by forming the volatile $SiF_4$. The etching model for Si etching in $SF_6$ plasma is shown in Figure 1(b). It is seen that the F* radicals can etch the sidewall under Al mask as well as the bottom. Therefore, the etching of Si using $SF_6$ is isotropic. Thus, because of its isotropic etching characteristics, $SF_6$ can be used as etchant to produce needle-like pyramid arrays.

However, since the reaction of $Si + F \rightarrow SiF_4$ takes place only on the surface, the surface of Si etched by $SF_6$ was quite rough. In order to improve the surface quality and to control the etched profiles, $O_2$ gas was added during the $SF_6$ gas etching process; the flow rate of the $SF_6$ was kept constant at 30 sccm and that of the $O_2$ gas was increased from 5 to 30 sccm. Figure 2(a), (b) show the side view SEM images of the etched Si structure under the capping of Al mask with the $O_2$ flow rates of 5 and 10 sccm, respectively. The chamber pressure and RF power were set at 4.0 Pa and 100 W, respectively.

As shown in the figure, the profiles of etched Si wafer under the Al mask can be clearly seen. The needle-like pyramids are obtained with addition of 5 sccm $O_2$ gas while the taper-like structures are formed with addition of 10 sccm $O_2$ gas. In an $SF_6$ and $O_2$ plasma, each gas has a known specific function and influence. Therefore, the etched

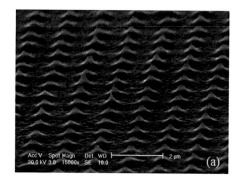

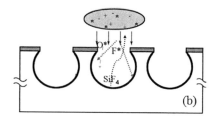

Figure 1.  SEM images of the Si pyramid by RIE under the patterned Al mask with $SF_6$: 30 sccm, 4.0 Pa, 100 W, 1 min. (b) Model for chemical etching Si in $SF_6$ plasma.

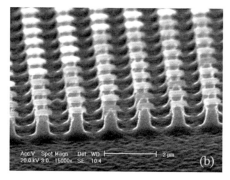

Figure 2.  SEM images after isotropic etching of silicon layer under the capping of the Al mask with additional $O_2$ of (a) 5 sccm and (b) 10 sccm.

profile is easily controlled by just changing the flow rate of one of these gases. It was reported that, under the influence of strong electric field, $SF_6$ and $O_2$ generate F* and O* free radicals. At the condition of low oxygen concentration, adding $O_2$ to $SF_6$ plasma can facilitate $SF_6$ to convert into F* by Equation (1):

$$O + SOF_x \rightarrow SO_2F_{x-1} + F \tag{1}$$

and prevent the recombination reaction of $SF_6$ by Equation (2):

$$F + SF_x \rightarrow SF_{x+1} \ (x = 3 \sim 5) \tag{2}$$

However, it does not mean that the F* radical concentration always increases with the increase of oxygen concentration. At the condition of high oxygen concentration, there is a competition between the F* and O* radicals for reaction with silicon. The reaction of F* radicals with silicon results in etching effect, but the etching of O* results in passivating process by forming a passivation layer of $SiO_xF_y$ on the Si surface. Furthermore, with more oxygen being added, O* radicals can further dilute F atom and the availability of F* is reduced. As a result, the horizontal etching rate began to decrease and partial anisotropic etching occurred, as shown in Figure 2(b). Our results suggest that the etching process can be finely controlled by controlling the ratio of $O_2$ gas flow to total gas flow to realize isotropic etching. The surface morphology of etched surface can be improved by appropriately adding $O_2$ to $SF_6$ gas.

Figure 3 shows a normal-incidence reflectivity measurement of the textured Si wafer at the wavelength range of 300–1100 nm. For comparison, the reflectivity of a polished Si wafer was also measured. It is evident from the figure that the average reflectance of polished silicon wafer and textured silicon wafer is about 35% and 12%, respectively. For polished silicon, reflection takes place only once at the silicon surface, whereas for the textured silicon surface, reflection can take place twice, each at surface 1 and surface 2, as shown in Figure 3(b). As a result, the reflectance of textured wafer reduced from 35% to 12%. Through the technique of textured surface modifications, the surface reflectivity can be effectively suppressed. The multiple reflections on the textured surface extended the optical path and increased the absorption of photons.

## 4 SUMMARY

In summary, we developed a simple but efficient technique for fabricating texture Si wafer with good regularity. The needle-like pyramids were fabricated using isotropic dry etching in $SF_6/O_2$ plasma under the patterned Al mask. The RIE process parameters such as chamber pressure, RF power, and the ratio of mixed gases are the key factors controlling the etching profile. Through the technique of textured surface modifications, the surface reflectivity can be effectively suppressed. In particular, Si-based RIE can be compatible with the mature ULSI technology, which is critical for potential applications in solar cells.

ACKNOWLEDGMENT

This work was financially supported by the Shenzhen Innovation Foundation of Fundamental Research (JCYJ20140901153335318).

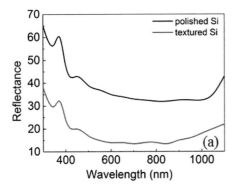

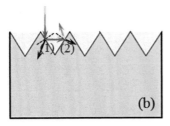

Textured Si wafer

Figure 3. Normal-incidence reflectivity measurement of the textured Si wafer.

REFERENCES

Dewan, R., J. I. Owen, D. Madzharov, V. Jovanov, J. Hüpkes, & D. Knipp (2012). Analyzing nano textured transparent conductive oxides for efficient light trapping in silicon thin film solar cells. *Appl. Phys. Lett.* 101: 103903-1-4.

Ferry, V. E., M. A. Verschuuren, H. B. T. Li, E. Verhagen, R. J. Walters, R. E. Schropp, H. A. Atwater, & A. Polman (2012). Light trapping in ultrathin plasmonic solar cells. *Opt. Express* 18: A237–A245.

Haase, C. & H. Stiebig (2007). Thin-film silicon solar cells with efficient periodic light trapping texture. *Appl. Phys. Lett.* 91: 061116-1-3.

Hitoshi, S., H. Fujii, K. Arafune, Y. Ohshita, M. Yamaguchi, Y. Kanamori, & H. Yugami (2006). Antireflective subwavelength structures on crystalline Si fabricated using directly formed anodic porous alumina masks. *Appl. Phys. Lett.* 88: 201116.

Hsu, C. M., C. Battaglia, E. Pahud, Z. Ruan, F. J. Haug, S. Fan, C. Ballif, & Y. Cui (2012). High efficiency amorphous silicon solar cell on a periodic nanocone back reflector. *Adv. Energy Mater.* 2: 628–633.

Jovanov, V., U. Palanchoke, P. Magnus, H. Stiebig, J. Hüpkes, P. Sichanugrist, M. Konagai, S. Wiesendanger, C. Rockstuhl, & D. Knipp (2013). Light trapping in periodically textured amorphous silicon thin film solar cells using realistic interface morphologies. *Opt. Express* 21: A595–A606.

Lewis, N.S. (2007). Powering the planet. *MRS Bull.* 32: 808–820.

Sai, H., H. Jia, & M. Kondo (2010). Impact of front and rear texture of thin-film micro-crystalline silicon solar cells on their light trapping properties. *J. Appl. Phys.* 108: 044505-1-9.

Shen, M. Y., C. H. Crouch, J. E. Carey, & E. Mazur (2004). Femtosecond laser-induced formation of sub-micro-spikes on silicon in water. *Appl. Phys. Lett.* 85: 5694–5696.

Tamang, A., A. Hongsingthong, P. Sichanugrist, V. Jovanov, H. T. Gebrewold, M. Konagai, & D. Knipp (2016). On the potential of light trapping in multi scale textured thin film solar cells. *Sol. Energy Mater. Sol. Cells* 144, 300–308.

Tan, H., E. Moulin, F. T. Si, J. W. Schüttauf, M. Stuckelberger, O. Isabella, F. J. Haug, C. Ballif, M. Zeman, & A. H. M. Smets (2015). Highly transparent modulated surface tex-tured frontelectrodes for high efficiency multi junction thin film silicon solar cells. *Prog. Photovolt. Res. Appl.* 23: 949–963.

Wan, D. H., H.-L. Chen, T.-C. Tseng, & C.-Y. Fang (2010). Antireflective nanoparticle arrays enhance the efficiency of silicon solar cells. *Adv. Funct. Mater.* 20: 3064–3075.

Yang, J., S. Y. Myong, & K. S. Lim (2015). Reduction of the plasmonic absorption in the nano textured back contact of *a*-Si:H solar cells by employing an *n*-SiOx:H/LiF interlayer. *Sol. Energy Mater. Sol. Cells* 132: 372–376.

*Environmental science and environmental engineering*

*Advances in Energy and Environment Research – Achour & Wu (Eds)*
© *2017 Taylor & Francis Group, London, ISBN 978-1-138-62682-9*

# Parametric analysis of a single basin solar still with a point-focus Fresnel lens in Shenzhen

Jun Dong & Xinhai Xu
*School of Mechanical Engineering and Automation, Harbin Institute of Technology, Shenzhen, China*

Shuyang Zhang
*Department of Aerospace and Mechanical Engineering, University of Arizona, Tucson, USA*

ABSTRACT:   A point-focus Fresnel lens was proposed to concentrate solar energy for a single basin solar still in the present study. Weather data in Shenzhen, China were used for the parametric study. The appropriate concentration ratio and receiver surface temperature were selected based on thermodynamic modeling of the receiver. The tilt angle of the receiver surface was determined by geometric calculation of the sun rays' angle and the focus area formed in the receiver surface. Concentration ratio of 2500 for the Fresnel lens, 350–400°C surface temperature and 30° tilt angle of the receiver are the optimized parameters. The fresh water production rate was calculated based on the heat transfer model of the solar still. It was proved that the Fresnel lens significantly improved performance of the solar still and increased the daily fresh water distillation rate. Daily fresh water production rate of 78.5, 37.1 and 68.8 L/m$^2$ water surface area can be obtained in Jan, Jun and Oct, respectively.

## 1 INTRODUCTION

With increased world population, the demand of fresh water increases considerably every year. Although 75% of earth's surface is covered by water, only 3% of the water is fresh water (Velmurugan 2008, Abdelmoez 2014, Reif 2015). Fortunately, fresh water can be obtained from seawater by desalination. Many desalination techniques have been developed mainly based on two different processes—membrane process and thermal process (Kalogirou 2004). However, most of these techniques such as Reverse Osmosis (RO), Multi-Stage Flash distillation (MSF), Multi-Effect Distillation (MED), and others consume a large amount of electricity or fossil energy to produce fresh water, which reduce the sustainability of desalination (Gude 2010).

Among all the sea water desalination techniques, solar still is the most simple and common device for low capacity fresh water production, because it is easy to fabricate, operate and maintain (El-Bialy 2016). Moreover, it is a more sustainable desalination technology since its major energy source is the most abundant and renewable solar energy. Basin solar still is the most frequently used type (Feilizadeh 2015). Application of the solar still system is mainly limited by its low productivity. Only one or two liters fresh water can be produced daily in a small solar still under normal solar radiation (Durkaieswaran 2015). Lot research works have

been conducted in order to improve the productivity of a solar still. Manokar et al. (2014) discussed the effects of various parameters on water evaporation rate in a solar still. Rajaseenivasan et al. (2013) reviewed the reported work of the multi-effect solar still system. Murugavel et al. (2008) summarized the improvement in a single basin solar still.

A solar concentrator can concentrate the incoming sun rays into a small focus area and significantly increase the evaporation temperature of the sea water in a solar still. If the temperature can be elevated to above 71°C for longer than 15 s, not only desalinated but also purified drinking water would be obtained (Beitelmal & Fabris 2015). Parabolic trough and Fresnel lens are the most industrial mature solar concentrators. Fresnel lens has the advantages of less susceptibility to wind, less occupied space, and easier maintenance compared to parabolic trough (Giostri 2013). Besides, Fresnel lens can be fabricated in any size as desired (Valmiki 2011). Line-focus and point-focus lens are the two types of Fresnel lens. Line-focus lens is generally used for low to medium concentrated applications with a geometrical concentration ratio less than 100 (Xie 2013). Pont-focus lens can have a much higher concentration ratio (Ryu 2006).

In the present study, a point-focus Fresnel lens was proposed to concentrate the solar energy for a single basin solar still. The concentrated sun rays were received by a tilted surface. The appropriate concentration ratio of the Fresnel lens, receiver surface temperature and tilt angle were determined by thermodynamic analysis based on the weather data of Shenzhen, China. The concentrated heat was absorbed and transported to the sea water in the solar still by a heat transfer fluid. Fresh water production rate at different evaporation temperatures were calculated based on a heat transfer model.

## 2 MATHEMATICAL MODELING

Figure 1 shows schematic of the proposed single basin solar still with a point focus Fresnel lens. The receiver is filled with a mineral oil (−20°C melting temperature and 400°C vaporization temperature), which acts as the heat transfer fluid to transport the concentrated heat energy to the solar still.

### 2.1 Receiver

Efficiency of the receiver is defined by Equation (1) as follows,

$$\eta_r = \frac{Q_{absorbed} - Q_{loss}}{Q_{solar}} \quad (1)$$

where $Q_{absorbed}$ is the heat energy absorbed by the receiver (W), $Q_{loss}$ is the energy loss from the receiver surface to the surroundings (W), and $Q_{solar}$ is the incoming energy from the solar (W). $Q_{absorbed}$ is the product of $Q_{solar}$ and absorptivity. $Q_{loss}$ is the sum of the radiation loss and the convection loss. The value of the convection heat transfer coefficient can be obtained based on Nusselt number, and $Nu$ is calculated by the characteristic number equation as shown in Equation (2).

$$Nu = 0.664 \, \text{Re}^{1/2} \, \text{Pr}^{1/3} \quad (2)$$

The obtained convection heat transfer coefficient $h$ is expressed as Equation (3).

$$h = 0.664 \frac{\rho^{1/2} V_{wind}^{1/2} c_p^{1/3} k^{2/3}}{\mu^{1/6} L_c^{1/2}} \quad (3)$$

The Fresnel lens and the receiver surface are not always parallel as the lens needs to rotate to track the sun. When they are parallel, the focus area formed in the receiver surface is a circle with the diameter "od" of $f \tan\theta$, where $f$ is the focal length of the Fresnel lens, and $\theta$ is half of the subtended angle of the sun viewed from the earth (Li 2013). $\theta$ has a value of 4.65 mrad (0.266°). When the lens and the receiver surface are not parallel, the focus area turns to an eclipse instead of a circle. Figure 2 shows the case in which the lens and the receiver surface are not parallel. The characteristic length of the eclipse "$od_1$" and "$od_2$" can be calculated by Equations (4) and (5) based on the geometry analysis as shown in Figure 2.

$$\frac{od_1}{od} = \frac{\sin(90° - \theta)}{\sin(90° - \theta - \varphi + \beta)} = C_1 \quad (4)$$

$$\frac{od_2}{od} = \frac{\sin(90° - \theta)}{\sin(90° - \theta + \varphi - \beta)} = C_2 \quad (5)$$

where $\beta$ is the tilt angle of the receiver surface, $\phi$ is the angle between the lens' optical axis and the direction perpendicular to the earth surface.

### 2.2 Solar still

Zheng et al. (2002) developed a new model to calculate the fresh water production rate in a single basin solar still as shown in Equation (6).

$$\dot{m}_e = \frac{h_c}{\rho C_p Le^{0.74}} \frac{M_w}{R} \left( \frac{P_w}{T_w} - \frac{P_g}{T_g} \right) \quad (6)$$

where $h_c$ is the convection heat transfer coefficient in the sea water surface (W/m²·K), $\rho$ is the sea water density (kg/m³), $C_p$ is the specific heat of sea water (J/kg·K), $Le$ is the Lewis umber, $M$ is molar weight (kg/mol), $R$ is the universal gas constant (8.31 J/mol.K), $T$ is temperature (K) and $P$ is pressure (Pa). The subscript $w$ and $g$ denote water and glass cover of the solar still, respectively. $h_c$ can be

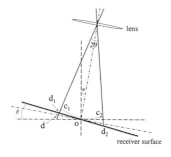

Figure 2. Geometry relationship of angles and dimensions.

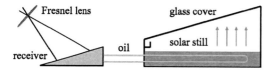

Figure 1. Schematic of the solar still system with a point focus Fresnel lens.

calculated based on the Rayleigh number as shown in Equation (7).

$$h_c = \frac{k}{5H_{w-g}} Ra'^{0.26} \qquad (7)$$

where $k$ is the thermal conductivity of sea water (W/m.K), $H_{w-g}$ is the average distance between the sea water free surface and the glass cover (m).

## 3 RESULTS AND DISCUSSIONS

Weather data in Shenzhen, China was used for a case study. Shenzhen is a coastal city of southern China, which locates in the Subtropical maritime climate zone. According to the historical monthly average data obtained from the NASA Langley Research Atmospheric Science Data Center, daily solar insolation in Shenzhen is the highest in Jun (905.5 W/m²), and the lowest in Jan (346.3 W/m²). However, May to Sep is the flood season in Shenzhen. Averagely there are 19.3 wet days in Jun, and only 6.1 wet days in Jan. Besides Jan and Jun, Oct was also selected as a reference month. The monthly average temperature is 12.6, 25.8, and 22.6°C in Jan, Jun, and Oct, respectively. And the average wind speed is 5.05, 3.82, and 5.13 m/s, respectively.

### 3.1 Concentration ratio, receiver temperature and tilt angle

Assuming the solar receiver is connected with a Carnot cycle heat engine, the ideal total efficiency is the product of the receiver efficiency and the Carnot cycle efficiency. Figure 3 shows the variation of the ideal total efficiency at different Fresnel lens concentration ratios and receiver surface temperatures in the three reference months. It can be seen that the ideal efficiency increases as the concentration ratio increases from 1500 to 2500. For each concentration ratio, various optimal receiver surface temperatures exist in different months. Thereafter a tradeoff has to be made to determine the receiver surface temperature. Considering the working temperature range of the mineral oil, it was concluded that the concentrating ratio of 2500, and the receiver surface temperature in the range of 350 to 400°C were the optimized operating parameters for the Fresnel lens.

Variation of factors $C_1$ and $C_2$ with different $\phi$ and $\beta$ were calculated according to Equation (4) and (5). It was observed that the trend of $C_1$ and $C_2$ with different angles were almost identical. As a consequence, only the relation of $C_1$ and different angles is shown in Figure 4. A small $C_1$ is expected to reduce the focus area so that the solar heat is more concentrated with less heat loss. It can be seen that with a horizontal surface the value of $C_1$ can increase to over 2.5 when the angle of $\phi$ is 5/12 rad. In contrast, $C_1$ is always in the range of 1–1.25 in the case of $\beta$ equals to 1/6 rad. Therefore the tilt angle of the receiver surface is optimized at 1/6 rad.

### 3.2 Fresh water production rate

Figure 5 shows the fresh water production rate at different seawater evaporation temperatures in the range of 60–100°C. Traditional single basin solar still cannot heat the seawater to a relatively high temperature under normal solar radiation condition. A Fresnel lens could concentrate the solar energy and heat the seawater in a solar still up to its boiling point. Apparently higher evaporation temperature leads to higher fresh water production rate. It can be seen that the fresh water distillation rate is the highest in Jan and the lowest in Jun, although the solar insolation is the lowest in Jan and the highest in Jun. This contradiction is attributed to the air temperature in Jun is much

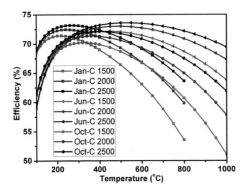

Figure 3. Ideal system efficiency at various concentration ratio and receiver surface temperature in Jan, Jun and Oct.

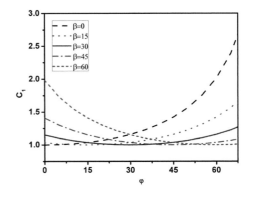

Figure 4. Change of $C_1$ with different $\phi$ and $\beta$.

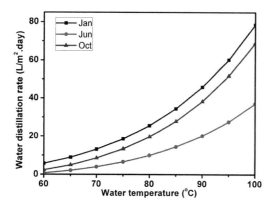

Figure 5. Fresh water production rate at different water heating temperatures.

higher than that in Jan, whereas the wind speed in Jun is only half of that in Jan. Higher air temperature and lower wind speed result in a higher glass cover temperature, which causes higher resistance for condensation of vapor fresh water. At the boiling point, daily fresh water distillation rate is 78.5, 37.1 and 68.8 $L/m^2$ water surface area in Jan, Jun and Oct, respectively. As a comparison, the average daily fresh water consumption per person in China is 84 L in 2006. Therefore the daily fresh water production rate of such a system in Shenzhen is able to satisfy the water demand of one person.

## 4 CONCLUSIONS

In this study a solar still with a point-focus Fresnel lens was proposed for seawater desalination. It was found that different optimal concentration ratios and receiver surface temperatures existed in different months. The concentrating ratio of 2500, and the receiver surface temperature in the range of 350 to 400°C were the optimized operating parameters for a Fresnel lens in Shenzhen. The tile angle of the receiver surface significantly influences the focus area of the sunlight. A 30° tilt angle of the receiver is able to restrict the focus area to its smallest size and keep the sunlight mostly concentrated. Fresh water distillation rate of a single basin solar still can be highly increased with a Fresnel lens due to the elevated evaporation temperature. Optimized daily fresh water production rate of 78.5, 37.1 and 68.8 $L/m^2$ water surface area could be obtained in Jan, Jun and Oct, respectively.

## ACKNOWLEDGMENT

Support from Harbin Institute of Technology (Shenzhen) (CA45001010) is gratefully acknowledged.

## REFERENCES

Abdelmoez, W., M. Mahmoud, & T. Farrag (2014). Water desalination using humidification/dehumidification (HDH) technique powered by solar energy: a detailed review. *Desalin Water Treat.* 52, 4622–4640.

Beitelmal, A., & D. Fabris (2015). Off-the grid solar-powered portable desalination system. *Appl Therm Eng.* 85, 172–178.

Durkaieswaran, P., & K. Murugavel (2015). Various special designs of single basin passive solar still—A review. *Renew Sust Energy Rev.* 49, 1048–1060.

El-Bialy, E., S. Shalaby, A. Kabeel, & A. Fathy (2016). Cost analysis for several solar desalination systems. *Desalination.* 384, 12–30.

Feilizadeh, M., M. Estahbanati, K. Jafarpur, R. Roostaazad, M. Feilizadeh, & H. Taghvaei (2015). Year-round outdoor experiments on a multi-stage active solar still with different numbers of solar collectors. *Appl Energy.* 152, 39–46.

Giostri, A., M. Binotti, P. Silva, E. Macchi, & G. Manzolini (2013). Comparison of two linear collectors in solar thermal plants: parabolic trough versus Fresnel. *J Sol Energy Eng.* 135, 011001-9.

Gude, V., & N. Nirmalakhandan (2010). Sustainable desalination using solar energy. *Energy Convers Manage.* 51, 2245–2251.

Kalogirou, S., (2004). Solar thermal collectors and applications. *Prog Energy Combust Sci.* 30, 231–295.

Li, P., P. Kane, & M. Mokler (2013). Modeling of solar tracking for giant Fresnel lens solar stove. *Solar Energy.* 96, 263–273.

Manokar, M., K. Murugavel, & G. Esakkimuthu (2014). Different parameters affecting the rate of evaporation and condensation on passive solar still—A review. *Renew Sust Energy Rev.* 38, 309–322.

Murugavel, K., K. Chockalingam, & K. Srithar (2008). Progresses in improving the effectiveness of the single basin passive solar still. *Desalination.* 220, 677–686.

Rajaseenivasan, T., K. Murugavel, T. Elango, & S. Hansen (2013). A review of different methods to enhance the productivity of the multi-effect solar still. *Renew Sust Energy Rev.* 17, 248–259.

Reif, H., & W. Alhalabi (2015). Solar-thermal powered desalination: Its significant challenges and potential. *Renew Sust Eng Rev.* 48, 152–165.

Ryu, K., J. Rhee, K. Park, & J. Kim (2006). Concept and design of modular Fresnel lenses for concentration solar PV system. *Solar Energy.* 80, 1580–1587.

Valmiki, M., P. Li, J. Heyer, M. Morgan, A. Albinali, K. Aihamidi, & J. Wagoner (2011). A novel application of a Fresnel lens for a solar stove and solar heating. *Renew Energy.* 36, 1614–1620.

Velmurugan, V., M. Gopalakrishnan, R. Raghu, & K. Srithar (2008). Single basin solar still with fin for enhancing productivity. *Energy Convers Manage.* 49, 2602–2608.

Xie, W., Y. Dai, & R. Wang (2013). Thermal performance analysis of a line-focus Fresnel lens solar collector using different cavity receiver. *Solar Energy.* 91, 242–255.

Zheng, H., X. Zhang, J. Zhang, & Y. Wu (2002). A group of improved heat and mass transfer correlations in solar stills. *Energy Convers Manage.* 43, 2469–2478.

*Advances in Energy and Environment Research – Achour & Wu (Eds)*
*© 2017 Taylor & Francis Group, London, ISBN 978-1-138-62682-9*

# China's rare earth markets: Value chain and the implications

Baolu Zhou, Zhongxue Li, Yiqing Zhao & Congcong Chen
*Beijing University of Science and Technology, Beijing, China*

ABSTRACT: Rare Earths Elements (REEs) impart important properties to modern technologies and clean energy technology. In 2010, the Chinese restrictions on rare earth exports caused their price to rapidly increase by several folds and there was worldwide panic over rare earth supply. Many rare earth deposits outside China were discovered and exploited. Substitution and recovery technologies that were developed in response to the crisis have improved resource efficiency. As a result, rare earth prices and demand have dropped and their supply has begun to diversify. In the meantime, China adopted a series of policies in the rare earth industry in order to improve environmental awareness. These policies have profoundly affected the rare earth industry. In this paper, the current status of the rare earth industry in China after the crisis is firstly reviewed and then, the lifecycle and value chain of the rare earth industry are analyzed; finally some measures to promote the development of China's rare earth industry are provided and its implications are discussed.

## 1 INTRODUCTION

China's monopoly on rare earth supply and restrictions on rare earth exports in 2010 have made the world aware of the unsustainability of the world's rare earth supply chain. Substitution and recovery technology for REEs have attracted many countries' attention in the subsequent years. This caused a significant impact on China's rare earth markets. The demand for China's rare earth exports has weakened; China's rare earth exports dropped to a 10-year decrease in 2011, which was even lower than the export quotas in 2011. Weakened demands drove rare earth prices down and this situation has not improved until 2016. These moves demonstrate that control of the export quotas of rare earths does not contribute to the healthy development of China's rare earth industry.

In this paper, we introduce the life cycle of China's rare earth industry, analyze the market value at every stage before ultimate application, and demonstrate that relocating industries is the most important driver of China's rare earth market, thereby building the strong rare earth downstream industry especially the permanent magnet material industry will be built to contribute to the improvement of China's rare earth industry.

This paper proceeds as follows. Section 2 provides an overview on rare earths. Section 3 introduces the current status on Chinese domestic rare earth markets. Section 4 analyzes the life cycle of the rare earth industry and assesses its value at every stage before ultimate application. Discussion and conclusion are presented in Sections 5 and 6.

## 2 RARE EARTH ELEMENTS: AN OVERVIEW

"Rare earth elements" is a general term for a group of 17 metals, including the 15 lanthanides (atomic number 57 to 71) along with scandium (atomic 21) and yttrium (atomic 39). Yttrium is regarded as an REE because of its chemical and physical similarities and affinities with the lanthanides, and yttrium generally occurs in the same deposits as REEs. Scandium is chemically similar to the REEs, but it does not occur in economic concentrations in the same geological settings as the lanthanides, and it is a dispersed element and so scandium is not generally discussed with other REEs (Gupta & Krishnamurthy 2005, USGS 2014).

In general, REEs could be divided into two groups according to atomic weight: light rare earth elements (LREEs, including the elements Lanthanum to Gadolinium) and heavy rare earth elements (HREEs, including Terbium to Lutetium and yttrium). Table 1 shows the classification and crustal abundance of REEs, the data are derived from USGS (USGS 2014). Yttrium, although its atomic weight is small, is included in the HREE group because of its similar chemical and physical properties with heavy rare earths. Typically, HREEs make up a much smaller share of the total rare earth content than LREE in rare earth minerals; as a result, they have higher economic value and status when compared with LREEs (Gupta & Krishnamurthy 2005). Meanwhile, some authorities divide it into the following three groups: light rare earth elements, middle rare earth elements,

Table 1. Rare earth elements: Classification and crustal abundance.

| Element | Symbol | Crustal abundance | Classification |
|---|---|---|---|
| Scandium | Sc | 17 | – |
| Lanthanum | La | 39 | LREE |
| Cerium | Ce | 51 | LREE |
| Praseodymium | Pr | 6.3 | LREE |
| Neodymium | Nd | 28 | LREE |
| Promethium | Pm | – | LREE |
| Samarium | Sm | 7.4 | LREE |
| Europium | Eu | 1.2 | LREE |
| Gadolinium | Gd | 7.4 | LREE |
| Terbium | Tb | 1.2 | HREE |
| Dysprosium | Dy | 4.6 | HREE |
| Holmium | Ho | 1.4 | HREE |
| Erbium | Er | 3 | HREE |
| Thulium | Tm | 0.3 | HREE |
| Ytterbium | Yb | 3 | HREE |
| Lutetium | Lu | 0.9 | HREE |
| Yttrium | Y | 25 | HREE |

and heavy rare earth elements. In this paper, we use the definition of the USGS and divide the REEs into the following two groups: LREE and HREE.

The lion's share of the known REEs' resources is mainly located in China, America, Australia, India, and Russia. The Bayan Obo Fe–Nb–REE deposit, located in the Chinese region of Inner Mongolia, is the world's largest rare earth deposit, followed by the deposit at Mountain Pass in USA, and then the deposit at Mt Weld in Australia. Rare earth minerals in these three regions are primarily composed of LREEs and they are the world's major suppliers of rare earth production. These places account for about 60% of the world's total output of rare earths in 2014 according to the data from USGS and Association of China Rare Earth Industry. Rare earth outputs in other deposits such as the Levozero mine in Russia is not high, ranging from hundreds to thousands of tons in recent years. However, outputs of Bayan Obo, Lynas (owns Mt Weld) and Molycorp (owns Mountain Pass) mainly consist of LREEs, with less than 5% of the mine output from Mt Weld and Mountain Pass consisting of heavy rare earths. When compared with LREEs, HREE production is mainly carried out by using ion-absorption clay deposits in South China (with Jiangxi and Guangdong provinces as the core). In spite of the low concentrations of REEs (from about 0.08 to 0.1% total rare earth oxides), the unique occurrence (rich in high-value HREEs) and the simple extraction process impart high production value to the ion-absorption clay deposits (CIITG 2011, Long

et al. 2010), which is the world's primary source of HREEs. Before other rare earth deposits, which are rich in HREEs, come into production, the world will remain heavily dependent on China's supply of HREEs.

## 3 CURRENT RARE EARTH MARKETS IN CHINA

Since the closure of Mountain Pass in 2002, China has become the largest and effective REE monopoly supplier to the world and it accounts for an average of 93% of global output of rare earths from 2000 to 2014. In 2010, China cut rare earth export quotas to 30,259 tons (see Table 2), which was decreased by about 40% when compared with 2009. This move drove REE prices to very high levels from the second half of 2010 to the first half of 2011 (see Figures 1 and 2), REE end users and businesses outside China begun to improve resource efficiency, including substitution and recovery technology for rare earths; meanwhile, the world has started to pay attention to the future rare earth supply chain and the price spike led to a global REE exploration boom, Mt Weld in Australia and Mountain Pass in the USA have recommenced, dozens of new rare earth projects being announced all over the world (TMR 2016).

In the context of the rare earth supply crisis, major rare earth consuming countries all over the world have taken the following measures: (1) constantly explore new resources and (2) strengthen substitution and recycling processes rare earths. Mountain Pass had started to produce rare earths again since 2012 (USGS 2013). Another project Bear Lodge, located in USA, is in the process of Environment Impact Statement (Rare

Table 2. Production and export quotas for rare earths in China (REO, tons).

| Year | China Output | China export quotas | ROW demand |
|---|---|---|---|
| 2004 | 95,000 | 65,580 | 52,000 |
| 2005 | 119,000 | 61,070 | 48,000 |
| 2006 | 119,000 | 59,643 | 50,000 |
| 2007 | 120,000 | 49,990 | 55,000 |
| 2008 | 120,000 | 34,156 | 54,000 |
| 2009 | 129,000 | 50,145 | 25,000 |
| 2010 | 130,000 | 30,259 | 49,000 |
| 2011 | 105,000 | 30,184 | 35,000 |
| 2012 | 100,000 | 30,996 | 51,000 |
| 2013 | 95,000 | 31,001 | 40,000 |
| 2014 | 105,000 | 30,610 | 45,000 |

Note: ROW is rest of world.

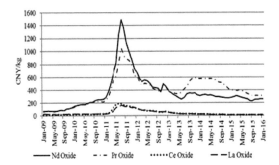

Figure 1. LREE Chinese domestic prices evolutions.

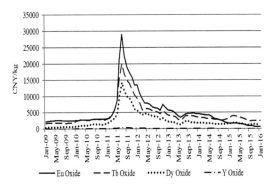

Figure 2. HREE Chinese domestic prices evolutions.

Element Resources 2016). Mt Weld has expanded its capacity to meet the increasing rare earth demand. Other countries, such as Japan, Canada, South Africa, Korea, and Germany, also have launched a series of rare earth projects all over the world. Korea exploited the Zandkopsdrift project with Frontier Rare Earths Ltd. in South Africa and Japan invested in a new rare earth project in Vietnam. These affairs significantly strengthen and diversify the global rare earth supply chain. As a result, China's share of world production has declined from over 95% in 2010 to 85% in 2015; some REE prices in China have dropped by over 90% when compared to their mid-2011 peak values.

At the same time, great efforts from downstream business enterprises and government agencies have been made to achieve rare earth substitution and recycling. Honda Corporation is developing a new motor using no permanent magnets. Philips is determined to design a new LED lighting technology by using less REEs. A critical materials institute established by the United States is sponsoring several rare earth research projects, including the recovery of rare earth oxides from lamp phosphor dusts and the rare earth metals from electronic wastes. The U.S. Geological Survey is continuing to evaluate potential domestic sources of rare

earths such as Mountain Pass tailing and Pea Ridge tailing (Grasso 2013). Substitution, recycling, and diversified supply chains all contributed to a significant reduce in the rare earth demand, and mitigate rare earth supply crisis.

On the other hand, rising rare earth prices suppressed the demand of downstream materials for rare earths. When compared with manufacturing costs of January 2011, the peak manufacturing cost of permanent magnets rose by 452% (pure Nd), 508% (Pr, Nd), 771% (high-grade NdFeB magnet containing Dy), the peak manufacturing costs of other new rare earth raw materials such as hydrogen storage metal, luminescent materials, and catalytic materials also rose by over 400% (Chen 2012). Manufacturers and industrial enterprises are looking for new technology and measures to reduce REE use or are searching for alternative materials and technology. Meanwhile, due to export restrictions and improper production technology, illegal mining and environmental damage have been rampant in South China. In 2015, the actual output of rare earths has reached over 150,000 tons, which far exceeds the scheduled output of 100,500 tons (ACREI 2014), the output of HREEs has reached over 40,000 tons, while its control output is just 179,000 tons. The black industry has begun to threaten the survival of the backbone of rare earth enterprises, to cause the oversupply of REEs, and thereby lead to a sharp decrease in rare earth prices.

After the 2010/2011 rare earth supply crisis, China has taken a series of measures to achieve several major long-term goals, including strengthening of the domestic rare earth industry, improve resources efficiency and environmental practices, and combat illegal production (SCIO 2012). In 2015, China has scrapped export quotas and export tariffs on rare earths after losing the case brought against it at WTO. The Ministry of Finance announced that the rare earth resource tax will be levied *ad valorem* instead of on weight and China continues to suspend granting of license to rare earth mining. After 2 years of operation, the integration of six rare earth groups has been basically completed, including China Aluminum Corporation, China Northern Rare Earth Group, China Minmetals Corporation, Xiamen Tungsten, Guangdong Rare Earth Group, and China Southern Rare Earth Group. More than 6 groups obtained 66 out of the total 67 rare earth mining licenses and integrate 77 out of the total 99 smelting separation enterprises until January 2016. However, the adjustment of the policy did not make the rare earth industry achieve its expected effects; prices of major REEs remain to be low, the operation of major rare earth enterprises encounters many difficulties and is in the situation of approaching loss.

## 4 CURRENT RARE EARTH MARKETS IN CHINA

REEs have unique and diverse magnetic, optical, and chemical properties that make them indispensable to traditional and modern high-technology industries. The rare earth industry began officially in 1883 (Jackson & Christiansen 1993) and subsequently underwent rapid development after the Second World War. In the 21st century, REEs are widely used as components in high-tech devices that surround us. They are also becoming increasingly more important in many defense and clean energy applications. The above-mentioned factors make the rare earth industry attract great attention from many countries.

Typically, the life cycle of rare earths can be divided into 4 processes from ore mining to the final application (see Figure 3). Generally, rare earth ore mining is regarded as an upstream industry; oxides and metals are recognized as the middle market, function devices is considered to downstream industry. From upstream to downstream, the value of rare earth products increases ceaselessly as the stage of processing proceeds. Before the 1980s, rare earth raw ore was the major transaction product in the international trade; in the 1990s, rare earth oxides and metals became the mainstream of international trade; since the 21st Century, the trade of rare earth function materials and devices are becoming more and more common.

In 2015, the output of China's rare earth amounts to 105,000 tons, including 87,100 tons of rock rare earths and 17,900 tons of ion-absorbed rare earths. According to the price of Inner Mongolia rare earth concentrates (TREO 70%) and Middle Y and Rich Eu mixed ores (TREO ≥ 92%), China's rare earth upstream has a market value of 6.185 billion CNY. The market value of rock rare earths and ion-absorbed rare earths is 2.489 billion CNY and 3.697 billion CNY, respectively. In the middle market, the output of Ce oxides is the biggest and reaches 44,602 tons. This is followed by La oxides, Nd oxides, and Pr oxides. According to the rare earth prices of 31 December 2015 in China, the market value of rare earth oxides is worth 8.53 (low prices) or 8.89 billion CNY (high prices). In low prices, Nd oxides worth 4.07 billion CNY, accounts for 47.7% of the total value and Pr oxides worth 1.51 billion CNY accounts for 17.7%, Pr and Nd are the main raw materials of Nd–Fe–B permanent magnetic and make the biggest

Table 3. The specific output and value of different oxides (tons, CNY).

| Oxide | Output | Market value |
|---|---|---|
| La | 29,179 | 335,558,500 |
| Ce | 44,602 | 468,321,000 |
| Pr | 4899 | 1,508,892,000 |
| Nd | 16,096 | 4,072,288,000 |
| Sm | 1667 | 23,338,000 |
| Eu | 236 | 153,400,000 |
| Gd | 1297 | 97,275,000 |
| Tb | 140 | 343,000,000 |
| Dy | 717 | 975,120,000 |
| Ho | 138 | 33,120,000 |
| Er | 394 | 78,800,000 |
| Yb | 267 | 49,395,000 |
| Lu | 53 | 257,050,000 |
| Y | 4798 | 134,344,000 |

contribution to the rare earth market value. The specific output and value of different oxides are shown in Table 3.

In the final market, various electric motors made of rare earth permanent magnets are very important components for wind turbine generators, automobiles, magnetic resonance imaging, and electric bicycle applications; rare earth catalytic materials are widely applied in automobile exhaust purification and gas desulfurization and denitration equipment; luminescent materials are indispensable raw materials for computer and TV screens and energy efficient illuminating applications; TV, mobile phone, glass, and ceramics polishing and coloring need rare earth polishing materials; rare earth alloys have been widely utilized in the aerospace industry, hubs, nodular cast iron, etc. These industries are generally recognized beyond the rare earth industry, which possesses immeasurable economic value.

## 5 DISCUSSION AND THE IMPLICATIONS

In fact, the REE market is a small market. China's rare earth market value is 26 billion CNY and accounts for less than 1% of China's steel market value in 2015 and the global output of rare earths is about 125,000 tons in 2014 and 2015, although they are indispensable elements to modern high-tech and clean energy industry. Meanwhile, high dependency on China often makes the global supply chain vulnerable to various disruptions; the supply crisis in 2010/2011 is not a random occurrence. The world responded promptly to the supply crisis, REE prices decreased sharply in the subsequent years, and the supply crisis has been

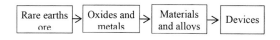

Figure 3. Rare earth life cycle.

alleviated. However, the violent fluctuation of prices has damaged the global rare earth industry including China's rare earth market.

At present, China has a huge advantage on the rare earth upstream, including rare earth concentrates and oxides, but it has no obvious advantage in the rare earth downstream, high-grade polishing powder need to import, the permanent magnet is facing patents "wall" from foreign companies such as Hitachi. China has been exploring ways to consolidate the rare earth industry and entered into the downstream industry link, while its output of rare earths is still controlled to 110,000 tons. At the same time, other rare earth mining operations outside China are suppressed by low prices and uncertain demands and the supply situation is quite similar to the previous demand in 2010. Along with the rising consumption in China, China could move from the net rare earth exporter to importer and the world should prepare for the new supply chain without China's exports.

## 6  CONCLUSION

The supply crisis of 2010/2011 served as a reminder to businesses and governments that future rare earth supply chains are no longer reliable. Shortly after the EU, the USA and Japan launched challenges against China's rare earth export policy to the WT. Businesses reopened old mines and started numerous new projects to strengthen the supply chain, thereby improving rare earth resource efficiency and increasing recovery and substitution of rare earths have become a hotspot in the world. All of these have reshaped the world's rare earth industry and dramatically decrease the REE prices.

Four years after the crisis, China's rare earth industry, although taking a series of measures, still faces many difficulties and challenges, including low prices, weakened demands, foreign diversified supply chains, overcapacity, environment issues, and illegal mining. Through analysis of the rare earth industry's life cycle and value chain, the proportion of value added from rare earth ore to oxides to function materials is 1:1.46:4.37. In the downstream market, the value of rare earth permanent magnets is the maximum and accounts for about 78% of the total value. The contribution of rare earths to high-tech industrials and clean energy is even immeasurable. As a result, China's rare earth industry should expand the industrial chain to downstream industries, especially the permanent magnet industry.

## REFERENCES

Association of China Rare Earth Industry (ACREI). 2014. China Rare Earth Industry Annual Report 2014. *Rare Earth Information* 4: 14–17. (in Chinese).

Commission of Industry and Information Technology of Ganzhou (CIITG). 2011. *The guidance of the development of rare earth industry in Ganzhou.* CIITG. (in Chinese).

Chen, Z. H. 2012. The effect of rising prices for rare earths raw materials on cost of new rare earths materials production. *Advanced Materials Industry* 11: 19–23. (in Chinese).

Grasso, V. B. 2013. *Rare earth elements in national defense: background, oversight issues, and options for Congress.* Library of congress Washington DC congressional research service.

Gupta, C. K. & Krishnamurthy, N. 2005. *Extractive metallurgy of rare earths.* Boca Raton Florida, CRC Press.

Jackson, W. D. & Christiansen, G. 1993. *International strategic minerals inventory summary report; rare-earth oxides.* U.S. Geological Survey.

Long, K. R., Van Gosen, B. S., Foley, N. K., et al. 2010. *The Principal Rare Earth Elements Deposits of the United States—A Summary of Domestic Deposits and a Global Perspective.* U.S Geological Survey.

Rare Element Resources. 2016. *Bear Lodge Project.* Rare Element Resources.

The State of Council Information Office of the People's Republic of China (SCIO). 2012. *Situation and Policies of China's Rare Earth Industry.* SCIO.

Technology Metals Research (TMR). 2016. *TMR Advanced Rare-Earth Projects Index.* TMR.

U.S. Geological Survey (USGS). 2013. *Mineral Commodity Summaries Rare Earths.* U.S. Geological Survey.

U.S. Geological Survey (USGS). 2014. *The Rare-Earth Elements—Vital to Modern Technologies and Lifestyles.* U.S. Geological Survey.

*Advances in Energy and Environment Research – Achour & Wu (Eds)*
*© 2017 Taylor & Francis Group, London, ISBN 978-1-138-62682-9*

# Research on the sensitivity of the marine industry to climate change

Hao Kong
*Fujian Provincial Key Laboratory of Coast and Island Management Technology, Xiamen, China*

Wei Yang
*APEC Marine Sustainable Development Center, Third Institute of Oceanography, State Oceanic Administration, Xiamen, China*

ABSTRACT:   Marine economic activities are mostly carried out in coastal zones with the most fragile ecological environment. Against the background of global climate change, the superposition of storm surges and sea level rise will pose great risks to land utilization and economic/social development in coastal zones and greatly undermine a sustainable marine economy. To evaluate the above-mentioned influence, this paper develops an evaluation system for sensitivity coefficient by quantifying specific indicators and finally confirms the sensitivity of different marine industries to climate change. The evaluation reveals that the marine salt industry is the most vulnerable to climate change followed by marine transportation, fishery, mining, shipbuilding, and coastal tourism. As observed from the proportion of all indicators, sea level rise and storm surge have the biggest influence on marine transportation and shipbuilding; extreme climatic conditions influence marine fishery and salt industries the most; high sea surface temperature and ocean acidification pose great threats to marine fishery and coastal tourism; the rising temperature and extreme weather conditions will endanger human health and labor-intensive production activities including marine salt industry, marine fishery, seawater utilization, and marine engineering.

## 1  INTRODUCTION

China's marine economy has maintained rapid and steady growth in the past decade, and GOP (Gross Ocean Product) in 2015 hit RMB 6466.9 billion, accounting for 9.6% of GDP (SOA, 2016). From 2006 to 2015, the marine economy grew at a rate of 13.34%, which is higher than that of the GDP in the same period. In this light, marine economy has become a new growth area of national economy and plays a key role in fuelling national economic development. The importance of marine economy will increase as China implements the strategy of building a maritime power and the "21st Century Maritime Silk Road".

However, marine economic activities are mostly carried out in coastal zones with the most fragile ecological environment. Against the background of global climate change, unique geographical conditions and their high relevance to human activities may amplify the influence of climate change and contribute to an increasingly fragile environment. Rising sea level, high SST (Sea Surface Temperature), ocean acidification, and extreme weather conditions caused by global warming will threaten the fragile marine ecosystem and the environment in coastal zones. Besides,

sea level rise and storm surges will pose great risks to land utilization and economic/social development in coastal zones and greatly undermine a sustainable marine economy.

To evaluate the above-mentioned influence, this work first analyzes qualitatively the influence of climate change on the marine industry. On this basis, it identifies major driving factors and key risks of such an influence, develops an evaluation system for sensitivity coefficient by quantifying specific indicators, and finally confirms the sensitivity of different marine industries to climate change.

## 2  IMPACTS OF CLIMATE CHANGE ON MARINE INDUSTRIES

Roessing et al. (2004) believe that climate change contributes to increasing oxygen expenditure of fishes, and further their foraging and migration in polar sea areas and sea areas of bleaching coral reefs, and estimated that even a tiny temperature change will alter fish distribution. Hunt et al. (2008) point out that ocean acidification has an influence on the calcification process of planktons, which is then transferred to other species via the food chain. According to Chen (2014), rising water temperature and changing environmental capacity caused by climate change will contribute to an obvious shift of anchovies from central south of the Yellow Sea to the north (about 2.5–2.7 latitudes) in following 30 years at an annual average rate of 0.09°. The latest IPCC (2014) report which carries out an estimation based on the SRES A1B scenario indicates that China's coastal areas will register a decrease of 6%–30% in fish catches from 2015 to 2016, and at worst more than 50% in some areas.

Coastal tourism, on one hand, catalyzes changes in functions of the ecosystem such as bleaching coral reefs, and on the other hand is highly vulnerable to frequent extreme weather conditions including storm surge and extreme rainfall. The vulnerability will become increasingly obvious as the sea level and temperature rise and the damage of the collapsing shoreline and store surge to coastal infrastructures exacerbate (Cai and Qi, 2014). Mclnnes et al. (2000) evaluate negative effects of the rising sea level and storm surges to seaside resorts in Queensland Australia. According to Amelung et al. (2007), climate change will result in re-allocation of tourism resources in time and space. As climate changes are anticipated, ideal climatic conditions for tourism will shift toward high altitudes. Besides, climate change also plays an obvious role in seasonal effects to tourism. Nurse et al. (2009) believe that climate change will give rise to sea level rise, frequent extreme weather conditions such as typhoon, shoreline erosion, and outbreak of climate-sensitive diseases, which are likely to undermine the appeal of islets as primary tourist destinations.

The negative effect of climate change to marine transportation is demonstrated in the following two aspects: loss from climatic disasters and the shock of emission reduction policies to the shipping industry. Climatic changes will bring loss to marine industries. For example, sea level rise may reduce the elevation of ports, wharfs, and warehouses and increase the frequency of floods caused by storm surge, weakens the functioning of ports, thereby preventing the marine industry from meeting requirements of economic development (Chinese Academy of Science, Division of Earth Sciences, 1994). Besides, extreme climatic events may interrupt port operation and transport routes, damage infrastructures, and cause loss of life and personal injury (Cai and Qi, 2014).

Climate change also has the potential influence to exploitation and utilization of marine resources, which is especially true for work platforms of oil and gas (Cai and Qi, 2014). On the other hand, climate change mitigation will promote the development of ocean energy (Allan et al., 2014), i.e. to realize the objective of carbon emission reduction by increasing the contribution of ocean energy to national energy mix.

## 3 CONSTRUCTION OF A SENSITIVITY COEFFICIENT EVALUATION SYSTEM

The above-elaborated analysis reveals the primary driving factors and key risks (Table 1) with regard to the influence of climate change to the marine industry. Sea level rise and storm surge will flood coastal lands and further paralyze normal production activities, which are subject to higher risk with greater difficulty in migration. Extreme climatic conditions such as extremely high rainfall, extremely high temperature, and extremely low temperature will greatly impede outdoor production activities. High SST and ocean acidification pose great threats to the marine ecosystem and marine biological resources (e.g. bleaching coral reef and decreasing fishery resources). What is more, rising temperature and extreme weather conditions will endanger human health and also normal production activities (especially labor-intensive activities).

Based on characteristics of key risks, this work defines four quantitative indicators to reflect the impact of such risks to the marine industry, including migration difficulty of production activities, protection of production activities (plant), the dependence of production activities to the marine ecosystem and marine biological resources and labor input per unit of GDP. See Table 1 for specific quantitative criteria of these indicators. The migration difficulty of production activities (A1), protection of production activities (plant) (A2), the dependence of production activities on the marine ecosystem and marine biological resources (A3) are confirmed via expert rating and labor input per unit of GDP (A4) is scored according to the China Marine Statistical Yearbook. In addition, the proportion of indicators is confirmed via expert rating.

Finally, the sensitivity coefficient of marine industries to climate change is calculated based on equation (1), which is as follows:

$$p_j = (\lambda_1 \cdot A1_j + \lambda_2 \cdot A2_j + \lambda_3 \cdot A3_j + \lambda_4 \cdot A4_j)/5 \quad (1)$$

where $p_j$ refers to sensitivity coefficient, $\lambda_1$ to $\lambda_4$ refer to the proportion of four quantitative indicators, and $A1_j$ to $A4_j$ refer to scores of marine industries in four indicators.

Table 1. Major components of sensitivity coefficient.

| Driving factor | Key risk | Quantitative index | Quantitative standard | Source |
|---|---|---|---|---|
| Rising sea level and extreme weather conditions (storm surge) | Floods in coastal lands and further, paralyzing of normal production activities. | Migration difficulty of production activities (A1) | The degree of dependence is rated from 0 to 5. The score increases positively with the degree of dependence. | Expert rating |
| Extreme weather | Extreme climatic events like extremely high rainfall, extremely high temperature, and extremely low temperature will greatly impede outdoor production activities. | Protection of production activities (A2) | The protection degree is rated between 0 and 5. The score increases negatively with protection degree. | Expert rating |
| SST rise and ocean acidification | Pose great threats to the marine ecosystem and marine biological resources (e.g. bleaching coral reef and decreasing fishery resources). | The dependence of production activities on the marine ecosystem and marine biological resources (A3) | The degree of dependence is rated from 0 to 5. The score increases positively with the degree of dependence. | Expert rating |
| Rising temperature and extreme weather conditions | Endanger human health and also normal production activities (especially labor-intensive activities). | Labor input per unit of GDP (A4) | The score is 0.5 when the labor input per unit of GDP (ten thousand yuan) is lower than 0.05; score is 1 when it is between 0.05 and 0.1; and score is 1.5 when it is between 0.1 and 0.15. Based on the principle, the score increases by 0.5 every time the range increases by 0.05; and the score is 5 when it is beyond 0.45. | Statistical yearbook of marine economy |

# 4 SENSTIVITY COEFFICIENT OF MARINE INDUSTRIES

The proportion of four indicators is identified via expert rating. The proportion of A1 is 40%, meaning that floods caused by sea level rise and storm surge to coastal lands are a leading key risk to marine economic activities. This is followed by A2 with a proportion of 30%. Both the proportion of A3 and A4 is 15%. The value of quantitative indicators for marine industries is summarized in Table 2.

The value of all indicators is analyzed below. First, marine transportation and shipbuilding have the highest value concerning the migration difficulty of production activities, because these two activities are the closest to the shoreline and the rising sea level will flood the premise of production activities like ports and wharfs. Besides, their operation is highly dependent on the deep coastline and water depth of the new shoreline caused by the rising sea level cannot be guaranteed. Therefore,

ports and wharfs should be elevated against sea level rise to guarantee normal production activities at the port and wharf.

Secondly, marine fishery and salt industry have the biggest value concerning the protection of production activities. Extreme weather conditions such as storm surge, extremely low temperature, and extremely high temperature have large influence on outdoor activities such as marine fishing, sea farming, and salt production. Storm surge will destroy facilities of these activities and extreme climatic events will shorten the period of normal production. Marine chemical industry, marine biopharmaceutics, and seawater utilization is subject to minimal effect. The shelter of dykes reduces the impact of storm surges and facilities such as plants also prevent these activities from extreme weather conditions.

Thirdly, the dependence of production activities is on the marine ecosystem and marine biological resources. Only marine fishery and coastal tourism are subject to direct influence. High SST

Table 2. Assignment and calculation of indicators.

| Industry\Proportion | A1 40% | A2 30% | A3 15% | A4 15% | $p_j$ |
|---|---|---|---|---|---|
| Marine fishery | 0.5 | 2.5 | 1.5 | 2 | 29.50% |
| Offshore oil and gas industry | 1.5 | 1 | 0 | 0.5 | 19.50% |
| Ocean mining | 1.5 | 2 | 0 | 0.5 | 25.50% |
| Marine salt industry | 1 | 2.5 | 0 | 3.5 | 33.50% |
| Marine chemistry industry | 1.5 | 0.5 | 0 | 0.5 | 16.50% |
| Marine biopharmaceutics industry | 1.5 | 0.5 | 0 | 0.5 | 16.50% |
| Marine electricity industry | 0.5 | 1.5 | 0 | 0.5 | 14.50% |
| Seawater utilization | 1.5 | 0.5 | 0 | 1.5 | 19.50% |
| Marine shipbuilding industry | 2 | 1 | 0 | 0.5 | 23.50% |
| Marine engineering | 0.5 | 1.5 | 0 | 1 | 16.00% |
| Marine transportation | 2.5 | 1.5 | 0 | 0.5 | 30.50% |
| Coastal tourism | 0.5 | 2 | 1.5 | 0.5 | 22.00% |

and ocean acidification will change the life cycle of marine organisms, threaten healthy ecosystem and reduce the quantity of species. The decreasing stock of fish thus caused will have an impact on marine fishery and marine ecosystem damage such as bleaching coral reef will undermine coastal tourism.

Fourthly, based on the China Marine Statistical Yearbook, the salt industry has the highest labor input per unit of GDP followed by marine fishery, seawater utilization, and marine engineering. Rising temperature and extreme weather conditions will endanger human health and normal production activities (especially labor-intensive activities). Production activities with greater labor input are subject to greater impact. In this sense, marine salt industry, marine fishery, seawater utilization, and marine engineering have a higher value in this indicator.

The sensitivity coefficient of these marine industries is calculated with equation (1) and summarized in Table 2. The marine salt industry has the highest sensitivity coefficient of 33.50% (or is the most vulnerable to climate change); the sensitivity coefficient of marine transportation is 30.50%, of marine fishery is 29.50%, of ocean mining is 25.50%, of marine shipbuilding is 23.50%, and of coastal tourism is 22.00%. Marine chemistry, marine bio-pharmaceutics, marine engineering, and marine electricity is exposed to less influence of climate change with a sensitivity coefficient of 16.50%, 16.50%, 16.00% and 14.50%, respectively.

## 5 CONCLUSION

By analyzing research findings at home and abroad, this work identifies major driving factors and key risks of the influence, and defines, based on characteristics of these key risks, four quantitative indicators to reflect the impact of such risks on the marine industry, including migration difficulty of production activities, protection of production activities (plant), the dependence of production activities on the marine ecosystem and marine biological resources and labor input per unit of GDP. The sensitivity coefficient of marine industries to climate change is thus calculated with the proportion of indicators.

The evaluation reveals that the marine salt industry is the most vulnerable to climate change followed by marine transportation, fishery, mining, shipbuilding, and coastal tourism; marine chemistry, bio-pharmaceutics, engineering, and electricity are not very sensitive to climate changes. As observed from the proportion of all indicators, sea level rise and storm surge have the highest influence on marine transportation and shipbuilding; extreme climatic events influence marine fishery and salt industry the most; high SST and ocean acidification pose great threats to marine fishery and coastal tourism industries; rising temperature and extreme weather conditions will endanger human health and also labor-intensive production activities including marine salt industry, marine fishery, seawater utilization, and marine engineering.

## REFERENCES

Allan G J, Lecca P, McGregor P G, et al. The Economic Impacts of Marine Energy Developments: A Case Study from Scotland [J]. Marine Policy, 2014, 43: 122–131.

Amelung B, Nicholls S, Viner D. Implications of Global Climate Change for Tourism Flows and Seasonality [J]. Journal of Travel Research, 2007, 45: 285–296.

Cai R S, Qi Q H. Key Points on Impact Assessment of Climate Change on the Ocean and Related Adaptation from the IPCC Working Group II Fifth Assessment Report [J]. Advances in Climate Change Research, 2014, 10(3): 185–190. (In Chinese).

Chinese Academy of Science, Division of Earth Sciences. The Influence and Solutions of Sea Level Rise to China's Delta region [M]. Beijing: Science Press, 1994. (In Chinese).

Chen Y L. Interannual Variations in Population Characteristics of Anchovy (Engraulis japonicus) and Redistribution of its Wintering Stock under Climate Change Scenarios in the Yellow Sea [D] Master Thesis of Ocean University of China, 2014. (In Chinese).

Hunt B P V, Pakhomov E A, Hosie G W, et al. Pteropods in Southern Ocean Ecosystem [J]. Progress in Oceanography, 2008, 78(3): 193–221.

IPCC. Climate Change 2014: Synthesis Report, Summary for Policymakers [R]. IPCC's Fifth Assessment Report, 2014.

Mclnnes K L, Walsh K J E, Pittock A B. Impact of Sealevel Rise and Storm Surges on Coastal Resorts [R]. A Report for CSIRO Tourism Research, 2000.

Nurse K, Niles K, Dookie D. Climate Change Policies and Tourism Competitiveness in Small Island Developing States [R]. NCRR Swiss Climate Research Conference on the International Dimensions of Climate Policies, 21–23 January 2009.

Roessing J M, Woodley C M, Cech J J, et al. Effects of Global Climate Change on Marine and Estuarine Fishes and Fisheries [J]. Reviews in Fish Biology and Fisheries, 2004, 14: 251–275.

SOA. 2015 Statistical Bulletin of China's Marine Economy [R]. Beijing: SOA, 2016. (In Chinese).

*Advances in Energy and Environment Research – Achour & Wu (Eds)*
*© 2017 Taylor & Francis Group, London, ISBN 978-1-138-62682-9*

# Research on the efficiency of regional environmental pollution abatement with consideration of centralized disposal pollution source in China—an analysis based on a model of undesirable outputs

Fanbin Kong & Yijun Yuan
*Faculty of Management and Economics, Dalian University of Technology, Dalian, China*

ABSTRACT: By using the model of undesirable outputs, in this paper, the discussion on the efficiency of regional environmental pollution abatement with consideration of centralized disposal of pollution sources in China is studied. The regional gap in efficiency, efficiency ranking, regional convergence, distribution, and changes of regional environmental pollution abatement over time are analyzed. It has been found that without considering the centralized disposal of pollution source or the efficiency promoted by centralized pollution abatement disposal, an underestimated efficiency of regional environmental pollution abatement exists in China. There is an obvious regional gap in the efficiency, efficiency rankings, and the convergence characteristics, but the distribution structure is relatively stable. By accelerating the construction of centralized pollution facilities, the auxiliary network project and auxiliary environment infrastructure projects have become policy choices to reduce the regional gap of environmental pollution abatement and to upgrade the distribution intervals of regional efficiency. However, enough attention should be paid to centralized disposal of pollution sources which have not yet constituted hazards, while China's local government strives to foster and develop the industry of centralized pollution disposal.

## 1 INTRODUCTION

The improvement of environmental quality is not significant though the efforts of strengthening the environmental pollution control have been enforced over the past few years. Therefore, research on the contact of these two aspects becomes very important and the efficiency of environmental pollution abatement is obviously one of the key factors to deal with these two aspects.

"Front-end prevention" and "end-of-pipe abatement" are two ways of environmental pollution governance. Aiming at the established fact of pollution emissions, the way of "end-of-pipe abatement" is chosen for this analysis. There are two solutions of "end-of-pipe abatement": internal pollution treatment by the enterprise itself and external treatment by centralized disposal facilities in China. Centralized disposal facilities are a company of centralized facilities or independent operation units, which provide compensable services to the society, and specialize in industrial wastewater (waste gas) treatment for Industrial Parks, joint industrial enterprises (including domestic sewage outside). In recent years, there has been an obvious trend of industrialization in centralized pollution disposal, which also leads to a new problem: a new pollution source resulted from centralized pollution treatment facilities or environmental pollution resulted from the governance process of pollution problems. According to the "China statistical yearbook on environment", by means of numerical computation, we can see that during 2011–2013, pollution sources from centralized disposal units produced an annual averaged 4620.33 million tons of waste water emissions, 2476 million tons of sulfur dioxide emissions, 38.68 million tons of nitrogen oxide emissions, and 2097.33 million tons of dust emissions, of which there was an average annual growth rate of over 10% for waste water emissions and nitrogen oxide emissions. Provincial emissions of pollution sources from centralized disposal units are quite different, for example, the annual growth rate of wastewater in Inner Mongolia and Heilongjiang reached more than 80% and the annual growth rate of sulfur dioxide emission in Fujian, Hubei, and Shaanxi reached more than 50%. An average annual growth rate of more than 55% is observed in Liaoning, Tianjin, Hubei, and Guangdong for nitrogen oxide emissions. An average annual growth rate of more than 45% was observed in Shanghai, Jiangxi, Guangdong, and Liaoning for dust emissions. Therefore, centralized pollution treatment facilities have become a substantial pollution source in all regions of the country.

Other pollution sources are directly related to the process of production, circulation, distribution, and consumption. In contrast, pollution sources of centralized disposal units result from the behavior on purpose of environmental protection and therefore, the pollution belongs to the category of undesirable outputs of environmental pollution abatement. It brings us naturally to think of questions such as "How to measure the efficiency of environmental pollution abatement while considering the undesirable outputs of centralized disposal?" "Under what conditions have the efficiency, efficiency ranking, regional convergence, and distribution existed and how would the conditions change with time if we consider undesirable outputs of centralized disposal units?" And we attempt to make a supplement to the margins.

## 2 A PRIORITY FOR CHINA: EFFICIENCY EVALUATION OF END-OF-PIPE POLLUTION ABATEMENT

There are two philosophies of evaluating the efficiency of environmental pollution abatement. One is to incorporate environmental factors into the framework of total factor productivity and the potential reduction of an absolute amount of pollutants or the relative amount of pollutants can be seen as the measure of the efficiency (Malin Song et al. 2013) or measure by the decomposition of losses of environmental production efficiency (Bing Wang et al. 2010). The other one is to govern pollution through the direct investment of human, capital, facility or so, to achieve the output effect of pollutant emissions minimization or pollutant removal maximization. The former forms front-end prevention of pollutants through technological innovation and its application to the production system, emphasizing production efficiency of the environmental technology adopted to reduce pollutants. The latter constructs a set of pollution abatement facilities for the end-of-pipe abatement of pollutants, thereby emphasizing the purification efficiency of the generated pollutants by independent governance of the production enterprise or centralized disposal unit of the third-party.

Although China put forward the environmental policy of the "giving priority to prevention, combining prevention with control", end-of-pipe abatement is dominant in actual pollution control. The so-called front-end prevention is mainly embodied in the preset of end-control technology of the production process. Färe et al. (2013) denote that production of good products and pollution governance are two subprocedures in industrial production processes. Because of ignorance of the environmental governance efficiency, the deviation in calculation is based on the traditional DEA method is significant. Therefore, we measure the efficiency of environmental pollution abatement in the second philosophy. Literature (Jun Yang

et al, 2012) concerning the efficiency of end-of-pipe abatement is relatively few, but there is no lack of representation. In general, more attention was paid to the efficiency of the front-end prevention (or produces efficiency of the environmental technology) than that of end-of-pipe abatement. However, efficiency evaluation of end-of-pipe pollution abatement is more suitable to China's condition, which should be considered as a priority.

## 3 METHOD AND DATA

### 3.1 *The directional distance function*

The DDF can be seen as a generalized form of the radial model

$$\max \beta$$
$$s.t. \ X\lambda + \beta g_x \le x_0$$
$$Y\lambda - \beta g_y \ge y_0 \qquad (1)$$
$$\lambda \ge 0$$
$$g_x \ge 0, g_y \ge 0$$

which is formulated appropriately when undesirable outputs exist.

$$\max \beta$$
$$s.t. \ X\lambda + \beta g_x \le x_k$$
$$Y\lambda - \beta g_y \ge y_k$$
$$B\lambda - \beta g_b \le b_k \qquad (2)$$
$$\lambda \ge 0$$
$$g_x \ge 0, g_y \ge 0, g_b \le 0$$

In equation (1), $g_x$ and $g_y$ denote the direction vectors associated with inputs $x$ and outputs $y$, respectively, and $\beta$ is the measure of inefficiency. Equation (2) differs from equation (1) in that it introduces a distinction between good outputs, denoted by $y$, and bad outputs, expressed by $b$. Accordingly, an additional direction vector $g_b$ is incorporated that refers to bad outputs $b$.

Cheng and Zervopoulos (2014) developed a generalized definition of the efficiency score for the DDF even if no simplification in the direction vectors is applied. This paper uses the form of which inputs–outputs are separate or the inputs and outputs are allowed to reach the frontier by using different spatial direction vectors in accordance with the different ratios and we assume the strong disposability of the undesirable outputs. Undesirable outputs in this paper are pollutants from centralized disposal units or pollutions generated by pollution governance. No matter in practice or in theory, pollutions that have resulted from the production, circulation, distribution, and consumption are the objects for environmental pollution abatement, but no attention is paid to the pollution that resulted from the process of pollution abatement. It means no regulation has been formed for this kind of undesirable outputs. And therefore, we define the directional distance function as follows:

$$\min \frac{1-\frac{1}{m}\sum_{i=1}^{m}\alpha g_{xi}/x_{ik}}{1+\frac{1}{s+p}\left(\sum_{r=1}^{s}\beta g_{yr}/y_{rk}+\sum_{q=1}^{p}\gamma g_{bq}/b_{qk}\right)}$$

$$s.t. \quad X\lambda+\alpha g_x \leq x_k \qquad (3)$$
$$Y\lambda-\beta g_y \geq y_k$$
$$B\lambda-\gamma g_b \leq b_k$$
$$\lambda \geq 0$$
$$g_x \geq 0, g_y \geq 0, g_b \leq 0$$
$$x_k \neq 0, y_k \neq 0, b_k \neq 0$$

$\alpha$, $\beta$, and $\gamma$ are respectively the measure of input inefficiency, desirable output inefficiency, and undesirable output inefficiency. And $m$, $s$, and $p$ are the amounts of input, desirable output, undesirable output, respectively. We denote that the directional vector is $(0, 1, -1)$ and the reason for $g_x = 0$ is the popular opinion of insufficient investment in environmental protection, and so we are not going to treat it as redundancy.

### 3.2 Data

Amount of pollutants discharged (produced) in the industrial waste water and volume of pollutants emission (produced) in the industrial waste gas data used in our analysis are obtained from the China Environmental yearbooks published in various years. Other data are obtained from the China Statistical Yearbook on Environment in various years. Our sample covers indicators of five inputs, three desirable outputs, and two undesirable outputs in China's 30 provinces between 2011 and 2013, which results in 90 individual observations (see in Table 1). Input indicators include investment in the treatment of environmental pollution ($X_1$), number of industrial waste water treatment facilities ($X_2$), number of industrial waste gas treatment facilities ($X_3$), annual expenditure of industrial waste water treatment facilities ($X_4$), and annual expenditure of industrial waste gas treatment facilities ($X_5$). Desirable output indicators include COD in industrial waste water removal rate ($Y_1$), AN (Ammonia Nitrogen) in industrial waste water removal rate ($Y_2$), and industrial SO2 removal rate ($Y_3$). Undesirable output indicators include waste water discharged of centralized pollution control facilities ($B_1$), SO2 emissions of centralized pollution control facilities ($B_2$).

Table 1. Descriptive statistics.

| Variable | Obs | Mean | St. d | Min | Max |
|---|---|---|---|---|---|
| $X_1$ (10000 Yuan) | 90 | 2614656 | 1827250 | 241000 | 8810000 |
| $X_2$ (set) | 90 | 2859.82 | 2461.73 | 176 | 10608 |
| $X_3$ (set) | 90 | 7510.71 | 5174.26 | 502 | 19392 |
| $X_4$ (10000 Yuan) | 90 | 226181.5 | 200714.8 | 9309 | 966723 |
| $X_5$ (10000 Yuan) | 90 | 502783 | 415861.5 | 36377 | 2243764 |
| $Y_1$ (%) | 90 | 81.37 | 10.72 | 44.71 | 98.36 |
| $Y_2$ (%) | 90 | 77.54 | 14.7842 | 26.45 | 95.65 |
| $Y_3$ (%) | 90 | 64.95 | 12.33 | 26.85 | 85.43 |
| $B_1$ (ton) | 90 | 1539778 | 1406059 | 70000 | 7130000 |
| $B_2$ (ton) | 90 | 82.57 | 125.57 | 1 | 551 |

*Aggregation and calculation based on data released from "China Statistical Yearbook on Environment" and "China's Environmental Yearbook".

Investment in the treatment of environmental pollution consists of three parts: investment in urban environment infrastructure facilities, investment in treatment of industrial pollution sources, and investments in environment components for new construction projects. And also, pollutant removal rate = (amount of pollutants produced—amount of pollutants discharged)/amount of pollutants produced.

In addition, it is worth noting that only a narrow sample has been used in this paper because statistical indicators of environmental data released by National Bureau of Statistics of the People's Republic of China are changing over time and the data of new pollution source we mentioned are really new and not so complete, which also proves the freshness of the practical problem we found. Therefore, we had to substantiate whether this sample is truly representative in a longer period of time. But the latest data for the longest period of time can be ensured in order to analyze the problem in this paper and the quantitative relationship between the samples and the number of input–output indicators is allowed in the DEA model or there is no estimation error because of the narrow sample.

## 4 REGIONAL COMPARISON AND DISTRIBUTION OF EFFICIENCY OF ENVIRONMENTAL POLLUTION ABATEMENT IN CHINA

### 4.1 Regional comparison of efficiency of environmental pollution abatement in China

The value of arithmetical average and variation coefficient of regional efficiency (rankings) of environmental pollution abatement in China with different models are shown in Tables 2 and 3.

99

Table 2. Efficiency of environmental pollution abatement by area.

| Area | E | Without undesirable output | | | With undesirable outputs | | |
|---|---|---|---|---|---|---|---|
| | | 2011 | 2012 | 2013 | 2011 | 2012 | 2013 |
| Eastern | | 0.32 | 0.36 | 0.32 | 0.41 | 0.49 | 0.44 |
| Central | | 0.20 | 0.22 | 0.18 | 0.41 | 0.54 | 0.43 |
| Western | AA | 0.51 | 0.57 | 0.51 | 0.76 | 0.79 | 0.71 |
| North-east | | 0.26 | 0.41 | 0.29 | 0.71 | 0.79 | 0.72 |
| Nation total | | 0.38 | 0.42 | 0.38 | 0.57 | 0.62 | 0.57 |
| Eastern | | 1.11 | 0.97 | 1.15 | 0.65 | 0.63 | 0.75 |
| Central | | 0.22 | 0.16 | 0.13 | 0.17 | 0.22 | 0.29 |
| Western | VC | 0.53 | 0.46 | 0.55 | 0.30 | 0.25 | 0.36 |
| North-east | | 0.64 | 0.52 | 0.44 | 0.51 | 0.45 | 0.43 |
| Nation total | | 0.70 | 0.61 | 0.74 | 0.48 | 0.43 | 0.46 |

*Aggregated and calculated by authors. E stands for efficiency, AA stands for arithmetical average, and VC stands for variation coefficient.

Table 3. Efficiency rankings of environmental pollution abatement by area.

| Area | ER | Without undesirable output | | | With undesirable outputs | | |
|---|---|---|---|---|---|---|---|
| | | 2011 | 2012 | 2013 | 2011 | 2012 | 2013 |
| Eastern | | 18.8 | 18.9 | 18.6 | 20.5 | 20.1 | 19.6 |
| Central | | 20.7 | 21.7 | 21.7 | 19.7 | 19.2 | 19.0 |
| Western | AA | 8.7 | 9.2 | 8.9 | 8.4 | 9.8 | 9.5 |
| North-east | | 18.7 | 14.0 | 15.0 | 11.7 | 8.7 | 11.7 |
| Nation total | | 11.4 | 11.3 | 11.2 | 10.3 | 10.5 | 10.6 |
| Eastern | | 0.57 | 0.56 | 0.60 | 0.49 | 0.49 | 0.57 |
| Central | | 0.16 | 0.10 | 0.12 | 0.16 | 0.26 | 0.27 |
| Western | VC | 0.53 | 0.60 | 0.65 | 0.77 | 0.71 | 0.82 |
| North-east | | 0.51 | 0.75 | 0.50 | 1.05 | 1.53 | 0.62 |
| Nation total | | 0.71 | 0.73 | 0.73 | 0.78 | 0.78 | 0.75 |

*Aggregated and calculated by authors. ER stands for efficiency rankings, AA stands for arithmetical average, and VC stands for variation coefficient.

From the perspective of the changes in the efficiency of regional environmental pollution abatement, the efficiency containing an undesirable output is 20% higher than that of the desirable output. It means that the efficiency of the regional environmental pollution abatement is enhanced by centralized pollution disposal facilities. The efficiency of environmental pollution abatement is significantly underestimated if we do not take centralized disposal pollution sources into account, which results in reducing the producers' positivity of pollution abatement and weakening of the supervision by the local government and it is not conducive to connotative development in national economy. In the period of 2011–2013, the regional efficiency of environmental pollution abatement calculated in two models show a trend of inverted V type, indicating that centralized disposal cannot change the trend of increase in the beginning and then decrease over trend.

From the perspective of the changes in the variation coefficient of regional efficiency of environmental pollution abatement, regional internal differences of efficiency without undesirable outputs are much greater than that of efficiency with undesirable outputs. If the government designs policies of narrowing the gap between the regional efficiency of environmental pollution abatement according to efficiency without undesirable outputs, there may be too much of a good thing. Results of two models show that the variation coefficients of eastern and northeastern regions are significantly higher than that of the central and western regions, which indicates that there exists a greater internal difference in the efficiency of environmental pollution abatement in the eastern and northeastern regions.

From the perspective of the changes in the efficiency ranking of regional environmental pollution abatement, when compared to the ranking of efficiency without undesirable outputs, the promotion of rankings in central and northeastern regions are greater than the corresponding reduction in eastern and western regions, which makes progress in the ranking of the national level. From the perspective of the changes in the variation coefficient of regional efficiency ranking in the West and East, efficiency ranking of environmental pollution abatement in the central region changes slowly, which represents a hesitation of improvement in the environmental pollution abatement to a certain extent.

In terms of $\alpha$-convergence, the results show that the convergence in the Eastern and Western regions is not significant, and there is a convergence condition in the Northeast region, but the choice of the method significantly influences the judgment of the convergence condition in the Central region. The efficiency with undesirable output in the Central region is convergent, while it is divergent without undesirable output. And therefore, the development of centralized pollution disposal units has become a way to catch up with other regions in terms of pollution abatement.

## 4.2 Distribution of efficiency of environmental pollution abatement in China

In order to present the relative efficiency of environmental pollution abatement in various regions and the efficiency distribution more directly, in this paper, the region (0, 1] I divided into four intervals from low to high as (0, 0.3), (0.3, 0.6), (0.6, 0.8)

[0.8, 1], respectively represented by gray, yellow, cyan, and green. And then, we draw the national distribution map of efficiency in two models during 2011–2013 (see Figures 1–6).

Efficiency of environmental pollution abatement in Northeast and West is higher than that in the Eastern and Central areas. Because the optimization space in factor allocation tends to be

gray=(0,0.3)
yellow=[0.3,0.6)
cyan=[0.6,0.8)
green=(0.8,0.1]

Figure 1. Picture showing the efficiency of environmental pollution abatement without undesirable output in the year 2011.

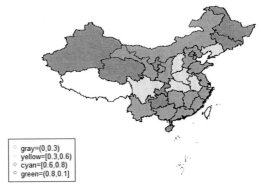

gray=(0,0.3)
yellow=[0.3,0.6)
cyan=[0.6,0.8)
green=(0.8,0.1]

Figure 4. Picture showing the efficiency of environmental pollution abatement with undesirable output in the year 2012.

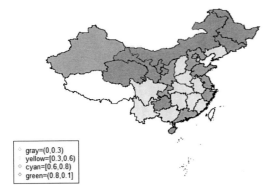

gray=(0,0.3)
yellow=[0.3,0.6)
cyan=[0.6,0.8)
green=(0.8,0.1]

Figure 2. Picture showing the efficiency of environmental pollution abatement with undesirable output in year 2011.

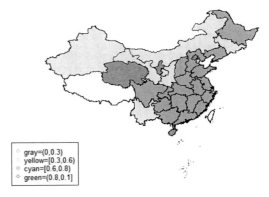

gray=(0,0.3)
yellow=[0.3,0.6)
cyan=[0.6,0.8)
green=(0.8,0.1]

Figure 5. Picture showing the efficiency of environmental pollution abatement without undesirable output in year 2013.

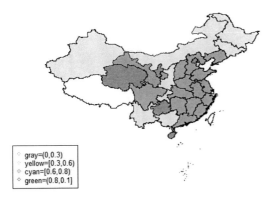

gray=(0,0.3)
yellow=[0.3,0.6)
cyan=[0.6,0.8)
green=(0.8,0.1]

Figure 3. Picture showing the efficiency of environmental pollution abatement without undesirable output in the year 2012.

gray=(0,0.3)
yellow=[0.3,0.6)
cyan=[0.6,0.8)
green=(0.8,0.1]

Figure 6. Picture showing the efficiency of environmental pollution abatement with undesirable output in year 2013.

saturated in the Northeast troubled by overcapacity and west troubled by the behindhand industrial technology under the long-term regulation by the environmental sector and the public supervision in China. Therefore, the improvement of space in the efficiency of environmental pollution abatement is limited; only the technology innovation and application of environmental pollution abatement becomes the sustainable increasing source of efficiency of environmental pollution abatement the Northeastern and Western areas. In the period of 2011–2013, there was the phenomenon of aging equipment in the eastern area and the capacity of industrial waste water (waste gas) treatment facilities decreased by 5.85% (8.03%), which is one of the reasons for the low efficiency of pollution abatement in the eastern area. Seeing in the specific provinces, with the change of time and calculation, the efficiency interval in some provinces do not change: Hebei, Tianjin, Shandong, Jiangsu, Shanghai, Zhejiang, and Guangdong have always been in the low interval of (0, 0.3), Qinghai, Ningxia, Hainan, and Beijing have always been in the high interval of [0.8, 1]. An unbroken distribution structure means that little difference on centralized pollution disposal units is shown in different provinces and in the industry of centralized pollution disposal, various provinces prefer "parallel advances" to "take the lead". Only through ways such as optimization of internal management and guarantee of safe operation of equipment can it improve the efficiency of pollution abatement by increasing the pollutant removal rate or reducing pollution from the source of centralized disposal. For provinces in the low interval band, there is large potential to enhance the efficiency by adopting the way mentioned above, but it had better inspired technological innovation to increase the efficiency for provinces in the high interval band.

## 5 CONCLUSION

An underestimated efficiency of regional environmental pollution abatement exists in China if not considering the centralized disposal pollution source. And also, there exists a greater internal difference between the efficiency of environmental pollution abatement in the eastern and northeastern regions. Ranking of efficiency with undesirable outputs changes faster, while the efficiency ranking of environmental pollution abatement in the central region is the slowest to change. The convergence in the Eastern and Western regions is not significant, and there is a convergence condition in

the Northeast region. When considering the effect of centralized pollution disposal, the efficiency interval has been upgraded in most regions of the country, but the distribution structure has not been broken.

We suggest that accelerating the construction of centralized pollution facilities, auxiliary network project and auxiliary environment infrastructure projects energetically foster and develop the industry of centralized pollution disposal. And also, invest more resources in technological innovation and achievement of transformation of environmental pollution abatement in the Northeast and Western areas. Strengthen facility maintenance and inspection of operation maintenance to solve the problem of equipment aging and the decreasing capacity of pollution treatment facilities. For the developed eastern area and central area undertaking industrial transfer, which are in the low efficiency interval, we should reduce the efficiency gap by accelerating the development of the industry of centralized pollution disposal. For the western and northeastern regions which are in the high efficiency interval, we should draw lessons of the "administration after the contamination", and the established system framework for environmental pollution abatement in advance, avoiding unnecessary pollution and economic losses effectively after substantive implementation of the policy lurches for regional development and transfer of industry resources.

## REFERENCES

Bing Wang, Yanrui Wu, Pengfei Yan. 2010. Environmental Efficiency and Environmental Total Factor Productivity Growth in China's Regional Economies. *J. Economic Research Journal, 2010(5)*:95–109.

Cheng G, Zervopoulos P. D. 2014. Estimating the technical efficiency of health care system: A cross-country comparison using the directional distance function. *J. European Journal of Operational Research, 2014(238)*:899–910.

Färe, Rolf, Shawna Grosskopf and Carl A. Pasurka Jr. 2013. Joint Production of Good and Bad Outputs with a Network Application. *J. Encyclopedia of Energy, Natural Resources and Environmental Economics, 2013(2)*:109–118.

Jun Yang, Yujia Lu. 2012. Environmental Investment Efficiency in China Based on Three-stage DEA Model. *J. Journal of Systems Engineering, 2012(10)*:699–711.

Malin Song, Shuhong Wang. 2013. Analysis of Environmental Regulation, Technology Progression and Economic Growth from the Perspective of Statistical Tests. *J. Economic Research Journal, 2013(3)*:122–134.

*Advances in Energy and Environment Research – Achour & Wu (Eds)*
© 2017 Taylor & Francis Group, London, ISBN 978-1-138-62682-9

# Preparation of phosphonic acid-functionalized silica magnetic microspheres for uranium(VI) adsorption from aqueous solutions

Limin Zhou, Hongbin Zou, Jieyun Jin, Zhirong Liu & Taian Luo

*Jiangxi Engineering and Technology Research Center for New Energy Technology and Equipment, East China University of Technology, Nanchang, P.R. China*

ABSTRACT: Phosphonic Acid-functionalized Silica Magnetic Microspheres (PA–SMMs) were synthesized for U(VI) adsorption from aqueous solutions. PA-SMMs were characterized by using various methods such as SEM, XRD, FTIR, and VSM. PA-SMMs exhibited a higher adsorption capacity for U(VI) when compared to unmodified silica magnetic microspheres due to the high affinity of phosphonic acid groups with U(VI). The adsorption was dominated by inner-sphere surface chelation, as evidenced by the slight influence of ionic strength. The adsorption isotherms fitted well to the Langmuir model, with the maximum adsorption capacity of 76.9 mg/g at 298 K and pH 5.0. PA-SMMs could be successfully regenerated using 0.2 M EDTA–0.5 M $HNO_3$ as the eluent.

## 1 INTRODUCTION

Magnetic sorbents represent an important category of sorbents for the separation of heavy metal ions due to their efficient collection from the extraction medium by using an external magnetic field. He et al. (He et al. 2013) studied the separation of Th(IV) by magnetic Th(IV)-ion imprinted polymers. And also, Wang et al. (Wang et al. 2011) investigated the adsorption of uranyl ions from aqueous solutions onto ethylenediamine-modified magnetic chitosan. The magnetic substance in the magnetic adsorbents should be encapsulated to avoid air oxidation or leaching under acidic conditions. In comparison with organic coating materials, $SiO_2$ can serve as a more ideal shell component due to its stability under acidic conditions, inertia to redox reactions, and abundance of surface hydroxyl groups which makes it easy for surface modification.

The grafting of new functional groups on sorbents may increase the density of adsorption sites and enhance the adsorption for target metals. Phosphonic acid groups can form stable complexes with lanthanides and actinides. Recently, uranium adsorption by phosphorus-containing silica (Milyutin et al. 2014) and phosphonate-functionalized mesoporous silica SBA-15 (Wang et al. 2012) was reported.

In this work, a novel Phosphonic acid-functionalized silica magnetic sorbent (PA-SMM) was synthesized. And the adsorption performance of PA-SMM for U(VI) adsorption from aqueous solutions was evaluated. PA-SMM has a potential application in the recovery of U(VI) from nuclear wastewaters.

## 2 EXPERIMENTAL

### 2.1 Chemicals

Diethylphosphatoethyltriethoxysilane (DPTS) was purchased from Gelest Inc., USA; $NH_3 \cdot H_2O$, tetraethyl orthosilicate (TEOS), isopropyl alcohol, and $UO_2(NO_3)_2 \cdot 6H_2O$ were obtained from Sinopharm Chemical Reagent Co., Ltd. All reagents were of analytical grade and were used as received.

### 2.2 Preparation of Silica Magnetic Microspheres (SMMs)

The nano-$Fe_3O_4$ particles were prepared according to literature previously reported study (Wang et al. 2011). SMMs were synthesized through a modified Stöber method (Stöber et al. 1968). In a typical process, 0.5 g of $Fe_3O_4$ was dispersed by ultrasonication in a mixture of 80 mL isopropyl alcohol and 25 mL water. Afterward, 4 mL TEOS and 5 mL $NH_3 \cdot H_2O$ was added dropwise and the reaction was allowed to proceed for 6 h under stirring. The product was collected by using a magnet and rinsed with distilled water and ethanol several times and then dried under a vacuum.

### 2.3 Preparation of Phosphonic Acid-functionalized SMM Microspheres (PA-SMMs)

2.5 mmol of DPTS was pre-hydrolyzed in concentrated acetic acid to obtain P-containing oligomers [≡Si($CH_2$)$_2$P(O)(OH)$_2$]$_n$ (the detailed procedure is described in reference previously reported study

(Dudarko et al. 2015). Preparation of PA-SMM was accomplished by a reaction of 1.0 g of SMM microspheres with the obtained P-containing oligomers in 50 mL of refluxing toluene for 2 h. The material was collected by using a magnet, washed with ethanol, and dried at 70°C in a vacuum for 12 h. The schematic presentation of the synthesis process is shown in Figure 1.

### 2.4 *Characterization of the sorbents*

The morphology of PA-SMM was observed by using a Leica Cambridge S360 Scanning Electron Microscope (SEM). FT-IR spectra were recorded by using a Nicolet, Magna-550 spectrometer. The X-Ray Diffraction (XRD) patterns were obtained by using a XRD-2000 X-ray diffractometer with Cu Ka radiation. Magnetic measurements were conducted in a Model 155 EG & G Princeton Vibrating Sample Magnetometer (VSM).

### 2.5 *Adsorption experiments*

50 mg of the silica-based sorbents (PA-SMMs or SMMs) was shaken with 50 mL of U(VI) solution for 4 h (the preliminary experiments showed that the adsorption equilibrium was achieved within 180 min). The initial pH of the U(VI) solution was adjusted by adding 0.2 M NaOH or 0.2 M HCl solutions. After adsorption, the sorbent was magnetically separated from the suspension. The concentration of U(VI) was determined by using the Arsenazo III spectrophotometric method at a wavelength of 650 nm. The adsorbed amount of U(VI) was calculated from the difference between initial and final U(VI) concentrations.

### 2.6 *Leaching of Fe and regeneration experiments*

50 mg of sorbents was dispersed in 50 mL of $HNO_3$ solutions with different concentrations. After a specific contact time, the sorbents were separated and the leached iron concentration in the supernatant was determined by using an atomic adsorption spectrophotometer. For desorption experiments, the U(VI)-loaded sorbents were mixed with acidified EDAT solution at 298 K under stirring for 6 h. After magnetic separation, the remaining U(VI) concentration in the supernatant was analyzed. The regenerated PA-SMM sorbents were washed and then reused for adsorption in the next cycle.

## 3 RESULTS AND DISCUSSION

### 3.1 *Characterization of sorbents*

The SEM images of PA-SMMs are shown in Figure 2a. The PA-SMM sorbents are observed to be spherical particles and their size is 0.4–0.6 μm. The encapsulation of silica on surfaces of $Fe_3O_4$ cores

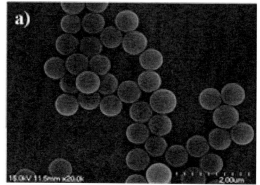

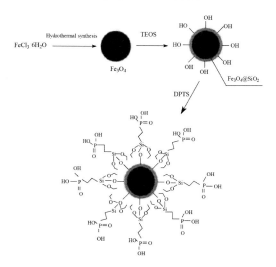

Figure 1. Schematic showing the synthesis process of PA-SMMs.

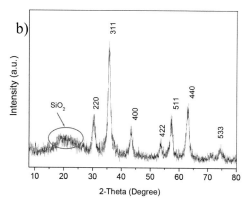

Figure 2. The SEM image (a) and XRD pattern (b) of PA-SMMs.

resulted in core–shell microspheres with black-colored $Fe_3O_4$ cores and grey-colored silica shells. The P content of PA-SMMs is 3.9% (determined by EDS elemental analysis) and the concentration of phosphonic groups in PA-SMMs is 1.26 mmol/g.

The XRD pattern (Figure 2b) of PA-SMMs exhibited seven diffraction peaks which could be indexed to (220), (311), (400), (422), (511), (440), and (533), respectively. These peaks are consistent with the database in the JCPDS file (JCPDS No. 75-0033) and confirmed the presence of pure $Fe_3O_4$ particles with a face-centered cubic structure (Wang et al. 2011) in PA-SMMs. The broad diffraction band at a $2\theta$ angle of about 20° can be attributed to amorphous silica. The results from VSM indicated that the saturated magnetization of PA-SMMs reached 42 emu/g. The PA-SMMs dispersed in the U(VI) solution could be completely separated by using a magnet within 1 min.

The FT-IR spectra of SMMs and PA-SMMs are shown in Figure 3. In the spectra for SMMs (Figure 3a), the peak at 586 cm$^{-1}$ is attributed to the stretching vibration of the Fe–O bond. The band at 1068 cm$^{-1}$ can be assigned to stretching vibrations of Si–O–Si. Apart from these bands, some new characteristic peaks have appeared in the spectra for PA-SMMs (Figure 3b), the peak at 1204 cm$^{-1}$ is attributed to the P = O stretching vibration of the P-containing groups. The peaks at 1646 cm$^{-1}$ and 2996 cm$^{-1}$ can be assigned to the deformation and stretching vibrations of –OH groups, respectively. The band that corresponds to vibrations (Si–CH$_2$) is identified at 1405–1456 cm$^{-1}$. Thus, it can be concluded that the surface layer of PA-SMMs contains phosphonic acid groups –(CH$_2$)$_2$P(O)(OH)$_2$, which have occurred from the hydrolyzation of DPTS.

### 3.2 Effect of pH and interfering ions on U(VI) adsorption

The impact of initial pH on U(VI) adsorption onto PA-SMMs is shown in Figure 4 (see the blank set). The adsorption capacity increases with the pH. The attachment of UO$_2^{2+}$ onto phosphonates is known to be a monodentate coordination through the phosphonyl oxygen groups (Vivero-Escoto et al. 2013). The increase of pH can neutralize the H$^+$ ions and decrease the competitive adsorption from H$^+$ ions. Meanwhile, increasing pH can also decrease the repulsion between UO$_2^{2+}$ and PA-SMMs through deprotonation of the functional groups, and thus enhance the adsorption amounts. The rapid increase of adsorption capacity at pH > 6.0 is probably due to the formation of UO$_2$(OH)$_2$ precipitation.

Figure 4 shows that the co-existing cations have negligible effects on U(VI) adsorption, indicating that the U(VI) adsorption is dominated by inner sphere surface chelation. It is also found that Cl$^-$ ions have no obvious effect on U(VI) adsorption over the entire pH range while the presence of SO$_4^{2-}$ reduces U(VI) adsorption at pH < 6.0. This can be attributed to the different chelation abilities of these anions. The chelation between Cl$^-$ and U(VI) is too weak to influence U(VI) adsorption, while SO$_4^{2-}$ can form stable soluble complexes with U(VI) at low pH values (e.g. UO$_2$(SO$_4$)$_2^{2-}$) (Bhalara et al. 2014).

### 3.3 Adsorption isotherms

Figure 5 shows the adsorption isotherms for U(VI) onto PA-SMMs and SMMs at an initial pH of 5.0 and different temperatures. The adsorption capacity (q$_e$) first sharply increases with increasing equilibrium U(VI) concentration and then slowly approaches saturation. The adsorption equilib-

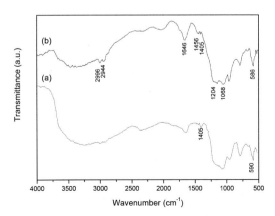

Figure 3.    FT-IR spectra of SMM (a) and PA-SMM (b).

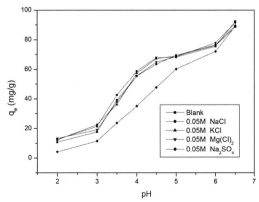

Figure 4.    Graph showing the effect of pH and ionic strength on U(VI) adsorption. (T = 298 K, C$_0$ = 90 mg /L, SD (sorbent dose) = 1.0 g/L).

rium data were analyzed by using the Langmuir model (equation [1]) and the Freundlich model (equation [2]), respectively.

$$\frac{C_e}{q_e} = \frac{C_e}{q_m} + \frac{1}{q_m K_L} \qquad (1)$$

$$\ln q_e = \ln k_F + 1/n \ln C_e \qquad (2)$$

where $q_m$ is the adsorption capacity at saturation of the monolayer of the sorbent (mg/g), $k_L$ is the Langmuir adsorption constant (L/mg), $k_F$ is the Freundlich isotherm constant, and n is the heterogeneity factor.

The parameters of the models are summarized in Table 1. The simulated curves (in Figure 5) are close to experimental points as a confirmation of the suitability of the Langmuir equation to fit adsorption isotherms. The value of the exponential term (i.e., 1/n) for the Freundlich model is less than unity, and thus the system can be considered as favorable; however, the correlation coefficient is much lower than that in the case of the Langmuir equation, suggesting that U(VI) absorbed forms a monolayer coverage and chemisorption is the predominant adsorption mechanism, which is consistent with the strong chelation between U(VI) ions and phosphonic acid groups on PA-SMMs.

The maximum adsorption capacity ($q_m$) of PA-SMMs for U(VI) is 76.9 mg/g, which was higher than MSPh-III ($q_m = 66.7$ mg/g) (Vivero-Escoto et al. 2013) and salicylaldehyde-modified magnetic silica ($q_m = 49.0$ mg/g) (Rezaei et al. 2012). These results indicate that PA-SMM exhibits a very competitive adsorption performance and thus, it exhibits a potential of real applications in the adsorption of U(VI) from aqueous solutions.

### 3.4  Stability and regeneration of PA-SMMs

The stability of PA-SMMs under acidic conditions was tested by monitoring the leached Fe content after contacting with $HNO_3$ solutions and the results are shown in Figure 6. Obviously, leaching of Fe in 24 h was greatly inhibited in acidic solutions due to successful coating of $SiO_2$. Desorption of U(VI)-loaded PA-SMMs was conducted with acidified EDTA solutions and the concentration was optimized. The results showed that the desorption percentage of U(VI) reached 96.5% by using 0.2 M EDTA–0.5M $HNO_3$ and a further increase in the eluent concentration only showed slight improvement of the desorption. Hence,

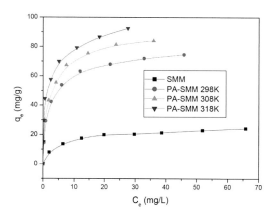

Figure 5.  Adsorption isotherms of U(VI) on SMMs and PA-SMMs (pH = 5.0, T = 298 ~ 308 K, $C_0 = 15 ~ 120$ mg / L, SD = 1.0 g/L).

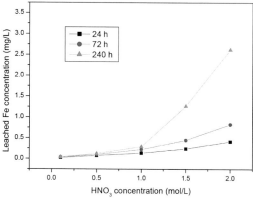

Figure 6.  Leaching of Fe in $HNO_3$ solutions with different concentrations.

Table 1.  The parameters for the Langmuir and Freundlich models of U(VI) adsorption on SMMs and PA-SMMs.

| Sorbents | T (K) | Langmuir | | | Freundlich | | |
|---|---|---|---|---|---|---|---|
| | 298 | $q_{max}$ (mg/g) | $K_L$ (L/mg) | $R^2$ | $k_F$ ($mg^{1-n}L^n/g$) | n | $R^2$ |
| MS | 298 | 25.70 | 0.170 | 0.9962 | 7.33 | 1.991 | 0.9463 |
| MS-DPTS | 298 | 76.92 | 0.492 | 0.9986 | 27.22 | 3.304 | 0.9268 |
| | 308 | 87.71 | 0.561 | 0.9984 | 31.34 | 3.445 | 0.9093 |
| | 318 | 94.34 | 0.785 | 0.9968 | 37.98 | 3.637 | 0.8780 |

0.2 M EDTA–0.5M HNO$_3$ solution was used as the eluent and the regenerated PA-SMMs were used for five consecutive cycles. After five consecutive adsorption/desorption cycles, the adsorption capacity of U(VI) slightly decreases by less than 10% ($q_{e,exp}$ decreases from 74.6 mg/g to 68.7 mg/g). The results indicate that U(VI)-loaded PA-SMMs could be efficiently regenerated by 0.2 M EDTA–0.5 M HNO$_3$ and that PA-SMMs can support long-term use in water treatment with low replacement costs.

## 4 CONCLUSIONS

The Phosphonic Acid-functionalized Silica Magnetic Microspheres (PA-SMMs) have been synthesized and tested for U(VI) adsorption from aqueous solutions. The maximum adsorption capacity was found to be 76.9 mg/g for U(VI) at pH 5.0 and 298 K. The Langmuir equation fits well with adsorption isotherms when compared to the Freundlich model. Finally, the adsorbent can be efficiently regenerated by using acidified 0.2 M EDTA–0.5 M HNO$_3$ as the eluent. The possibility to recover the sorbent by using an external magnetic field and efficient adsorption performance contributes to make EDA-MCCS very interesting for applications in wastewater treatment.

## ACKNOWLEDGMENTS

This work was financially supported by the National Natural Science Fund Program (21366001; 11375043) and Opening Funding from Jiangxi Engineering and Technology Research Center for New Energy Technology and Equipment (JXNE2014-12).

## REFERENCES

Bhalara, P. D., Punetha, D. & Balasubramanian, K. (2014). A review of potential remediation techniques for uranium(VI) ion retrieval from contaminated aqueous environment. *J. Environ. Chem. Eng.,* 2, 1621–1634.

Dudarko, O. A., Gunathilake, C., Wickramaratne, N. P., Sliesarenko, V. V., Zub, Y. L., Górka, J., Dai, S. & Jaroniec, M. (2015). Synthesis of mesoporous silica-tethered phosphonic acid sorbents for uranium species from aqueous solutions. *Colloid & Surf. A Physicochem. Eng. Aspects,* 482, 1–8.

He, F., Wang, H., Wang, Y., Wang, X., Zhang, H., Li, H. & Tang, J. (2013). Magnetic Th (IV)-ion imprinted polymers with salophen schiff base for separation and recognition of Th (IV). *J. Radioanal. Nucl. Chem.,* 295, 167–177.

Milyutin, V. V., Gelis, V. M., Nekrasova, N. A., Melnyk, I. V., Dudarko, O. A., Sliesarenko, V. V. & Zub, Y. L. (2014). Sorption of actinide ions onto mesoporous phosphorus-containing silicas. *Radiochemistry,* 56, 262–266.

Rezaei, A., Khani, H., Masteri-Farahani, M. & Rofouei, M.K. (2012) A novel extraction and preconcentration of ultra-trace levels of uranium ions in natural water samples using functionalized magnetic-nanoparticles prior to their determination by inductively coupled plasma-optical emission spectrometry. *Anal Methods* 4: 4107–4114.

Stöber, W., Fink, A. & Bohn, E. (1968). Controlled growth of monodisperse silica spheres in the micron size range. *J. Colloid Interface. Sci.,* 26, 62–69.

Vivero-Escoto, J. L., Carboni, M., Abney, C. W., Dekrafft, K. E. & Lin, W. (2013). Organo-functionalized mesoporous silicas for efficient uranium extraction. *Micropor. & Mesopor. Mat.,* 180, 22–31.

Wang, J. S., Peng, R. T., Yang, J. H., Liu, Y. C. & Hu, X. J. (2011). Preparation of Ethylenediamine-modified magnetic chitosan complex for adsorption of uranyl ions. *Carbohydr. Polym.,* 84, 1169–1175.

Wang, X., Yuan, L., Wang, Y., Li, Z., Lan, J., Liu, Y., Feng, Y., Zhao, Y., Chai, Z. & Shi, W. (2012). Mesoporous silica SBA-15 functionalized with phosphonate and amino groups for uranium uptake. *Sci. China Chem.,* 55, 1705–1711.

Yang, S., Zong, P., Ren, X., Wang, Q. & Wang, X. (2012). Rapid and highly efficient preconcentration of Eu (III) by core–shell structured Fe$_3$O$_4$@ humic acid magnetic nanoparticles. *ACS Appl. Mater. Interface,* 4, 6891–6900.

*Advances in Energy and Environment Research – Achour & Wu (Eds)*
© 2017 Taylor & Francis Group, London, ISBN 978-1-138-62682-9

# A new scenario for enhancing phosphorus removal in SBR$_S$-AA and SBR$_S$-AO systems during the winter season

Sa'ad Abu-Alhail Al-Khalil & Ahid Zuhair Hamody
*Department of Civil Engineering, College of Engineering, University of Basrah, Basra City, Iraq*

ABSTRACT: This paper deals with a new scenario for enhancing phosphorus removal in pilot plant SBRs (AA) and SBRs (AO) through improvement of PAOs and DNPAOs during the winter season when temperature ranges from 8C to 10C. It is observed that the SBRs (AO) can achieve acceptable Phosphorus Removal Efficiencies (PREs) after 39 days of their operation. This acceptable Phosphorus Removal Efficiency (PRE) can be achieved in the SBRs (AA) after 79 days. This means that the improvement of Phosphorus Accumulating Organisms (PAOs) can be reached within a short time period than that of denitrifying phosphorus accumulating organisms during the winter season. The experimental results showed that Denitrifying Phosphorus Accumulating Organisms (DNPAOs) can immediately achieve good phosphorus removal under aerobic conditions more than anoxic conditions. Phosphorus Accumulating Organisms (PAOs) can achieve low phosphorus removal under anoxic conditions. Therefore, there are two different types of Accumulibacter enriched in SBRs (AA) and SBRs (AO). Scanning Electron Microscopy (SEM) is used in this study, whereas SEM analysis shows that Accumulibacter is prevailing in SBRs (AA) and SBRs (AO) whereas, a long-rod morphology is observed. The scenario proposed in this study was confirmed to be effective in increasing the enrichment of Accumulibacter during the winter season.

## 1 INTRODUCTION

As phosphorus is the limiting nutrient in algal blooms, phosphorus removal from wastewater, hence, has become an important requirement to protect public health and reduce ecological risk. In order to meet a stringent limit for phosphorus, Enhanced Biological Phosphorus Removal (EBPR) is generally regarded as an economical and environmentally friendly technology for the removal of phosphorus from wastewater due to its advantages relative to conventional chemical precipitation method, such as ferric chloride and aluminium oxide. The discharge of nutrient materials (i.e. nitrogen and phosphorus) from wastewater to soil and waters may adversely affect water resources in some ways, especially potential contributions to eutrophication. Nitrogen and phosphorus-polluted surface waters often need to be pretreated prior to use in drinking water systems, such as Tigers River, Euphrates River, and Shatt AL-Arab, Iraq. In the EBPR process, a group of bacteria is generally enriched through sequential anaerobic–aerobic conditions, known as Polyphosphate Accumulating Organisms (PAOs) responsible for phosphorus removal. During the anaerobic period, PAOs take up carbon sources, particularly Volatile Fatty Acids (VFA), such as acetate and propionate, store them as Poly-β-Hydroxyalkanoates (PHA), supply with energy from the hydrolysis of polyphosphate (poly-P) (resulting in the release of phosphorus from the cells of PAOs) and glycolysis of glycogen. In the subsequent aerobic period, PAOs take up phosphorus in excess of the anaerobic release to store them as poly-P, which is usually called as the luxury phosphorus uptake, simultaneously accompanying the growth of biomass and the regeneration of glycogen, with the required energy from the oxidation of PHA stored in cells of PAOs under anaerobic conditions [1, 2]. There is no information available on the running performance of the EBPR system at even lower temperatures such as around 10°C, often presenting in winter. Thus, a new strategy for obtaining microorganisms responsible for phosphorus removal (i.e., PAOs and denitrifying Phosphorus Accumulating Organisms at 8–11°C was developed in this study. This proposed strategy was performed by using two lab-scale EBPR reactors, which were operated under both anaerobic-aerobic and anaerobic-anoxic conditions. The results obtained from this study, linking running performance with microbial population, may serve as a new suggestion for the design and operation of the EBPR system, especially during the winter season.

## 2 MATERIALS AND METHODS

### 2.1 *Experimental reactor*

Two lab-scale Sequencing Batch Reactors (SBR) with a working volume of 3.3 L (see Fig. 1) were used for phosphorus removal; one operated under a sequence of anaerobic-aerobic conditions and anaerobic-anoxic. The cycle time consisted of a 0.5 h filling period, a 2 h anaerobic period, a 4 h aerobic or anoxic period, a 1 h settling period, and a 0.5 h decant period. In each cycle, 1.9 L of synthetic wastewater (composition is detailed in Table 1) was fed to the reactor during the filling phase, resulting in a 13.9 h Hydraulic Retention Time (HRT) and an effluent of the same amount as the influent (1.9 L) was discharged at the end of one cycle. In the AA reactor, sodium nitrate solution was pumped in the first 1 min of the anoxic period to create anoxic conditions. Volumes of 330 mL and 115 mL mixed liquor were removed per day from the AO reactor and AA reactor to maintain the Solids' Retention Time (SRT) at 10 and 20 days, respectively. Air was supplied at a flow rate of 1.5 L/min to maintain the Dissolved Oxygen (DO) levels at greater than 2 mg/L during the aerobic period. The pH in two reactors was maintained at 7.0 ± 0.2, with the addition of 0.5 M HCl or 0.5 M NaOH when the pH was above or below this setpoint. Two reactors were operated at room temperature, ranging from 8°C to 11°C. Two reactors responsible for the PAOs and denitrifying phosphorus accumulating organisms' enrichment were operated for 80 days.

### 2.2 *Batch tests*

For the batch tests, 0.5 L of activated sludge was taken from both the AA and AO reactors under aerobic and anoxic conditions at 80 days and was immediately washed twice with the nutrient solution (see Table 1) not containing the basic medium. The activated sludge treated from the AA reactor responsible for anoxic phosphorus removal was divided into two parts and filled into two 1 L test devices; one part was operated with an anaerobic-aerobic mode and the other under anaerobic-anoxic conditions.

## 3 RESULTS

The EBPR performance of anaerobic-aerobic and anaerobic-anoxic systems was investigated throughout the enrichment experiments of PAOs and denitrifying phosphorus accumulating organisms during the winter season.

### 3.1 *Performance of AA and AO reactors*

The phosphorus removal performance throughout the microorganisms' acclimatization process in both EBPR reactors, namely AA and AO proposed in this study, at low temperatures (8~11C) over the 80 days' time period is shown in Fig. 2. During the AO reactor operation, under anaerobic-aerobic conditions, a stable phosphorus removal performance was presented after 40 days, as given by the variation of the phosphorus concentration in the effluent, while the AA reactor reached the similar stable-state phase after approximately 80 days of running under anaerobic-anoxic conditions. With the same operation parameters, each reactor responded differently throughout the PAOs and denitrifying phosphorus accumulating organisms' enrichment experiments, in which denitrifying phosphorus accumulating organisms' acclimatization to attain stable state required one times more time than PAOs, indicating the higher activities of PAOs at low temperatures than denitrifying phosphorus accumulating organisms. At the stable-state phase, both the AA and AO reactors exhibited a good phosphorus removing performance and the effluent phosphorus concentrations

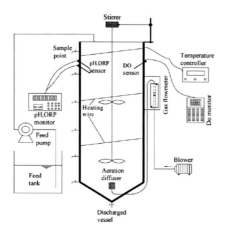

Figure 1.   Schematic diagram of the SBR system.

Table 1.   Composition of inlet wastewater.

| Feeds | Weight g/L | Nutrient | Weight g/L |
|---|---|---|---|
| $CH_3COONa$ | 41.00 | $Fecl_3 \cdot 6H_2O$ | 1.50 |
| $KH_2PO_4$ | 7.04 | $H_3BO_3$ | 0.15 |
| $(NH_4)_2SO_4$ | 18.84 | $CuSO_4 \cdot 5H_2O$ | 0.03 |
| $Cacl_2$ | 0.85 | KI | 0.18 |
| $MgSO_4 \cdot 7H_2O$ | 7.20 | $Mncl_2 \cdot 4H_2O$ | 0.12 |
| Nutrient | 0.60 mL/L | $Na_2MoO_4 \cdot 2H_2O$ | 0.06 |
| | | $ZnSO_4 \cdot 7H_2O$ | 0.12 |
| | | $Cocl_2 \cdot 6H_2O$ | 0.15 |
| | | EDTA | 10.00 |

Notes: COD: P = 20 : 1; pH: 7.0 ± 0.2.

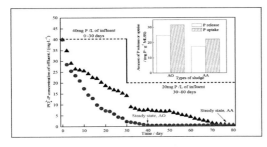

Figure 2. Graph showing phosphorus removal with respect to AA and AO reactors.

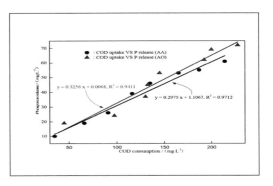

Figure 3. Graph showing the COD consumption versus phosphorus release.

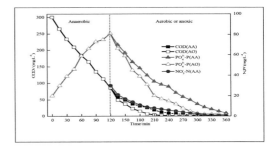

Figure 4. Nutrient profiles of DNPAOs.

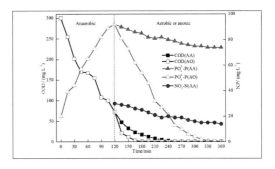

Figure 5. Graph showing the nutrients of PAO sludge-exposed AA or AO reactors.

were both less than 0.5 mg P/L. Consistently, the stable concentrations of MLSS, MLVSS, phosphorus release and uptake and the constant ratios of MLVSS to MLSS and phosphorus release to phosphorus uptake were also observed both in the AA and AO reactors. The amount of phosphorus stored in the microorganisms such as PAOs and denitrifying phosphorus accumulating organisms can be generally implied according to the ratio of MLVSS to MLSS, and the lower the ratio, the greater amount of phosphorus may be stored in microbes responsible for phosphorus removal. At the end of acclimatization of PAOs and denitrifying phosphorus accumulating organisms in the two reactors, the average MLVSS and MLSS concentrations were 2.6 g/L and 3.7 g/L, 3.5 g/L and 4.5 g/L, respectively, exhibiting that their ratios were 0.70 and 0.78, respectively. These results implied that a higher amount of phosphorus was stored in the AO sludge than AA sludge per gram of biomass. A typical key phosphorus biochemical transformation responsible for EBPR was observed both in the AA and AO reactors through a cycle batch test performed at the end of the acclimatization study, as given in Fig. 2, Fig. 3, Fig. 4 and Fig. 5, strongly suggesting that PAOs and denitrifying phosphorus accumulating organisms were predominantly present in their respective reactors proposed in this study. However, a significant difference in the amount of phosphorus release and uptake per MLSS between the AO sludge and AA sludge was monitored (see Fig. 2), probably due to the fact that Accumulibacter exhibits a different metabolic process based on the different running modes, namely anaerobic-aerobic and anaerobic-anoxic conditions. For AO sludge, the anaerobic phosphorus release rate and aerobic phosphorus uptake rate were 19.46 mg P/(g MLSS) and 24.74 mg P/(g MLSS), respectively, both higher than the phosphorus release rate and anoxic phosphorus uptake rate of AA sludge, which were 13.56 mg P/(g MLSS) and 17.33 mg P/(g MLSS), respectively. These results demonstrated that the PAOs and denitrifying phosphorus accumulating organisms' phenotypes are responsible for

phosphorus removal from wastewater. Although the significant difference existing in the amount of phosphorus release/uptake between PAOs and denitrifying phosphorus accumulating organisms was observed, the ratio of the phosphorus release to the phosphorus uptake (0.786) in the AO sludge was quite consistent with that in the AA sludge (0.782), further suggesting that both PAOs and denitrifying phosphorus accumulating organisms were dominant in their respective reactors at the end of the enrichment period. This explanation was also demonstrated by the linear relationship between the amount of COD consumption and that of phosphorus release under anaerobic conditions, as given in Fig. 3 (discussed later).

### 3.2 Anaerobic-aerobic batch test with denitrifying phosphorus accumulating organisms sludge

The phosphorus release and uptake capacities of denitrifying phosphorus accumulating organisms during two different cycles (anaerobic-aerobic and anaerobic-anoxic) at the end of the acclimatization phase were investigated through two batch tests proposed here. Typical profiles (variation of carbon, nitrogen, and phosphorus with time) monitored in these batch tests are shown in Fig. 4. Under the anaerobic conditions, sodium acetate was mostly taken up, which was accompanied by phosphorus release, additionally showing a good correlation between the sodium acetate uptake and phosphorus release (see Fig. 3). The phosphorus anaerobic release rate of denitrifying phosphorus accumulating organisms obtained here was 13.56 mg P/g MLSS which is lower than that of PAOs (19.46 mg P/g MLSS), and which is likely due to the less amount of Accumulibacter enriched in the AA reactor when compared to the AO reactor (see Fig. 6). After a two-hour anaerobic phase, denitrifying phosphorus accumulating organisms sludge exhibited a good phosphorus uptake performance both under anoxic and aerobic conditions. The phosphorus uptake rates obtained in these batch tests were 17.33 mg P/g MLSS in the anoxic mode and 17.76 mg P/g MLSS under aerobic conditions, indicating that denitrifying phosphorus

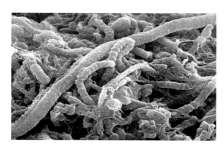

(A) AA sludge, ×5000

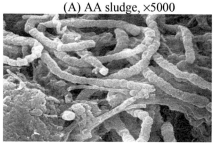

(B) AO sludge, ×5000

Figure 6.   SEM images of activated sludge at the end of the activation period.

accumulating organisms were able to immediately use oxygen as the electron acceptor when exposed to aerobic conditions, as evidenced by the rapid phosphorus uptake rate (see Fig. 4). Concurrently, residual sodium acetate from the anaerobic phase was completely consumed by the denitrifying bacteria or by the heterotrophic bacteria. Obviously, nitrate added in the anoxic phase was removed from wastewater by the denitrifying phosphorus removing bacteria, namely Accumulibacter, with the function of simultaneous denitrification and phosphorus removal.

### 3.3 Anaerobic-anoxic batch test with PAO sludge

Two batch tests similar to those elaborated in section 3.2 were performed to compare the phosphorus uptake capacity of PAOs from the AO reactor in anaerobic-anoxic and anaerobic-aerobic modes at the end of the acclimatization phase. The results obtained in these batch tests are shown in Fig. 5. During a 2 h anaerobic phase, both phosphorus release and sodium acetate uptake exhibited a good performance for the PAO sludge from the AO reactor, as also illustrated in Fig. 3, where the phosphorus release rate was 19.46 mg P/g MLSS and the sodium acetate uptake rate was 61.49 mg COD/g MLSS, both higher than that of denitrifying phosphorus accumulating organisms (13.56 mg P/g MLSS, 47.56 mg COD/g MLSS, as shown in Fig. 4). In contrast with denitrifying phosphorus accumulating organisms (see Fig. 4), however, a significant difference of phosphorus uptake performance of PAOs between aerobic and anoxic conditions was clearly present in Fig. 5. Here, the aerobic phosphorus uptake rate was 24.74 mg P/g MLSS, while that was 4.86 mg P/g MLSS under anoxic conditions, suggesting that the phosphorus uptake ability of the PAO sludge was inhibited when exposed to anoxic conditions, as also evidenced by the less nitrate reduction in this batch test.

### 3.4 SEM analysis

Throughout the entire start-up period of AA and AO reactors, the different phosphorus removal performances observed in the two reactors could be due to the variations in the microbial population. For this, Scanning Electron Microscopy (SEM) was adopted here to obtain a better understanding of the microbial community shift both in AA and AO reactors during the acclimatization phase. From SEM images (see Fig. 6), it can be clearly observed that significant differences in microbial morphologies were observed between the seed sludge from A2O and the AA or AO sludge from EBPR systems studied here. Long-rod morphology microbes were abundantly enriched both in the AA and AO reactors, while the seed sludge has a higher proportion of cocci or short-rod morphology microorganisms. Similar microbes were enriched in the two reactors during the acclimatization process suggested that the long-rod morphology of Accumulibacter responsible for phosphorus removal may preferably take up sodium acetate, as supplied in the influent in this study, regardless of the types of electron acceptors used.

## 4 DISCUSSION

### 4.1 *Operational strategy of phosphorus organisms*

The strategy of enrichment of PAOs and denitrifying phosphorus accumulating organisms under anaerobic-aerobic and anaerobic-anoxic conditions, respectively was developed based on the previous reports that Accumulibacter, a known PAO, contains two different types: one is capable of not only aerobic phosphorus uptake by using oxygen as the electron acceptor, but also anoxic phosphorus uptake by using nitrates, namely denitrifying phosphorus accumulating organisms, and the other type only uses oxygen instead of nitrate as the electron acceptor for phosphorus removal [3, 4] and that temperature seems to be one of the most important influence factors on wastewater systems containing EBPR in practical operations, particularly at low temperature [panswad et al. 2003]. It can be observed from Fig. 1 that both AO and AA reactors operated under anaerobic-aerobic and anaerobic-anoxic conditions, respectively confirmed the phenotypes of PAOs and denitrifying phosphorus accumulating organisms responsible for phosphorus removal and reached the similar stable-state, as evidenced by the effluent phosphorus concentrations, MLSS, MLVSS, the ratio of MLVSS/MLSS, phosphorus release rate and uptake rate and their ratio. The AO and AA reactors reached the stable-state after 40 days and 80 days respectively, suggesting the higher activities of PAOs to low temperatures than denitrifying phosphorus accumulating organisms, probably due to the fact that the amount of energy generated from the oxidative phosphorylation with $NO_3^-$ is about 40% lower than that with $O_2$ [Ahn et al. 2002]. The ratio of MLVSS to MLSS from 0.86 (3.6/4.2) of the start-up phase (namely, seed sludge collected from an aerobic basin within an $A_2O$ process) decreased to 0.70 (2.6/3.7) of the stable-state phase in the AO reactor and to 0.78 (3.5/4.5) in the AA reactor, which are in agreement with the reports [Lu et al. 2006], indicating that the higher amount of phosphorus was stored in PAOs or denitrifying phosphorus accumulating organisms than in the seed sludge, suggesting that this strategy studied here dramatically promoted the growth of PAOs and denitrifying phosphorus accumulating organisms in their respective

reactors. The specific phosphorus release and uptake rates for PAOs were estimated to be 19.46 and 24.74 mg P/g MLSS both higher than that for denitrifying phosphorus accumulating organisms, 13.56 and 17.33 mg P/g MLSS, respectively. This was likely due to the following two reasons: (1) the energy produced by PAOs with oxygen was higher than that of denitrifying phosphorus accumulating organisms with nitrate [Ahn et al. 2002]; (2) the size of denitrifying phosphorus accumulating organisms was bigger than that of PAOs, causing the limited transfer of carbon, nitrogen, and phosphorus to the active biomass [Zeng et al. 2003]. Indeed, SEM conducted in this study showed that denitrifying phosphorus accumulating organisms grow with the aggregation of biomass into similar granules, while PAOs grow with flocs. Overall, these results obtained here demonstrated that the operational strategy at low temperatures proposed in this study is rather effective in acclimatization of PAOs and denitrifying phosphorus accumulating organisms in EBPR systems, thereby providing a practical strategy for the stable-state operation of this process at low temperatures such as in winter.

### 4.2 *Phosphorus organisms during switching tests*

Figs. 4 and 5 show the correlation between the types of electron acceptor and the phosphorus removal performance. For denitrifying phosphorus accumulating organisms, no considerable difference in the phosphorus uptake rate with nitrate and oxygen as electron acceptors was observed by switching the mode from normal anoxic to aerobic, namely that when denitrifying phosphorus accumulating organisms exposed to aerobic conditions it can take up phosphorus immediately, which agrees well with the report [Gebremariam et al. 2011]. However, the phosphorus uptake performance of PAOs was obviously inhibited when it was exposed to anoxic rather than aerobic conditions. These results obtained through the switching batch tests suggested that denitrifying phosphorus accumulating organisms can readily produce the quality of enzymes for aerobic metabolisms similar to anoxic metabolisms, while PAOs lack the enzymes required for anoxic metabolisms using nitrate instead of oxygen as an electron acceptor [Martin et al. 2011]. From Fig. 5, the phosphorus uptake rate of PAOs was very low under anoxic conditions as compared with the aerobic conditions. Interestingly, some studies [7, 10] have demonstrated that when the anoxic phase was extended to 30 h rather than 4 h, the phosphorus uptake rate of PAOs can be obviously improved, suggesting that a several hours lag phase may be in existence in phosphorus uptake when PAOs are exposed to anoxic conditions. From these studies, we hypothesize that PAOs could gradually develop the required amount of enzymes for anoxic metabolisms during the lag time, which may be in agreement with the explanation mentioned above (PAOs lack the enzymes required for anoxic metabolisms using nitrate). Through four batch tests, comparison of the phosphorus removal performance of PAOs between aerobic and anoxic conditions and a similar comparison of denitrifying phosphorus accumulating organisms were conducted, demonstrating that Accumulibacter has, at least, two different types, which supports the reports [3, 4] described above and nevertheless, based on Carvalho et al.'s findings [Carvalho et al. 2007].

### 4.3 Identification of accumulibacter in AA and AO

An analysis of SEM images show that identical microbial morphologies (long-rod microbes) were present in the two reactors at the end of the acclimatization period, suggesting that this kind of Accumulibacter may display good affinities for sodium acetate as the carbon source. Indeed, two different types (rods or cocci) of Accumulibacter were found by Martin et al. in two EBPR systems, in which one was fed with sodium acetate and the other with propionate. Similarly, He et al. [He et al. 2011] also found the distribution of the different types of Accumulibacter in one lab-scale reactor and six full-scale reactors both presenting good phosphorus performance. These studies may support the hypothesis that different carbon sources fed to the phosphorus removal microorganisms could promote the growth of different types of Accumulibacter, probably regardless of electron acceptors, which is also partially supported by the results obtained in this study. Overall, the combination of chemical analysis with microbial analysis suggested that Accumulibacter, both PAOs and DNPAOs, with a long-rod morphology was more preferably enriched with sodium acetate when compared with other carbon sources, such as sodium acetate.

## 5 CONCLUSIONS

A new scenario for enhancing phosphorus removal during the winter season is investigated in this study. The results showed the following points:

The COD/P ratio of the two EBPR systems operated in Anaerobic-Aerobic (AA) and anaerobic-anoxic (AO) modes respectively, from 20:1 to 15:1; with the decrease of concentration COD, $PO_4^{3-}$–P and $NO_3$–N in the influent from 800 mg/L, 40 mg/L, and 50 mg/L to 300 mg/L, 20 mg/L, and 30 mg/L, respectively.

Different Solid Retention Times (SRTs) were adopted for enrichment of different types of Accumulibacter; no excess activated sludge was wasted at the beginning of the start-up and then 10 days SRT for PAOs and 20 days SRT for denitrifying phosphorus accumulating organisms were adopted.

Maintaining high and different MLSS concentrations in AA and AO reactors, based on the metabolic characteristics of Accumulibacter supplied the different electron acceptors as follows: 4.5 g/L MLSS for the AA reactor and 3.7 g/L MLSS for the AO reactor.

This proposed strategy is shown to be effective in achieving a very high enrichment of Accumulibacter at low temperatures by linking chemical analysis with microbial observation. Chemical analysis of batch tests indicated the existence of two types of Accumulibacters, in which one uses oxygen as the electron acceptor and the other uses nitrate. Through microbial observation, a high abundance of Accumulibacter was present in both AA and AO reactors. Although the strategy may not be the unique method for the enrichment of phosphorus removal microorganisms during winter, it is recommendable for future studies in practical applications.

## REFERENCES

Ahn J, Daidou T, Tsuneda S, Hirata A. 2002. Characterization of denitrifying phosphate-accumulating organisms cultivated under different electron acceptor conditions using polymerase chain reaction-denaturing gradient gel electrophoresis assay. Water Research, 36(2): 403–412.

Carvalho G, Lemos P C, Oehmen A, Reis M A M. 2007. Denitrifying phosphorus removal: linking the process performance with the microbial community structure. Water Research, 41(19): 4383–4396.

Gebremariam S Y, Beutel M W, Christian D, Hess T F., 2011. Research advances and challenges in the microbiology of enhanced biological phosphorus removal—A critical review. Water Environment Research, 83(3): 195–219.

He S, Gu A Z, Mcmahon K D., 2006. Fine-scale differences between Accumulibacter-like bacteria in enhanced biological phosphorus removal activated sludge. Water Science and Technology, 54(1): 111–118.

Lu H B, Oehmen A, Virdis B, Keller J, Yuan Z G., 2006. Obtaining highly enriched cultures of Candidatus Accumulibacter phosphates through alternating carbon sources. Water Research, 40(20): 3838–3848.

Martín H G, Ivanova N, Kunin V, Warnecke F, Barry K W, Mchardy A C, Yeates C, He S, Salamov A A, Szeto E., 2006. Metagenomic analysis of two enhanced biological phosphorus removal (EBPR) sludge communities. Nature Biotechnology, 24(10): 1263–1269.

Panswad T, Doungchai A, Anotai J. 2003. Temperature effect on microbial community of enhanced biological phosphorus removal system. Water Research, 37(2): 409–415.

Zeng R J, Saunders A M, Yuan Z, Blackall L L, Keller J. 2003. Identification and comparison of aerobic and denitrifying polyphosphate accumulating organisms. Biotechnology and Bioengineering, 83(2): 140–148.

*Advances in Energy and Environment Research – Achour & Wu (Eds)*
© *2017 Taylor & Francis Group, London, ISBN 978-1-138-62682-9*

# Chaotic analysis of metallogenic element concentration

L. Wan, X.Y. Xiong & J.J. Lai
*School of Mathematics and Information Science, Guangzhou University, Guangdong, China*
*Key Laboratory of Mathematics and Interdisciplinary Sciences of Guangdong Higher Education Institutes,*
*Guangzhou University, Guangdong, China*

ABSTRACT:  Identifying the mechanisms and interactions that influence the spatial structure of metallogenic element concentration distribution is important for both metallogenic prognosis and quantitative assessment. In this paper, the techniques of phase space reconstruction and chaos theory are applied to study the chaotic characteristics of metallogenic element concentration for Au-Cu-Pb-Zn in Shangzhuang ore deposit, Jiaodong Gold Province, China. The results show that the correlation dimension, the corresponding saturated embedding dimension, and the maximum Lyapunov exponent of Au in intensely mineralized area are 2.73, 24 and 0.02, respectively, indicating an obvious chaotic dynamic feature of Au distribution. The correlation dimensions of Cu-Pb-Zn are unsaturated, suggesting a strong random distribution. Our conclusion can provide new approaches for research on ore-forming dynamics process.

## 1 INTRODUCTION

Characterization of metallogenic element concentration distribution is essential for economic planning in the mining industry. Metallogenic element distribution often shows a highly irregular structure, and exhibits scale-dependent changes in structure, and needs nonconventional statistical methods. For a better comprehension of the element of concentration volatility function properties, it is necessary to ascertain the element concentration changes by investigating the structure of latency at the microscopic level (Yu 2007). In recent years, non-linear study of the geological and metallogenic data is mainly focusing on fractal or multifractal (Zhu 2012, Cheng 2007, 2012, Deng et al. 2009, Agterberg 2007, Wan et al. 2010). However, the ore-forming process might be better understood through chaotic analysis that belongs to nonlinear deterministic model (Yu 2006, Wan et al. 2015).

In this paper, the concentration series of along drifts were selected from Shangzhuang Deposit, Jiaodong Gold Province, China, by using the non-linear analysis method, to verify that metallogenic element concentration series, Au-Cu-Pb-Zn, are deterministic chaotic phenomena. These results are to investigate the dynamical behavior of geochemical element concentration series or the influenced domain of the complex mineralization system.

## 2 METHODS AND MATERIALS

### 2.1 *Analysis methods of chaos characteristics*

The complex behavior of a dynamic system can be reflected by univariate time series, which contains the information of all characteristic variables concerned with the evolution of a system. The univariate time series should be expanded in to a high-dimensional space to reveal its dynamic characteristics (Mañé 1981). This is called the phase–space reconstruction which can be performed through the time delay technique proposed by Takens (1981). The details are shown as follows.

Given a time series $\{x_i\}$, where $i = 1, 2, \ldots\ldots$, $N$, the phase space can be reconstructed using the method of the standard delay-coordinate embedding. In this higher dimension space, the vectors of phase space reconstruction can be expressed as below:

$$X(t_i) = [x(t_i), x(t_i + \tau), \cdots, x(t_i + (m-1)\tau] \quad (1)$$

$$i = 1, 2, \ldots, M$$

where $\tau$ is the time delay, $m$ is the embedded dimension, i.e. the coordinate number of the phase space, and the constants $m$, $N$, $M$ and $\tau$ are related as $M = N - (m-1)\tau$. The dynamic properties of systems could be studied by reconstruction of the phase space if $m \geq 2D_c + 1$, where $D_c$ is the fractal dimension of the system. $D_c$ is estimated from the correlation integral $C(N, r)$, which is the number of points in the phase space of dimension $m$ that are closer than $r$.

$$C(N, r) = \frac{2}{N(N-1)} \sum_{1 \leq i \leq j \leq N} \theta(r - d_{ij}), r > 0 \qquad (2)$$

and $d_{ij} = \left\| X_i - X_j \right\|$

where $\theta(r - d_{ij})$ is the Heaviside function, with $\theta(r - d_{ij}) = 1$ for $r > d_{ij}$, and $\theta(r - d_{ij}) = 0$ for $r \leq d_{ij}$. $N$ is the size of the data set. $\|\cdot\|$ is the maximum distance.

$$D_C = \lim_{r \to 0} \frac{\ln C(N, r)}{\ln r} \qquad (3)$$

For different values of $r$, the least squares regression is introduced to calculate the points $(\ln r, \ln C(N, r))$ in the non-scale region, and the slope is considered as the correlation dimension. Practically one computes the correlation integral for increasing embedding dimension $m$ and calculates the related $D_c(m)$ in the scaling region. If the $D_c(m)$ reaches saturation value $D_c$ for relatively small $m$, this gives an indication that an attractor with dimension $D_c$ exists underlying the analyzed time series. On the contrary, if the $D_c(m)$ increases without bound with increase in the embedding dimension $m$, the system under investigation is considered as stochastic (Grassberger & Procaccia 1983, Liebert & Schuster 1989).

As the system evolves, the sum of a series of attractor point values (in each dimension) will converge or diverge (Eckmann & Ruelle 1985). Lyapunov exponents measure the rate of convergence or divergence in each dimension, and a chaotic system will exhibit trajectory divergence in at least one dimension. Thus, a positive Lyapunov exponent is a strong indicator of chaos. It is ensured that the time series has chaos if only the largest Lyapunov exponent is larger than zero and it indicates the chaos degree of the system directly. The method used in calculating the largest Lyapunov exponent is based on averaging the local divergence rates or the local Lyapunov exponents. Rosenstein et al. (1993) proposed a new method to calculate the largest Lyapuno exponent from an observed series, which is small data method. The largest Lyapunov exponent, the definition is:

$$\lambda_1(i, k) = \frac{1}{k \Delta t (M - k)} \sum_{j=1}^{M-k} \ln \frac{d_j(i+k)}{d_j(i)} \qquad (4)$$

where $\Delta t$ is the sample period, $k$ is a constant, $d_j(i)$ is the distance of the $j$th couple nearest position after $i$ time steps pass.

The calculated positive largest Lyapunov exponents prove that the attractor of the mean delay spread time is a strange attractor which is sensitive to the initial condition. Such sensitivity indicates that the mean delay spread time series of flights has apparent chaos phenomena.

## 2.2 Study area and data acquisition

The Shangzhuang ore deposit is located in Jiao-jia goldfield in the Jiaodong gold province occurs on the Jiaodong Peninsula of eastern China. The main rock types in the field are the metamorphic rocks of Jiaodong group and the Yanshaninan granitoids. The main gold orebodies, controlled by NNE- and NE-trending Wangershan fault zone, strike N25°~45°E with dips 30°~40° towards NW. The major ore zone is 1800 m long, 2–4 m thick, and continues to approximately 750 m depth. The alteration includes strong silicification, sericitization, sulphidation and K-feldspar alteration. Gold grades are variable, ranging from 2 to 20 g/t with the average about 5 g/t. Au occurs mainly as gold and electrum in pyrite, with minor free gold and silver in the altered rocks (Deng et al. 2006, 2008).

The data of metallogenic element concentration, Au-Cu-Pb-Zn, are obtained from the continuous channel sample with 1 m length in Shangzhuan Deposit, Jiaodong Gold Province, China. These samples were assayed. The composition of alteration rock in the central of alteration belt has the largest degree of variation, a large number of permeability fluid carry active components into the dilation space, which has significance to the further enrichment of the ore-forming element (Au) and the remobilization of correlative elements (Cu, Zn and Pb).

Figure 1 shows the change curve of Au elements concentration as the change of sampling location.

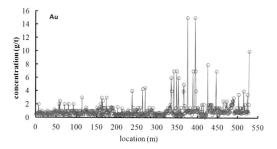

Figure 1. Change curve of Au elements concentration as the change of sampling location.

116

# 3 RESULTS AND DISCUSSION

## 3.1 Descriptive statistics

To describe the main features of a collection of data in different samples, statistical treatment of the data was performed using to excel. We implement a Jarque-Bera (JB) test for normality with a relative large sample here. The statistical quantity for the JB test is

$$JB = n\left[\frac{S^2}{6} + \frac{(K-3)^2}{24}\right] \quad (5)$$

where $S$ is skewness, $K$ is kurtosis, $n$ is the volume of sample, and JB is statistical quantity. The main descriptive statistics of the Au-Cu-Pb-Zn concentration series are given in Table 1.

According to data in Table 1, the kurtosis values lower than 3 are an indication of the presence of platykurtosis in the probability distribution, and the skewness values greater than 0 are an indication of the positive skewness. The results of the the Jarque-Bera test show that the values of statistic are more than the significance at the 5% level. So we can reject the null hypothesis of following a normal distribution to four samples. Thus, there is a clear departure from Gaussian normality.

## 3.2 Chaotic parameters

In order to reconstruct the original phase space, we first estimate reconstruction parameters, the delay time $\tau$ and embedding dimension $m$. We calculate both the autocorrelation function with time lags of 1–40 m, and the cures have dropped to (1–1/e) of its initial value of the autocorrelation function (Liebert 1989). The delay time $\tau$ for Au-Cu-Pb-Zn, respectively (Fig. 2), and the delay time $\tau$ are 2 for all elements.

Subsequently, we calculate the embedding dimensions for our dataset using Eq. 2–Eq. 4, $m$, from 2 to 26. Fig. 3 shows the relationship between the correlation exponent values $D_c(m)$, and the embedding dimension values $m$. It can be seen that the correlation exponent value increases with the embedding dimension up to a certain value, and embedding dimension of Au shows saturation,

Table 1. Descriptive statistics for the elements concentration.

| Element | Avg | Std | K | S | J-B |
|---|---|---|---|---|---|
| Au | 1.08 | 1.36 | 48.62 | 6.06 | 55148.18* |
| Cu | 8.85 | 8.38 | 148.57 | 11.21 | 130553.12* |
| Pb | 23.54 | 13.07 | 76.45 | 7.07 | 54865.72* |
| Zn | 30.79 | 15.97 | 29.62 | 3.69 | 20495.95* |

Note: *Statistical significance at the 5% level.

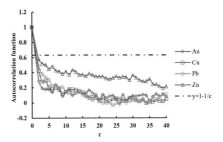

Figure 2. Autocorrelation function of the metallogenic elements.

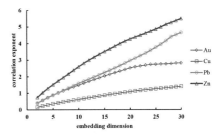

Figure 3. Relation between correlation exponent $D_c(m)$ and embedding dimension $m$.

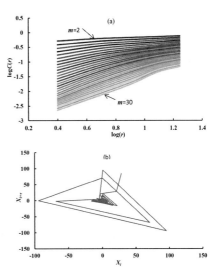

Figure 4. (a) $\log C(r)$ versus $\log(r)$ plots of the elements concentration of Au; (b) two-dimensional phase portrait of Au.

but none of Cu-Pb-Zn. The saturation of the correlation exponent beyond a certain embedding dimension value is an indication of the existence of deterministic dynamics.

Fig. 4(a) (b) shows the relationship between the correlation function $\ln C(r)$ and the radius $\ln r$ for

Table 2. Chaotic parameters for the elements concentration.

| Element | $\tau$ | $m$ | $D_c$ | $\lambda_1$ |
|---------|--------|-----|-------|-------------|
| Au | 2 | 24 | 2.73 | 0.02 |
| Cu | 2 | 2~30 | 0.16~1.40 | 0.56~0.07 |
| Pb | 2 | 2~30 | 0.40~4.65 | 0.02~0.01 |
| Zn | 2 | 2~30 | 0.74~5.49 | 0.00 |

increasing $m$, and two-dimensional phase portrait of Au, respectively.

The Lyapunov exponent are estimated using the small data sets method, Eq. 6. The results of chaotic parameters, Au-Cu-Pb-Zn, are listed in Table 2.

The largest Lyapunov exponent of Au is $\lambda_1 = 0.02 > 0$, saturation dimension is $m = 24$, and correlation dimension is about $D_c = 2.73$. The result indicated the existence of chaotic characteristic in mineralization processes of Au, and the decisive factor of dynamics of the geology processes is not more than 3. The correlation dimensions of Cu-Pb-Zn are unsaturated, indicating strong random distribution.

## 4 CONCLUSIONS

The chaotic characteristics of the metallogenic element concentration series at the continuous channel samples in Shangzhuang ore deposit, Jiaodong Province, China, are investigated by the phase space reconstruction technology and chaos theory in this study. The empirical results show that Au, the largest Lyapunov exponents $\lambda_1$ is above zero, and correlation fractal dimensions $D_c$ is no-integer, which suggests the existence of chaos in the element enrichment processes.

The deposition of ore-forming material is a complex nonlinearity dynamics process. Through all kinds of coupling and feedback processes among the tectonic stress, fluid flow, fluid-rock reaction and minerals deposition at the appropriate site in fault structures at last and then form epithermal deposits with chaotic character. The study provides positive evidence regarding the existence of chaotic behavior in metallogenic elements concentration series, leading to better understanding of the dynamics of the geology processes.

## ACKNOWLEDGMENTS

This research is supported by the National Natural Science Foundation of China (Grant No. 41172295).

## REFERENCES

Agterberg, F. P. (2007). Mixtures of multiplicative cascade models in geochemistry. *Nonlinear Proc. Geoph. 14,* 201–209.

Cheng, Q. M. (2007). Mapping singularities with stream sediment geochemical data for prediction of undiscovered mineral deposits in Gejiu, Yunnan Province, China. *Ore Geo. Rev. 83–87,* 314–324.

Cheng, Q. M. (2012). Singularity theory and methods for mapping geochemical anomalies caused by buried sources and for predicting undiscovered mineral deposits in covered areas. *J. Geochem. Explor. 122,* 55–70.

Deng, J., Yang, L. Q., Ge, L. S., Wang, Q. F., Zhang, J., Gao, B. F., Zhou, Y. H. & Jiang S. Q. (2006). Reasearch advances in the Mesozoic tectonic regimes during the formation of Jiaodong ore cluster area. *Prog. Nat. Sci. 16,* 777–784.

Deng, J., Wang, Q. F. Yang, L. Q. Zhou, L., Gong, Q. J., Yuan, W. M., Xu, H. Guo, C. Y. & Liu, X. W. (2008). The Structure of Ore-controlling Strain and Stress Field in the Shangzhuang Gold Deposit in Shandong Province, China. *Acta Geo. Sin-Engl. 82,* 769–780.

Deng, J., Wang, Q. F. Wan, L., Yang, L. Y. Gong, Q. J, Zhao, J. & Liu, H. (2009). Self-similar fractal analysis of gold mineralization of Dayingezhuang disseminated-veinlet deposit in Jiaodong gold province, China. *J. Geochem. Explor. 102,* 95–102.

Deng, J., Wang, Q. F. Wan, L., Liu, H., Yang, L. Y. & Zhang, J. (2011). A multifractal analysis of mineralization characteristics of the Dayingezhuang disseminated-veinlet gold deposit in the Jiaodong gold province of China', *Ore Geo. Rev. 40,* 54–64.

Eckmann, J. P. & Ruelle, D. (1985). Ergodic theory of chaos and strange attractors. *Rev. Mod. Phys. 57,* 617–656.

Grassberger, P. & Procaccia, I. (1983). Measuring the strangeness of strange attractors. *Physica D. 9,* 189–208.

Liebert, W. & Schuster, H. G. (1989) Proper choice of the time delay for the analysis of chaotic timeseries. *Phys. Lett. A. 142,* 107–111.

Mañé, R. (1981). On the dimension of compact invariant sets of certain nonlinear maps. In: Dynamical systems and turbulence, lecture notes in mathematics. *Berlin: Heidelberg: Springer,* 898: 230–242.

Rosenstein, M. T., Collins J. J. & De luca, C. J. (1993). A practical method for calculating largest Lyapunov exponents from small data sets. *Physica D. 65,* 117–134.

Takens, F. (1981). Detecting strange attractors in turbulence. *Lect. Notes Math. 898,* 366–381.

Wan, L., Wang, Q. F., Deng, J., Gong, Q. J., Yang, L. Q. & Liu, H. (2010). Identification of Mineral Intensity along Drifts in the Dayingezhuang Deposit, Jiaodong Gold Province, China. *Resour. Geol. 60,* 98–108.

Wan, L., Liu, H.,Yang, L. & Zhu, Y. Q. (2015). Chaotic mechanisms of the ore-forming element accumulation: Case study of porphyry and disseminated-veinlet gold deposits. *Acta Petrol. Sin. 31,* 3455–3465.

Yu, C. W. (2006). Fractal growth of mineral deposits at the edge of chaos. *Education Press, Anhui, China.*

Yu, C. W. (2007). Complexity of Geosystem. *Beijing: Geological Publishing House.*

Zhu, Y. F. (2012). Introduction Geochemistry of Mineral Deposits. *Beijing: Beijing University Press.*

*Advances in Energy and Environment Research – Achour & Wu (Eds)*
© 2017 Taylor & Francis Group, London, ISBN 978-1-138-62682-9

# Test study of the degradation ability of the coated sludge microbe-nanometer iron system on perchlorate

Yujie Tong & Bingguang Liu
*College of Biology and Environmental Engineering, Tianjin Vocational Institute, Tianjin, China*

Tielong Li
*College of Environmental Science and Engineering, Nankai University, Tianjin, China*

ABSTRACT: In biological degradation, perchlorate is the ultimate electron receptor of the electron transport chain and is reduced to chloride. Microbe-Nanometer Zero-Valent Iron (NZVI) is oxidized and $H_2O$ reacts to give hydrogen as an electron donor to perchlorate. In order to preserve NZVI, agar is used as a medium to make coated microbe-nanometer zero-valent iron and dried to be a composite. The test studied the composite's degradation ability on perchlorate, and also the factors, including preservation period, activation, and adding trehalose. The contrast group includes NZVI and activated sludge. Perchlorate measuring 50.0 mg/L declines to detection limits in 12 days. When trehalose is added to activated composite preserved for 7 days, 50.0 mg/L perchlorate declines to detection limits in 18 days. The dried composite's degradation ability is lower than the activated sludge. When the preservation period is 30 days, the composite has about the same degradation ability as that preserved for 7 days. It suggests that the dried composite can be preserved effectively, which provides a basis for its industrial production and practical application. If trehalose is not added to NZVI, the composite is slightly different from that added with trehalose. The reflection point is delayed to 12 days, which proves that trehalose can protect microbial activity of the composite to a certain degree. If the composite is not activated, perchlorate only declines by 8% in 18 days. It can be concluded that activation plays an important role in the degradation on perchlorate.

## 1 INTRODUCTION

Perchlorate is a strong oxidizing agent and can persist endocrine disruption, which can restrain the absorption of iodine and affect the function of thyroid. Pollution of perchlorate has recently caused serious concerns and the processing technology of perchlorate pollution has been further developed. In biological degradation, perchlorate is the ultimate electron receptor of the electron transport chain and is reduced to chloride finally (Li, 2016).

Immobilized microbes can be applied in the treatment of wastewater (Ye, 2014). Materials like gel can be used to couple microbe and nanometer-iron. It is added to groundwater for treatment of pollution. As nanometer-iron can be oxidized easily when it encounters oxygen in water (Liu, 2013), this kind of immobilized gel is not beneficial for the preservation of nanometer-iron (Chibata, 1979; Nussinovitch, 1994). When dry preservation technology is applied to produce the composite of nanometer-iron and microbes, it can be preserved in dry environment and cannot be easily oxidized (Sarathy, 2008; Numi, 2005; Hoerle, 2004).

Agar is a kind of poly sugars and it begins to melt at 95°C in water. Its molecular chain spreads in liquid and freezes at 37°C. The freezing agar has strong fastness properties and is difficult to dissolve or decompose, and hence, it keeps the composite stable for a long period (Pelegrin, 2005; Soriano, 2001).

Agar is used as a medium to make coated microbes with nanometer zero-valent iron and dried to be a composite. This paper studies the perchlorate degradation ability of the composite and also its test condition, including preservation period, activation, and adding trehalose.

## 2 RESEARCH METHOD

### 2.1 *Preparation of dried composite*

Activated sludge is a microbial community with high vitality and bacteria that exist is Zoogloea. It is often used for bacteria acclimation. Microbes were collected from activated sludge of city sewage treatment plant and were acclimatized for 4 months. Nanometer Zero-Valent Iron (NZVI) is made by the method of liquid-phase reduction and

$KHB_4$ is used to reduce $FeSO_4 \cdot 7H_2O$. The reaction equation is as follows:

$$Fe^{2+} + 2BH_4^- + 6H_2O \rightarrow Fe^0 + 2B(OH)_3 + 7H_2\uparrow$$

The product NZVI is cleaned thrice with deionized water (Wang, 2006).

In the anaerobic condition, 0.056 g NZVI is added to liquid agar (20 mL, 3500 mg/L) and heated to boil. After cooling the liquid, the upper liquid is removed to get NZVI which is coated by agar. Activated sludge (0.50 mL) is centrifuged at 5000 r/min for 1 min and then the upper liquid is removed, 0.050 g trehalose is added and it is diluted to 1 mL using deionized water.

In anaerobic operation box, NZVI mud is squeezed and mixed with activated sludge. Being dried for 24 hours under the condition of nitrogen at room temperature, the product is a black chip solid which is microbe-NZVI (Nanometer Zero-Valent Iron) coated by agar. Then, the composite is preserved in the air at room temperature.

## 2.2 Composite's degradation test on perchlorate

The composite is used to test its degradation ability on perchlorate (50.0 mg/L, 100 mL). No. 0 is the contrast group to study its autotrophic degradation ability, including NZVI (0.056 g) and activated sludge (0.5 mL). No. 1 includes dried composite (0.090 g), trehalose is added in the making of NZVI (0.056 g), and the preservation period is 7 days, activated by beef extract peptone broth. No. 2 is the same as No. 1, but the preservation period is 30 days; No. 3 is the same as No. 1, but trehalose is not added; No. 4 is the same as No. 1, but is not activated.

Added 0.090 g composite and beef extract peptone broth (50 mL) to 120 mL plasma bottle, and activated in an incubator at 35°C for 12 hours. Then, the upper liquid is removed, adding perchlorate at a concentration of 50.0 mg/L. This is an oscillatory reaction and the concentrations of perchlorate and chloride are detected regularly.

## 2.3 Detection of perchlorate and chloride

Perchlorate and chloride are detected by ion chromatography (DIONEX ICS-1500). The separator column and the guard column are, respectively, IonPac AS20 (4 × 250 mm) and IonPac AG20 (4 × 50 mm).

## 3 RESULTS AND DISCUSSION

Transmission Electron Microscopic (TEM) image of NZVI indicates that NZVI is sphere particulate with a diameter of 62 nm (Fig. 1). The particles are easy to polymerize together as they have magnetic properties and high specific surface area.

## 3.1 Degradation ability of the composite on perchlorate

The system degrades 50.0 mg/L perchlorate to detection limits in 18 days and there is a reflection point on day 9 (Fig. 2). The reaction has two dynamics stages: the first stage is from the beginning to day 9 and the degradation rate is −0.274 (mg/L · d) and the second stage is from day 9 to day 18 and the degradation rate is −0.726 (mg/L · d).

Compared with the contrast group, which degradation time is 12 days (Fig. 3), and it demonstrates that the drying procedure affects the ecological system of bacterial community in activated sludge, although it has been activated.

## 3.2 Factors of degradation ability on perchlorate

### 3.2.1 Effect of preservation period on composite's degradation ability on perchlorate
The composite preserved for 30 days has similar degradation dynamics to that preserved for 7 days (Fig. 4). The composite is stable and can be preserved for long time without losing its degradation ability. Dried microbes can have high activity after some years. If trehalose is used as a protective agent, the microbe's survival has essentially no change in 7 weeks (Zayed G et al., 2004).

### 3.2.2 Effect of adding trehalose to composite's degradation ability on perchlorate
Both No. 3 group and No. 1 group degrade perchlorate's concentration to detection limits in

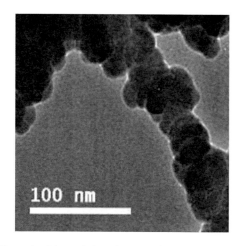

Figure 1. Transmission electron microscopic image of NVZI.

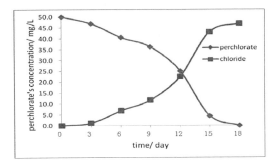

Figure 2. Composite's degradation ability on perchlorate.

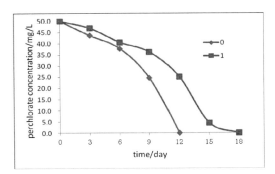

Figure 3. Comparison of perchlorate degradation ability of the contrast group and the composite.

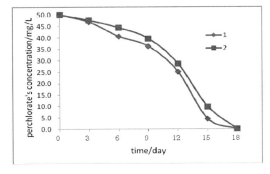

Figure 4. Comparison of the perchlorate degradation ability of the composite preserved for 7 days (1) and 30 days (2).

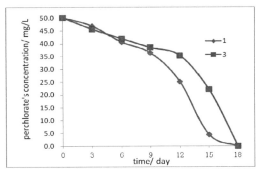

Figure 5. Comparison of perchlorate degradation ability of the composite with trehalose (1) and without trehalose (3).

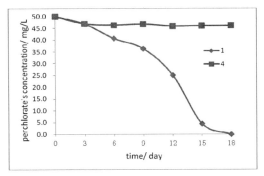

Figure 6. Comparison of perchlorate degradation ability of composite activated (1) and non-activated (4).

hydrogen bond with the peptone of microbe's cell membrane (Zavaglia et al. 2003).

### 3.2.3 Effect of activation of composite's degradation ability on perchlorate

In No. 4 group, the concentration of perchlorate on the day 18 is 46.1 mg/L, which means that perchlorate drops only by 8% in 18 days (Fig. 6). The reaction graph does not have the characteristic of degradation dynamics. There is little reduction in the concentration of perchlorate after day 3 and has a lowest point on day 12, and finally, there is a little rise in the last 6 days. It suggests that the composite has only adsorption and desorption effects on perchlorate.

When the composite is not activated, the inorganic environment cannot supply enough carbon source and nitrogen source to sustain the physiological function of microbes. Consequently, it does not have the ability to degrade perchlorate. Agar has many oxhydryl and other activated groups, which can absorb and desorb ion easily[9–10] (Pelegrin Y F et al. 2005 and Soriano E M 2001).

18 days (Fig. 5). The reflection point in No. 3 group is achieved on day 12; whereas in No. 1 group it occurred on day 9. Therefore, the degradation ability of No. 3 group is a bit lower than that of No. 1 group. The reflection point of reaction dynamics is related to the physiological activity of microbes. Trehalose is used as a drying protective agent in the processing of production of microbe-NZVI. Fructose, saccharose, and trehalose are used instead of the function of water in dry environment and bring

## 4 CONCLUSIONS

Microbiology has the enzyme system of degradation on perchlorate. One is perchlorate reductase and chlorate reductase, and the other is chlorite dismutase. First, perchlorate is reduced to chlorate and then chlorite by perchlorate reductase and chlorate reductase, respectively. Subsequently, chlorite is reacted to be chloridion (Hatzinger, 2005; Bardiya, 2011).

The microbes from activated sludge are coated by agar and then mixed with Nanometer Zero-Valent Iron (NZVI) to be a dried composite (Microbe-NZVI). The composite is activated by adding beef extract peptone broth in the procedure of making microbe-NZVI. Activation plays an important role in its degradation on perchlorate. Adding trehalose as a protective agent also promotes the reaction. Preservation period has less impact on its degradation ability, which suggests that the dried composite is stable and can be preserved for long time.

## REFERENCES

Bardiya N, Bae J H. Dissimilatory perchlorate reduction: a review [J]. *Microbiological Research,* 2011, 166(4): 237–254.

Chibata I. Immobilized microbial cells with polyacrylamide gel and carrageenan and their industrial applications [J]. *Immobilized Microbial Cells,* 1979, 13: 187–202.

Hatzinger P B. Perchlorate biodegradation for water treatment [J]. *Environ Science Technology,* 2005, 39(11): 239A–247A.

Hoerle S, Mazaudier F, Dillmann P, et al. Advances in understanding atmospheric corrosion of iron II: mechanistic modeling of wet-dry cycles [J]. *Corrosive Science,* 2004, 46(6): 1431–1465.

Li Haixiang, Zhang Huan, Yao Yi, Overview of advances in research of microbial reductive degradation for perchlorate [J], Water Purification Technology, 2016, 35(1): 16–20, 42.

Liu Jiang-hong, Xu Rui-dan, Pan Yang, Lu Yan, On treating sewage with multiple contents by immobilized Microorganisms [J], Journal of Safety and Environment, 2013, 13(2): 36–39.

Numi J T, Tratnyek P G, Sarathy V, et al. Characterization and properties of metallic iron nanoparticles: spectroscopy, electrochemistry, and kinetics [J]. *Environmental Science Technology,* 2005, 39(5): 1221–1230.

Nussinovitch A. Resemblance of immobilized *Trichoderma viride* fungal spores in an alginate matrix to a composite material [J]. *Biotechnology Progress,* 1994, 10(5): 551–554.

Pelegrin Y F, Murano E. Agars from three species of *Gracilaria* (*Rhodophyta*) from Yucatán Peninsula [J]. *Bioresource Technology,* 2005, 96(3): 295–302.

Sarathy V, Tratnyek P G, Nurmi J T, et al. Aging of iron nanoparticles in aqueous solution: effects on structure and reactivity [J]. *Physical Chemistry,* C, 2008, 112(7): 2286–2293.

Soriano E M. Agar polysaccharides from *Gracilaria* species (*Rhodophyta, Gracilariaceae*) [J]. *Journal of Biotechnology,* 2001, 89(1): 81–84.

Wang W, Jin Z H, Li T L, et al. Preparation of spherical iron nanoclusters in ethanol–groundwater solution for nitrate removal [J]. *Chemosphere,* 2006, 65(8): 1396–1404.

Ye Lin-jing, Guan Wei-sheng, Lu Xun, et al., Adsorbing power of modified nano-$Fe_3O_4$ in removing tetracycline from the aqueous antibiotic pollutants [J], *Journal of Safety and Environment,* 2014, 14(1): 202–207.

Zavaglia G, Tymczyszyn A, Antoni E D, et al. Action of trehalose on the preservation of Lactobacillus delbrueckii ssp. bulgaricus by heat and osmotic dehydration [J]. *Journal of Applied Microbiology,* 2003, 95(6): 1315–1320.

Zayed G, Roos Y H. Influence of trehalose and moisture content on survival of *Lactobacillus salivarius* subjected to freeze-drying and storage [J]. *Process Biochemistry,* 2004, 39(9): 1081–1086.

*Advances in Energy and Environment Research – Achour & Wu (Eds)*
© *2017 Taylor & Francis Group, London, ISBN 978-1-138-62682-9*

# Assessment of the impact of cropland loss on bioenergy potential: The case of Yangtze River Delta, China

Fengsong Pei
*School of Geography, Geomatics and Planning, Jiangsu Normal University, Xuzhou, China*

Xiaoping Liu
*School of Geography, Geomatics and Planning, Jiangsu Normal University, Xuzhou, China*
*School of Geography and Planning, Sun Yat-sen University, Guangzhou, China*

Jie Guo
*College of Land Management, Nanjing Agricultural University, Nanjing, China*

Gengrui Xia & Changjiang Wu
*School of Geography, Geomatics and Planning, Jiangsu Normal University, Xuzhou, China*

ABSTRACT: Bioenergy plays a critical role in the world's energy consumption and climate change adaption. However, large uncertainties exist on the potential of bioenergy from crop residues and the impacts of urban expansion on the bioenergy in the process of urbanization. In this paper, the bioenergy of crop residue in the Yangtze River Delta in China was estimated for the period 2001–2010 by using satellite-based modeling. The impacts of land use change on the bioenergy were further analyzed. We found that a large amount of cropland was encroached for other uses in this period, approximately 8% of the total cropland in 2010. The cropland loss reduced the bioenergy of crop residues at a rate of 15 PJ (1 PJ = $10^{15}$ J) per year, approximately 3% of annual total bioenergy from crop residues. In this context, further research and policies are needed for cropland protection and sustainable urban planning in future decades.

## 1 INTRODUCTION

The concentration of atmospheric greenhouse gases, like carbon dioxide ($CO_2$), has increased from 278 ppm in 1750 (Raynaud et al., 2005) to 407 ppm by 2016 (Dlugokencky and Tans, NOAA/ESRL), mainly due to burning of fossil fuels. The anthropogenic increase in greenhouse gas accounted for more than half of the observed increase in global mean surface temperature from 1951 to 2010 (Collins et al., 2013). The large-scale production of renewable heat, electricity, and gaseous, solid, and liquid fuels from biomass is an important strategy for climate change mitigation and energy supply (Metz et al., 2007; Taylor et al., 2010). Land-based mitigation strategies, especially bioenergy development, play a potentially significant role to reduce the carbon emission from fossil fuel (Rose et al. 2012). Particularly, bioenergy from crop residues is one of the important sources to obtain low carbon energy to reduce carbon emission. However, this bioenergy received little attention than deserved, particularly in the context of global change.

As one of the largest emitters of fossil-fuel $CO_2$ into the atmosphere (Gregg et al., 2008), China experiences rapid and unprecedented industrialization and urbanization in recent decades, accompanied by massive urban land development at a cost of large cropland losses. At the same time, the country produces a large amount of agricultural residues that can be used for generating renewable energy. Jiang et al. (2012) analyzed bioenergy potential from crop residues in China. Wang et al. (2013) evaluated the spatial distribution of cereal bioenergy potential in China. However, large uncertainties still exist on the potential of bioenergy from crop residues in the process of human-induced land use change, especially urbanization.

This paper focuses on the potential assessment of crop residue as an energy resource in the Yangtze River Delta in China experiencing fast urban expansion by using a satellite-based method. First, the terrestrial Net Primary Productivity (NPP) was calculated using the Carnegie-Ames-Stanford Approach (CASA). The bioenergy production potential from crop residues was then discussed in the process of cropland loss from 2001 to 2010.

## 2 MATERIALS AND METHODS

### 2.1 *Study area*

The study area, which covers approximately 100,000 km², is located in the Yangtze River Delta (YRD) in eastern China (Figure 1). The predominant climate in this region is subtropical monsoon climate, with an annual rainfall of 1,158 mm and average temperature of 16°C. Owing to the distribution in the same seasons for the rainfall and heat energy, such climate provides a favorable environment for plant growth. The major vegetation is subtropical evergreen broadleaf forest. Since the economic reform in 1978, the YRD has witnessed both rapid industrialization and unprecedented urbanization, accompanied by massive urban land development.

### 2.2 *Data preparation and preprocessing*

Meteorological data, including temperature, precipitation, and solar radiation data, were obtained in monthly scale from the Chinese National Metrological Information Center. Continuity and consistency were validated by screening and eliminating the suspicious and missing records. Land use data for 2000 and 2010 were obtained from the GlobeLand30 dataset at 30 m resolution, which was interpreted using satellite images from Landsat and Chinese HJ-1 (Chen et al., 2014). Monthly Normalized Difference Vegetation Index (NDVI) images were derived from the Moderate Resolution Imaging Spectroradiometer (MODIS) product (i.e., MOD13) with a spatial resolution of 1 km².

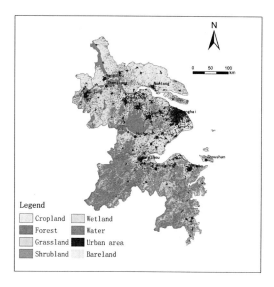

Figure 1. Location of study area.

Vegetation data were compiled from a vegetation map at a scale of 1:1,000,000 (Editorial Board of Vegetation Map of China, 2001). Soil data, including soil texture classes and the associated particle sizes, were obtained from the harmonized world soil database (Freddy et al., 2008). Field-based biomass/NPP data were derived from Luo's study (1996). Many site-dependent records such as biomass and NPP for most of the plant components were mainly compiled from national forest inventories during 1989–1993, and some other published literatures (Ni, 2004; Jin et al., 2007).

### 2.3 *The CASA model for the NPP estimation*

In the past few years, a wide range of models have been developed to estimate the terrestrial NPP, including the Carnegie-Ames-Stanford Approach (CASA) (Potter et al., 1993), the BIOME-BGC (Running et al., 1991), the LPJ-DGVM (Sitch et al., 2003), and the SEIB-DGVM (Sato et al., 2007). The CASA, which is one of the satellite-based models (Potter et al., 1993; Field et al., 1995), has been widely employed in evaluating the terrestrial NPP in the United States (Lobell et al., 2002), in China (Piao et al., 2005), as well as at a global scale (Potter et al., 1993). In this paper, the distributions of terrestrial NPP in the Yangtze River Delta in China were modeled by using the CASA model. According to the algorithm of the CASA, the NPP are calculated as the product of the amount of Photosynthetically Active Radiation (PAR) absorbed by green vegetation ($APAR$) and the light use efficiency ($\varepsilon$) that the radiation is converted into plant biomass increment:

$$NPP(x,t) = APAR \times \varepsilon \qquad (1)$$

where $NPP(x, t)$ is the NPP fixed by vegetation at a grid cell $x$ in month $t$, and $APAR$ is the amount of photosynthetic active radiation. $APAR$ is calculated by using the data on solar surface irradiance ($S$) and the fraction of photosynthetic active radiation absorbed by green vegetation ($FPAR$). $\varepsilon$ for each grid cell can be determined as the product of $\varepsilon_{max}$, and scalars representing the availability of soil moisture ($W$) and the suitability of temperature ($T_1$, $T_2$). Thus, the NPP in location $x$ and time $t$ becomes:

$$NPP(x,t) = S(x,t) \times FPAR \times 0.5 \times \varepsilon^* \\ \times T_1(x,t) \times T_2(x,t) \times W(x,t) \qquad (2)$$

where the factor 0.5 accounts for the fact that approximately half of the incoming solar radiation is in the photosynthetic active radiation waveband (0.4–0.7 $\mu m$). $FPAR$ is defined as a linear function of the NDVI simple ratio. $\varepsilon_{max}$ is determined by a calibration with field data from Luo's (1996) study (Zhu et al., 2006; Pei et al., 2013; Pei et al., 2015). The details of the CASA model can be found in the studies by Potter et al. (1993) and Field et al. (1995).

For verifying reliability of the performance of the CASA model, we validated the model results based on Luo's (1996) investigation data and other literatures. The plot sites that have the same vegetation types as the vegetation map were selected from Luo's (1996) data.

### 2.4 Bioenergy potential of crop residue

The bioenergy potential of crop residue (*B*1) is calculated as the products of the CASA-based NPP (*NPP*), Residue Product Ratio (*RPR*), conversion ratio for Dry Matter (*DM*), and Returned Field factor (*RF*).

$$BI = NPP \times RPR \times DM \times (1 - RF) \qquad (3)$$

It is estimated that approximately 15% of straw and stalk resource are returned to the field as fertilizer or left in the field in China (i.e., $RF = 0.85$). In addition, the *RPR* was set according to Li et al. (2005). To convert CASA NPP to dry biomass, we assume a mass ratio of 2.2 for dry biomass to carbon (*DM*), and that 50% of the NPP is above ground. For conversion efficiencies from biomass to bioenergy, we consider a biomass heating value of 20 kJ per gram of dry biomass (Zumkehr, 2013).

## 3 RESULTS AND DISCUSSION

### 3.1 Validation of the NPP

For validating the NPP simulation, correlation between our CASA-based NPP and some field-based biomass/NPP records was studied. A significant correlation is found between our simulated NPP and observation-based data (r = 0.854, P < 0.030, n = 6). In addition, comparison was made between our estimated NPP and other published summaries of NPP studies. According to our study, the average annual NPP in China is 42 Tg C during 2001–2010 (1 Tg C = $10^{12}$ g C), close to the result of Dai et al. (2012) using the CASA as well. These indicate that the CASA is applicable to the modeling of the NPP in the YRD.

### 3.2 Cropland losses from 2001 to 2010

As shown in Figure 2, net cropland losses were found between 2001 and 2010, approximately 8% of the total cropland. In particular, 65% of the cropland loss was due to fast urban expansion in the past decade. However, some increases in cropland were also found (Figure 2), primarily due to reclamation

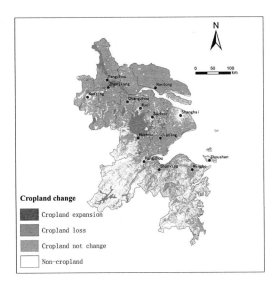

Figure 2. Variations of Cropland in the YRD during 2001–2010.

of grassland and deforestation (Liu et al., 2005). As a whole, a total of 4, 423 km² cropland loss was estimated in this 10-year period. Furthermore, the encroachments are accelerating, and they have potentially important implications for bioenergy.

### 3.3 Variations in the NPP

The average NPP in the YRD, which decreased markedly from northern parts to southern parts of the YRD, showed an obvious geographical heterogeneity in either the amounts or the spatial distributions (Figure 3). This can be associated with the wide distribution of forest in the southern YRD. As for the individual grid cell, annual NPP ranged from the values less than 10 gCm⁻² year⁻¹ in urbanized regions to 1,016 gCm⁻² year⁻¹ in broadleaf forests in the southern YRD.

Besides the spatial pattern of the average NPP, interannual variation in the NPP was investigated as well. The terrestrial NPP reduction in YRD reached 2.61 TgC (R2 = 0.108, P = 0.355) over the period 2001–2010, approximately 6% of regional totals. Particularly, the NPP in YRD was anomalously low in 2006 and 2010 (Figure 4), associated with the cumulative effects of the droughts and serious flooding that occurred in the corresponding period, respectively (Gao et al., 2014).

### 3.4 Bioenergy potential of crop residues in the YRD

Figure 5 shows the temporal-spatial distribution of bioenergy of crop residues in the YRD in 2001. We

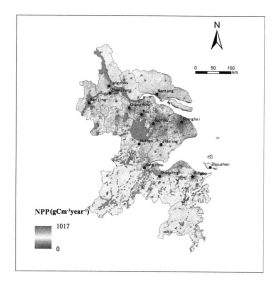

Figure 3. Spatial pattern of average terrestrial NPP during 2001–2010 in the YRD.

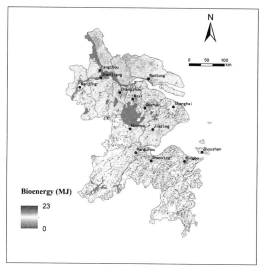

Figure 5. Distribution of bioenergy from crop residues in the YRD in 2000.

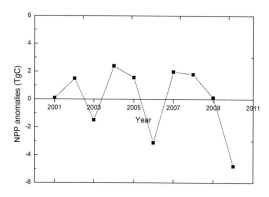

Figure 4. Interannual variations in terrestrial NPP from 2001 to 2010.

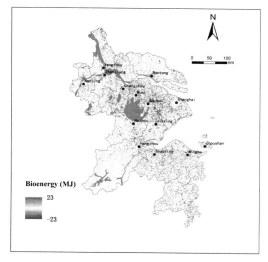

Figure 6. Changes in bioenergy from crop residues in the YRD between 2001 and 2010.

estimated that the bioenergy of crop residues in the YRD ranging from less than 0 to 23 MJ m$^{-2}$, with an average value of 9 MJ m$^{-2}$ in this year. Spatially, the bioenergy of crop residues was mainly concentrated in the northern YRD. This phenomenon could be associated with the wide distribution of the cropland in the northern YRD, instead of forest in the southern YRD.

The changes in the bioenergy of crop residues were further analyzed between 2001 and 2010 (Figure 6). Similar to the cropland conversions, the bioenergy of crop residues in the YRD showed either reductions or increases locally. However, taking it as a whole, the bioenergy of crop residues shows an average decrease of 15 PJ per year over the period 2001–2010, approximately 3% of the annual total bioenergy from crop residues in this region.

## 4 CONCLUSIONS

In this paper, cropland change, as well as the changes in bioenergy potential from crop residues in YRD, was estimated for the period 2001–2010 by using a satellite-based model. We found that approximately 12% of the total cropland was encroached between 2001 and 2010. In particular, urban expansion accounted for most of the cropland loss (65%) in the past decade. Consequently, this loss reduced the bioenergy of crop residues, at a rate of 147 kJ m$^{-2}$ year$^{-1}$. Considering the expected dramatic increase in urban land use (Seto et al., 2012), additional research and policies should be carefully considered for cropland protection and sustainable urban planning in future decades.

## ACKNOWLEDGMENTS

This study was supported by the National Natural Science Foundation of China (Grant No. 41401438).

## REFERENCES

Chen, J. and J. Chen, et al. (2014). "Global land cover mapping at 30 m resolution: A POK-based operational approach." ISPRS Journal of Photogrammetry and Remote Sensing.

Collins, M. and R. Knutti, et al. (2013). "Long-term climate change: projections, commitments and irreversibility.".

Dai, L. and S. L. Zhou, et al. (2012). "Temporal-spatial variation of terrestrial net primary productivity in the Yangtze river delta over the past decade." Resources and Environment in the Yangtze Basin 21 (10): 1216–1222.

Dlugokencky, E. and P. Tans NOAA/ESRL (www.esrl. noaa.gov/gmd/ccgg/trends/, Accessed 7–4–2016).

Editorial Board Of Vegetation Map Of China (2001). Vegetation Atlas of China (1:1000000). Beijing, Science Press.

Field, C. B. and J. T. Randerson, et al. (1995). "Global net primary production: Combining ecology and remote sensing." Remote Sensing of Environment 51 (1): 74–88.

Freddy, N. and V. V. Harrij, et al. (2008). Harmonized World Soil Database, Food and Agriculture Organization of the United Nations.

Gao, T. and L. Xie (2014). "Multivariate regression analysis and statistical modeling for summer extreme precipitation over the Yangtze River basin, China." Advances in Meteorology 2014.

Gregg, J. S. and R. J. Andres, et al. (2008). "China: Emissions pattern of the world leader in CO2 emissions from fossil fuel consumption and cement production." Geophysical Research Letters 35 (8).

Jiang, D. and D. Zhuang, et al. (2012). "Bioenergy potential from crop residues in China: Availability and distribution." Renewable and Sustainable Energy Reviews 16 (3): 1377–1382.

Jin, Z. and Y. C. Qi, et al. (2007). "Storage of biomass and net primary productivity in desert shrubland of Artemisia ordosica on Ordos Plateau of Inner Mongolia,China." Journal of Forestry Research 18 (04): 298–300.

Junfeng, L. and H. Runqing, et al. (2005). "Assessment of sustainable energy potential of non-plantation biomass resources in China." Biomass and Bioenergy 29 (3): 167–177.

Liu, J. and M. Liu, et al. (2005). "Spatial and temporal patterns of China's cropland during 1990–2000: An analysis based on Landsat TM data." Remote Sensing of Environment 98 (4): 442–456.

Lobell, D. B. and J. A. Hicke, et al. (2002). "Satellite estimates of productivity and light use efficiency in United States agriculture, 1982–98." Global Change Biology 8 (8): 722–735.

Luo, T. X. (1996). Patterns of net primary productivity for Chinese major forest types and its mathematical models. Beijing, Chinese Academy of Sciences. Ph.D.

Metz, B. and O. Davidson, et al. (2007). Mitigation of climate change, Cambridge University Press Cambridge.

Ni, J. (2004). "Estimating net primary productivity of grasslands from field biomass measurements in temperate northern China." Plant Ecology 174 (2): 217–234.

Pei, F. and X. Li, et al. (2013). "Assessing the differences in net primary productivity between pre- and post-urban land development in China." Agricultural and Forest Meteorology 171–172(0): 174–186.

Pei, F. and X. Li, et al. (2015). "Exploring the response of net primary productivity variations to urban expansion and climate change: A scenario analysis for Guangdong Province in China." Journal of environmental management 150: 92–102.

Piao, S. and J. Fang, et al. (2001). "Application of CASA model to the estimation of Chinese terrestrial net primary productivity." Acta Phytoecologica Sinica 25 (5): 603–608.

Potter, C. S. and J. T. Randerson, et al. (1993). "Terrestrial ecosystem production: a process model based on global satellite and surface data." Global Biogeochemical Cycles 7 (4): 811–841.

Raynaud, D. and J. Barnola, et al. (2005). "Palaeoclimatology: The record for marine isotopic stage 11." Nature 436 (7047): 39–40.

Rose, S. K. and H. Ahammad, et al. (2012). "Land-based mitigation in climate stabilization." Energy Economics 34 (1): 365–380.

Running, S. W. and S. T. Gower (1991). "FOREST-BGC, a general model of forest ecosystem processes for regional applications. II. Dynamic carbon allocation and nitrogen budgets." Tree Physiology 9 (1–2): 147.

Sato, H. and A. Itoh, et al. (2007). "SEIB-DGVM: A new dynamic global vegetation model using a spatially explicit individual-based approach." Ecological Modelling 200 (3–4): 279–307.

Seto, K. C. and B. Güneralp, et al. (2012). "Global forecasts of urban expansion to 2030 and direct impacts on biodiversity and carbon pools." Proceedings of the National Academy of Sciences 109 (40): 16083–16088.

Sitch, S. and B. Smith, et al. (2003). "Evaluation of ecosystem dynamics, plant geography and terrestrial carbon cycling in the LPJ dynamic global vegetation model." Global Change Biology 9 (2): 161–185.

Taylor, P. (2010). "Energy Technology Perspectives 2010." Scenarios & strategies to 2050.

Wang, W. and Y. Liu, et al. (2013). "The spatial distribution of cereal bioenergy potential in China." GCB Bioenergy 5 (5): 525–535.

Zhu, W. and Y. Pan, et al. (2006). "Simulation of maximum light use efficiency for some typical vegetation types in China." Chinese Science Bulletin 51 (4): 457–463.

Zumkehr, A. and J. E. Campbell (2013). "Historical US cropland areas and the potential for bioenergy production on abandoned croplands." Environmental science & technology 47 (8): 3840–3847.

*Advances in Energy and Environment Research – Achour & Wu (Eds)*
© *2017 Taylor & Francis Group, London, ISBN 978-1-138-62682-9*

# The sixth technological revolution in construction industry: Noospheric paths

M.V. Shubenkov
*Development of Education of Urban Planning and Urbanism, Department of Urban Development, Moscow Institute of Architecture, State Academy, Moscow, Russian Federation*

S.D. Mityagin
*JSC, NIIP Gradostroitelstva» Research and Design Institute of Urban Development, St. Petersburg, Russian Federation*

Z.A. Gaevskaya
*State Polytechnical University, St. Petersburg, Russian Federation*

ABSTRACT: Human civilization is on the threshold of the sixth technological stage. In the heart of it there will be an energetic structure shift conditioned by transition to the society and economy not capitalizing on carbon technologies. Construction industry develops in close relation with general trends of economy development. Energy efficiency of using natural resources in the sphere of industry, transportation and constructional production remarkably influences energy balance, climate, and environmental change. In biospheric context, transformation of nature function of human society is regulated by professional apparatus of city planning activity in a system of management as it has been developed historically. Thus, nowadays, city planning may be considered as one of the major integrated means of management of social evolution of biosphere. City planning as a means of social evolution and transformation of biosphere through construction may become a basis for innovative development of the whole construction industry.

## 1 INTRODUCTION

The prevailing life styles that modern humankind is leading based on insufficient knowledge of laws running the Earth have resulted in global social, economic and ecological catastrophes.

At the end of the 20th Century, the issue of seeking alternative models of development of means of production, management, and consumption securing processes of societal life was put on the agenda. In 1992, the UN Conference on Environment and Development (held in Rio de Janeiro) declared a thesis about necessity to provide a balanced solution to socioeconomic problems and problems of preserving environmental public goods and natural-resources potential in order to meet the needs of present and future generations. In the conference final document, this statement came to be known as «sustainable development» (AGENDA 21).

Today, green building means, to a considerable extent, «sustainable development» (self-maintenance, balance, self-sufficiency). Construction sector is one of the largest sources of environmental pollutions due to the increase in the amount of natural resources used, degradation of vulnerable ecological zones, etc. Necessity to adopt the mandatory requirement to enable the construction sector to meet human settlement development goals, while avoiding harmful side-effects on human health and on the biosphere is obvious. However, despite being clear, this condition of future development of society is not fulfilled. This requirement was first declared in «Agenda XXI» UN report in 1992. Ch. 7 noted the necessity « to adopt standards and other regulatory measures which promote the increased use of energy-efficient designs and technologies and sustainable utilization of natural resources in an economically and environmentally appropriate way» (AGENDA 21, ch.7, article 7.69).

The combination of energy efficiency, ecology, and economy is possible only if solutions in all industrial sectors will be based on scientific knowledge. Cosmic scientific vision of the Earth which became practical in the second part of the 20th Century. brought a system of concepts about the unity of organic and inorganic worlds, their interdependence and social control of their evolution to modern commonplace sense. The information about spatial, material, and energetic limits of social development and the role of human in global changes of biosphere has become widespread. The growth of green movements in the world can be considered as the demonstration of social attitude towards seeking forms, methods, and techniques of controlling the quality of the environment in order to create and maintain the conditions for progressive co- evolution of two interacting parts of the biosphere: nature and society (Mityagin 2011).

It is possible to achieve goals of this kind using a professional apparatus of city planning which aims to find a solution to the problems of formation of spaces of life activities with pre-planned properties. City planning should not be understood pragmatically as support building complex but should have a biospheric component. Noospheric territory planning will allow infrastructure capital construction objects to be effectively incorporated into ecologically balanced framework of engineering and transport infrastructure, real estate development, and natural complexes.

Technological revolution in resource saving and alternative energy requires the construction sector to develop an entirely new methodology of design and management aiming at constructive responses to the 21st Century challenges.

Y. P. Mukhin, T.S. Kuzmina, and V.A. Baranov distinguish two main directions of sustainable development (Mukhin 2002):

- The first direction is based on solving ecological problems (improving the environmental quality) by technical means, e.g., applying modern innovative treatment technologies, wasteless technologies, etc.
- The second direction focuses on conservation of majority of natural or near-natural ecosystems in order to prevent environmental disaster. Nowadays, when the 1% consumption threshold has been exceeded (by man), Le Châtelier's principle, according to which, whenever a system in equilibrium is disturbed, the system will adjust itself in such a way that the effect of the change will be nullified, is not true for ecosystems.

The history of humankind development at the end of the 21st Century showed that the first direction is not efficient enough to solve all environmental problems. The concept of sustainable development of the second direction continues to develop the theory of noosphere genesis first formulated in V.I. Vernadsky's works. Philosophical and scientific foundation of noosphere genesis was laid in works by Russian cosmist scientists of the 20th Century: V.I. Vernadsky (Vernadsky 2004), N.N. Moiseev (Moiseev 1987, 1985), and N.F. Glasovsky (Glazovsky 2005). The theory of noosphere genesis related to the development of

Vernadsky's doctrine is in progress at the moment (Doctor of Architecture Mityagin S. D., adviser of Russian Academy of Architecture and Construction Sciences (RAACS) (Mityagin 2011, 1989,1986, 2016), Doctor of Philosophy Ursul A.D. (Ursul 2005), Doctor of philosophy and Economics Subetto A.I. (Subetto 2009), Doctors of Geography Shalnev V. A. (Shalnev 2006) and Sdasyuk G.V. (Sdasyuk 2005), academician of RAMS Kaznacheev V.P. (Kaznacheev 1989), PhD in Architecture Gaevskaya, Z.A. (Gaevskaya 2012, 2014), and foreign scientists Norgaard, R. B (Norgaard 1994) and Hödl, Elisabeth (Hödl 2012).

Thus, scientific foundations of noospheric development of the construction industry are at the early stage of their formation and need further research and extension.

According to these, proposals on defining noospheric directions of construction and city planning sectors development based on the foreseen challenges of the VI technological revolution are very relevant at present.

## 2 METHODS

The UNEP report prepared for Rio+20 «Towards a Green Economy: Pathways to Sustainable Development and Poverty Eradication» continues to develop the concept of sustainable development adopted at the 1992 UN conference in Rio de Janeiro in modern conditions. The report argues that «A green economy substitutes clean energy and low carbon technologies for fossil fuels, addressing climate change, … creating decent jobs and reducing import dependencies. New technologies promoting energy and resource efficiency provide growth opportunity in new directions, compensating for «brown economy» job losses (UNEP, 2011).

Transition of the world civilization to the sustainable development based on «green economy» requires modernizing all spheres of production, consumption, and environmental management. Space-based observations of dynamics of global processes show continuity of systematic unity of all natural landscape components as well as human economic and industrial activity and the environment. Life is not a purely chance phenomenon on this one little planet, the Earth. According to Vernandsky (Vernadsky 2004), the face of the Earth reveals the surface of our planet, its biosphere, and outer area dividing it from the cosmic force. Academician Ursul A.D. suggested the following definition of sustainable development: « It may be defined as a forthcoming form of co-evolution of society and nature providing their mutually safe coexistence and noosphere formation» (Ursul 2005).

In cognition and control of biosphere, scientific knowledge functions as the main tool. Academician Glasovsky N.F. stressed that... «noosphere is a scientifically conscious space-time domain of the world» (Kotlyakov 2006).

Sdasyuk G.V. thinks that the fundamental difference of the concept of sustainable development from conventional views and management practices is in its holistic (integral) approach to development as a holistic process (Sdasyuk 2005).

Noospheric thinking should be built on Vernadsky's prediction: nature is not amorphous and shapeless as people thought for centuries but has very fine organization which as such should be reflected and taken into account in all conclusions and judgments related to nature (Vernadsky 2004). This is the most important condition of sustainability of nature.

This idea is formulated in Article 39 of the Resolution adopted by the UN General Assembly July, 27, 2012 as follows: «We recognize that planet Earth and its ecosystems are our home and that the «Mother Earth» is a common expression in a number of countries and regions, and we note that some countries recognize the rights of nature in the context of the promotion of sustainable development. We are convinced that in order to achieve a just balance among the economic, social and environmental needs of present and future generations, it is necessary to promote harmony with nature» (UN General Assembly 2012).

Humankind is bound to degrade if it fails to organize rationally its life environment, protect environment, and form a convenient infrastructure. For many centuries, activities on planning, development, and management of settlements were realized without taking into account scientific knowledge and were connected to the formation of their own theoretical knowledge based on practice, development of their own language of setting, and solving problems. There is a mediation link between scientific knowledge and practical activity in city planning: planning, development, and technical implementation tools. The methodology of planning, development, and management is known to be principally different from research methodology. Fundamental and applied R&D requires expanding the range of problems addressed, changing a way of professional thinking, and elaborating a principally new approach to the phenomena under study since a quantitative accumulation of knowledge and conventional approaches of problem solving are not sufficient today (Shubenkov 2015)

The problem of efficiency is especially urgent in planning and development activity. Human practice is systemic. Systematicity of human action consists of following a certain plan or algorithm.

It is necessary to compare consequences of all possible steps by modeling them rather than taking them in reality. The sixth scientific and technical revolution raises a question about the systematicity of the world, which, in turn, results in necessity to understand development as a creative process aiming at perspective, connected with decision making, developing new technical systems and technologies, new optimal in perspective models of systems (Peregudov 1987).

Algorithmization of action is an important means of its development. The algorithm is a mode of future action, its model. Modeling is the major way of building theories and goal-oriented representation of the original. Therefore, it is vital that the organization of the whole system was subordinated to the specific goal because it is an image of the desired future. The sixth technological revolution determines the following goal for construction industry. It must become a competitive industry forming safe environment for human life and action which meets high standards of quality and effective management in order to provide sustainable development of the environment based on biospheric laws.

Thus, algorithmization of a city planning activity is necessary since development depends on the ability to use resources, and not just their availability. In our view, it should consist of consequent elaboration of the above-mentioned goal at different levels of hierarchy.

1. Development of a city planning theory as Architecture of the Earth. Modern city planning must use fundamental notions of science, regularities of biosphere functioning, and development (Mityagin 2011). Spatial arrangement of the territory is characterized by natural resources location (woods, water bodies, minerals), organization of residents' settlement, the arrangement of working places, transport and engineering public facilities, and cultural and natural heritage sites. This definition allows us to consider it as a complex system, consisting of anthropogenic infrastructure and the environment. Building, industrial, and communication infrastructures cause a considerable impact on the climatic system of the Earth. According to prof. Gorshkov S.P., making the Earth abiotic environment, turning the Green Earth into Gray one due to anthropogenic impact for a short historic period changes the structure of the hydrologic system resulting in a decrease in evapotranspiration, which in turn leads to an increase in heating ground air. On the Gray Earth (which is featured by deforestation, desertification, substitution of overwetting, and aquatic habitats for fields, plantations,

and man-made grazing land; fast expansion of anthropogenic infrastructure, especially through urbanization), natural disasters are becoming more intensive (Kotlyakov 2006). Thus, for the mankind to survive, it is necessary to adapt the anthropogenic environment to local ecosystems, not vice versa.

2. Development of multidimensional systems of project management integrated with systems of modeling objects and their life cycle management based on their integration into the natural environment. Copenhagen Declaration (of International Union of Architects), adopted at 2011 General assemble of IUA, considers Sustainable by Design concept which, in particular, recognizes that «all architecture and planning projects are part of a complex interactive system, linked to their wider natural surroundings...» (Copenhagen Declaration, 2009). Architectural design lags behind engineering design, which has already experienced remarkable changes. System engineering of construction, machine-building, instrument-making, etc., are new fields of corresponding professional activity. Architectural design will have to adopt new design methodologies in the nearest future. New principles and techniques will be worked out. They will be built on the results of modern applied development and research on scientific basis of design, building, and functioning of integrated interactive complexes of analysis and synthesis of architectural and spatial systems and related design and technological documentation. Improvement of design processes based on the wide use of computer aids, information technologies, and computer networks will become the earnest of solving such problems. Principally, new methods and means of «designer-computer» interaction based on new interfaces will appear (Shubenkov 2006).

3. Considering a single building and complexes of building objects as a complex unified power aerohydrodynamic system. Such understanding lays scientific foundations of designing really «green» objects. Concept of «green architecture» is to include concepts such as power efficient, cost-effective, ecological, and ergonomic architecture.

4. Development or renovation of buildings should imply solve a threefold problem taking into account the following aspects: 1) creation of healthy indoor climate, comfort, and ecology; 2) fossil fuel and primary energy saving, using alternative energy sources; and 3) construction cost-effectiveness, reducing the consumption of material and technical resources (Belyaev 1991). Simultaneously, taking energy intensity, durability, level of thermal protection, and

construction site location into account is possible only if we consider an energy-efficient building as a complex system which is an element of a more complex power-efficient architectural and landscape system (Gaevskaya 2014). Such an innovative approach will require updating standard technical documents regulating construction industry as the concept of sustainable development implies a new natural evolutionary stage of society development.

New methods of waste water treatment, human waste recycling, development of technologies of cheap autonomous shelter construction, energy generation from renewable sources, and clean industrial manufacturing are among the most promising directions in construction industry.

Mutual influence of natural and anthropogenic structures and processes, lack of solid knowledge about the variety of manifestations, and degree of disrupting impact of economic activity make development of management system of social evolution of the biosphere and its consequent formation on the global, continental, national, regional, and local scale considerably more difficult.

## 3 RESULTS AND DISCUSSION

The goal of modern city planning activity is to harmonize relationship and mutual influence of society and its life environment. In our opinion, this is the most reasonable way of investigating city planning sites.

Life environment artifacts are imprinted in the territory's sketch plan. We would suggest the following principles of biospheric optimization of construction industry:

1. More comprehensive and insightful study and consideration of horizontal relations among morphological components of the landscape. Thus, mutual arrangement industrial premises, residential quarters, green zones, and water bodies should be in agreement with a wind rose, geomagnetic structure of the territory, effects of groundwater flow and overland runoff, and the history of its development.

2. Creation of the «Green Earth» in order to provide continuity of evolution of natural processes. To prevent secondary gravigenic processes and loss of soil particles, it is important to provide required forest square not only along waterways and ravins but also and especially on watershed divides and slopes regardless of the value of these lands for other types of land use (Muhin, 2002).

3. Finding agreement between economic benefit and ecological balance. Capital construction

objects should not create anthropogenic load destroying nature but ought to support ecological balance resulting in maximal ecological-social--economic synergistic effect.

4. Buildings and their complexes are included into an interactive system linked to their environment based on different types of energy-efficient organization of their properties (e.g. architectural, technological, landscape, and typological) (Gaevskaya, 2014).

5. Search for modern energy-efficient building materials and technologies of building erection and renovation should be in agreement with the concept of energy efficiency of landscape and typological object organization. Careful and considered design of buildings' shape and geometry, choice of spatial strategy together with relevant materials, equipment, and functional space arrangement can reduce resource consumption, green gases emission, and total negative impact on the environment by 50–80%» (Remizov, 2016).

## 4 CONCLUSIONS

New directions of noospheric development of construction industry taking into account forthcoming transition to the sixth technological revolution epoch can be suggested today. They include the following directions:

- Development of «Architecture of Earth» theory. Its methodological basis consists of integration of capital construction objects into biosphere within its industrial capacity, with horizontal relations among morphological components of natural and anthropogenic landscapes being taken into account as fully as possible.
- Development of new generation program interfaces to determine properties and quality of a socio-functional and natural system in order to run multidimensional systems of design, modeling, and managing life cycle of construction objects.
- Development of the scientific basis of «green building» and «green architecture» based on the theory of noosphere genesis.

All above-mentioned directions raise questions about working out a principally new methodology of design thinking and management in construction industry aiming at a constructive response to the 21st Century challenges.

## REFERENCES

Agenda 21 (1992). Adopted by United Nations Conference on Environment & Development Rio de Janerio, Brazil. Retrieved from:/Available from: http://daccess-dds-ny.un.org/doc/UNDOC/GEN/N92/836/55/PDF/N9283655.pdf?OpenElement.

Belyaev, V.S., Khokhlova, L.P. (1991). Design of power-efficient and energy active civilbuildings. Moscow: Vysshaja shkola. 255, 244.

Copenhagen Declaration (Dec.7, 2009). International Union of Architects. Available from http://www.uia.archi/sites/default/files/COP15_Declaration_EN.pdf.

Gaevskaya, Z.A. (2012) Town-Planning Sustainable Rural Area Development as a New Theoretical Direction // Academia. Arhitektura i stroitel'stvo. No 2, pp. 106–110.

Gaevskaya, Z.A., Mityagin S.D. (2014) Capital construction and noosphere genesis// Applied Mechanics and Materials. No 587–589. pp. 123–127.

Gaevskaya, Z.A., Rakova X.M. (2014) Modern bulding materials and the concept of «sustainability project» // Advanced Materials Research. No 941–944. pp. 825–830.

Glazovsky N.F., Tishkov, A.A. (2005). Lider of Russian Geography (instead of preface)]. In: Multifaceted Geography, pp. 5–14. Moscow. T KMK. pp. 5–14.

Hödl, E. (2012). Die Noosphäre als Bezugsrahmen für das Recht [The noosphere as a framework for the conception of law] in: E. Schweighofer, F. Kummer/ Transformation juristischer Sprachen, Tagungsband des 15. Internationalen Rechtsinformatik Symposions, pp. 639–648.

Kaznacheev, V.P. (1989). Vernadsky's doctrine about biosphere and noosphere. Novosibirsk: Nauka. 245.

Kotlyakov, V.M., Alexeenko N.A., Tishkova A.A., Sdasyuk G.V. (2006). Environmental management and sustainable development. World eco-systems and problems of Russia]. Moscow: KMK. 448, 24–413.

Mityagin, S.D. (1986). Biosphere development and city planning. 1986. Iss.6.

Mityagin, S.D. (1989). Architecture and city planning in noosphere genesis . Leningrad. 1989, p. 24.

Mityagin, S.D. (2011) Economics of the Biosphere and City Planning // Biosphere. V. 3(2). pp. 264–276.

Mityagin, S.D. (2011) Earth Architecture as constructive landscape design, the basis of rational and ecologically balanced environmental management and promising methodology of spatial planning. Moscow: MGU Publ. 54–58.

Mityagin, S.D. (2016). Spatial planning as a tool of sustainable development. Available from State research and design center of Saint Petersburg Master Plan Web site. Available from: http://www.gugenplan.spb.ru/index.php?id=133&l=RU.

Moiseev, N.N. (1987) Algorithms of Development. Moscow. Nauka. 304.

Moiseev, N.N., Aleksandrov, V.V., Tarko, A.M. (1985). Man and Biosphere. The Case of System Analysis and Experiments with Models: Moscow. Nauka. 1985. 304.

Muhin, Ju.P., Kuz'mina, T.S., & Baranov V. A. (2002) Sustainable development: ecological optimization of agro-and urban landscapes: Volgograd: Vol-GU., 124, 9–26.

Norgaard, R. B. (1994) Development betrayed: the end of progress and a coevolutionary revisioning of the future. London-NewYork, Routledge.

Peregudov, F.I., Tarasenko, F.P (1987). Introduction to system analysis: Moscow: Vysshaja shkola. 360.

Remizov, A.N. (2016) Strategy of development of sustainable architecture in Russia. Available from: Arcitekturnye sezony web site http://old.kpfu.ru/f2/bin_files/trofimovrubcovermolaev_regionanaliz!133.pdf

Sdasyuk, G.V. (2005) Objective necessity of transition to SARD: role of information, knowledge and management. Moscow: KMK., 615, 88.

Shalnev, V.A. (2006). Problems of interrelations of society and nature: geographer's view. Stavropol: SGU. 2006. 110.

Shubenkov M.V. (2006). Issues of architectural activities in the context of computer technologies development // Arhitekton: Available from: http://archvuz.ru/2006_3/14

Shubenkov, M.V. (2015) Single issues of Russian city planning theory development // Arhitektura i sovremennye informacionnye tehnologii (AMIT). MArchI. Available from: http://www.marhi.ru/AMIT/issues.php.

Subetto, A.I. (2009). Noospherism. St. Petersburg. Vol. 8. 709.

UN General Assembly (2012). The future we want // Resolution adopted by the UN General Assembly. Available from: https://worldwewant.de/worldwewant/de/home/file/fileId/54

UNEP (2011). Towards a Green Economy: Pathways to Sustainable Development and Poverty Eradication— A Synthesis for Policy Makers. UNEP, 2011. Available from: www.unep.org/greeneconomy

Ursul, A.D. (2005) Sustainable development: Conceptual model//Nacional'nye Interesy. No. 1 Archive. Available from:http://ni-journal.ru/archive/2005/n1_05/5324690e/d93f12df/index.htm

Vernadsky, V.I. (2004) Biosphere and Noosphere. Moscow. Ajris-Press. 576, 35, 385–386.

*Advances in Energy and Environment Research – Achour & Wu (Eds)*
© 2017 Taylor & Francis Group, London, ISBN 978-1-138-62682-9

# Database design of the Qingyi River basin for multi-objective and multi-department integrated management

Zhihui Tian & Jingjing Peng
*College of Water Conservancy and Environmental Engineering, Zhengzhou University, Zhengzhou, China*

Xiaolei Wang
*Smart City Institute, Zhengzhou University, Zhengzhou, China*

ABSTRACT: As affected by the conditions of population, economy, and water resources, the water environment is gradually being destroyed. At the same time, the overlapping functions, poor cooperation, and the different system construction standards increase some difficulties for water environmental management. Therefore, it is urgent to access the comprehensive data from the management department and build a database. In this paper, the database of the Qingyi River basin in Xuchang is designed, which plays an important role in water environment management. In order to achieve the database construction of the Qingyi river basin, the paper, based on ArcSDE + Oracle, makes a deep research on the conceptual, logic, and storage design for database, and then proposes a data integration processing method to solve all various types of data sharing problem. Finally, the database can achieve the functions of data query and statistical analysis. This research shows that it provides a database standard and data support for the integrated management platform, and realizes the multi-department integration application.

## 1 INTRODUCTION

With the rapid development of social economy and the impact of human production, there are severe problems of water environment. At present, the pollution situation of the river basin is not optimistic; a sharp decline of water environmental quality has become the key factor that influences the sustainability of the water resources and the social and economic development. Each region has paid more attention on the issue of water environment management and constructed the corresponding pollution source monitoring system, such as water quality evaluation system and flood disaster management system. However, owing to the different construction standards, the sharing of data is difficult and cannot provide effective information for water environment management to decide. Therefore, it is an urgent task for the environmental protection and other departments that how to realize the effective management, storing, and sharing of the mass data.

In recent years, the GIS, database technology, integration technology, and remote sensing technology have rapid development (VA Tsihrintzis 1997), and the scholars tend to study the water environmental management by using the relational model (Tan & Chen 2002, Krause et al. 2005, Sreedevi et al. 2013). There are some researches on the information management for the watersheds. For example, based on Oracle+ArcSDE, Yang (2013) establishes the database of Dianchi north shore and realizes the efficient storage and management of spatial data and attribute data, but business attribute data lacks consideration. Li (2012) makes a detailed design in the database structure of the Dianchi River basin, but the data processing is slightly deficient. By the research results of relevant academic studies and characteristics of Qingyi River, this paper proposes the exploring process of the attribute data and the integration processing to realize the data storage and sharing.

In this paper, we use the ArcGIS, database technology, and integration technology to establish the database for integrating the different sources of data. The database meets the needs of the water departments for the daily work, and solves the poor coordination. The database will improve the utilization efficiency of data resource, and realize the management of vast amounts of geographical information and attribute information.

## 2 RESEARCH AREA

Qingyi River originates from the ditch grass garden of Xinzheng, and then flows into Xuchang from Guanting; the river covers an area of

2362 Km$^2$ and there is an area of 1585 Km$^2$ in the Xuchang city, which occupies about 67%. It belongs to the warm temperate humid monsoon climate, and has a clearly four seasons, rich heat energy, concentrated rainfall, abundant sunshine, etc. Qing Yi River is an important water body for flood control, pollution prevention, and landscape water of Xuchang, and therefore, the water quality is of great significance for the people's living and the urban healthy development. However, the river is experiencing severe water problems. According to the investigation, the main pollution source of Qingyi Basin is due to the pollutant emissions from the leather, hair products, and paper industry. Currently, the water environment management of Qingyi is not a single department, but involves multi-sector coordination and cooperation, including environmental protection, water conservancy, agriculture, and hydrology.

## 3 DATABASE DESIGN FRAMEWORK

### 3.1 *Purpose*

The purpose of database is to solve the using and sharing problem of the multiple heterogeneous data in the information construction process. It could provide the basic data for the integrated management platform application.

### 3.2 *Database design framework*

The general design of database framework includes the following aspects: requirement demand analysis, concept design, logic design, and physical store design. In this paper, we study the method of data organization and management based on geodatabase (Song & Wan 2004) and the relational data model. The ArcSDE (Spatial Database Engine) (Robert 2004) is considered as the passage that will connect GIS to Oracle. It also allows the users to edit, view, and modify other commands to manage geographic information in the different management system. The database supports all the applications of ArcGIS (SH Mahmoud et al. 2016) to access the data resources by the server configuration link.

#### 3.2.1 *Data source*
The requirement of data analysis stage is treated as the beginning of the database design. The purpose of analysis includes: 1) the user needs 2) the business relationships in different sectors, and 3) the data sorting out. In this study, the data is obtained mainly from some departments, such as environmental protection, water department, meteorological department, hydrology and water resources department, and land and resources department.

All available data are assembled into a complex database, which stores spatial and attribute information. The composition of the Qingyi River basin database includes geographic data, socioeconomic data, water quality monitoring data, pollution source data, allocation of water resource, water ecological data, sewage outfall data, hydrology, and meteorological data.

#### 3.2.2 *Conceptual design of database*
The task of conceptual design is to express the relationship, attribute, and key word among the entities in the real world. The entities are described by the E-R model with the specialization and abstraction way. A description of the entities involved in the database is provided in the Figure 1.

#### 3.2.3 *Logical design of the spatial database*
Graphic entity data are abstracted as point, line, and polygon data and assembled into a spatial database. It is expressed as different types of data organizational structure by the geometric network, feature class, relationship, feature dataset, and raster dataset. Main contents are presented as follows: (1) selected layers; (2) how to express the layer information; and (3) how to organize the layers as well as control the spatial scale and precision. Spatial database is positioning reference base for all kinds of applications; meanwhile, it supports and controls the expression of various thematic map. In this paper, according to the national standard and the application requirements of the management platform, the spatial information is divided into some layers. Finally, the spatial data is stored in the Oracle database.

#### 3.2.4 *Logical design of attribute database*
The attribute database consists of the water quality monitoring information, reservoirs, dam information, hydrological information, water information, water ecological information, and socio-economic information. Main design contents are as follows: (1) classification of the attribute data; (2) how to use the E-R model to express; and (3) designing of the attribute table that contains the table name, attribute fields, data type, precision, length, and primary keys; the primary key, as the unique identifier of the table, cannot be empty and is the basis of related spatial layers and attribute tables. The structure of each table must correspond to the needs of paradigm to reduce the redundant data and improve the computer processing, data query, calculation, and update rate; and (4) understanding the element coding. The elements coding refers to the relevant national, provincial, and industry standards, which can reduce the computing time and processing time of the system.

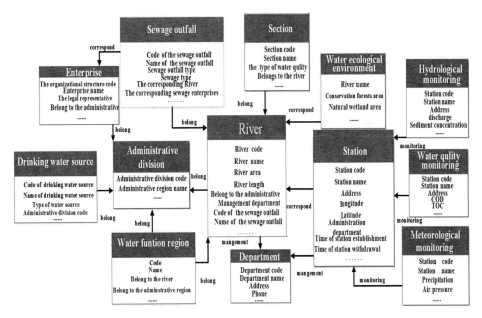

Figure 1. The E-R model of the database. The E-R method is the abbreviation of "Entity-Relationship Approach".

### 3.2.5 *Database application*

The Database application is mainly reflected in two aspects: 1) the heterogeneous multivariate data from the various sectors integrated and 2) the decision-making platform of the Qingyi River basin could use the database to achieve the data query, statistical analysis, and other functions.

## 4 REALIZATION AND APPLICATION OF THE DATABASE

### 4.1 *Data integration applications*

Data integration is a technology that integrates all kinds of data with different sources, formats, and standards by the processes of data preprocessing, matching, conversion, and loading, which can achieve the data sharing and break the plight of the "information island". According to a set of business rules, the converted data were installed in the database. The contents of spatial data integration (Assaf & Saadeh 2008; J Noh 2015) include data form, data format, the basis of mathematics, layered feature, the expression of attribute values, and topology. The contents of the attribute data integration include data format and table structure. Figure 2 displays the data integration process.

1. The applications of multi-department data integration

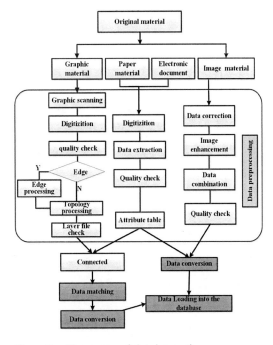

Figure 2. The process of data integration.

Owing to the different construction standards and data formats, which make data sharing difficult and cannot provide effective information for decision of water environment manage-

137

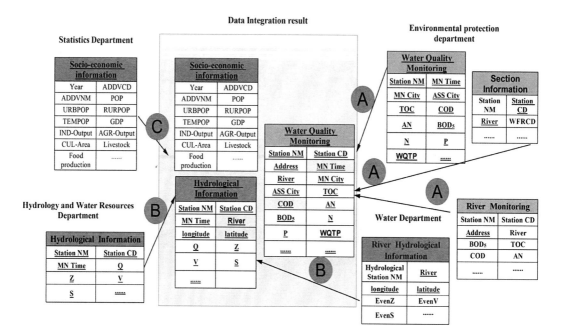

Figure 3.  The multi-department data integration.

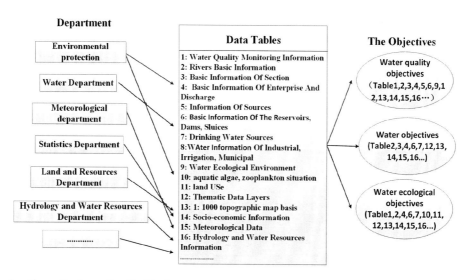

Figure 4.  The applications of multi-objective.

ment, and therefore, data integration technology is applied to solve this problem. The three forms of data integration (Figure 3) are listed below:

A. Integration of the environmental protection department and others. An integration principle is that the environmental protec-

tion is the standard, while other data are a reference.

B. Integration of the other departments except the environmental protection. An integration principle is that the charge of the water sectors is the standard, while other data look as a reference.

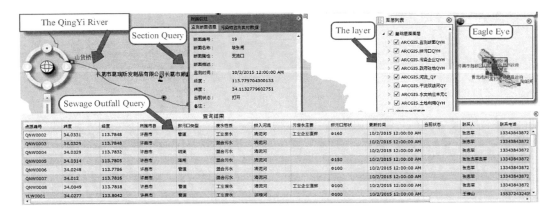

Figure 5.    The query result of sewage outfall information.

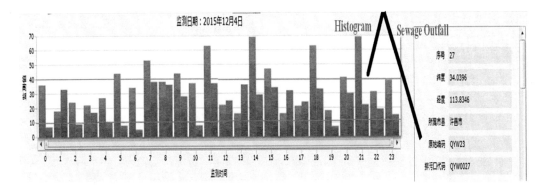

Figure 6.    The pollutant emission statistics of water section.

C. Integration of the department with only information source. The principle is based on the department's data as the standard.

2. The applications of multi-objective management
The water environment management involves many departments, which made the water environment management sank into a complex dilemma. By integrating the sectors business information, we can achieve the multi-objectives of water quality, water quantity, and water ecology management (Figure 4).

### 4.2    Realization of database

Based on the management decision-making platform, we achieve data query, statistical analysis, and layer management functions. The query result of sewage outfall information is shown in Figure 5. The pollutant emission statistics of water section is shown in Figure 6.

## 5    CONCLUSION

This paper, by applying ArcGIS, database, and data integration technology, has established the multi-objective and multi-department of the integrated management database. By integrating spatial data and non-spatial data, we realized the data storage and management of multiple heterogeneous. This paper not only improves the utilization efficiency of data resource, but also realizes the vast amounts of geographical information and attribute information integration management, and solving the disadvantages of manual management.

## ACKNOWLEDGMENTS

This work was supported by grants from the China National Critical Project for Science and Technology on Water Pollution Prevention and Control (2015ZX07210-002-005).

# REFERENCES

Assaf, H. & Saadeh, M. (2008). Assessing water quality management options in the Upper Litani Basin, Lebanon, using an integrated GIS-based decision support system. J. Environmental Modelling & Software. 23 (10–11), 1327–1337.

DeBao Tan, & XueJun Chen (2002). The Construction and application of integrated database for flood control and disaster reduction based on ArcSDE+Oracle9i. J. Journal of wuhan university (information science edition). 01: 90–93.

JingShan Li, & Kun Shi (2012). Design of the pollution source database in Dianchi Basin based on ARC-SDE. J. Science Technology and Engineering. 08: 1977–1980.

Krause P., Kralisch S., & Flugel WA (2005). Model integration and development of modular modelling systems. J. Adv Geosci. 4: 1–2.

Mahmoud, S.H. & A.A. Alazba (2016). Integrated remote sensing and GIS-based approach for deciphering groundwater potential zones in the central region of Saudi Arabia. J. Environmental Earth Sciences. 75(4), 1–28.

Min Yang, XiaoMei Xu, & HongBin Zhou (2013). Design and application of Dianchi north shore database based on Oracle and ArcSDE. J. Environmental Science and Technology. 36 (7): 171–168.

Noh, J., H. Choi, & S. Lee (2015). Water quality projection in the Geum River basin in Korea to support integrated basin-wide water resources management. J. Environmental Earth Sciences. 73(4), 1745–1756.

Robert West (2004). Understanding ArcSDE. M. USA: Esri Press.

Sreedevi, P.D., Sreekanth PD, Khan HH, & Ahmed S (2013). Drainage morphometry and its influence on hydrology in an semi arid region: using SRTM data and GIS. J. Environ Earth Sci. 70: 839–848.

Tsihrintzis, V.A., H.R. Fuentes, & K.G. Rao (1997). GIS-Aided Modeling of Nonpoint Source Pollution Impacts on Surface and Ground Waters. J. Water Resources Management. 11(3), 207–218.

Yang Song, & YouChuan Wan (2004). A new spatial data model Geodatabase. J. Bulletin of surveying and mapping. 11: 31–33.

*Advances in Energy and Environment Research – Achour & Wu (Eds)*
*© 2017 Taylor & Francis Group, London, ISBN 978-1-138-62682-9*

# The design and implementation of the Qingyi River water environment basic information system

Hengliang Guo, Yuanyuan Wei & Xiaohui He
*College of Water Conservancy and Environmental Engineering, Zhengzhou University, Zhengzhou, China*

ABSTRACT: According to the demand of multi-objective and multi-department integrated management platform construction of the Qingyi River basin, the database and system of the Qingyi River basin water environment are designed and implemented. The basic function of geo-information and water environment thematic information service has been built. Based on ASP.NET framework and B/S architecture, the water environment information system has been designed and developed to realize online management and sharing of watershed spatial information, water environment monitoring information, multimedia information, user information, and other basic information.

## 1 INTRODUCTION

As fundamental and strategic resources, water resources are an important support for sustainable economic and social development. Currently, the lack of water resources and pollution problems has emerged seriously in China. Therefore, strengthening water environment management of river basin has become a strategic requirement for water ecological civilization construction. However, in the water environment management, the loss of information management of basin environment persists for a long time (Rojanamon et al. 2009). Besides this, the management of basin ecological environment has complex, multi-scale, and multi-objective characteristics. All of these factors lead to water environment management because of problems that long-term plagued the people.

Qingyi river basin is an important water area along the river; to control flood, drainage, receiving pollution, and urban landscape water, its water environment pollution problem has to be immediately attended. However, at present, many problems exist in the basin management, such as water environment information sharing and connectivity. These problems are urgent and rely on the information means to build a management platform. The multi-objective integrated and multi-sectoral decision-making of river basin could break the barriers to "data" and make water environment information openness and sharing come true (Tillman et al. 2012, Guo et al. 2015). Water environment basic information system based on Web and GIS technology is the basic condition.

## 2 SYSTEM RESEARCH AND DESIGN

The overall analysis and design stage of basic information system need to be based on many principles such as economic, reliable, flexible, and systematic. Under the premise of satisfying the requirements of the system, the R&D costs need to be reduced as much as possible. In addition, it is more important to avoid the complexity of the system design process, and fully guarantee the simplicity of each module in order to shorten the processing flow of the system (Kaewboonma et al. 2013, Ruddell et al. 2014).

The overall design of the Qingyi River watershed basic information system includes five aspects. They are user characteristics and needs analysis, system analysis, system architecture design, function design, and programming environment of the system design.

The Qingyi River water environment basic information system is mainly oriented to three types of users that include decision of users, enterprise users, and the public users of water environment management. The rights of users satisfy full consideration in the process of building the system because of different users' demand characteristics. Management users are concerned about the river basin information sharing, interoperability, and scientific and efficient management of the water environment (Volk et al. 2007). Enterprise user's demand mainly means that enterprise's business can dock system, enhance the enterprise's production efficiency, and reduce the cost of environmental management. Public users are concerned about their right to know and supervise the water environment of the river basin, and it is convenient for the public to participate in the management of water environment (Düpmeier & Geiger 2006).

## 2.1 Framework design

The logical structure of the Qingyi River water environment basic information system is mainly divided into three levels: data layer, services layer, and presentation layer (Fig. 1).

Data layer provides the database for the operation of the system that includes spatial information database and water environment thematic database. Moreover, ArcSDE is used to achieve the link of the two databases. The service layer provides water environment map service, query and statistics service, spatial analysis service, watershed environmental monitoring service, etc. For the water environment management of the river basin, ArcGIS Server is used for the release of these services and it is contributed to conveniently call these services for front-end users. The presentation layer is committed to providing a friendly, good interactive, and beautiful front-end interface in order to improve the effect of application.

## 2.2 Database design

Watershed data is divided into two categories: geographic information data and attribute data (Fig. 1). The basin geographic information data mainly involves basic geographic information (the scope of administrative divisions, roads, rivers, and towns) and all kinds of thematic geographic information (land use, control unit, and point and non-point source pollution and relates to the spatial position information). Watershed attributes database mainly includes the sewage outfall, enterprises, and other point source pollution of environment monitoring data and water quality data; moreover, it also relates to unit management, hydrology, meteorology, and user information, and other aspects of the attribute data (Dar et al. 2010).

ArcGIS Geodatabase is used for the management of basic geographic information database of the

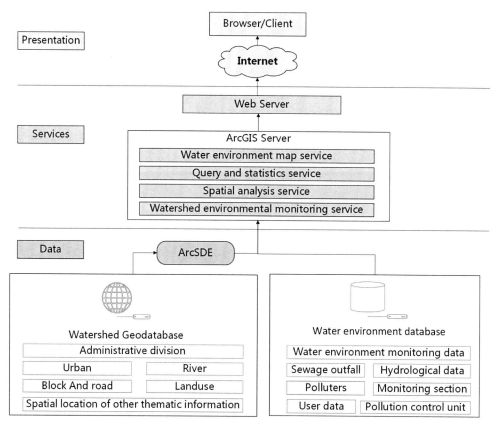

Figure 1.   System framework and structure diagram.

Qingyi River basin. Arc catalog is used for spatial data management. It plays a spatial data resource manager role in order to achieve the operation of the geographic information data organization and management, additions and deletions, and other operations. Oracle, currently one of the very popular B/S structure databases, is used to organize and manage all kinds of water environment thematic attribute information of the river basin in this paper.

### 2.3 Function design

The function of the Qingyi River water environment basic information system mainly consists of three modules: basic geo-information service, environment project information service, and user management service (Fig. 2). The system functions in design are divided into seven items that include map management, information query, geographic analysis, data editing, basin detection, statistics and analysis, and user management.

1. Basic geo-information service
   In these functions, map management mainly aims to achieve some of the basic map operations that including map display, additions and deletions, layer management, online map overlay display, zoom in, zoom out, roaming, view switching, and hawk-eye. Information query is used to achieve attribute query and spatial query for all kinds of water environment information that include the sewage outfall, sewage enterprises, section, and pollution control unit.
2. Environment project information service
   Data editing functions need to achieve spatial graphics editing, such as point, line, surface and other types, and water environment thematic attribute information input. Basin detection can be implemented on the management and monitoring of some objects, such as sewage outfall,

sewage companies, and monitoring sections and the monitoring content related to their live photos, videos, and other related information. Simultaneously, it can realize the real-time monitoring and display of water environment information, such as discharge and pollutant concentration. Statistics and analysis functions required to implement various types of charts display, such as line chart, histogram, and pie chart. The statistical content includes the flow hydrograph, pollutants content comparison, and pollutant source analysis. It is more important to use the results of statistics to realize the water level evaluation. In addition, all kinds of statistical results can be printed online or output in the form of JPG, PNG, or PDF, and other forms. Some of the important spatial analysis tools include measurement and analysis tools, impact analysis, and viewshed range of analysis.

3. User management
   User management functions to achieve the user's registration, authentication, login, query rights management, and log management.

## 3 SYSTEM IMPLEMENTATION

### 3.1 Study area

The Qingyi River basin belongs to Shaying River. The economy of the Qingyi River basin has been developed and has a rapid process of industrialization and urbanization in Henan province. The industry accounted for a large proportion in the economy and water pollution in the industry such as hair products, food, textiles, paper, and leather industry. In terms of natural conditions, Qingyi River lacks natural runoff, which causes the river water ecological function degradation and water environment pollution to become serious. It has

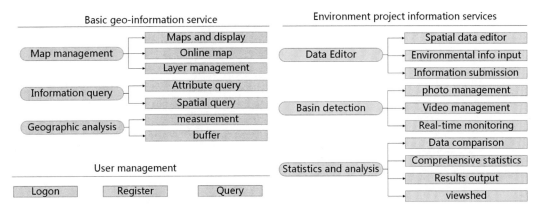

Figure 2. System function diagram.

become the most heavily polluted river in the Shaying River system causing serious problems.

Therefore, it is critically necessary to build the water environment basic information system for Qingyi River, which serves as the basic supporting system of multi-department and multi-objective integrated management decision-making platform. The system can lay the foundation for the sharing and opening of the basic geographic and water environment information of the river basin.

### 3.2 System development

The development and implementation of the system mainly follow four main steps: establishing river basin basic database, creating and publishing watershed map service, developing system function, and designing client. Each step has its own characteristics. However, they are not isolated, and these steps are complement and cooperate with each other. Besides this, the functions need to be considered and gradually improved at different stages of the system development.

First, all kinds of spatial and environmental data of the Qingyi River basin are acquired. Then, the watershed spatial information collation is finished and imported database according to the program which has been designed in the watershed spatial database. Map service is the foundation of the river basin basic information service and other kinds of functions are built on the basis of the map service. In addition, the design of client not only can significantly enhance the user experience,

but is also conducive to strengthening the function of the watershed based information system. The Qingyi River basin-based map service development and Web services publishing are completed, respectively, based on ArcGIS Desktop and ArcGIS Server. The system is implemented based on the Asp.net framework in the Visual Studio2013 development environment and is developed through the C# and JavaScript based on the DOJO framework. In addition, the front-end development is also used for JQuery, Bootstrap framework, and many third-party tools, such as Highcharts, layer, and others.

### 3.3 Function realization

The Qingyi River basin water environment basic information system has realized a series of functions that include online acquisition, storage, display, query, editing, analysis, statistics, output, and other functions about the basin spatial information, the water environment monitoring information, multi-media information, user information, and other various types of basic information (Fig. 3). It provides a series of basic information services for river basin, such as geographic information, and the project information of water environment management. The system is the basis for building mechanism of information online management, analysis, and sharing. It is beneficial to realize water environment management information sharing of river basin, connectivity, and business synergy. In addition, it could provide a support system

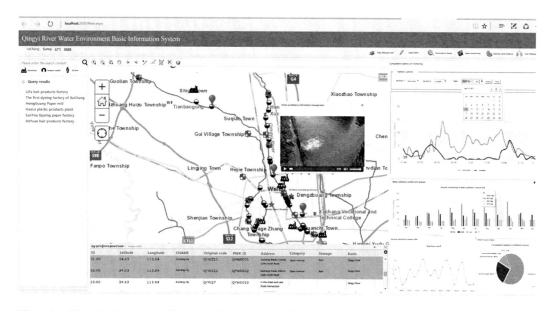

Figure 3.    Watershed-based map display and function example.

for building the management platform with multi-objective integrated and multi-sectoral decision-making of the river basin.

## 4 CONCLUSIONS

A series of research and practice has been completed about water environment basic information service system and it realizes the design and development of the basic functions by ArcGIS API for JavaScript based on Web GIS technology and B/S architecture. This paper proposes the design of database and basin water environmental information system. It includes in-depth research and detailed design before the development of the system implementation and is constantly perfect in the follow-up of the system development process.

The function of system mainly consists of three modules that are basic geo-information service, environment project information services, and user management service. It covers the basic information collection, management, analysis, statistics, output, and open sharing of the water environment management and can provide effective information support for the management of the water environment of the river basin.

## ACKNOWLEDGMENTS

This work was supported by grants from the China National Critical Project for Science and Technology on Water Pollution Prevention and Control (2015ZX07210-002-005).

## REFERENCES

Dar I A, Sankar K, & Dar M A (2010). Remote sensing technology and geographic information system modeling: an integrated approach towards the mapping of groundwater potential zones in Hardrock terrain, Mamundiyar basin. J. Journal of Hydrology. 394(3):285–295.

Düpmeier C, & Geiger W (2006). Theme park environment as an example of environmental information systems for the public. J. Environmental Modelling & Software. 21(11):1528–1535.

Guo Y T, Wang H W, & Nijkamp P, et al. (2015). Space–time indicators in interdependent urban–environmental systems: A study on the Huai River Basin in China. J. Habitat International. 45:135–146.

Kaewboonma N, Tuamsuk K, & Kanarkard W (2013). Knowledge acquisition for the design of flood management information system: Chi river basin, Thailand.J. Procedia-Social and Behavioral Sciences.73:109–114.

Rojanamon P, Chaisomphob T, & Bureekul T (2009). Application of geographical information system to site selection of small run-of-river hydropower project by considering engineering/economic/environmental criteria and social impact. J. Renewable and Sustainable Energy Reviews. 13(9), 2336–2348.

Ruddell B L, Zaslavsky I, & Valentine D, et al. (2014). Sustainable long term scientific data publication: Lessons learned from a prototype Observatory Information System for the Illinois River Basin. J. Environmental Modelling & Software. 54(3), 73–87.

Tillman F D, Callegary J B, & Nagler P L, et al. (2012). A simple method for estimating basin-scale groundwater discharge by vegetation in the basin and range province of Arizona using remote sensing information and geographic information systems. J. Journal of Arid Environments. 82:44–52.

Volk M, Hirschfeld J, & Schmidt G, et al. (2007). A SDSS-based ecological-economic modelling approach for integrated river basin management on different scale levels–The project FLUMAGIS. J. Water Resources Management. 21(12):2049–2061.

*Advances in Energy and Environment Research – Achour & Wu (Eds)*
*© 2017 Taylor & Francis Group, London, ISBN 978-1-138-62682-9*

# Flotation performance measurement and the DFT study of salicylhydroxamic acid as a collector in niobite flotation

Hao Ren, Zihuan Wang, Lifang Hu, Feifei Ji, Yufen Zhang & Taizhong Huang
*School of Chemistry and Chemical Engineering, University of Jinan, Shandong, China*

ABSTRACT: In this paper, Salicylhydroxamic Acid (SA) has been investigated as a new collector for niobite by mineral flotation tests, zeta potential determination, and the Density Functional Theory (DFT) calculation. The results of flotation tests and zeta potential determination validated that SA showed the optimal flotation effect for single niobium atoms in the condition that the preferable pH for flotation of SA for niobite was equal to 9 and the preferable background solution is calcium ions solution. Simultaneously, SA for artificial mixed mineral also had a very good separation effect. In addition, the DFT computation results indicated that the dianion of SA exhibited stronger chemical reactivity than their anions and neutral molecules. In alkaline aqueous solutions, the two oxygen atoms in the dianions of SA charged more negative charges than the other atoms, and hence became their reactive center. The dianion of SA exhibited higher atomic charge value, HOMO (Highest Occupied Molecular Orbital) energy, and bigger dipole moment, and thus, it had stronger collecting ability for niobite, which achieved excellent agreement with the flotation test.

## 1 INTRODUCTION

Niobium plays an important role in the industrial development of our country, and it is a kind of rare earth metal with a wide range of industrial applications. The world's proven reserves of niobium resources is approximately 25 million tons, mainly in Brazil, China, and Canada (Yu, 2007), and China's niobium reserves rank second in the world. However, from the perspective of recycling and utilization, there are very significant differences from the world's existing mining of niobium deposits. In China, owing to the characteristic of multitudinous niobium minerals, low grade and high degree of dispersion and embedded fine, it is difficult to achieve the effective recovery of niobium resources. With the rapid development of economy, the niobium resources requirement of the industry is increasing rapidly and the niobium output is far from meeting the needs of the national economic growth. Therefore, it is significant to carry out the research on the recovery of niobium from Nb-bearing minerals.

Bayan Obo deposit has rich resources of niobium in China. It has been found that niobium minerals are a total of 20 species of 7 families, of which iron ores of niobium, niobium iron rutile, yellow stone, and aeschynite mainly contain niobium minerals, and their niobium mineral contents are 69.49%, 11.42%, 60.87%, and 27.92%, respectively (Zhang, 1994). In this paper, we mainly study the amount of Nb being rich in columbite.

The mineral is scattered all over the banded fluorite-type rare earth iron ore, white stone-type niobium rare earth iron ore, and dolomite-type rare earth iron ore. The niobium ore symbiosis minerals include hematite, fluorite, barite, apatite, rare earth minerals, and ilmenorutile. Therefore, the flotation has become an effective means for the recovery of the mineral and the separation between niobium and other minerals. It has been widely used in the separation of niobium-bearing minerals.

Collector is one of the key factors in the process of flotation. The flotation process can be controlled flexibly and it realize separation of minerals effectively by using a flotation agent with good flotation effect and studying the mechanism of flotation. The structure and performance of the collector is a key factor for the flotation effect. Scientists have put forward the method and principle of the structure design and selection of the collector (wang, 1991). Computational chemistry has been widely used in the study of the structure and performance of collector, which has become a powerful complement for the search of high-efficiency flotation reagents.

This paper mainly explores the floatation efficiency of Salicylhydroxamic Acid (SA) and diphosphonic acid in different conditions, aimed at finding the best recovery conditions of columbite and the optimal recovery agent, realizing the recycling and utilization of niobium resources efficiently. The B3 LYP method of Density Functional Theory (DFT) was used to study the interaction of collector and niobium minerals. HOMO (Highest Occupied Molecular Orbital) energy of the flotation agent and niobium mineral interactions is used to determine the search capacity of collectors. The theoretical explanation and experimental results are in agreement.

## 2 RESEARCH METHODS

### 2.1 *Experimental methodology*

The flotation tests were carried out in a flotation tube with a volume of 40 mL. In each test, one sample (1.0 g) was dispersed in the tube with distilled water for 1 min. The pH was adjusted by a pH regulator for 2 min. Then, the collector was introduced and the pulp was conditioned for 3 min, followed by flotation for 5 min.

Artificial mixed minerals flotation process was concordant with the single ore flotation process. Only the mass ratio of the niobium ore, hematite, barite, and fluorite is 1:1:1:1, and other operating conditions remain unchanged (Zhong, 1993).

### 2.2 *Computational analyses*

In this paper, all the computational analyses were performed using Gaussian 03. The molecular geometry obtained was further optimized by Gaussian using the DFT methods at B3 LYP/6 level. The energy was calculated at the same basis set. The atomic charge values were determined by Mulliken Population Analysis (MPA) and Natural Population Analysis (NPA) to describe the molecular reactivity.

## 3 RESULTS AND DISCUSSION

### 3.1 *Pure mineral experiments*

#### 3.1.1 *Flotation tests of SA*
The effect of pH on niobite flotation with SA as collectors (50 mg/L) is shown in Fig. 1. The flotation results indicated that the floatation rate of niobite by SA was higher than that by deionized water in the experiment pH 2–12. The preferable pH for recovery of niobite was 8–10.

#### 3.1.2 *Determination of zeta potential on niobite surface*
The zeta potential of niobite in the presence and absence of SA is presented in Fig. 2. The change in zeta potential in the presence of SA was bigger than that of the primitive zeta potential of niobite, indicating that more SA collectors were adsorbed onto the niobite surface. Therefore, SA exhibited superior improvement in flotation performance to

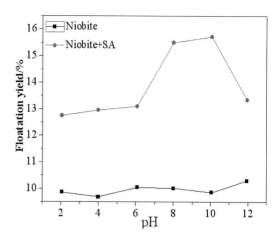

Figure 1. Effect of pH on niobite floatability at 50 mg/L SA.

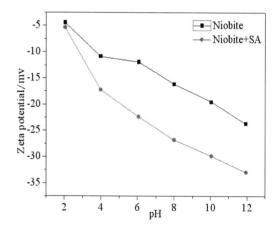

Figure 2. The zeta potential on niobite with and without the addition of SA.

niobite, which was in agreement with the results of flotation tests.

#### 3.1.3 *Flotation test of separation of mixed ores*
It was shown in Fig. 4 that flotation yield of hematite was far greater than the other three kinds of minerals. However, the floating rate of niobite was the highest in the whole range of pH. When the condition was set at pH 8, the flotation yield of niobite was much larger than that of the other three minerals. Therefore, Salicylhydroxamic Acid (SA) could achieve effective separation of niobite and other minerals.

### 3.2 *Computational analysis*

#### 3.2.1 *The stability analysis of two tautomers*
Salicylhydroxamic Acid (SA) is a kind of organic compound containing –CONHOH groups, which

mainly have two tautomers: keto-form and oxime-form, as shown in Fig. 4. The total energy of the optimized molecular structures of two tautomers is demonstrated in Table 1. From Table 1, it can be observed that the total energy of the oxime form was lower than that of the corresponding keto form, indicating that the oxime tautomer was more stable than the keto tautomer for SA. Hence, only this tautomer was considered in further study.

### 3.2.2 The frontier molecular orbital analysis

There are oxime form of Salicylhydroxamic Acid (SA) and their ionization equilibrium in aqueous solutions. The molecule of SA and their ionic species are listed in Fig. 5. The calculated dipole moment and HOMO energy of SA are given in Table 3. We followed this order: dianion > anion > molecule, hence their dianions had the strongest electron donation ability. It can be assumed that SA was adsorbed on niobite surfaces mainly through the dianion configuration at alkaline conditions.

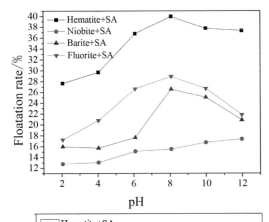

Table 2. The calculated HOMO energy and dipole moment of SA tautomers in different forms.

| Forms | HOMO energy/a.u. | Dipole moment/D |
|---|---|---|
| Molecule | −0.23003 | 1.6859 |
| Anion | −0.03165 | 6.8083 |
| Dianion | 0.17042 | 11.3539 |

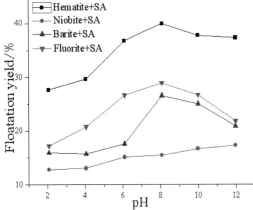

Figure 5. Ionization of salicylhydroxamic acid (SA) in aqueous solutions.

Table 3. Natural charges of atoms on SA groups.

| Binding site | Natural charges | | |
| | Molecule | Anion | Dianion |
|---|---|---|---|
| O1 | −0.63259 | −0.67898 | −0.65150 |
| N2 | −0.30297 | 0.42908 | −0.20540 |
| C3 | −0.20411 | −0.23259 | 0.42207 |
| O4 | −0.70341 | −0.78253 | −0.85269 |

Figure 3. Flotation test of separation of mixed ores with the addition of SA.

Figure 4. Two tautomers of salicylhydroxamic acid (SA).

Table 1. The calculated total energy of two tautomers of SA in different forms.

| Tautomers | Keto | Oxime |
|---|---|---|
| Total energy/a.u. | −551.3560053 | −551.3566935 |

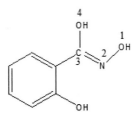

Figure 6. Schematic representation of SA.

149

The results given in Table 2 also indicated that the dipole moment of SA ions was much bigger than their molecules, especially for dianion. Therefore, the dianion species of SA had stronger electrostatic and Van der Waals interaction to metal atoms in mineral surfaces, which was in agreement with the results of flotation tests.

### 3.2.3 *Natural charges*

The selected structure parameters of SA with the numbering scheme are shown in Fig. 3. The calculated natural charge values are given in Table 3. Table 3 indicates that, for both SA ions and molecule, the negative charges were mainly focused on O1 and O4 atoms in the SA group. Moreover, the two atoms of SA dianions had more negative charges. Hence, O1 and O4 were of electron-donating center, chemical reactivity center, and the stronger electron-donation ability. Therefore, it was inferred that the metal ion can chelate SA dianions to form five-membered ring complexes (Bátka, D. 2006, Ali, O. 2011).

## 4 CONCLUSIONS

In this paper, SA has been investigated as a new collector for niobite by mineral flotation tests, zeta potential determination, and the Density Functional Theory (DFT) calculation.

The flotation results indicated that the preferable pH for recovery of SA was 8~10. Moreover, when the preferable pH was set at 8, SA could achieve the effective separation between niobite and other minerals. The SA exhibited superior improvement in flotation performance for niobite.

In aqueous solution, the oxime tautomer was more stable than the keto tautomer. In alkaline solutions, the dianion of SA exhibited higher atomic charge value and HOMO energy, bigger dipole moment, and hence stronger collecting ability for Niobite, which was in agreement with the results of flotation tests.

As a result, SA has been proven to have stronger collecting ability in niobite flotation by theory and experimental investigation in this work. Consequently, we believe that SA is a justified candidate collector in niobite flotation and the separation with other minerals.

## ACKNOWLEDGMENTS

This work was financially supported by the 2015 Shandong Provincial Key Research and Development Program (2015GSF116005). Special thanks are given to the research team for providing support to this study.

## REFERENCES

Ali, O.Y., Fridgen, T.D.. Structures of electrosprayed Pb (Uracil-H) + complexes by infrared multiple photon dissociation spectroscopy. International Journal of Mass Spectrometry, 2011, 308 (2–3): 167–174.

Bátka, D., Farkas, E. Pb (II)-binding capability of aminohydroxamic acids: Primary hydroxamic acid derivatives of a-amino acids as possible sequestering agents for Pb (II). Journal of Inorganic Biochemistry, 2006, 100 (1), 27–35.

Dianzuo wang, Xiangyun Long, Shuiyuo Sun. The oxidation of sulfide ore and the flotation mechanism of quantum chemistry research [J]. Chinese journal of nonferrous metals, 1991, 11 (1): 15–23.

Guozhong Zhang, Qun Bi, Guoan Zeng, et al. Baiyun obo ore mining and metallurgy technology [M]. Inner Mongolia: baosteel printing plant, 1994, 10: 507–508.

Kangnian Zhong, Jing shen. With double our acid separation collophane and dolomite in the [J]. Journal of wuhan institute of chemical technology, 1993, 15 (1): 8–13.

Yongfu Yu. Baiyun obo ore, niobium rare earth resources comprehensive utilization is of great significance to study [J]. Rare earth information, 2007, (8): 8–9.

# Study of the preparation of a magnetic recyclable emulsifier

Hao Ren, Zihuan Wang, Feifei Ji, Zhen Li, Yufen Zhang & Taizhong Huang
*School of Chemistry and Chemical Engineering, University of Jinan, Shandong, China*

ABSTRACT: A magnetic emulsifier was synthesized using the two-step method. First, $Fe_3O_4$ nanoparticles were synthesized and then modified with silane coupling agent 3-(methacryloyloxy) propyltrimethoxysilane (KH-570). Second, the final composite particles were synthesized by the reaction between emulsifier octylphenol polyoxyethylene ether OP-10 ($C_8H_{17}C_6H_4O(CH_2CH_2O)_{10}H$) and modified $Fe_3O_4$ nanoparticles by silane coupling agent KH-570. Then, the products were characterized by X-Ray Diffraction (XRD), Raman spectroscopy, and Fourier transform infrared spectroscopy (FT-IR). The result indicated that the mean diameter of $Fe_3O_4$ nanoparticles was approximately 28.2 nm and the silane coupling agent KH-570 can be effectively modified $Fe_3O_4$ nanoparticles.

## 1 INTRODUCTION

Oil and water are mutually insoluble. In order to make the aqueous solution distribute into the oil, an emulsifier is usually used to make a two-phase immiscible oil emulsion from a stable emulsion (Liu, 2012). The emulsifier plays an extremely important role in the emulsification process (Solans, 2005; Zhang, 2006; Ren, 2014). Nanosized $Fe_3O_4$ was surface modified with silane coupling agent KH-570 hydrolyzed in ethanol solution (Mohapatra, 2009; He, 2014; Wang, 2011; Iijima, 2007; Stewart, 2013; Mohammad-Beigi, 2011; Iijima, 2009). Then, OP-10 was compounded with the modified $Fe_3O_4$, which can be recycled by magnetic separation.

Generally, when the emulsifier dispersed in the surface of the dispersoid, a thin film or a double layer can be formed, and thus, the dispersed phase resulted with a charge. The droplets of the dispersed phase can prevent condensation from each other, so that a stable emulsion is formed. Emulsifiers have two different kinds of groups: a hydrophobic group and a hydrophilic group.

Nano-sized $Fe_3O_4$ was surface modified with silane coupling agent KH-570 hydrolyzed in ethanol solution. Then, OP-10 when applied magnetic field was compounded by the modified $Fe_3O_4$, which can be recycled by magnetic separation.

## 2 EXPERIMENTAL METHODOLOGY

### 2.1 Chemicals and measurements

Ferrous sulfate ($FeSO_4 \cdot 7H_2O$), ferric chloride ($FeCl_3 \cdot 6H_2O$), Sodium hydroxide (NaOH), methylbenzene ($C_7H_8$), ethyl alcohol absolute ($C_2H_6O$), silane coupling agent (KH-570), emulsifier OP-10, distilled water, and oxalic acid ($H_2C_2O_4$) for pH variations are the chemicals used in this experiment.

FT-IR spectra were recorded on a Bruker Tensor 27 spectrometer, using KBr pellets for solids. XRD patterns were recorded using an X-Ray Diffractometer. The scanning range of $2\theta$ was set between 10 and 80° with a scan rate of 0.2 s-1 (Masoumi, 2014).

### 2.2 Synthetic procedures

The steps are given in detail in Figure 1. The following process could contribute to the synthesis of the composite particles:

1. Preparation of $Fe_3O_4$:
   To 200 mL of deionized water, 7 g of $FeSO_4 \cdot 7H_2O$ and 7 g $FeCl_3 \cdot 6H_2O$ were added in a clean beaker and then stirred to dissolve completely.
2. Preparation of aqueous oxalic acid solution:
   To 200 mL of deionized water, 1 g of $C_2H_2O_4$ was added in a clean beaker and then stirred

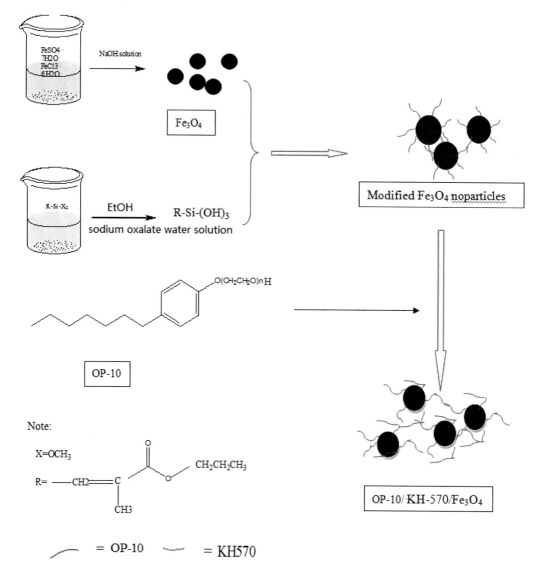

Figure 1. Schematic illustration of the reaction.

to dissolve completely; 0.04 mol/L of aqueous solution of oxalic acid was formulated, which was employed in the follow-up experiments.

## 3 RESULTS AND DISCUSSION

### 3.1 *X-Ray Diffraction (XRD)*

Figure 2 shows the crystal structure of the $Fe_3O_4$ particles. The characteristic peaks of $Fe_3O_4$ particles were at $2\theta$ of $18.438°$, $30.144°$, $35.633°$, $43.169°$, $53.624°$, $57.180°$, and $62.593°$. The dif-

fraction peak intensity is high and sharp that crystallization is good (Zhou, 2011). When compared with the XRD standard card, it conforms to the $Fe_3O_4$ standard card card PDF # 65–3107. The crystal size of $Fe_3O_4$ was 28.2 nm calculated by using the Scherrer equation.

$$D = K\lambda/b\cos\theta$$

### 3.2 *FTIR spectroscopy*

The spectra of $Fe_3O_4$ nanoparticles are shown in Figure 3a. The characteristic absorption bands

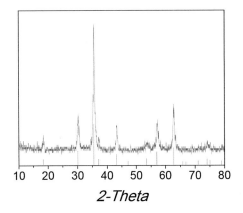

Figure 2. XRD patterns of Fe$_3$O$_4$ particles.

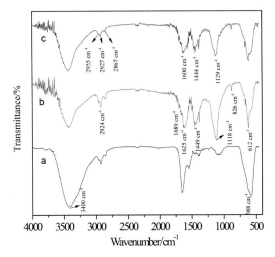

Figure 3. Infrared spectra (IR) of synthesized particles. (a: Fe$_3$O$_4$ prepared by co-precipitation; b: KH-570/Fe$_3$O$_4$ particles; c: OP-10/KH-570/Fe$_3$O$_4$ particles).

The weak absorption peak at 2924 cm$^{-1}$ was related to the stretching vibrations of −CH$_2$− of KH-570 (Taghvaei, 2009). The bands at 1689 and 1625 cm$^{-1}$ were due to the stretching vibration of C=O and bending vibrations of O−H of KH-570. In addition, the band near 1449 cm$^{-1}$ was assigned to Si−C−H.

In addition, bands at 1118 cm$^{-1}$ were due to O-H stretching vibrations. The weak absorption peaks for Si−O−Si bending vibrations were 826 cm$^{-1}$. The absorption bands at around 612 cm$^{-1}$ resulted from the Fe−O−Si stretching vibration of Fe$_3$O$_4$. This reveals that KH-570 was chemisorbed into Fe$_3$O$_4$ nanoparticles to form the covalent bonds of Fe−O−Si.

In Figure 3c, the new absorption bands at 2955 cm$^{-1}$ were due to −CH$_3$ stretching vibration. The bands at 2927 cm$^{-1}$ and 2867 cm$^{-1}$ were due to stretching vibration of −CH$_2$−. Peaks for the aromatic ring stretching vibrations were observed at 1600−1444 cm$^{-1}$. In addition, the band near 1129 cm$^{-1}$ was assigned to C−O−C.

### 3.3 *Raman spectrum of Fe$_3$O$_4$/KH-570 particles*

To further verify successful synthesis of particles, Raman spectrum of particles is also displayed in Figure 4.

Figure 4 shows Raman spectrum of Fe$_3$O$_4$/ KH-570 particles. Characteristic Raman bands were displayed at 1308 cm$^{-1}$, resulting from the vibrations of C−C. The peak at 689 cm$^{-1}$ was due to symmetrical stretching vibrations of −Si(CH$_3$)$_3$ groups and the band at 487 cm$^{-1}$ was assigned to Si−O stretching vibration.

Raman spectrum of Fe$_3$O$_4$/KH-570 particles is shown in Figure 5. The characteristic Raman peak at 1582 cm$^{-1}$ was associated with the stretching vibrations of benzene ring of OP-10. Another band assigned to the stretching vibrations of

of Fe$_3$O$_4$ nanoparticles were nearby 3400 cm$^{-1}$ and 588 cm$^{-1}$, which were assigned to the O−H absorbed on the surface stretching vibration and characteristic absorption of Fe$_3$O$_4$ nanoparticles. Moreover, the absorption bands nearby 588 cm$^{-1}$ were due to the stretching vibration of Fe-O.

Figure 3b shows that the peak at 3462 cm$^{-1}$ was associated with the stretching vibrations of O−H absorbed on the surface of Fe$_3$O$_4$. Owing to a small amount of the coupling agent absorbed on the surface of Fe$_3$O$_4$, a characteristic peak of Fe$_3$O$_4$ can still be reflected. However, the characteristic absorption bands showed some changes. This suggests that dehydration reaction occurred between O−H obtained from KH-570 hydrolyzed and O−H absorbed on the surface of Fe$_3$O$_4$ nanoparticles.

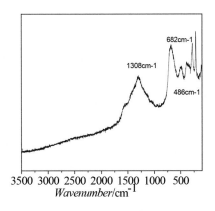

Figure 4. Raman spectrum of Fe$_3$O$_4$/KH-570 particles.

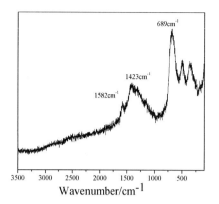

Figure 5. Raman spectrum of Fe$_3$O$_4$/KH-570/op-10 particles.

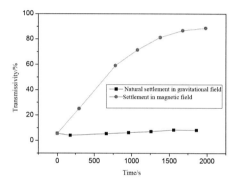

Figure 6. Settlement curve of OP-10/KH-570/Fe$_3$O$_4$ particle.

−CH$_2$− was also observed at nearly 1423 cm$^{-1}$. In addition, the band near 689 cm$^{-1}$ was assigned to −Si(CH$_3$)$_3$ groups.

### 3.4 *Settlement of OP-10/KH-570/ Fe$_3$O$_4$ particle*

Determination of process: OP-10/KH-570/Fe$_3$O$_4$ particle was diluted with water, divided into two parts. One with a magnet on the bottom of the beaker, and the other with natural subsidence. After a certain time interval, transmittance of the supernatant liquid was measured (580 nm).

Figure 6 shows that transmittance of particles changes very slightly in gravity field. However, in the magnetic field, with the increase in time, transmittance rises rapidly. After 1800 s, the transmittance is nearly unchanged. This shows that modified particles have a good magnetic field.

## 4 CONCLUSIONS

Based on the above test results, the following conclusions can be drawn:

1. Ferroferric oxide particles are nanoscale particles.
2. Nanoscale ferroferric oxide particles were surface modified with silane coupling agent KH-570. The result indicated that the silane coupling agent KH-570 can be modified ferroferric oxide nanoparticles successfully.
3. OP-10/KH-570/Fe$_3$O$_4$ composite particles were synthesized successfully.

## ACKNOWLEDGMENTS

This work was financially supported by the 2015 Shandong Provincial Key Research and Development Program (2015GSF116005). Special thanks are given to the research team for providing support to this study.

## REFERENCES

He, F., S. Chao, X. D. He, et al, Ceramics International. 2014; 40, 6865–6874.
Iijima, M., M. Tsukada, H. Kamiya, Journal of Colloid and Interface Science. 2007; 305, 315–323.
Iijima, M., N. Sato, I.W. Lenggoro, et al, Colloids and Surfaces A: Physicochem. Eng. Aspects. 2009; 352, 88–93.
Liu, X. H., Z. H. Bai, Y. H. Shuang, et al, Applied Mechanics and Materials. 2012; 217–219, 1363–1367.
Masoumi, A., M. Ghaemy, A. N. Bakht, Ind. Eng. Chem. Res. 2014; 53, 8188–8197.
Mohammad-Beigi, H., S. Yaghmaei, R. Roostaazad, et al, Physica E: Low-dimensional Systems and Nanostructures. 2011; 44, 618–627.
Mohapatra, S., P. Pramanik, Colloids and Surfaces A: Physicochem. Eng. Aspects. 2009; 339, 35–42.
Ren, Z. H., Ind. Eng. Chem. Res. 2014; 53, 10035–10040.
Solans, C., P. Izquierdo, J. Nolla, et al, Current Opinion in Colloid & Interface Science. 2005; 10, 102–110.
Stewart, A., B. Schlosser, E. P. Douglas, Appl. Mater. & Interfaces. 2013; 5, 1218–1225.
Taghvaei, A. H., H. Shokrollahi, A. Ebrahimi, et al, Materials Chemistry and Physics. 2009; 116, 247–253.
Wang, C. X., H. Y. Mao, C. X. Wang, et al, Ind. Eng. Chem. Res. 2011; 50, 11930–11934.
Zhang, D., X. M. Song, F. X. Liang, et al, J. Phys. Chem. B. 2006; 110, 9079–9084.
Zhou, Z. H., J. Wang, X. Liu, et al, J. Mater. Chem. 2001; 11, 1704–1709.

*Advances in Energy and Environment Research – Achour & Wu (Eds)*
© *2017 Taylor & Francis Group, London, ISBN 978-1-138-62682-9*

# Analysis of the variation of porosity and permeability in different periods of Lamadian Oilfield

Liping Wang
*No. 6 Oil Production Plant of Daqing Oilfield Limited Company, Daqing, China*

ABSTRACT: After more than 30 years of development and adjustment, the reservoir microstructure of Lamadian Oilfield has a certain change. Based on coring well data and oilfield test data in different periods of Lamadian Oilfield, this paper studies the variation law of porosity and permeability of the reservoir. Correct understanding of the variation in these parameters and the characteristics of reservoir parameters at the present stage and the factors influencing the change in reservoir parameters has a very important practical significance for further tapping oil, tertiary oil recovery, and water-injection development adjustment.

## 1 INTRODUCTION

Lamadian Oilfield has experienced several periods of basic well pattern, one encryption, two encryption, PuI1-2 polymer flooding, and flooding upwards in thin and poor layers. Owing to the development of water injection for many years, there have been some changes in the microstructure characteristic parameters of the reservoir (including the physical parameters of the reservoir, rock composition and properties, and pore structure parameters). The porosity and effective permeability of the same layer in different periods are very different, the most prominent being the PuI2 oil layer (Wang, 1999). Majority of the basic well patterns are several hundred millidarcy, while the adjacent infill well networks are mostly over one thousand millidarcy. We made the geological modeling and numerical simulation, the residual oil analysis, as well as the development and adjustment of the preparation of the program difficult to carry out smoothly (Zhao, 1992).

## 2 STUDY ON THE VARIATION LAW OF POROSITY AND PERMEABILITY PARAMETERS OF THE RESERVOIR

It is an effective means to study the variation in physical property parameters of the reservoir by analyzing the data of the closed core well (Zhang, 2003). To examine the effects of development and adjustment in different periods, there are 18 closed coring inspection wells from early development to the present. According to the time and purpose, the core well is divided into six stages: 1981–1982,

1988–1990, 1992–1995, 1997–2001, 2004–2009, and 2011–2012 (Table 1).

### 2.1 *Variation in porosity and permeability parameters of different types of reservoirs in different periods*

With the increase in mining time, the water injection multiple gradually increases, the oil saturation decreases, and the reservoir parameters change. According to the statistical results of the inspection well data, from 1981 to 2012 (Table 2).

The effective porosity of class I oil layer increased from 28% to 29.56%, with an average annual increase of 0.04%/a; the air permeability of class I oil layer increased from 1.975 $\mu m^2$ to 2.349 $\mu m^2$, which increased by 18.9%, and the average annual increase range is 0.5%/a.

The effective porosity of class II oil layer increased from 28.1% to 28.67%, with an average annual increase of 0.01%/a; the air permeability of class II oil layer increased from 1.026 $\mu m^2$ to 1.258 $\mu m^2$, which increased by 22.6%, and the average annual increase range is 0.6%/a.

The effective porosity of class III oil layer increased from 27.5% to 27.52%, with an average annual increase of 0.0%/a; the air permeability of class III oil layer reduced from 0.405 $\mu m^2$ to 0.316 $\mu m^2$, which reduced by 18.9%, and the average annual reduction range is 0.6%/a.

From the above data, it can be observed that the air permeability and porosity of the class I and class II oil layers are increasing; the porosity of the class III oil layer is not changed, but its permeability is decreased slightly.

Table 1.　Basic condition of closed coring inspection well.

| Stages | Well no. | Core time | Core length (m) | Sampling number | Drilling reservoir conditions Layer number | Drilling reservoir conditions Sandstone thickness (m) | Drilling reservoir conditions Effective thickness (m) |
|---|---|---|---|---|---|---|---|
| 1981–1982 | X1 | 198107 | 243.39 | 731 | 53 | 108.6 | 85.4 |
| | X2 | 198207 | 292.15 | 839 | 66 | 115 | 84.9 |
| 1988–1990 | X3 | 198807 | 248.36 | 484 | 32 | 74.9 | 54.2 |
| | X4 | 198906 | 310.18 | 923 | 78 | 137 | 91.4 |
| | X5 | 199005 | 362.36 | 974 | 69 | 118.9 | 83.8 |
| | X6 | 199001 | 169.24 | 400 | 18 | 38.5 | 27.7 |
| 1992–1995 | X7 | 199206 | 238.24 | 757 | 54 | 78.5 | 59.8 |
| | X8 | 199507 | 272.22 | 835 | 41 | 81.9 | 66.2 |
| 1997–2001 | X9 | 199701 | 251.71 | 713 | 52 | 85.9 | 63.3 |
| | X10 | 200106 | 363.67 | 1253 | 71 | 101.4 | 74.4 |
| 2004–2009 | X11 | 200409 | 337.42 | 1163 | 69 | 134.5 | 105.3 |
| | X12 | 200510 | 55.87 | 198 | 8 | 27 | 23.9 |
| | X13 | 200511 | 312.36 | 910 | 22 | 66.6 | 63 |
| | X14 | 200610 | 132.02 | 433 | 23 | 53.5 | 39.9 |
| | X15 | 200908 | 394.75 | 1446 | 66 | 88.9 | 63 |
| | X16 | 200909 | 77.52 | 424 | 8 | 44.6 | 37 |
| 2011–2012 | X17 | 201103 | 304.26 | 837 | 56 | 94.7 | 73.4 |
| | X18 | 201111 | 305.98 | 929 | 49 | 104.9 | 86.5 |
| Total | | | 4671.7 | 14249 | 835 | 1555.3 | 1183.1 |

Table 2.　Characteristics of porosity and permeability in different periods.

| Stages | Effective porosity (%) Class I oil layer | Effective porosity (%) Class II oil layer | Effective porosity (%) Class III oil layer | Air permeability ($\mu m^2$) Class I oil layer | Air permeability ($\mu m^2$) Class II oil layer | Air permeability ($\mu m^2$) Class III oil layer |
|---|---|---|---|---|---|---|
| 1973–1974 | 28 | 28.1 | 27.5 | 1.975 | 1.026 | 0.405 |
| 1981–1982 | 28.4 | 28.33 | 26.43 | 2.065 | 1.098 | 0.423 |
| 1988–1990 | 28.7 | 28.38 | 27.07 | 2.149 | 0.989 | 0.558 |
| 1992–1995 | 28.42 | 27.83 | 26.87 | 2.169 | 0.831 | 0.412 |
| 1997–2001 | 29.42 | 28.92 | 26.9 | 2.24 | 0.994 | 0.448 |
| 2004–2009 | 29.72 | 29.16 | 27.33 | 2.386 | 1.112 | 0.292 |
| 2011–2012 | 29.56 | 28.67 | 27.52 | 2.349 | 1.258 | 0.316 |

## 2.2　Variation in porosity and permeability parameters of different sedimentary sand bodies at different stages

### 2.2.1　Variation law of air permeability

The air permeability of channel sand body increased from 1.185 $\mu m^2$ to 1.593 $\mu m^2$, which increased by 34.4%, and the average annual increase range is 1.1%/a. S,P oil reservoir river increased by 17.4% and the average annual increase rate is 0.6%/a; the G oil reservoir river increased by 4% and the average annual increase rate is 0.1%/a.

The air permeability of main sand sheet increased from 0.561 $\mu m^2$ to 1.228 $\mu m^2$, which increased by 119%, and the average annual increase range is 4%/a. S,P oil reservoir river increased by 39.6% and the average annual increase rate is 1.3%/a; the G oil reservoir river increased by 156% and the average annual increase rate is 5.2%/a.

The air permeability of minor sand sheet reduced from 0.448 $\mu m^2$ to 0.427 $\mu m^2$, and the average annual reduction range is 0.05%/a. S,P oil reservoir river average annual decreased by 0.01%/a, while G oil reservoir average annual increased by 0.01%/a.

As seen from table, in the process of water injection development, air permeability increased

156

Table 3. Variation characteristics of air permeability in different sand bodies (unit: $\mu m^2$).

| Stages | S,P,G | | | S,P | | | G | | |
|---|---|---|---|---|---|---|---|---|---|
| | Channel sand | Main sand sheet | Minor sand sheet | Channel sand | Main sand sheet | Minor sand sheet | Channel sand | Main sand sheet | Minor sand sheet |
| 1981–1982 | 1.185 | 0.561 | 0.448 | 1.533 | 0.983 | 0.701 | 0.328 | 0.302 | 0.416 |
| 1988–1990 | 1.198 | 0.446 | 0.34 | 1.539 | 0.504 | 0.705 | 0.23 | 0.322 | 0.212 |
| 1992–1995 | 1.157 | 0.69 | 0.555 | 1.236 | 0.877 | 0.68 | 0.38 | 0.462 | 0.334 |
| 1997–2001 | 1.302 | 0.672 | 0.544 | 1.392 | 1.111 | 0.797 | 0.571 | 0.377 | 0.395 |
| 2004–2009 | 1.581 | 0.78 | 0.455 | 1.665 | 0.975 | 0.56 | 0.301 | 0.352 | 0.253 |
| 2011–2012 | 1.593 | 1.228 | 0.427 | 1.799 | 1.372 | 0.424 | 0.341 | 0.773 | 0.448 |

Table 4. Variation characteristics of effective porosity in different sand bodies (unit: %).

| Stages | S,P,G | | | S,P | | | G | | |
|---|---|---|---|---|---|---|---|---|---|
| | Channel sand | Main sand sheet | Minor sand sheet | Channel sand | Main sand sheet | Minor sand sheet | Channel sand | Main sand sheet | Minor sand sheet |
| 1981–1982 | 28.1 | 27 | 26.6 | 28.3 | 28 | 27.3 | 27.2 | 26.1 | 26.3 |
| 1988–1990 | 28.1 | 26.9 | 25.9 | 28.6 | 27.5 | 26.5 | 26.4 | 21.9 | 21.5 |
| 1992–1995 | 27.9 | 27.4 | 27.1 | 28 | 28.4 | 27 | 26.6 | 26.1 | 27.1 |
| 1997–2001 | 28.8 | 27.9 | 27.4 | 29 | 29.4 | 28.6 | 27.3 | 26.9 | 26.4 |
| 2004–2009 | 29.3 | 28.4 | 27.2 | 29.4 | 28.7 | 27.1 | 28.2 | 21 | 26.6 |
| 2011–2012 | 29.2 | 28.5 | 27.3 | 29.4 | 28.6 | 27.7 | 28.2 | 28.3 | 26.6 |

significantly in the channel sand body and main sand sheet with large thickness and good physical properties; while in minor sand sheet with poor deposition characteristics, the increase in air permeability was smaller, and even reached a negative increase (Table 3).

### 2.2.2 Variation law of effective porosity

The effective porosity of channel sand bodies increased from 28.1% to 29.2%, and the average annual increase range is 0.03%/a. The average annual increase of S,P oil reservoir is 0.03%/a and the average annual increase of G oil reservoir range is 0.03%/a.

The effective porosity of main sand sheet increased from 27% to 28.5%, and the average annual increase range is 0.05%/a. The average annual increase of S,P oil reservoir is 0.02%/a and the average annual increase of G oil reservoir range is 0.07%/a.

The effective porosity of minor sand sheet increased from 26.6% to 27.3%, and the average annual increase range is 0.02%/a. The average annual increase of S,P oil reservoir is 0.01%/a and the average annual increase of G oil reservoir range is 0.01%/a.

Through data analysis, the porosity of channel sand and main sand has an increasing trend, especially, after polymer flooding, the porosity and permeability increased obviously, while the porosity change of the minor sand is not obvious (Table 4).

## 3 ANALYSIS OF THE MECHANISM OF THE CHANGE IN POROSITY AND PERMEABILITY PARAMETER IN THE RESERVOIR

Relevant research results indicate that the main mechanism of the microstructure change in water-drive reservoir is the hydration, expansion, dispersion, migration of clay minerals, and the migration of other formation particles. The migration of clay minerals and formation particles increases the permeability and pore throat radius, and clogging of the throat during migration leads to a decrease in permeability and pore throat radius (Qian, 2001).

The effect of water injection on the dissolution of rock minerals promotes the detachment and migration of particles and enlarges the circulation channels, while precipitation interaction reduces the permeability and pore throat radius. The injected water flow in the pore throat channel with small flow resistance, the scouring action of clay minerals and formation particles, and the dissolution and precipitation of the skeleton minerals and cements mainly occurred in the large pore throat passage. As a result, the larger the pore throat radius, the smaller the pore throat radius, and formation particles are easy to be blocked here, leading to the decrease in pore throat radius and the decrease in permeability.

With the development of oil field, for coarse lithology, high-permeability reservoir, with pore throat, is more uniform, and long-term water washing can cause an increase in the porosity and permeability; for fine lithology, low-permeability reservoir, due to its micropore throat, is small, the micro throat is easy to be blocked after water flooding, resulting in a small increase in porosity and permeability, and even lower (Wang, 2004).

## 4 CONCLUSION

The air permeability and porosity of the class I and class II oil layers are increasing; the porosity of the class III oil layer is not changed, but its permeability is decreased slightly.

The permeability and porosity of channel sand and main sand sheet are generally increasing, especially after polymer flooding, the permeability and porosity increased obviously; the change in permeability and porosity of the minor sand sheet is not obvious.

According to the laboratory experiment, after long-term water scouring and soaking, the minerals and clay will be changed, and the pore structure parameter of the rock will be better.

## REFERENCES

Qian Zhi-hao, Cao Yin: Geochemical effects of crude oil migration in North Tarim basin, Xinjiang [J]. Petroleum Geology & Experiment, 2001, 23(2): 186–190, In Chinese.

Wang Yun-long: Chacracterisctics and exploration potential of Southern Deep Oil-gas Accumulation Zone in Songliao basin [J]. Petroleum Geology & Oilfield Development in Daqing, 2004, 23(4): 5–6, In Chinese.

Wang Zhizhang, Cai Yi, Yang Lei: Variation law and mechanism of reservoir parameters in the middle and late stage of development (The petroleum Industry Press, Beijing 1999), In Chinese.

Zhang Feng, Liu li, Yang Hua: Application of the statistical analysis in mother rock study in the west strata of Songliao basin [J]. Geology & Oilfield Development, 2003, 22(3): 332–324, In Chinese.

Zhao Hanqing: Study on sedimentary model and heterogeneity of sand bodies in the large fluvial delta system in the north of Songliao Basin, Research Report on oil and gas reservoir evaluation in China, 1992, In Chinese.

*Advances in Energy and Environment Research – Achour & Wu (Eds)*
© 2017 Taylor & Francis Group, London, ISBN 978-1-138-62682-9

# Research on greenhouse gas carbon emission reduction in the public bicycle service system

Dong-Xiang Cheng
*Nanjing Vocational Institute of Transport Technology, Nanjing, China*
*Nanjing University Business School, Nanjing, China*
*Jiangsu Engineering Technology Research Center for Energy Conservation and Emission Reduction*
*of Transportation, Nanjing, China*

Ya-Nan Huang
*Hohai University College of Civil and Transportation Engineering, Nanjing, China*

Jing Chen
*Nanjing Vocational Institute of Transport Technology, Nanjing, China*
*Jiangsu Engineering Technology Research Center for Energy Conservation and Emission Reduction*
*of Transportation, Nanjing, China*

ABSTRACT:   As transportation is one of the main sources of carbon emissions, and now people pay more attention to the environmental issues and the problem of traffic, the public bicycle service system construction has been carried out rapidly in major cities. As the public bicycle system is a zero emissions zero pollution green traffic, it has attracted the attention and affection of more and more people. In this paper, through the analysis of domestic and foreign traffic greenhouse gas emissions, research status, and the existing carbon emission reduction calculation model, research on the environmental benefits is brought about by the public bicycle service system in order to establish a model of greenhouse gas carbon emission reduction, which is unique to the public bicycle service system.

## 1 INTRODUCTION

As transportation is one of the main sources of carbon emissions, and now people pay more attention to the environmental issues and the problem of traffic, the public bicycle service system construction has been carried out rapidly in major cities. With the concept of low carbon deeply rooted in the hearts of people, the green and healthy way to travel will attract the attention and favor of more and more people. Although many researches have been conducted on public bicycle, the studies to quantify its environmental benefits are few. This paper helps us understand the impact of greenhouse gas carbon emission factors and carbon emission reduction accounting method referring to a large number of domestic and foreign literatures, in order to establish a model of greenhouse gas carbon emission reduction, which is unique to the public bicycle service system.

## 2 RESEARCH STATUS OF DOMESTIC AND FOREIGN TRAFFIC GREENHOUSE GAS CARBON EMISSIONS

### 2.1 *Research status of foreign traffic greenhouse gas carbon emissions*

#### 2.1.1 *Study on the calculation method of foreign traffic greenhouse gas carbon emissions*

According to the literature research (IPCC2006), it can be known that the IPCC measurement algorithm is one of the main methods of the internationally recognized traffic greenhouse gas carbon emission calculation. The IPCC measurement algorithm includes the "top-down" and "bottom-up" methods. The carbon emission of the former is based on the sales of transport fuels in a certain area multiplied by fuel carbon emission coefficient. The carbon emission of the latter is based on the range of all kinds of transportation multiplied by fuel consumption in unit distance and then multiplied by the sales of transport fuels.

Based on the research of two kinds of IPCC measurement algorithm, the foreign scholars have made many researches on greenhouse gas carbon emissions in different regions.

Lu IJ (2007) calculated and analyzed the greenhouse gas emissions of road transport in Germany, China, Taiwan, and Korea. Taking Antwerp area as the research object, Mensink (2000) constructed a city traffic carbon emission measurement model, which can provide all the streets and even the sections per hour of $CO_2$, NO, and other gas emissions.

### 2.1.2 Study on the foreign transport greenhouse gas emissions factors

With the continuous increase in the urban motor vehicle, the factors of the influence of the greenhouse gas emissions have been further studied.

Some of the views were the same, for example: Lisa Göransson (2010) studied the effects of parallel hybrid cars using the wind and thermoelectric integrated power supply system. Ou Xunmina (2010) comparatively analyzed the greenhouse gas emissions resulting from the use of different energy electric vehicles, and they have pointed out that the use of hybrid power can reduce carbon dioxide emissions. The others were the opposite, for example: Chung (2004) pointed out that urbanization is a factor of influencing the increase in traffic carbon emissions. York (2007) also believed that urbanization brings about the growth of energy consumption and more carbon emissions. However, Chen H (2008) believed that the increase in the population of the city has forced people to use public transport, which greatly reduced transport greenhouse gas emissions.

There are other aspects of research, for example, Schipper (1997) analyzed and studied the factors influencing the increase in freight traffic carbon emissions from 1973 to 1982 in OECD countries. It has been concluded that economic factors are the main factors that affect the rise of the carbon emission.

In summary, foreign scholars analyzed the factors from different angles that influence the traffic carbon emissions, and laid a good foundation for the further study of scholars.

### 2.2 Research status of domestic traffic greenhouse gas carbon emissions

### 2.2.1 Study on the calculation method of domestic traffic greenhouse gas carbon emissions

Now, it has not formed a complete greenhouse gas emissions accounting system in China, but there are many methods to study the measurement of greenhouse gas carbon emission. Liu Jiancui (2011) used the linear regression method that forecasts the Chinese passenger and freight turnover after 10, 20, and 40 years with relevant data of Chinese transportation department from 1980 to 2008, estimated the energy consumption and $CO_2$ emission in the future transportation range, and analyzed the potential of energy saving and emission reduction in the range of transportation department. Chen Fei (2009) used the scenario analysis method to predict $CO_2$ emissions in Shanghai from now to 2020. At the same time, he also analyzed the important factors that influence the $CO_2$ emissions.

Now, Chinese scholars are still in the use of foreign advanced accounting methods and form their own accounting model after a large number of analyses. However, I believe that China will have its own transport greenhouse gas carbon emissions accounting system in the near future.

### 2.2.2 Study on the domestic transport greenhouse gas emission factors

Many domestic scholars have made many researches on the factors influencing the carbon emission. Li Zheng (2008) established a two-stage decomposition model, analyzed the factors that influence the energy consumption of passenger transport quantitatively taking Guangzhou, Shanghai, and Beijing as examples, and found the main factors that influence the energy consumption of these three cities.

Wu Cuifang (2015) used the ridge regression statistical methods based on the STIRPAT model to make a qualitative and quantitative analysis of the factors influencing the greenhouse gas emissions in Gansu province.

Xu Yanan (2011) used the existing STIRPAT model to analyze the factors influencing the greenhouse gas carbon emissions and emphatically carried on the secondary decomposition to demographic factors. It is concluded that the increase in population will lead to the increase in carbon emissions and economic growth; at the present stage of China, it will cause further damage to the environment.

### 3 THE DEVELOPMENT STATUS OF THE PUBLIC BICYCLE SERVICE SYSTEM AND THE RESEARCH STATUS OF THE GREENHOUSE GAS CARBON EMISSION REDUCTION CALCULATION MODEL

With the accelerating process of urbanization, the number of motor vehicles is increasing, and greenhouse gas emissions are also getting more and more serious. With traffic as a major source of greenhouse gas emissions, domestic and foreign experts have put forward many ways to solve the greenhouse gas carbon emissions, such as giving exclusive priority to public transport, optimizing the energy structure, and improving the automotive process. The public bicycle as a kind of public transport has been actively promoted and widely concerned.

### 3.1 Development status of the public bicycle service system

In order to reduce the environmental pollution, Amsterdam implemented the "white bicycle plan", which was called "the first generation of public bicycle system", but it was not long before it was destroyed. After 1995, Copenhagen launched "the second generation of public traffic system" with coin-operated. Since the end of the Twentieth Century, the rapid development of various emerging technologies drive the rapid development of public bicycles, and the public bicycles are known as the third-generation public bicycle system (Wang Zhigao, 2009), for which users use their real names. With the continuous improvement, the public bicycle system has been more successful in foreign countries, such as Paris, Copenhagen, and London.

1. Paris
   To ease the excessive energy consumption, environmental pollution, and traffic congestion problems, Paris in 2007 launched a bike rental business making a great contribution to Paris (Sun Ying, 2010). There are thousands of rental points and tens of thousands of bicycles in the city and the distance between the rental points of Paris is not more than 300 meters, which can facilitate the public to borrow a bike. In Paris, bicycle travel has already become a kind of fashion, not only because the city offers bike lanes, but also because the scenery on both sides of the road is very beautiful. The implementation of the public bicycle system has added an infinite luster to Paris.
2. Copenhagen
   Copenhagen is famous as "bicycle city" in the world. Not only because bicycling is the tradition of Copenhagen, but also because the government supports the development of public bicycle, the public bicycle network distributes in the region, and hundreds of services for public using (Ye Yifang, 2011). In Copenhagen, everywhere we can feel the priority of bicycle and carry out frequent bicycle movement, greatly improving people's interest rate and the usage rate of bicycle. Public bicycle made a non-negligible contribution to Copenhagen.
3. London
   In London, public bicycle which is as popular as taxi and bus has become a landmark of London (Huigang, Chinese bike). The major bright spot of the London bicycle system is that it can be rented by text messages (Li Zhenghao, 2010), the public network is intensive in the central city, and the public bicycle offers 24 hours of service, facilitating the public and visitors to travel.

However, the development of public bicycle of domestic users started late and has reached only better development in Hangzhou and Wuhan.

1. Hangzhou
   In China, Hangzhou is the first city to promote public bicycle, and invested 2500 public bicycles, 30 mobile service points, and 31 fixed points of service for the first time (Shi Xiaofeng, 2011). With the continuous advancement of the time, public bicycle system in Hangzhou is improving continuously, and now it is more user-friendly, reflected not only in the bicycle network but also in the system of borrow and lend; what is more, it embodies the people-oriented design concept at each place in Hangzhou.
2. Wuhan city
   In 2009, the first free public bicycle system for domestic users was established in Wuhan. It was funded by Wuhan Xin Fei Da Group Company. The government gives enterprise advertising rights, and the enterprise enjoyed benefits by advertising for the construction, operation, and management of public bicycle service system (Deng Lequan, 2013). With the transport development, public bicycle has become an important part of the green public transport in Wuhan, which cannot be lacked.

### 3.2 Research status of the measurement model of greenhouse gas carbon emission reduction in domestic public bicycle

Although now the domestic and foreign studies on public bicycle system are very deep, the environmental benefits brought by the public bicycle service system are not quantified. It is not intuitive to see the benefits brought by the public bicycle and cannot be rational implementation of using the public bicycles.

The following model is the only model of public bicycle greenhouse gas carbon emission reduction calculation:

Wu Anqi (2013) calculation of greenhouse gas emissions from traffic in Hangzhou used the IPCC calculation method. Moreover, based on full investigation and study on the operation mode of public bicycle, the travel sharing rate of public bicycle to motor vehicle was calculated. Then, a dynamic model of public bicycle service system was established to calculate the contribution rate of public bicycle to the environment.

Finally, she put forward some suggestions for solving the problem of the city public bicycle system development.

Zhou Qicong (2011) investigated the public bicycle service system in Guicheng, Foshan, deeply. He found out the alternative of public bicycle travel tools, statistically studied the times and the number of people using public bicycle, built the carbon emission reduction model of urban public bicycle service system, and quantified the contribution of the public bicycle service system to environment, and he also comprehensively analyzed the benefits of the public bicycle service system.

Guo Yuzhu (2013) mainly researched the public bicycle system on achieving the reduction in greenhouse gas carbon emissions related to other public transports. He analyzed and determined the measurement method of greenhouse gas carbon emission reduction in urban public bicycle through many researches. Then, taking Hangzhou as an example, he also calculated the carbon emission reduction in greenhouse gas in the public bicycle service system, and quantified the benefits to the environment in Hangzhou based on traffic data in 2010.

Wu Anqi and Shao Mengyun used the average annual total discharge of various means of transport multiplied by public bicycle transportation sharing ratio to calculate the Hangzhou public bicycle greenhouse gas carbon emission reductions, while Zhou qichung and Guo Yuzhu selected the object of public bicycle, which can reduce emissions to calculate total carbon emission reductions after providing the baseline. These two models have some subjectivity, and therefore, the result has a certain error in comparison with actual.

## 4 COMPARATIVE ANALYSIS OF DOMESTIC AND FOREIGN RESEARCH PROGRESS

### 4.1 *Shortcomings of existing models*

Domestic and foreign scholars of public bicycle system research are numerous, such as public bicycle station selection, development model, and vehicle scheduling. However, the environmental benefits of the public bicycle system can only be stopped to know, few people go to quantify the benefits of public bicycle, so that there is no perfect public bicycle greenhouse gas carbon emission reduction effect assessment method.

Nowadays, some people have put forward the quantitative model of public bicycle benefit in China. According to different places, different calculating formulas which have a great effect are given; it has opened up a precedent to quantify the public bicycle system's contribution to the transportation energy conservation and emissions reduction.

### 4.2 *Intended research content*

This paper helps us understand the composition and operation status of the public bicycle service system, and a variety of ways for people to travel, and determines the project boundary and applicable conditions of the greenhouse gas carbon emission reduction method of public bicycle service system through the study of domestic and foreign literatures and field survey.

Based on the abroad carbon emission model, this paper analyzes the key factors that affect carbon emissions one by one, and establishes a model of greenhouse gas carbon emission calculation, through investigation and study on the sharing rate of public bicycle to motor vehicle, and greenhouse gases emitted by the construction and operation of public bicycle service systems.

## 5 CONCLUSION

With the continuous development of the transportation industry, environmental problems will become more serious increasingly, prompting all around to step up the pace of building a public bicycle service system. Problems will follow, and some places are not suitable for the construction of the public bicycle service system. If building, it has not played the role of energy-saving emission reduction. Therefore, this paper studies a large number of domestic and foreign literatures, in order to build a greenhouse gas carbon emission reduction calculation model of the public bicycle service system. When a public bicycle service system is not built in a place, it can give the environmental benefits resulting from the assumption of the construction of the public bicycle service system. After the public bicycle service system has been built in a certain place, it will give the effect evaluation and suggestions of the public bicycle service system.

## ACKNOWLEDGMENTS

This paper was supported by the following three fund projects: the Jiangsu Transportation Science and Technology Project "Research on The Slow Traffic Guarantee System" (No. 2014 N03-1), the Jiangsu University Philosophy and Social Science Research Fund Project "Study on Evaluation System of Innovative City—Taking the Medium and Large Cities of Jiangsu Province for Example" (No. 2014SJB276), and the Postdoctoral

Research Funding Plan of Jiangsu Province (No. 1501147 B).

## REFERENCES

Chen Fei, Zhu Dajian, and Xu Kun. (2009). Urban low carbon traffic model, problems and Strategy—Taking Shanghai city as an example. *J. Urban Planning Journal. 6*, 40–46.

Chen H., Jia B., and Lau S.S.Y. (2008). Sustainable urban form for Chinese compact cities: Challenges of a rapid urbanized economy. *J. Habitat International. 32*, 28–40.

Chung U., Choi J., and Yun J.I. (2004). Urbanization Effect on the Observed Change in Mean Monthly Temperatures between 1951–1980 and 1971–2000 in Korea. *J. Climate Change. 66*, 127–136.

Deng Lequan. (2013). The study on feasibility of promoting the healthy development by using the public bicycle system in Wuhan. *D. Shanghai: Huazhong Normal University.*

Guo Yuzhu, Zhu Yanan, Weiming Yu, and Huang Jie. (2013). A preliminary study on the measurement of emission reduction of urban public bicycle traffic system. *J. Resource Conservation and Environmental Protection.*

IPCC. 2006 IPCC Guidelines for National Greenhouse Gas Inventories [EB/OL]. *http://www.ipcc-nggip.iges.or.jp/Public/gl/invs1.html.*

Li Zheng, Fu Feng, and Gao Dan. (2008). Analysis on the factors of energy demand of passenger transport in Chinese mega cities. *J. Journal of Tsinghua University (Natural Science Edition). 4811*, 1945–1948.

Li Zhenghao. (2010). Analysis on the scale of the long term development of urban public bicycle rental station. *J. Traffic Energy Saving and Environmental Protection. 2*, 44–46.

Lisa Göransson, Sten Karlsson, and Filip Johnsson. (2010). Integration of plug-in hybrid electric vehicles in a regional wind-thermal power system. *J. Energy Policy. 38*, 5482–5492.

Liu Jiancui. Energy conservation potential and carbon emission prediction of Chinese transportation sector. *J. Resource Science. 334*, 640–646.

Lu I.J., Lin S.J., and Lemis C. (2007). Decomposition and decoupling effects of carbon dioxide emissions from highway transportation in Taiwan, Germany, Japan and South Korea. *J. Energy Policy. 356*, 3226–3225.

Mensink C., I. De Vlieger, and J. Nys. (2000). An urban transport emission model for the antwerp area. *J. Atmospheric Environment. 327*, 4595–602.

Ou Xunmina, Xiaoyu Y., and Zhang Xilianga. (2010). Using coal for transportation in China: Life cycle GHG of coal-based fuel and electric vehicle and policy implications. *J. International Journal of Greenhouse Gas Control. 4*, 878–887.

Shan Huigang. (2014). The introduction and Inspiration of London balek public bicycle system. *J. Chinese Bike.*

Shao Mengyun, Wu Anqi. (2013). The contribution of public bicycle to traffic energy conservation and emission reduction in Hangzhou city. *J. Operation and Management.*

Shi Xiaofeng, Cui Dongxu, and Wei Wei. (2011). The research of the planning and construction and use of the public bicycle system in Hangzhou. *J. Urban and Rural Planning.*

Shipper L., Scholl L., and Price L. (1997). Energy Use and Carbon Emissions from Freight in 10 Industrialized Countries: An Analysis of Trends From 1973 to 1992. *J. Transportation Research Part D: Transport and Environment. 21*, 57–76.

Sun Ying. (2010). France Paris bicycle rental business and Its Enlightenment to China. *J. Journal of Traffic and Transportation Engineering and Information. 82*, 75–80.

Wang Zhigao, Kong Zhe, Xie Jianhua, and Yin Lie. (2009). The case and enlightenment of the third European public bike system. *J. City Traffic. 7.*

Wu Cuifang, Xiong Jinhui, Wu Wancai, Gao Wenqi, and Liu Xuebin. (2015). Analysis of the traffic carbon emission and its influencing factors in based on STIRPAT model in Gansu. *J. Glacier Frozen Soil. 37 3.*

Xu Yanan, and Du Zhiping. (2011). Measurement and factor decomposition of carbon emission in Chinese transportation industry. *J. Logistics Technology. 306*, 16–18.

Ye Xiafei, and Ye Yifang. (2011). Analysis of typical urban bicycle development mode in foreign countries and Its Enlightenment. *J. Comprehensive Transportation.*

York R. (2007). Demographic trends and energy consumption in European Union Nations: 1960–2025. *J. Social Science Research. 363*, 855–872.

Zhou Qicong, Xu Mingjian, Gao Jun, and Jian Wei. (2011). Analysis of the urban public bicycle to reduce carbon emissions by value model. *J. Association Forum.*

*Advances in Energy and Environment Research – Achour & Wu (Eds)*
© 2017 Taylor & Francis Group, London, ISBN 978-1-138-62682-9

# Three mining ideas put forward by the Chinese and their typical applications in the early 21st century

Qing-fa Chen, Zhong Yu, Qiong-ying Zhong, Chun-hui Feng & Wen-jing Niu
*College of Resources and Metallurgy, Guangxi University, Nanning, China*

ABSTRACT: Under the guidance of the sustainable development idea, three mining ideas, namely green mining, mining environment reconstructing, and synergetic mining, were put forward by the Chinese mining scientists and engineers in the early 21st century, which could coordinate the relationship between mining and environment. This paper systematically summarized the fundamental concepts, the connotations, and the technology systems of the three mining ideas, and analyzed the differences and connections among them. In addition, some applications of these ideas were introduced in detail. The examples show that the three mining ideas had far-reaching impact on the sustainable development of the Chinese mining. Meanwhile, they played a promoting role in the international mining industry.

## 1 INTRODUCTION

Mineral resources are the material foundation of the national economy, social development, and people's life. Statistics show that 95% of the energy in national economy development and 80% of the industrial raw materials are derived from mineral resources (Huang 1999). The exploitation of mineral deposits has polluted the environment. Resources shortage and environmental pollution have become a common challenge which human society would face during the sustainable development in the current world. How to balance the relationship between mineral resources exploitation and environmental protection, reduce waste emissions, and implement high efficiency mining has become the most important issue for sustainable development in mining industry.

Since the 1970s, people have gradually realized that the mining industry must take the road of sustainable development. In recent years, the main mining ideas are wasteless mining, harmonic mining, etc. (Adisa 2004). These ideas treat mining and environment as different systems and improve their relationship using the method of system engineering.

## 2 FUNDAMENTAL CONCEPTS OF THE THREE MINING IDEAS

### 2.1 *Green mining*

Under the guidance of the sustainable development of mining, the green mining idea was put forward by Qian (2003). The concept of the theory was that, through using the coal mining method, the rock strata control, and relative technologies to improve, ecological problems were caused by traditional coal mining technology. The green mining had three aspects of content. First, the mine castoff, such as groundwater, gas, land, and other resources, which were wrong treatments, were reused elsewhere. Second, the idea suggested that environmental damages could be reduced at the beginning of the mining rather than after being destroyed by adjusting mining methods. Third, controlling the rock strata movement to reduce the coal mining influence on the environment as much as possible.

### 2.2 *Mining environment reconstructing*

In order to achieve the safe and efficient mining of soft and broken ore body under the complex condition, the "mining environment reconstructing" was put forward by Gu (Zhou et al. 2004). The theory holds that the essence of mining was a reconstructing process of mining environment. Through some means of regulation, such as the ecological compensation system and the ecological compensation system, the environment for mining could be restored. It also paid attention to protect environment while mining. Moreover, mine roadways could be used as the common underground space to improve the ground transportation, save the land, etc.

### 2.3 *Synergetic mining*

A new scientific idea of synergetic mining was put forward by Chen & Su (2013). The idea suggested that the negative effects of risk factors should be treated by taking some effective measures (rock control technologies, disaster control technologies, and other related technologies) while developing resources. Overall, synergetic mining was the coordination and synchronization among the resources mining, the disaster handling, and other related technologies in the process of mining. Thus, the higher synergistic effects could be exported from the mining system.

## 3 TECHNOLOGY SYSTEMS OF THE THREE MINING IDEAS

### 3.1 Coal green mining technology system

The green mining technology mainly consists of water-preserved mining, simultaneous extraction of coal and gas, filling and strip mining, grouting and reducing subsidence, coal roadway supporting and partial waste rock underground processing, and underground gasification. The water-preserved mining referred to the protection of the groundwater resources and the utilization of the mine drainage. The coal and gas simultaneous extraction held that the coal and gas existing in the coal layer were all mine resource, and it achieved the simultaneous mining of the two resources. The filling and strip mining and the grouting and reducing referred to the reduction of the surface subsidence caused by mining and protecting the land resources and the buildings on the ground. The coal roadway supporting and partial waste rock underground processing referred to the reduction of the discharge and accumulation of the coal gangue (Figure 1).

### 3.2 Mining environment reconstructing technology system

The mining environment reconstructing technology system included the stress control technology, the mine microclimate control technology, the strata control technology, the limited space security early warning technology, the groundwater

environment regenerating technology, and the cement filling technology. The stress control technology included the induced caving technology, pressure relief groove technology, and hydraulic fracturing technology. The mine microclimate control technology consisted of the mine dust prevention technology, the mine ventilation technology, and the mine cooling technology. The strata control technology contained the technology of shotcreting support, anchor spray net support, and combined support by long anchor and short bolting. The limited space security early warning technology included the technology of megatons detection, acoustic emission testing, and stress monitoring. The groundwater environment regenerating technology included underground reservoir storage technology, water disaster prevention and control technology, underground water disaster early warning technology, groundwater resources configuration technology, and groundwater purification technology. The cemented filling technology mainly included the tailing-cemented filling technology, the cemented rock fill technology, and the block stone cemented filling concrete technology (Figure 2).

### 3.3 Synergetic mining technology system

Combined with the spirit of synergetic mining conception and the application background of mining engineering, the general goal of synergetic mining is formed in the analysis of risk factors associated with solid deposit and with engineering purposes. According to the time table of realizing the general goal, synergetic mining was divided into three periods, namely early synergy, middle synergy, and late synergy. Each period of research tasks and the corresponding technical requirement were different. The synergetic mining technology system was composed of the integrate technologies of different periods, as shown in Figure 3.

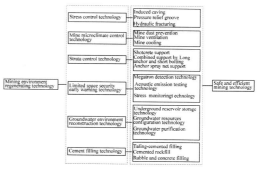

Figure 2. Reconstructed mining environment technology system.

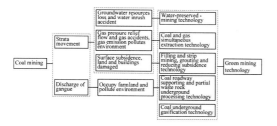

Figure 1. Green mining technology system.

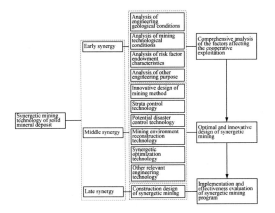

Figure 3. Synergetic mining technology system.

# 4 DIFFERENCES AND RELATIONS AMONG THE THREE MINING IDEAS

The three mining ideas had their own characteristics. The green mining emphasized the importance of environmental protection caused by traditional mining technology. It suggested that the destruction should be reduced through the coal mining methods, strata control, and other related technologies of coal mining. Mining environment reconstructing emphasized to build a good condition for the ore rock mining for the complex and soft broken ore body. It focused on the reconstructing the existing mining environment structure through taking effective engineering technologies, which could protect the mineral resources under poor mining conditions. Synergetic mining was an integrated technology, which emphasized the improvement of technologies for exploitation of resources in the destroyed environment. The final goal of the three is to achieve the efficient exploitation of mineral resources and realize the sustainable development of mining industry.

# 5 TYPICAL APPLICATIONS OF THREE MINING IDEAS IN CHINESE MINES

## 5.1 Study on green mining technology in the Huainan coal mine

The Huainan coal mine is a typical high-gas coal mine located in the southern of Huaibei plain of Anhui province in China. The geological structure of the mining area is complex and the surrounding rock of coal seams is soft rock with high stress, which makes roadway support difficult. In order to promote the mine sustainable development, a series of pressure relief and permeability increasing technologies had been developed successfully through the theoretical analysis and the field trials based on the theory of "critical layer theory of strata control" (Qian et al.

2003). The gas content in coal seam was reduced and the coal seam outburst dangerousness was eliminated. Then, the green mining technology of "coal and gas simultaneous extraction" was formed.

## 5.2 Study on mining environment reconstructing technology in the Kelatongke copper nickel mine

The Kelatongke copper-nickel mine is one of the famous copper nickel resource bases in China. It is a typical complex difficult mining ore body in which the joint fissure are well developed, the surrounding rocks are in poor stability, and the ground stress is high. It took the No.1 ore body as the test section, which had a large of fissures, rich fissure water, and little major fault. According to the conditions, the technical scheme of mining environment reconstructing deep hole-induced caving filling mining method was proposed by Zhou (2008). It suggested to regenerate a new safety mining environment which could satisfy the need for the mining engineering through the traditional horizontal descending stopping cemented filling mining technology, strata control technology, and grouting technology. Then, a large number of mining methods were used to replace the stratified filling mining method, and the low-cost filling material is used to replace the high-cost concrete (shows in Figure 4). The new mining method can improve the working conditions, reduce the intensity of labor, and keep the costs down.

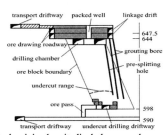

(a) Ore block mining longitudinal planar graph

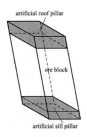

(b) Schematic diagram of mining environment reconstructing

Figure 4. Mining environment reconstructing deep hole-induced caving filling mining method.

### 5.3 Synergetic mining technology in the Gaofeng mine

The Gaofeng mine is located in Dachang Town of Nandan County. The ore grade was high and its value was precious in the fractured ore section, and therefore, a new method with high recovery and low dilution was needed. Zhou et al. (2008) proposed the deep hole shrinkage synergetic mining method under the guidance of synergetic mining, as shown in Figure 5.

The main resources were divided into three 10 m- thick ore layers. This paper only introduced the synergetic mining scheme of the layered ore body ranging from –130 m to –140 m. The pillar continuous mining method was used to mining in the layer. The layer was divided into parts A, B, and C

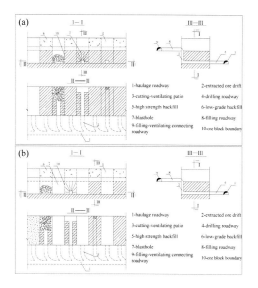

Figure 5. Layering and stripping medium-length hole caving synergetic mining method.

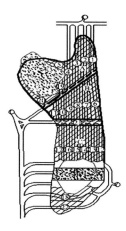

Figure 6. Synergetic mining scheme of layer No. 7.

by two of the branch roadways and 22 stripes, as shown in Figure 6.

The application effect showed that a new synergetic mining technology not only successfully handled the hidden danger with cavities and achieved the hidden danger resources safety mining, but also greatly reduced the cost of processing cavities, or even zero cost, and exported the best synergic effect; the arrangement and design of the inclined stripe was one bright spot in the synergetic mining design and embodied the essence of synergetic mining technology intensively.

## 6 CONCLUSIONS

The proposed mining ideas of green mining, mining environment reconstructing, and synergetic mining have greatly promoted the progress of theory and technology in the Chinese mining industry and obtained the highly appraisal by the mining academia. These ideas are in line with the international mining technology to the high efficiency, safety, and low-cost direction of the development of the big trend. Moreover, this would provide a further positive reference to promote the sustainable development of the global mining industry.

### ACKNOWLEDGMENTS

The authors appreciate the support provided by the National Natural Science Foundation of China (51464005) and the Scientific Research and Technological Development Projects of Guangxi (AE130063).

### REFERENCES

Adisa, A. (2004). Developing a frame work for sustainable development indicators for the mining and minerals industry. *Journal of Cleaner Production* 12: 639−662.

Chen, Q.F., Su, J.H. (2013). Synergetic mining and its technology system. *Journal of Central Southern University (Science and Technology)* 44: 732−736.

Huang, J.M. (1999). Situation and countermeasures of nonferrous metal mining in China. *World Nonferrous Metals* 14: 17−19.

Huang, Y.M. (2002). A modification on the mining method for the deep orebody in guangxi gaofeng mining ltd. *Mining Research and Development* 22: 12−14.

Qian M.G. (2003) Technology system and green mining concept. *Coal Science and Technology Magazine* 24: 1−3.

Qian, M.G., Xu, J.L. & Miao, X.X. (2003). Green technique in coal mining. *Journal of China University of Ming & Technology* 32: 343−348.

Zhou, K.P., Gao, F. & Gu, D.S. (2004). Ming environment regenerating and new thoughts on the development of mining industry. *China Ming Magazine* 16: 34−36.

Zhou, K.P., Zhu, H.L. & Xiao, X. et al. (2008). Study on the minimum safety thickness of artificial roof based on reconstructing mining environment. *Ming and Metallurgical Environment* 28: 13−17.

*Advances in Energy and Environment Research – Achour & Wu (Eds)*
© *2017 Taylor & Francis Group, London, ISBN 978-1-138-62682-9*

# Study on the application of fire-resistant cloth to prevent fire spreading between buildings

Xingshao Wu & Zhigang Song
*Faculty of Civil Engineering and Architecture, Kunming University of Science and Technology, Kunming, China*

Guangyuan Yu
*Kunming Architectural Design and Research Institute Co. Ltd., Kunming, China*

ABSTRACT: Technical feasibility and relevant requirements of the Fire-Resistant Clothes (FRC) used to prevent fire spreading between wooden buildings have been studied. Heat transfer process of buildings covered with FRC was simplified as a one-dimensional model. Heating curves under different protective conditions, air gap conditions, cloth thickness, and thermal radiation intensity are obtained using the finite difference method. The equations used for determining the thickness of FRC and calculating the temperature curve of exterior walls were obtained, and example buildings in the ancient city of Lijiang were taken to show the process of determining the thickness and the temperature curve. The results indicate that: (1) FRC can significantly reduce the surface temperature of wooden exterior wall, so that it can extend the time that the wall reaches the ignition temperature. (2) A better protective effect can be obtained if the FRC is hanging from those extruding building components to form a vertical inter-layer.

## 1 INTRODUCTION

China has a large number of historical buildings. Those buildings composed by the wooden frames and the clay or brick walls, in general, have lower fire-resistant ratings but higher fire loads. Further, these buildings are often crowdedly built with little Fire Prevention Spaces (FPS), which often makes the fire easily spread from buildings to buildings. Thus, the fire fighters have to cool the surrounding buildings while extinguishing the fire of the ignition buildings. This will require more water and more FPS. However, poor water supply and limit FPS are often encountered in the ancient city because of the crowded buildings, the narrow streets, the poor fire fighting equipment, and the inadequate water supply. Local residents created a fire-spread prevention method by using the wet blankets to cover adjacent buildings in the long term of fire fighting practices. The method is proven to be effective by the past fire extinguishing practices. Inspired by this method, the study investigates an improved method by alternating the wet blankets with the Fire-Resistant Clothes (FRC). It is feasible because the sizes of historic buildings are generally small. If this alternative method is proven to be effective, the water supply requirements and FPS can be significantly reduced.

## 2 PROTECTION OF THE BUILDINGS BY FRC

### 2.1 Problem description

The main way of heat transfer between buildings is generally thermal radiation. In this case, the radiation sources, for example, the wooden exterior wall and the unprotected windows can be treated as surface radiation sources and the corresponding radiation flux can be determined as

$$q''_{rad} = \varphi_{total} \varepsilon \sigma T^4 \tag{1}$$

where $T$ is the temperature of the radiation source (K). For most residential buildings, a maximum T can be chosen as 1187.15 K. $\varepsilon = 0.8$ is the emissivity. $\varphi_{total}$ is the angle factor. If radiation source is windows, doors, or wooden exterior wall, it can be considered as a rectangular surface radiation source (see Fig. 1). The angle factor $\varphi_{total}$ can be obtained by Eq. (2).

$$\begin{cases} X = \dfrac{a}{c}, Y = \dfrac{b}{c} \\ \varphi_{total} = 4 \times \dfrac{1}{2\pi} \times \\ \left( \dfrac{X}{\sqrt{1+X^2}} tg^{-1} \dfrac{Y}{\sqrt{1+X^2}} + \dfrac{Y}{\sqrt{1+Y^2}} tg^{-1} \dfrac{X}{\sqrt{1+Y^2}} \right) \end{cases} \tag{2}$$

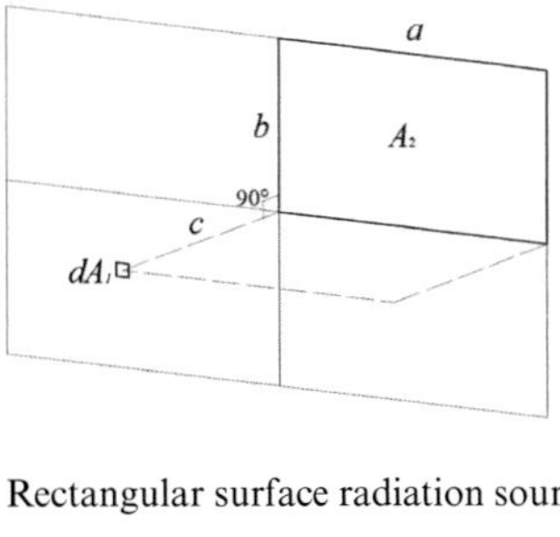

Figure 1. Rectangular surface radiation source.

## 2.2 Heat transfer from the FRC to the protected buildings

The heat transfer process can be simplified to one-dimensional, as shown in Fig. 2.

The heat transfer model consists of three layers: fire-resistant cloth, wooden exterior wall, and air inter-layer. There is no air inter-layer if the clothes and wall stick together. $q''_{rad}$ is the thermal radiation received by FRC, which can be determined by Eq.(1) and Eq.(2). For a one-dimensional heat transfer model, it has

$$\rho c \frac{\partial T}{\partial t} = \frac{\partial}{\partial x}\left(\lambda \frac{\partial T}{\partial x}\right) \quad (3)$$

where $\rho$ is the density, $c$ is the specific heat, and $\lambda$ is the thermal conductivity. For fire-resistant cloth, the values of these parameters are $\rho_{fab} = 38.3$ kg/m³, $c_{fab} = 3528$ J/kg·K, and $\lambda_{fab} = 0.125$ W/(m·K) and for wooden exterior wall, the values of these parameters are $\rho_w = 527$ kg/m³, $c_w = 2850$ J/kg·K, and $\lambda_w = 0.35$ W/(m·K). The one-dimensional heat transfer model has the following initial conditions and boundary conditions.

$$
\begin{cases}
t=0 & T = 20\ ^\circ C \\
x = L1 & -\lambda_{fab}\dfrac{\partial T}{\partial n} = \varepsilon_{Al}\sigma T_1^4 - q''_{rad}\varepsilon_{Al} \\
& \quad + h_{convair}(T_1 - T_{air}) \\
x = L2 & -\lambda_{fab}\dfrac{\partial T}{\partial n} = \dfrac{\sigma\left(T_2^4 - T_3^4\right)}{\dfrac{1}{\varepsilon_{Al}} + \dfrac{1}{\varepsilon_w} - 1} \\
& \quad + h_{convgap}(T_2 - T_3) \\
x = L3 & -\lambda_w\dfrac{\partial T}{\partial n} = \dfrac{\sigma\left(T_3^4 - T_2^4\right)}{\dfrac{1}{\varepsilon_{Al}} + \dfrac{1}{\varepsilon_w} - 1} \\
& \quad + h_{convgap}(T_3 - T_2) \\
x = L4 & -\lambda_w\dfrac{\partial T}{\partial n} = \varepsilon_w\sigma\left(T_4^4 - T_{air}^4\right) \\
& \quad + h_{convroom}(T_4 - T_{air})
\end{cases} \quad (4)
$$

where $t = 0$ is the initial condition of the model, and initial temperatures of cloth and wall are 20°C. $L1$ is the boundary condition on the side that cloth is exposed to fire. $n$ is the outward normal direction of the cloth. $q''_{rad}$ is the cloth receiving thermal radiation. $T_1$ is the surface temperature of the cloth. $\varepsilon_{Al}$ is the emissivity of the aluminum foil attached on the cloth surface. According to the study on the aluminum foil by Licai Hao, $\varepsilon_{Al} = 0.2$ is a reasonable value. $\sigma = 5.67 \times 10^{-8}$ W/m²·K is the StefanBoltzman constant. The ambient temperature, $T_{air}$, is 20°C. $h_{convair}$ is the convective heat transfer coefficient of the side that the cloth is exposed

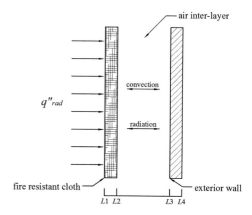

Figure 2. One-dimensional heat transfer model of using fire-resistant cloth to protect buildings.

to fire. When the temperature of clothes on the side exposed to fire reached 600°C, the ambient temperature is 20°C, and the height of clothes is 3.0 m, the convective heat transfer coefficient is $h_{convair} = 11.2$ W/(m²·K).

$L2$ is the boundary condition on the side that fire-resistant cloth is unexposed to fire. $T_3$ is the surface temperature on the side of the exterior wall. $\varepsilon_w = 0.8$ is the emissivity of the wall. $h_{convgap}$ is the convective heat transfer coefficient of air inter-layer. When the temperature on the side that the cloth contacts with the air inter-layer reached 400°C, the temperature on the side that the wall contacts with the air inter-layer reached 300°C, and the air inter-layer width is 0.1 m, the convective heat transfer coefficient is $h_{convair} = 1.7$ W/(m²·K).

$L3$ is the boundary on the side that wooden exterior wall contacts with the air inter-layer. When the cloth and the exterior wall stick together, there is no air inter-layer, so that the $L2$ boundary condition and $L3$ boundary condition should be replaced by the interface continuity condition, that mutation of heat flux density and temperature cannot occur at the interface of cloth and exterior wall.

$$\left(\lambda_{fab}\frac{\partial T}{\partial x}\right)_- = \left(\lambda_w\frac{\partial T}{\partial x}\right)_+, \quad T_- = T_+ \quad (5)$$

$L4$ is the condition of the interior side of the exterior wall, $T_4$ is the surface temperature of the exterior wall, and $h_{convroom}$ is the convective heat transfer coefficient of indoor. When the temperature of the exterior wall reached 200°C, the room temperature is 20°C, and the exterior wall height is 3.0 m, the convective heat transfer coefficient is $h_{convair} = 7.6$ W/(m²·K).

## 3 SOLVING THE HEAT TRANSFER EQUATIONS BY THE FINITE DIFFERENCE METHOD

Solving the heat transfer equations can obtain the temperature of fire-resistant cloth and wooden exterior wall. A finite difference method is used to get the numerical solution of the equations. An explicit finite difference scheme obtained by the Taylor expansion method is listed in Eq. (6). The explicit finite difference scheme of $L1$ boundary condition was given in Eq. (7).

$$T_n^{(i+1)} = \frac{\lambda \Delta \tau}{\rho c \Delta x^2}\left(T_{n+1}^{(i)} + T_{n-1}^{(i)}\right) + \left(1 - 2\frac{\lambda \Delta \tau}{\rho c \Delta x^2}\right)T_n^{(i)}; \quad (6)$$

$$1 < n < N;$$

$$
\begin{aligned}
T_1^{(i+1)} = &T_1^{(i)}\left(1 - \frac{2h\Delta \tau}{\rho c \Delta x} - \frac{2\lambda \Delta \tau}{\rho c \Delta x^2}\right) + \frac{2\lambda \Delta \tau}{\rho c \Delta x^2}T_2^{(i)} \\
&+ \frac{2h\Delta \tau}{\rho c \Delta x}T_{air} + \frac{2q\Delta \tau}{\rho c \Delta x}\left(q_{rad}''\varepsilon_{Al} - \varepsilon_{Al}\sigma T_1^4\right)
\end{aligned} \quad (7)
$$

where subscript $n$ and superscript (i) represent the location and time, respectively. $\Delta x$ is the space step and $\Delta \tau$ is the time step. The $\Delta x$ and $\Delta \tau$ should be specially chosen according to Eq. (8) to ensure convergence. In this paper, $\Delta x = 1/3$ mm and $\Delta \tau = 0.02$ s are chosen.

$$1 - 2\frac{\lambda \Delta \tau}{\rho c \Delta x^2} \geq 0, \ 1 - \frac{2h\Delta \tau}{\rho c \Delta x} - \frac{2\lambda \Delta \tau}{\rho c \Delta x^2} \geq 0 \quad (8)$$

## 4 ANALYSIS OF NUMERICAL RESULTS

### 4.1 Effect of FRC

A burning traditional timber building can generally project a radiation flux to the surrounding building as high as 20 kW/m² ~ 60 kW/m². To confirm that FRC can resist the radiation flux, six different cases are considered (see Table 1), and the corresponding temperature rising of the exterior wall is shown in Fig. 3.

As shown in Fig. 3, compared with the cases 4 to 6 (no FRC), temperatures of the outer surface of exterior wall in case 1 to case 3 (with FRC) show a significant drop up to 257°C, 323°C, and 350°C respectively, in 8000 s. There are two reasons for this phenomenon. First, the surface of the fire-resistant clothes is attached with aluminum foil. The emissivity of the aluminum foil $\varepsilon_{Al}$ is 0.2. According to Kirchoff's law, the fire-resistant clothes can absorb only 20% energy of irradiation. Second, fire-resistant clothes have lower heat conductivity, lead to heat transfer from fire-resistant clothes to wooden exterior wall have to overcome a higher thermal resistance.

### 4.2 Effect of air gap

The FRC can be installed stick together with the surface of the components or hang from the outstanding components of the protected buildings. For the later case, there is an air gap between the FRC and the surface of the building components. To analyze the effects of air gaps, six different cases are considered (see Table 2), and the corresponding temperature rising of the surface of exterior wall is shown in Fig. 4.

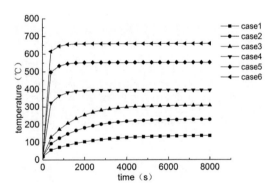

Figure 3. Effect of fire-resistant cloth.

Table 1. Working conditions of fire-resistant cloth.

| Case no. | FRC | Thermal radiation kW/m² | Thickness of the cloth mm |
|---|---|---|---|
| Case 1 | Yes | 20 | 2 |
| Case 2 | Yes | 40 | 2 |
| Case 3 | Yes | 60 | 2 |
| Case 4 | No | 20 | – |
| Case 5 | No | 40 | – |
| Case 6 | No | 60 | – |

Table 2. Working conditions of air inter-layer.

| Case no. | Air inter-layer | Thermal radiation kW/m² | Width of Air gap cm |
|---|---|---|---|
| Case 1 | Yes | 20 | 10 |
| Case 2 | Yes | 40 | 10 |
| Case 3 | Yes | 60 | 10 |
| Case 4 | No | 20 | – |
| Case 5 | No | 40 | – |
| Case 6 | No | 60 | – |

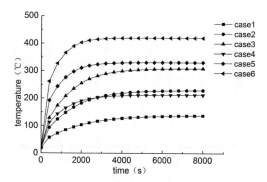

Figure 4. Effect of air inter-layer.

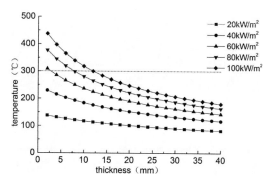

Figure 5. Cloth thickness-wall surface temperature curve.

From Fig. 4, it is clear that the temperatures in 8000 s further drop to 74.5°C, 101.1°C, and 111.6°C, respectively. This indicates that the air gap can increase the thermal resistance and decrease the wooden exterior wall temperature. In order to achieve better effect of FRC, the FRC should be hung from the prominent components, such as cornices.

## 5 DETERMINING THE THICKNESS OF FRC

A critical thickness is the minimum cloth thickness that can ensure that the wall surface temperature is always lower than the minimum ignition temperature. Li's research indicated a minimum of 3h fire duration in timber structure fire. The maximum surface temperatures of the exterior wall in 3h fire exposure time when received radiation flux are, respectively, 20 kW/m², 40 kW/m², 60 kW/m², 80 kW/m², and 100 kW/m², as shown in Fig. 5.

300°C can be viewed as a conservative critical temperature of the igniting wood. According to the critical temperature, the critical thicknesses under the different radiation fluxes are obtained (see Fig. 6).

The fitting formula of the critical thickness is obtained, which has a determination coefficient as high as 0.99, see Eq. (9).

$$h = 0.2309 \times q''_{rad} - 10.89 \qquad (9)$$

where $q''_{rad}$ is the thermal radiation flux received (kW/m²) and $h$ is the critical thickness of fire-resistant cloth (mm). The thickness of FRC is generally greater than 1 mm, and therefore, if the result of $h$ is less than 1 mm, it is still taken as 1 mm.

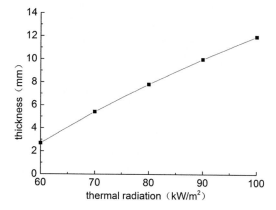

Figure 6. Critical thickness-thermal radiation intensity curve.

## 6 EXAMPLE

A typical wooden building in the ancient city of Lijiang is taken as an example. The outer walls of the building are soil and have the windows with a maximum size up to 3.0 m in width and 1.5 m in height. The width of the street and alley is 3.0 m and 1.5 m, respectively. According to Eq. (2), the $\phi_{total}$ for street and alley is 0.13 and 0.54, respectively. The thermal radiation flux across the street and alley is calculated as 11.4 kW/m² and 46.7 kW/m², respectively according to Eq. (1). From Eq. (9), the critical thicknesses of FRC are both 1 mm for street and alley. When the thickness of FRC is 1 mm, the temperature rising curve of exterior wall is calculated by numerical calculation (see Fig. 7). It is clear that the FRC with a thickness of 1 mm can well protect the surrounding buildings.

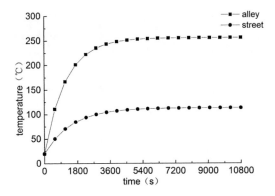

Figure 7. Surface temperature curve of the exterior wall.

## 7 CONCLUSION

The feasibility and effects of using FRC to protect the surrounding buildings from fire ignition are analyzed in this paper by one-dimensional heat transfer process analysis. The conclusion can be drawn as follows.

1. FRC can significantly reduce the highest temperature of exterior wall and extend the time of reaching the maximum temperature. Using FRC can effectively prevent fire spread between buildings.
2. An air gap can reduce the highest temperature of the exterior wall by increasing the thermal resistance of the heat transfer process. In practical applications, the FRC should be hung from the prominent components.
3. The equations determining the critical thickness of FRC have been proposed. For a typical case in the ancient city of Lijiang, a minimum thickness of 1 mm of FRC can prevent fire spread across street and alley.

ACKNOWLEDGMENTS

This study was sponsored by the National Natural Science Foundation of China (grant no. 51168021) and the National Key Technology Support Program (grant no. 2013BAJ07B01).

REFERENCES

Chen Yongming. (2009). A number of technicalfire prevention measures for historic district protection and renovation project. *J. China city*. 5, 48–50. (in Chinese).

David a. Torvi j. Douglas dale. (1999). Heat transfer in thin fibrous materials under high heat flux. *J. Fire technology*. 3.

Fan Weicheng, Wang Qingan, Jiang Fenghui, et al. (1995). Fire science concise guide. 1st edition. Hefei: *University of science and technology of china press*. (in Chinese).

Gao Ye, Huo Yan, Wu Hongmei, et al. (2010). Numerical simulation of outside temperature influence on the indoor fire. *J. Journal of Tianjin University*. 9, 777–782. (in Chinese).

Hao Licai. (2011). The Invalidation behaviour of fire protective and heat insulating flexible composite fabrics. *Donghua university*. (in Chinese).

Li Peijian. (2015). A historic district building fire duration calculation. *J. Fire Science and Technology*. 1, 69–71. (in Chinese).

Song Guowen, Patirop Chitrphiromsri, Dan Ding. (2008). Numerical simulations of heat and moisture transport in thermal protective clothing under flash fire conditions. *J. International journal of occupational safety and ergonomics*. 1, 89–106.

Song Zhigang, Chen Shuo, Bai Yu. (2008). Statistic analysis of fire accidents in urban villages of Kunming. *J. Journal of safety and environment*. 6, 112–116. (in Chinese).

Yang Xianrong, Ma Qingfang, Yuan Gengxin, et al. (1982). Radiation heat transfer angle factor manual. Beijing: *National defense industry Press*. 106–111. (in Chinese).

Yang Shiming, Tao Wenquan. (2006). Heat Transfer. 4th edition. Beijing: *Higher education press*. (in Chinese).

*Advances in Energy and Environment Research – Achour & Wu (Eds)*
© *2017 Taylor & Francis Group, London, ISBN 978-1-138-62682-9*

# Study on coal-bed methane isothermal adsorption at different temperatures based on the monolayer adsorption theory

Tingting Cai, Dong Zhao & Zengchao Feng
*Institute of Mining Technology, Taiyuan University of Technology, Taiyuan, Shanxi, China*

ABSTRACT: Adsorption volume is not only an important parameter to evaluate the porous characteristics of coal, but can also provide basic data for adsorption capacity calculation. The apparent and true adsorption volumes were introduced and the calculation of adsorption volume was modified. The differences between apparent and true adsorption volumes were compared and analyzed. The fitted formula between adsorption constant and temperature was established in this work. The results indicated that: the differences between apparent and true adsorption volumes become much more significant with the rising temperature. The adsorption ability of coal is in close relationship with specific surface, and the Langmuir constant *a* presents a slow growth due to the increase in specific surface area when the temperature was raised. Such research results not only can help understand the adsorption properties of coal mass, but also can provide appropriate guidance in exploitation of Coal-Bed Methane (CBM).

## 1 INTRODUCTION

Coal-Bed Methane (CBM) is an unconventional natural gas, with the core composition of methane. It is formed in the coal-forming process and reserved in pores and fractures of coal and surrounding rocks, mainly in adsorption state. It is of substantial guidance to study its adsorption regularity to get more information about CBM resource storage, exploration, and gas prevention and control in mines (Fu X.H. et al. 2011).

In order to utilize CBM resource more efficiently, many domestic and international scholars weigh in its adsorption regularity inquiry. Many common results indicate that, except temperature, pressure, metamorphism degree and particle size of coal samples, water, ash, and pore structure inside the coal all have different effects on its adsorption regularity. In terms of experimental test options, temperature and pressure are two main factors. The increasing pressure has a significant promotion in adsorption volume, but in the same pressure gradient, the promotion varies. Concretely, adsorption volume increases lineally with pressure in low pressure range, while it presents a slow growth in high pressure range (Zhang Q. et al. 2008, Zhang Q.L. et al. 2004). Similarly, several isothermal experiments have been carried out to study the effect of temperature on the adsorption mechanism of coal (Zhang Q. et al. 2008, Zhang Q.L. et al. 2004, and Wang G. et al. 2012). Azmi A.S. (2002) proved that adsorption volume is inversely proportional to temperature by

his adsorption experiments at low temperatures. Wang Z.J. et al. (2014) studied the influence of temperature on the adsorption characteristics of coal and came to the conclusion that under the same pressure, methane adsorption volume of coal with identical degree of metamorphism decreased as the temperature was raised. Thus, the inhibition of temperature has been proved and unified. Furthermore, Cui Y.J. (2005), Zhang Q. (2008), and Zhong L.W. (2002a) established a comprehensive temperature-pressure adsorption model of coal under adsorption potential theory and proposed a new CBM storage prediction method, which made experiment options close to the original CBM reservoir conditions quite well.

A series of widely unified achievements have been obtained in the study of factors that affect the adsorption of coal, and most of them were achieved through isothermal adsorption experiments. In China, the majority of isothermal adsorption experiments were conducted according to the national standard (GB/T19560-2008, 2008), in which the experimental material is crushed into tiny coal samples with different particle sizes ranging from 60 to 80 mesh members. Actually, coal contains many highly developed pores and fractures, and in the exploitation of CBM resource, gas and moisture all migrate along the fractures inside it; however, many of the fractures are destructive in the crushed coal samples with such tiny particle sizes, then the specific surface area of coal and the paths of molecular migrate are affected, too. Consequently, adsorption differences between experimental tests and practical engineering application resulted. These will be limited and not objective to determine the critical desorption pressure and other parameters in coal-bed methane exploitation through the data tested in laboratory with coal powders, because scale affects a lot and cannot be neglected. Thus, it is of significant importance to study the adsorption characteristics of coal with large scale to be close with engineering application.

## 2 APPARENT ADSORPTION VOLUME AND TRUE ADSORPTION VOLUME

Referring to the national standard GB/T19560–2008, in most of the isothermal experiments, the adsorption equilibrium pressure was measured, and then the free gas state equation was combined to calculate the free methane volume before and after adsorption at certain temperature and pressure spot. The difference value is the volume of adsorbed methane, and such volume is the apparent adsorption volume. In this calculation, the free space volume in adsorption instrument with coal samples inside was artificially assumed to be unchanged, and the free space volume contains three parts: the spare space after the placement of coal samples in adsorption instrument, pore and fracture volume in coal, and the air volume in connecting pipes and valves. However, in practical measurement, the free space volume was found to be changing; it varied with different gas injection pressures. Therefore, the calculation of apparent adsorption volume is not objective based on unchanged free space volume.

Actually, in methane adsorption of coal, after adsorption, there are some methane molecules absorbed on the surface of coal, which take up free space volume. However, such volume is considered as the spare space volume in instruments and gets subtracted in usual calculation, resulting in a relatively larger free gas volume, while the adsorption volume is relatively smaller. As a consequence, the apparent adsorption calculated is relatively smaller than its true adsorption volume (Yang Z.B. et al. 2011).

The critical temperature of methane is –81.6°C, and its critical pressure is 4.6 MPa, which means that it turns into super-critical fluid state under the condition that temperature and pressure both exceed the critical point. The density of the super-critical fluid quite approaches that of the liquid, and it is easily affected by temperature and pressure due to its compressibility. Under the above isothermal adsorption experiment conditions, there are obvious differences between the apparent adsorption volume and the true adsorption volume.

The relation between the true adsorption volume and the apparent adsorption volume can be obtained as follows in Moffat's study (1955):

$$V = \frac{V'}{1 - \dfrac{\rho_f}{\rho_{ad}}} \tag{1}$$

where $V$ is the true adsorption volume, cm$^3$/g; $V'$ is the apparent adsorption volume, cm$^3$/g; $\rho_f$ is the density of free methane/(g/cm$^3$), and can be calculated by the following equations: $PV = nRTZ$, $V = m/\rho_f$, $n = m/M$ (for methane, the relative molecular mass M = 16); and $\rho_{ad}$ is the density of adsorbed methane, g/cm$^3$. Generally, 0.375 g/cm$^3$ is taken.

In the study of physisorption apparatus (Li X. et al. 2009) and methane adsorption (Yang Z.B. et al. 2011), some scholars found that the apparent adsorption volume did not make much difference with the true adsorption volume in low pressure range, but once the pressure exceeded 2 MPa, their difference would increase with the growing pressure and become much more significant, and the error between them could reach 10% when the pressure reached 6 MPa. The adsorption constant is an important indicator of the adsorption ability of coal, as the error between apparent adsorption volume and true adsorption volume is set, the adsorption constant calculated by apparent adsorption volume is much different to reflect the genuine adsorption characteristic of coal. Thus, it is necessary to further understand and investigate the true adsorption volume and adsorption constant in isothermal experiments theoretically.

## 3 SATURATED ADSORPTION CAPACITY $A$

The adsorption volume can be expressed as follows based on the Langmuir monolayer adsorption theory (Langmuir I. 1916, 1918):

$$V = \frac{a_{bp}}{1 + bp} \tag{2}$$

where $V$ is the methane adsorption volume per gram in coal samples, cm$^3$/g; $p$ is the equilibrium adsorption pressure, MPa; $a$ is the adsorption constant, the saturated methane adsorption volume per gram in coal samples, cm$^3$/g; and $b$ is the adsorption constant, MPa$^{-1}$.

Constant $a$ is the saturated adsorption volume or limiting adsorption volume at a certain temperature and pressure spot, and it can reflect the adsorption ability of the adsorbent. The value of constant $a$ gets restricted by properties of both adsorbent and adsorbate. As for methane adsorption of coal, it ranges from 15 to 55 cm³/g. Based on the Langmuir monolayer adsorption theory (Langmuir I. 1916, 1918), $a$ can be calculated as:

$$a = \frac{V_0 s}{\delta N_a} \qquad (3)$$

where $V_0$ is the gas volume per mole under the standard situation, 22.4 L/mol; $s$ is the specific surface area of coal, m²/g; $\delta$ is the area a methane molecule takes up in monolayer adsorption, the diameter of methane molecular is 0.48 nm, and its sectional area is generally taken as 0.18 nm²; and $N_a$ is the Avogadro constant, $6.02 \times 10^{23}$ mol⁻¹.

From equation (3), we can see, the property of coal itself is the only factor that affects the adsorption constant. The saturated adsorption volume is proportional to the specific surface area of coal. Wen X.H. & Ma D.M. (2007) found, in their controlled trial, that an increase in the specific surface area can augment methane adsorption of coal samples significantly. Zhong L.W. (2002b, 2004) proved in her microexperiments that the adsorption ability of coal presents a positive correlation trend with the specific surface area of total pores and micropores and the larger the specific surface area, the stronger the adsorption ability of coal to storage gas. Besides, by the method of mercury intrusion porosimetry, Zhang Y.T. et al. (2007) measured the fractal pores in coal at different temperatures, and came to the conclusion that pore fractal structures increase lineally with the rising temperature. Therefore, a linear relation between the specific surface area and temperature is assumed as follows:

$$s = mT + n \qquad (4)$$

then equation (3) can be transformed as:

$$a = \frac{V_0 m}{\delta N_a} T + \frac{V_0}{\delta N_a} n \qquad (5)$$

The above equation shows that the saturated adsorption volume increases with the rising temperature. There are some preliminary conjectures as follows (Liu G.F. et al. 2009): On the one hand, as the temperature rises, coal samples are heated, keyholes, micropores, and other pore structures inside it expand or get oxidative slowly, and therefore, there are more effective vacancies left to adsorb more methane in free state. On the other hand, moisture and ash may escape from the coal by heat, and thus, new keyholes and micropores reproduce. Both facets produce more keyholes

and micropores with porosity and specific surface area larger than before. In addition, keyholes and micropores are the main places where methane adsorption of coal occurs, and hence, these explanations make it quite clear that the Langmuir constant $a$ presents a slow growth with the rising temperature, which is consistent with equation (5).

Former studies on the relationship between saturated adsorption volume and temperature are mainly fitted and analyzed based on apparent adsorption volume. Liang B. (2002) and Zhao D. (2012) elucidated a negative linear relation between the apparent adsorption volume and the rising temperature with a good correlation. As analyzed before, when the temperature rises, the saturated adsorption volume increases theoretically, while on the contrary, the apparent adsorption volume decreases. With such discrepancy, one can infer that the difference between apparent adsorption volume and true adsorption volume would become more and more significant on the same temperature gradient when the temperature rises, which means that the adsorbed volume on the surface of coal is becoming larger and larger, and such volume cannot be neglected.

In conventional isothermal adsorption experiments, the apparent adsorption volume calculated based on the unchanged free space volume is not the authentic adsorption volume. Studying the relation between apparent adsorption volume and true adsorption volume is quite beneficial to correctly understand the adsorption phenomena. In view of this, large-scale coal samples are taken as experimental material, and a series of isothermal adsorption experiments under the condition of six sets of pressure and eight temperature options were conducted to investigate the adsorption regularity of lump coals in this work. Besides, the apparent adsorption volume and the true adsorption volume were compared and analyzed, and the fitted formulas of saturated adsorption volume calculated based on true adsorption volume and temperature were established to further understand the adsorption characteristics of lump coals and offer a new reference about the proven reserves of CBM resource.

## 4 EXPERIMENTS

### 4.1 Experimental coal samples and instruments

The lump coal materials were taken from Gucheng mine and Gaohe mine in Qinshui coalfield in North China. After field sampling, the coal materials were wax sealed, and then be machined into cylindrical specimen with the size of ø100 mm × 100 mm by the core-taking drilling machine, and the two specimens were numbered as 1#, 2#, respectively. Both specimens are lean coal ($R_{o,max1}$ = 2.26, $R_{o,max2}$ = 2.02). Owing to the core-taking direction which is along the joint, both specimens have a smooth surface without large fractures. In this work, the coals with identical degree of metamorphism were chosen for mutual verification.

The main experimental equipment includes an adsorption instrument, a high-temperature adsorption platform, and a GW-1200 A temperature controller. The temperature controller has a high sensitivity in temperature rising and controlling with error range less than 1°C, and such accuracy can meet the requirement of the experiments quite well. There are also some auxiliary experimental devices, such as precise digital pressure gauge, vacuum pump, thermocouple thermometer, methane gas cylinder, and water drainage device. The accuracy of the precise digital pressure gauge is 0.001 MPa, and it can record and store the real-time gas pressure value inside the adsorption instrument in the whole experimental process. As for the water drainage device, in order to make sure that the exhaust methane volume was equivalent each time, the graduate with measurement range 1 L and the least calibration 10 ml was used to drain water for collecting methane. The experimental system is shown in Figure 1.

### 4.2 *Isothermal adsorption experiments*

After the experimental system was assembled and fixed well, a series of isothermal adsorption experiments were conducted, and these experimental steps can be divided into three stages. The

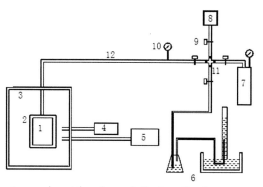

1. experimental coal sample 2.adsorption instrument
3. high-temperature adsorption platform
4. thermocouple thermometer 5.temperature controller
6.water drainage device 7.methane gas cylinder
8. vacuum pump 9. valve 10.digital pressure gauge
11. four directions device 12.pipeline

Figure 1.   Experimental system.

Table 1.   Test scheme.

| No. | Pressure/MPa | Temperature/°C |
|---|---|---|
| 1# | P1 | 20, 30, 40, ..., 90 |
| 2# | Exhaust $\Delta P$, $P_2 = P_1 - \Delta P$ | 20, 30, 40, ..., 90 |
| | Exhaust $\Delta P$, $P_3 = P_2 - \Delta P$ | 20, 30, 40, ..., 90 |
| | Exhaust $\Delta P$, $P_4 = P_3 - \Delta P$ | 20, 30, 40, ..., 90 |
| | Exhaust $\Delta P$, $P_5 = P_4 - \Delta P$ | 20, 30, 40, ..., 90 |
| | Exhaust $\Delta P$, $P_6 = P_5 - \Delta P$ | 20, 30, 40, ..., 90 |

first-stage dealt with system airtightness by the method of high-pressure helium. After accomplishment, the vacuum pump was taken on to degas the free air in the adsorption instrument. After 24 hours, it was taken off with the vacuum degree lower than 70 Pa, and then the methane gas cylinder was taken on to inject methane to certain volume. In the second stage, the isothermal adsorption experiments were conducted. In this work, eight temperature spots were set, ranging from 20 to 90°C at the interval of 10°C. In the whole adsorption process, the precise digital pressure gauge was used to record the real-time pressure of free methane inside the adsorption instrument. At the time when pressure was quite stable and the reading change was no more than 0.002 MPa in 20 minutes, it was considered that the adsorption process had reached equilibrium, and then the final equilibrium pressure was recorded. It took about 6 hours for each adsorption equilibrium. After the equilibrium, the temperature controller was set to the next temperature spot to continue adsorption. Last but not least, after the equilibrium of 90°C, the adsorption instrument was cooled naturally to room temperature, and then the device, which was used to drain water for collecting methane, was opened to exhaust some free methane by certain volume to make the free methane gas inside the adsorption instrument get a new pressure state. In the third stage, the isothermal adsorption experiments at the temperature of 20–90°C in the second stage were repeated under the new gas pressure. The above operational steps were repeated 6 times and ensured that the exhaust free methane gas volume was still the same each time. The concrete test scheme is shown in Table 1.

## 5   RESULTS

Both specimens were processed with isothermal adsorption experiments under six pressure spots and eight temperature spots. The apparent adsorption volume (absorbed methane volume per gram, $cm^3/g$) was calculated on the combination of equilibrium adsorption pressures, free gas state equa-

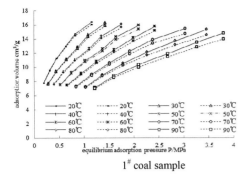

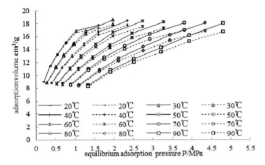

Figure 2. Contrast diagram of apparent adsorption curves and true adsorption curves in isothermal adsorption from 20°C to 90°C.

Table 2. Errors between apparent and true adsorption volume at different temperatures.

| Temperature (°C) | $\Delta V_{max}$ (cm³/g) | Error (%) |
|---|---|---|
| 1# coal sample | | |
| 20 | 0.334 | 2.09 |
| 30 | 0.400 | 2.53 |
| 40 | 0.475 | 3.05 |
| 50 | 0.544 | 3.55 |
| 60 | 0.603 | 3.98 |
| 70 | 0.719 | 4.88 |
| 80 | 0.793 | 5.47 |
| 90 | 1.058 | 7.59 |
| 2# coal sample | | |
| 20 | 0.503 | 2.92 |
| 30 | 0.634 | 3.55 |
| 40 | 0.731 | 4.13 |
| 50 | 0.827 | 4.73 |
| 60 | 0.956 | 5.55 |
| 70 | 1.070 | 6.30 |
| 80 | 1.222 | 7.27 |
| 90 | 1.335 | 8.01 |

tion, and the Langmuir monolayer adsorption theory (Langmuir I. 1916, 1918). Correspondingly, the true adsorption volume was also calculated by equation (1), and then the contrast diagram of apparent adsorption curves and true adsorption curves in isothermal adsorption at different temperatures are shown in Figure 2. Full lines refer to true adsorption curves, while the dashed ones refer to apparent adsorption curves.

The difference values and errors between adsorption volumes and true adsorption volumes at certain pressure and temperature spots were calculated and shown in Table 2. Table 2 shows that the difference between apparent adsorption volume and true adsorption volume is becoming larger and larger with the increasing temperature, and the error between them is gradually becoming much more significant, which is coherent to the curves in Figure 2. In light of actual reservoir conditions, temperature and pressure both increase with the burial depth of coal, and thus, one can infer that with the comprehensive effect of both temperature and pressure, the difference between apparent adsorption volume and true adsorption volume will become much more significant than this in deeper burial depth.

## 6 VERIFICATION OF RELATION BETWEEN SATURATED ADSORPTION CAPACITY AND TEMPERATURE

In equation (2), let $y = \frac{1}{V}$, $x = \frac{1}{p}$ then the equation (2) can be transformed as follows:

$$y = \frac{1}{ab}x + \frac{1}{a} \quad (6)$$

The saturated adsorption capacity $a$ at different temperatures were simulated based on true adsorption volume and equation (7), as can be seen in Table 3.

The relation curves of saturated adsorption capacity $a$ changing with temperatures are shown in Figure 3. Clearly, the saturated adsorption capacity grows gradually with the increasing temperatures. Equation (5) was used to fit the two curves, and the results are given in Table 4.

As the fitted formulas show, the saturated adsorption capacity calculated based on true adsorption volume is in positive linear correlation with the temperature, and the correlation coefficient is greatly near 1, which manifests that equation (5) can be used to describe the relationship between methane adsorption ability of coal and temperature quite well, in concrete terms, when the temperature rises, the specific surface of coal becomes larger, and the saturated adsorption capacity grows. Such results corroborate the former statement that the true adsorption volume increases with the increasing temperature. As the decrease in apparent adsorption volume

Table 3. Simulated results.

| Temperature/°C | Fitted formulas | $R^2$ | $a$ |
|---|---|---|---|
| (a) 1# coal sample | | | |
| 20 | y = 0.016x + 0.051 | 0.9843 | 19.608 |
| 30 | y = 0.024x + 0.049 | 0.9834 | 20.408 |
| 40 | y = 0.036x + 0.044 | 0.9915 | 22.727 |
| 50 | y = 0.048x + 0.042 | 0.9928 | 23.890 |
| 60 | y = 0.064x + 0.040 | 0.9935 | 25.000 |
| 70 | y = 0.084x + 0.037 | 0.9991 | 27.027 |
| 80 | y = 0.106x + 0.036 | 0.9997 | 27.778 |
| 90 | y = 0.128x + 0.035 | 0.9988 | 28.571 |
| (b) 2# coal sample | | | |
| 20 | y = 0.012x + 0.050 | 0.9843 | 20.000 |
| 30 | y = 0.020x + 0.047 | 0.9834 | 21.277 |
| 40 | y = 0.029x + 0.044 | 0.9915 | 22.727 |
| 50 | y = 0.041x + 0.041 | 0.9928 | 24.390 |
| 60 | y = 0.053x + 0.040 | 0.9935 | 25.000 |
| 70 | y = 0.071x + 0.037 | 0.9991 | 27.027 |
| 80 | y = 0.091x + 0.036 | 0.9997 | 27.778 |
| 90 | y = 0.115x + 0.033 | 0.9988 | 30.303 |

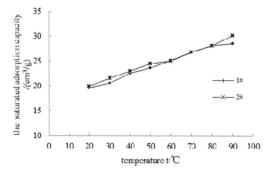

Figure 3. Relationship curves of saturated adsorption capacity $a$ changing with temperatures of both coal samples.

Table 4. Simulated results of saturated adsorption capacity $a$.

| No. | Fitted formulas | Coefficient/$R^2$ |
|---|---|---|
| 1# | $a = 0.0955t + 17.432$ | 0.9682 |
| 2# | $a = 0.0689t + 18.629$ | 0.9524 |

is compared, the conclusion that the difference between apparent adsorption volume and true adsorption volume is becoming larger and larger with the temperature is proved.

Under the actual conditions of coal reservoir, pressure and temperature will be much larger than the experiment value in certain burial depth, and the difference between apparent adsorption volume and true adsorption volume will be quite larger

than the experimental error, too, which means that the apparent adsorption volume measured in the laboratory with coal powder is not a genuine reflex of true adsorption volume, accordingly, the CBM resources reservoir prediction based on the apparent adsorption volume is not objective, and the critical desorption pressure and gas saturation calculated by the isothermal adsorption curves may not coincide with practical application very well. Thus, it is of great necessity to take the difference between apparent adsorption volume and true adsorption volume into consideration in the exploitation of coal-bed methane, and it may make theoretical analysis approach to engineering application much better.

## 7 CONCLUSIONS

1. The apparent adsorption volume and true adsorption volume were introduced and analyzed, and the difference between apparent adsorption volume and true adsorption volume becomes more and more significant with the increasing temperature. The adsorbed phase volume on the surface of coal is getting larger and larger and cannot be neglected. The true adsorption volume can be used to reflect the practical engineering situation.
2. The adsorption capacity of coal can be characterized by its specific surface, and the saturated adsorption capacity has a slow but linear growth because of the increase in specific surface area when the temperature increases.

## REFERENCES

Azmi A. S., Yusup S., Muhamad S. 2002. The influence of temperature on adsorption capacity of Malaysian coal. Chemical Engineering and Processing, 45:392–396.

Cui Y.J., Li Y.H., Zhang Q., et al. 2005. The methane adsorption character curves of coal and its use in coal-bed methane reservoir study. Chinese Science Bulletin, 10(1):76–81.

Fu X.H., Qin Y., Wei C.T., 2011, Coal-bed gas geology. Xuzhou: China University of Mining and Technology Press.

Langmuir I. 1916. The constitution and fundamental properties of solids and liquids, Journal of the American Chemical Society, 38(11):2221–2295.

Langmuir I. 1918. The Adsorption of Gases on Plane Surfaces of Glass Mica and Platinum. Journal of the American Chemical Society, 38:1361–1403.

Li X., Zhang X.H., Cong X.Z., et al. 2009, Study on adsorption capacity of coal based on low-temperature nitrogen absorption and high-pressure adsorption. Safety in Coal Mines, 5:9–11.

Liang B., 2002. Experimental study on influence of temperature on coal adsorption ability, Journal of Heilongjiang Institute of Mining, 10(1):20–22.

Liu G.F., Zhang Z.X., Zhang X.D., et al. 2009. Pore distribution regularity and absorption–desorption characteristics of gas coal and coking coal. Chinese Journal of Rock Mechanics and Engineering, 28(8):1586–1592.

Moffat D.H., Weale K.E., 1955. Sorption by coal of methane at high pressures, Fuel, 34:449–462.

Standardization Administration of the People's Republic of China. GB/T19560-2008, Methods of isothermal adsorption experiments of coal at high pressures. 2008.

Wang G., Cheng W.M., Pan G. 2012. Influence of temperature on coal's adsorption ability. Journal of Safety and Environment. 12(5):231–234.

Wang Z.J., Song W.T., Ma X.T., et al. 2014. Analysis of experiments for the effect of temperature on adsorbing capability of coal. Journal of Henan Polytechnic University (natural science), 33(5):553–557.

Wen X.H., Ma D.M. 2007. Preparation of methane sorbing material and study on adsorption/desorption characteristics. China Coal-bed Mathane, 4(2):33–37.

Yang Z.B., Qin Y., Gao D., et al. 2011. Differences between apparent and true adsorption quantity of coal-bed methane under supercritical conditions and their geological significance. Natural Gas Industry, 31(4):13–16.

Zhang Q., Cui Y.J., Zhong L.W., et al. 2008. Temperature pressure comprehensive adsorption model for coal adsorption of methane, Journal of China Coal Society, 33(11):1272–1278.

Zhang Q.L., Cui Y.J., Cao L.G. 2004. Influence of pressure on adsorption ability of coal with different deterio-ratio level. Natural Gas Industry, 1:98–100.

Zhang Y.T., Wang D.M., Zhong X.X. 2007, Features of fissure sharp in coal borehole and variation law with temperature. Coal Science and Technology, 35(11):73–76.

Zhao D. 2012. Study on coal-bed methane adsorption & desorption mechanism influenced by coupling of water injection and temperature. Taiyuan: Taiyuan University of Technology.

Zhong L.W., 2004. Adsorptive Capacity of Coals and Its Affecting Factors. Earth Science-Journal of China University of Geosciences, 29(3):327–332.

Zhong L.W., Zhang H., Yun Z.R. 2002. Influence of specific pore area and pore volume of coal on adsorption capacity. Coal Geology and Exploration, 30(3):26–28.

Zhong L.W., Zheng Y.Z., Yun Z.R., et al. 2002. The adsorption capability of coal under integrated influence of temperature and pressure and predicted of content quantity of coal bed gas. Journal of China Coal Society, 27(6):581–585.

*Advances in Energy and Environment Research – Achour & Wu (Eds)*
© *2017 Taylor & Francis Group, London, ISBN 978-1-138-62682-9*

# Evaluation of carbonate reservoir fracture on Shun Nan district in the northern slope of the central Tarim area

Bing Zhao, Yudi Geng, Yang Wang & Pandeng Luo
*Research Institute of Petroleum Engineering Northwest Oilfield Company, Urumqi, Xinjiang, China*

Meiheng Duan & Xinwei Lu
*School of Oil and Natural Gas Engineering, Southwest Petroleum University, Chengdu, Sichuan, China*

ABSTRACT: Accurate evaluation of reservoir fractures is a key factor to implement the effective development of oil and gas productivity prediction and volume fracturing. The formation of Shun Nan district in the northern slope of the central Tarim area is mainly due to carbonate rocks, controlled by strike slip fault and present stress, and therefore, fracture distribution is extremely complex. Combining with geology, core, well logging, and other materials, the paper analyzes features of geological structure, crustal stress, and reservoir natural fracture, and gets the qualitative relationship among geological structural trend, crustal stress direction, and attitude of reservoir natural fracture. In addition, based on the multiple regression theory, the fitting model of crustal stress size and natural fracture parameters is established. This study reveals that the reservoir of Shun Nan district in the northern slope of the central Tarim area is a microfracture development formation with a mass of high angle fracture. The direction of the in-situ stress in this area approximately parallels the fault (tectonic) tendency of NEE. The fracture of this formation is affected by the early geological formation movement. The present in-situ stress presents the characteristics of complex fracture distribution. In general, fracture occurrence is more affected by current stress than the early geological formation movement, correlation between crack width and in-situ stress is the biggest, and the width, length, and porosity of fracture reflect a negative contribution to the value of the in-situ stress, but the fracture density reflects a positive contribution to the value of the in-situ stress.

## 1 INTRODUCTION

As the main storage space and percolation channel of the reservoir, the fracture is a necessary condition for the formation of industrial production capacity, and the evaluation of the fracture in the reservoir is one of the important means to understand the law of oil and gas (refer Hui Zhao et al. 2008). The conventional fracture evaluation mainly includes the identification of the fracture zone and the non-fracture zone, and the identification of the type of fracture. The common well logging techniques in fracture identification have all borehole Formation Microresistivity Imaging (FMI), Formation Microresistivity Scanning Imaging (FMS), azimuth lateral imaging, and acoustic imaging (Junsheng Dai et al. 2011). The accuracy and effectiveness of the fracture are poor by using conventional logging techniques. In addition, it is difficult to identify the fracture angle, direction, density, etc. Acoustic imaging logging and electric imaging logging have obvious advantages, can accurately identify fracture width, inclination angle, and density parameters (KJ Smart & DA Ferrill 2009,

A. H. Awdal 2013), combined with core analysis, numerical simulation, seismic data, and geological theory, has been widely applied to the evaluation of reservoir fracture (Houyi Luo et al. 2004, Hengmao Tong 2006). In addition, in order to reduce the high cost of imaging logging, some scholars have put forward establishment of dual lateral logging evaluation model to estimate the fracture porosity and joint opening (Yan Jin 2002).

Tectonic process is one of the main factors of formation and expansion of formation fracture. Establishing the relationship between formation geological parameters and fractures is an important means to effectively evaluating the characteristics of reservoir fractures. Wei Ju et al. (2014) proposed the "fracture evaluation index" based on the two parameters of tectonic fracture porosity and young modulus. Huiliang Zhang et al. (2014) established the method of characterization and evaluation of fracture pore-type ultra deep reservoir under the double action of rock forming compaction and tectonic compression. However, due to the complexity of structural fractures and the particularity of the region, the systematic and quantitative comprehensive evaluation of structural fracture development is still a frontier problem to be solved (Ke Wang et al. 2015).

The northern slope of the central Tarim basin of the Ordovician top surface is a slope structure, and fathered nine major NE trending fault zones, greatly influenced by complex geological processes and the reservoir is characterized by low porosity, anisotropy, and heterogeneity, which is a typical complex reservoir of fractured carbonate rocks. In this paper, based on the geological and imaging logging data of Shun Nan district in the northern slope of the central Tarim basin, we analyzed the regional geological tectonics, in-situ stress and natural fracture feature. Further, the qualitative relationship between the geological characteristics of the area and natural fracture occurrence was build. Furthermore, the fitting model of in-situ stress and fracture parameters is established. These will provide the basis for higher yield of the reservoir fracturing.

## 2 ANALYSIS OF RESERVOIR GEOLOGICAL CHARACTERISTICS

The northern slope of the central Tarim basin is located in the NO.1 fault footwall of the central Tarim basin. On the whole, the structure is constructed by the gently inclined slope of NW treading. Shun Nan district is located in Shuntuoguole fault depression and the western part of Guchengxu uplift, and the tectonic position is shown in Figure 1 (Taizhu Huang 2014). In the process of long evolution, the Shun Nan district has been on the slope, and the northeast is adjacent to the Majiaer hydrocarbon depression, which is the favorable direction of oil and gas migration. The northern slope of the central Tarim basin developed 9 NE strike slip faults from west to east, belonging to small- and middle-scale slip faults. NE series of left lateral strike slip faults will cut the northern

slope into stripe blocks and combine features of Domino's bone plane. Moreover, there is segmentation of the Shun Nan fault, mainly controlled by two groups of NE trending strike slip faults, respectively, NE 65 degrees and NE 30 degrees; the latter one is larger, the main activities of the time is the late Caledonian S-D1-2.

According to several wells' drilling data analysis, the formation of Shun Nan district is dominated by carbonate rocks, bedding feature is obvious, formation structure angle is low, the average structural angle is about 1.2~2 degrees, the inclination is consistent, the main tendency is north and northeast. Moreover, the direction of the principal in-situ stress is determined by three methods, which are the cross-borehole short-axis azimuth, the fast-shear azimuth, and drilling-induced fracture; the analysis shows that the direction of the horizontal principal stress is mainly NEE. Owing to the limited space, only take the SN6 well as an example, Figures 2, 3 respectively, for the log analysis diagram of the tectonic for SN6 well, and the log analysis diagram of the in-situ stress orientation for SN6 well.

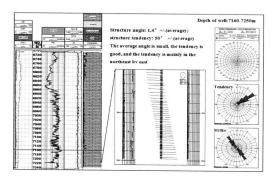

Figure 2. Log analysis diagram of the tectonic for SN6 well.

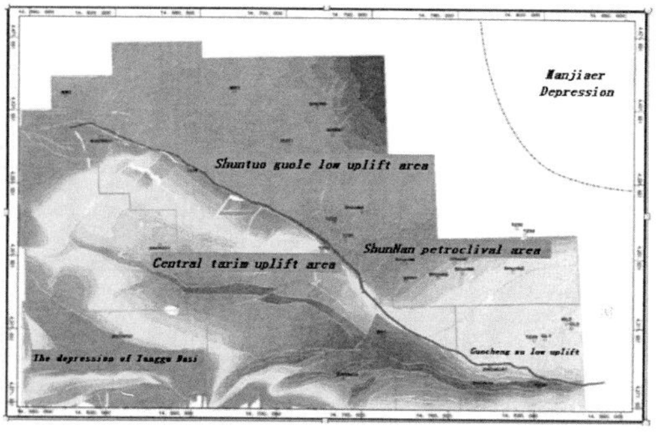

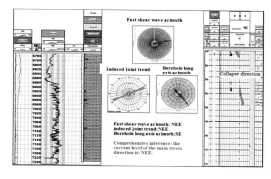

Figure 1. Tectonic map of the north slope in the central part of the Tarim basin.

Figure 3. Log analysis diagram of the in-situ stress orientation for SN6 well.

# 3 ANALYSIS OF NATURAL FRACTURE CHARACTERISTICS OF THE RESERVOIR

## 3.1 Natural fracture identification and evaluation method

1. Calculation of fracture width

The resistivity of the FMI detection range is $R_{xo}$, the sidewall has a width of $W$, and the fracture is filled with the resistivity of $R_m$ drilling fluid, which is assumed to be $R_m \ll R_{xo}$. When the measuring electrode of FMI is near the fracture, the abnormal low resistance of the drilling fluid in the fracture will cause an increase in the FMI measuring electrode current, and the phenomenon of current increases will continue to increase, until the measuring electrode is far from the fracture and is not affected by the low resistivity anomaly. S. Mluthi and P. Souhaite established the correction formula about fracture width calculation.

$$W = CAR_m^b R_{xo}^{1-b} \tag{1}$$

where $W$ is the fracture width, m; $C$ and $b$ are the constant; $R_m$ is the mud resistivity, $\Omega \cdot$ m; and $R_{xo}$ is the formation resistivity for the FMI electrode detection range, $\Omega \cdot$ m. A is the added current value caused by fracture. The following formula can be used for calculating the value:

$$A = \frac{1}{V_e} \int_{h_0}^{h_n} \left[ I_b(h) - I_{bm} \right] dh \tag{2}$$

where $V_e$ is the potential difference between the measuring electrode and the upper return electrode, V; $I_b(h)$ is the current value of the electrode at the depth of $h$, $\mu$A; $I_{bm}$ is the current measurement of natural fracture, $\mu$A; $h_o$ is the depth of the fracture when beginning to affect electrode measurements, m; and $h_n$ is the depth of the fracture, when the impact on the electrode measurement ended, m.

When the fracture dip is 0~40 degrees, the value A is basically independent of the fracture dip. When the fracture dip is more than 40 degrees, with the increase in the fracture dip, the value A has a decreasing trend. The level of the value A and the electrode are not related to the contact degree of the sidewall.

With the comprehensive application of formulas (1) and (2), the average width of the fracture can be obtained by using the weighted hydrodynamic formula:

$$W_a = \left[ \sum_{i=1}^n L_i W_i^3 \left( \sum_{i=1}^n L_i \right)^{-1} \right]^{1/2} \tag{3}$$

where $W_a$ is the average width of the fracture, mm; $W_i$ is the width of the $i$ section of the fracture on the FMI image, mm; and $L_i$ is the length of the $i$ section of the fracture on FMI image, mm.

2. Calculation of fracture length and density

Formula of the fracture length is

$$F_a = \frac{1}{2\pi RHC} \sum_{i=1}^n L_i \tag{4}$$

where $F_e$ is the cumulative fracture length of the unit area in the well section, m; $R$ is the radius of the borehole, m; $C$ is the borehole cover, dimensionless; and $L_i$ is the length of the $i$ section of fracture on FMI image, m.

Using FMI data, the fracture density can be calculated, generally expressed as a linear density:

$$F_d = \frac{1}{H} \sum_{i=1}^n L_i \tag{5}$$

where $F_d$ is the apparent fracture density, strip/m; $\sum_{i=1}^n L_i$ is the total coefficient of the well section.

## 3.2 Characteristics of regional reservoir fracture

By analyzing the core data (Figure 4) and imaging log interpretation, it can be observed that many wells can be picked up to the obvious natural opening fracture, the number of fracture is uneven, the fracture opening and porosity are very low, and the development of fracture is mainly microfractures, which indicate that there is no strong tectonic activity in the present structure. The open joint cumulative thickness of a single well is about 20.7 m. The average angle of the fracture is about 65 degrees. The main type of fracture is middle- and high-angle fracture, and the tendency is mainly North West.

Taking SN6 well as an example, the cumulative thickness of the opening fracture is 27. 9 m, which is mainly in the low-angle microfractures, and the tendency of the fracture is scattered, and has the characteristics of multicomponent fracture. A lot of induced fractures were generated due to the

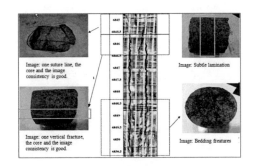

Figure 4. Rock core and imaging logging graphic of SN6 well

induction; the average dip angle is 73 degrees, the average inclination is 157 degrees, and the main tendency is North West. Table 1 is the fracture characteristics obtained from the fracture distribution compared with the normal curve; Figure 5 is the imaging logging graphic of the fracture character for SN6 well.

Table 1. Types of fractures at different depths of SN6 well.

| Depth range (m) | Types of fractures |
|---|---|
| 6799–6882 | The middle- and low-angle microfracture |
| 6885–6957 | High-angle fracture and the development of vertical fracture, localized development of microfracture in low angle |
| 6971–7030 | Characteristics of joint and middle- and low-angle microfracture |
| 7034–7043 | Characteristics of joint and low-angle microfracture |
| 7046–7060 | Joint and middle- and low-angle microfracture |
| 7122–7157 | Middle- and low-angle microfracture |
| 7179–7220 | Low-angle microfracture, localized high-angle microfracture, and the bad opening |

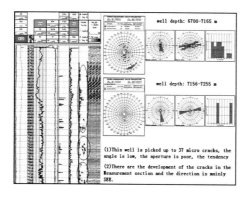

Figure 5. Imaging logging graphic of the fracture character for SN6 well.

Table 2. Orientation of the geology and fracture.

## 4 RELATIONSHIP BETWEEN NATURAL FRACTURE MORPHOLOGY AND GEOLOGICAL CHARACTERISTICS OF THE RESERVOIR

### 4.1 Establishment of the qualitative relationship between the geological features and the occurrence of natural fractures

As we know that the Shun Nan district is affected by the two groups of strike slip faults. They are NE60 degrees and NE30 degrees, namely NNE in direction and NEE in direction. Table 2 is the basic data of geology and fracture of five wells in Shun Nan district. In addition, the direction of minimum horizontal principal stress is the direction of in-situ stress.

As we can see in Table 2: ①The direction of in-situ stress is approximately parallel with the fault (structure) dip of the area for the majority of wells, such as SN1 and SN6. Some wells have some deviations, but the difference is not too large, which indicates that the fault tendency is approximately the same as the minimum horizontal principal stress due to the release of tectonic stress. ②There is a large angle deviation between the direction of the minimum horizontal stress and the direction of the fracture, which are SN1, SN6, and SN501, the difference of angle is greater than 45 degrees, and the SN5 and SN401 are approximately 45 degrees. The consistency of the tectonic tendency and fracture strike is poor, where the SN1 and SN501 are more than 45 degrees, SN5 and SN401 are less than 45 degrees, and the SN6 is roughly 45 degrees. This may be influenced by the early tectonic compression and the present in-situ stress. The early tectonic compression makes fracture growth along the same direction with the structural tendency. In the later stage, the horizontal in-situ stress direction is changed due to the in-situ stress released by the fault. The direction of fracture is affected by the present in-situ stress and perpendicular to the direction of the growth of the minimum horizontal principal stress direction. In addition, the in-situ stress is determined by a variety of methods, and each well has the fracture and the fault. The rock of

| Name | In-situ stress direction | Tectonic characters | | Fracture | | |
|---|---|---|---|---|---|---|
| | | Angle (°) | Tendency | Number and angle | Tendency | Direction |
| SN1 | NEE | 2.8 | NEE | 2-middle and high (56°) | NW | SE |
| SN5 | NNE | 1.8 | NWW | 5-middle and high (85°) | NNW | NEE |
| SN6 | NEE | 1.4 | NEE | 37-low angle | NW | SE |
| SN401 | SN | 2 | NEE | 3-middle and high (close fracture) | SW | NE |
| SN501 | NEE | 1.4 | NNW | 41-low angle (50°) | NW | SE |

sidewall is broken. The influence of the formation of elliptical hole can cause the deviation in the measurement of the in-situ stress and the occurrence of the fracture. Moreover, there will be a slight difference between the analysis and the actual situation. From the statistical analysis of the relationship between the direction of the fracture and the in-situ stress direction and the structural tendency, it can be found that the influence of the present in-situ stress is larger than the early tectonic action.

### 4.2 Establishment of the quantitative fitting relationship between the in-situ stress and natural fracture parameters

Based on the analysis of the fracture occurrence, which is greatly influenced by the present in-situ stress, on Shun Nan district in the northern slope of the central Tarim basin is discussed in section 4.1. Using SPSS, the fracture parameters were defined, including the fracture density $x_1$, fracture length $x_2$, fracture width $x_3$, and fracture porosity $x_4$ (show in Table 3) as prediction factors, and the minimum principal stress $y_1$, the maximum horizontal direction $y_2$ as the dependent variables. We can explore the relationship between the prediction factors for the contributions to the value of the dependent variable. The relationship between y and $x_i$ was analyzed using the multiple linear regression method, which is shown in formulas (6) and (7).

The correlation coefficients of R2 in the formula are 0.81 and 0.89, which show that the numerical value of the in-situ stress is closely related to the fracture parameters. We can know from formulas (6) and (7) that the fracture width and in-situ stress have the maximum correlation. The contribution variation trend of fracture width, fracture length, and fracture porosity to in-situ stress is nearly uniform, which has a negative value. In addition, the contribution of the fracture density to the in-situ stress is positive. The wider fracture and longer fracture cause larger porosity. The smaller the fracture density, the smaller the stress:

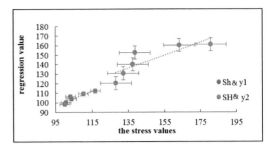

Figure 6. Contrast analysis diagram between the measured value and the regression value.

$$y_1 = 26.4x_1 - 6.8x_2 - 5914.1x_3 \\ - 2916.2x_4 + 60.7 (R^2 = 0.89) \quad (6)$$

$$y_1 = 77x_1 - 21x_2 - 28270x_3 - 8598x_4 \\ + 21 (R^2 = 0.81) \quad (7)$$

The above analysis indicates the multiple regression equation can be used to predict the in-situ stress in the well section with some known fracture parameters, as shown in Figure 6. The study shows that the Shun Nan district in the northern slope of the central Tarim basin is used to predict the direction of minimum horizontal principal stress accuracy, which was significantly higher than that in the direction of maximum horizontal principal stress.

## 5 CONCLUSION

Combined with geology, core, and well logging data, the tectonic, in-situ stress, and the characteristics of natural fracture in the reservoir have been analyzed. The relationship among them is built, the quantitative fitting relationship between the in-situ stress and natural fracture parameters is established, and the following conclusions are obtained.

1. The formation of Shun Nan district in the northern slope of the central Tarim basin is mainly dominated by carbonate rocks, and the fracture is mainly in the middle and high angle. The main types are microfracture. In addition, the area is controlled by the strike-slip fault, and there is no strong tectonic activity in the structure.
2. The direction of in-situ stress in this area is approximately parallel to the fault (tectonic) tendency, and the main direction is the North East by East. The occurrence of fracture is affected by the early tectonic and the present

Table 3. In-situ stress and fracture parameters.

| Name | Sh | SH | Density (1/m) | Length (1/m) | Width (cm) | Porosity (pu) |
|------|------|-------|------|------|------|------|
| | | | Fracture property | | | |
| SN1 | 112.4 | 161.2 | 2.74 | 2.32 | $8 \times 10^{-4}$ | 0 |
| SN5 | 104.2 | 140.2 | 2.18 | 1.55 | $7 \times 10^{-4}$ | $3 \times 10^{-5}$ |
| SN6 | 98.8 | 120.6 | 2.10 | 1.88 | $5 \times 10^{-4}$ | $9 \times 10^{-5}$ |
| SN7 | 106.3 | 152.5 | 2.21 | 2.03 | $3 \times 10^{-4}$ | $1 \times 10^{-4}$ |
| SN401 | 109.4 | 160.5 | 2.43 | 2.14 | 0 | 0 |
| SN501 | 100.2 | 130.9 | 2.12 | 1.95 | $3 \times 10^{-4}$ | $3 \times 10^{-4}$ |

in-situ stress, which is characterized by complex fracture distribution. In general, fracture occurrence affected by the in-situ stress is larger than the early tectonic action.

3. The quantitative fitting relationship between the fracture parameters and the stress can be established, and the maximum correlation between fracture width and in-situ stress can be obtained using the SPSS software. The width, length, and porosity of the fracture reflect a negative contribution to the value of the in-situ stress. However, the fracture density reflects a positive contribution to the value of the in-situ stress.

## ACKNOWLEDGMENTS

The authors thank the National Natural Science Foundation of China (Grant No. 51474185) for the financial support.

## REFERENCES

Awdal, A. H., A. Braathen, O. P. Wennberg, G. H. Sherwani. The characteristics of fracture networks in the Shiranish Formation of the BinaBawi Anticline; comparison with the TaqTaq Field, Zagros, Kurdistan, NE Iraq. Petroleum Geoscience, 2013, 19(2): 139–155.

Hengmao Tong. Application of imaging well logging data in prediction of structural fracture [J]. Natural Gas Industry, 2006, 26(9): 508–61.

Houyi Luo, Tao Guo, Yaoting Sun, et al. The evaluation of fractures on the metamorphic formation in Well CCSD-PPII [J]. Petroleum Exploration and Development, 2004, 30(2): 81–83+86.

Hui Zhao, liqiang Si Ma, Qibin Yan, etal. Daanzai stage fracture evaluation and reservoir productivity prediction method [J]. Well Logging Technology, 2008, 32(3): 277–280.

Huiliang Zhang, Zhang Ronghu, Yang Haijun, et al. Characterization and evaluation of ultra-deep fracture-pore tight sandstone reservoirs: A case study of Cretaceous Bashijiqike Formation in Kelasu tectonic zone in Kuqa foreland basin, Tarim, NW China[J]. Petroleum Exploration and Development, 2014, 41(2): 158–167.

Junsheng Dai, Zhendong Feng, Hailei Liu, et al. Some conditions apply Reservoir Fracture Evaluation Method Analysis [J]. Advances in Geophysics, 2011, 26(4): 1234–1242.

Ke Wang, Huiliang Zhang, Ronghu Zhang, etal. Comprehansive assessment of reservoir structural fracture with multiple methods in Keshen-2 gas field, Tarim Basin [J]. ActaPetroleiSinica, 2015, 36(6): 673–687.

Smart, K. J., D. A. Ferrill, A. P. Morris. Impact of interlayer slip on fracture prediction from geo-mechanical models of fault related folds [J]. American Association of Petroleum Geologists Bulletin, 2009, 93(11): 1447–1458.

Taizhu Huang. Tazhong North Slope structural analysis and hydrocarbon exploration [J]. Petroleum Geology and Experiment, 2014, 36 (3): 257–267.

Wei Ju, Guiting Hou, Shaoying Huang, et al. Evaluation of the the Sandstone Tectonic Fractures in the Yinan-Tuzi Area, Kuqa Depression [J]. ActaScientiarum Naturalium Universitatis Pekinensis, 2014, 50(5): 859–866.

Yan Jin, Xu Zhang. Research on log fracture parameter estimation and reservoir fracture evaluation method [J]. Natural Gas Industry, 2002, 22(Supp.): 64–67.

*Advances in Energy and Environment Research – Achour & Wu (Eds)*
© *2017 Taylor & Francis Group, London, ISBN 978-1-138-62682-9*

# Structure and performance of *in situ* $ZrH_{1.8}$ coatings in amounts of urea

Guoqing Yan, Yang Chen, Ming Wu, Jiaqing Peng, Jiandong Zhang & Lijun Wang
*Division of Rare Metals Metallurgy and Materials, General Research Institute for Nonferrous Metals, Beijing, China*

ABSTRACT: Coatings for preventing hydrogen escaping were formed on disk-type $ZrH_{1.8}$ by the *in situ* oxidation method. Urea was placed in a vacuum chamber and decomposed into complicated chemical compounds, including carbon, nitrogen, and oxygen source. Under experimental temperature, chemical reaction was processed. The phase structure and morphology of coatings were analyzed by X-Ray Diffraction (XRD) and Scanning Electron Microscopy (SEM). With the composition and depth of elements, bond states were tested by Auger Electron Spectroscopy (AES) and X-ray Photoelectron Spectroscopy (XPS). The results indicate that the oxidation weight gains increase with the increasing urea concentration. XRD shows that the phase structure of coatings consists mainly of monoclinic $ZrO_2$ ($m$-$ZrO_2$) and tetragonal $ZrO_x$ ($t$-$ZrO_x$). SEM depicts carbon, nitrogen, oxygen and zirconium distributions uniformly. Oxide particles accumulate densely. XPS shows that the coating included $ZrO_2$, $Zr_2N_2O$, and C-N bonds.

## 1 INTRODUCTION

Zirconium hydride materials have been found to have main application in the field of fusion reactor engineering, having high stability and hydrogen density, low neutron absorption cross-section, high neutron scattering cross-section, and good thermal conductivity (Konashi et al. 2003, Lee et al. 1962). However, the problem of hydrogen loss from zirconium hydride exists at a working temperature range, generating higher hydrogen equilibrium partial pressure causing hydrogen loss and reducing its moderating efficiency (Shoji & Inoue 1999). To prevent or slow down the hydrogen loss in a moderator from zirconium hydride, this has become the primary key issues in moderating materials applied to space reactor. The coatings that are formed on the surface of these materials are meant to be the barrier to prevent hydrogen releasing from the base materials. For hydrogen permeation barrier coatings, some ceramic materials seem promising because of their low hydro-gen diffusivity and solubility, including oxides, nitrides, and carbides (Hollenberg et al. 1995, Causey et al. 2012, Serra et al. 2005).

The effective way to prevent hydrogen permeation and diffusion is adopted surface modification technology to prepare metal oxide, nitride, and carbide coating, that the research is focused on stainless, heat-resistant alloy materials. Recent reports on hydrogen permeation barriers of zirconium hydride are mostly concentrated in the following techniques. Electroplating technique, *in situ* oxygen methods, micro-arc oxide technique, sol-gel methods, etc. (Monev 2012, Ye et al. 2013, Yan et al. 2013, Fan et al. 2014), have been used in coatings. Lee studied that the hydrogen permeation resistance of metal oxides is that H atoms permeate in organic coatings needed to break bonds of metal and oxygen, then forming O-H bonds (Aroutiounian 2007). Some researchers have verified that the carbon can capture H atoms that formed C-H bonds, preventing hydrogen penetration effectively (Zhao et al. 2005). The existence of the chemical bonding, for C-H and O-H bonds, plays an effective part in preventing hydrogen permeation.

In this paper, coatings were fabricated by utilizing urea decomposed into complicated chemical compounds, including carbon, nitrogen, and oxygen source, looking forwarding to form composite coatings.

## 2 EXPERIMENTAL METHODOLOGY

### 2.1 *Specimen preparation*

Disc-shaped zirconium hydride specimens (diameter of 20 nm; thickness of 10 nm) were prepared, then polished by models of abrasive paper, cleaned by acetone, alcohol, and de-ionized water, and finally dried. Weigh 0.1, 0, 2, 0.3, 0.4, and 0.5 g of urea, and place it with disc-shaped specimens into a vacuum chamber, vacuuming to $10^{-5}$ Pa. Heat treatment was proceeding in a tubular muffle furnace at a rate of 276 K/min up to 773 K. Coatings were fabricated on keeping the experimental temperature for 20 hours.

## 2.2 Surface examination

The phase of the coatings was analyzed by X-Ray Diffraction (XRD; D/max-2550HB, Rigaku Corp., Japan). The morphology of the coatings was examined using the Hitachi S4800 cold field emission Scanning Electron Microscope (SEM; Hitachi Corp., Japan). The chemical binding states in the coatings were investigated using X-Ray Photoelectron Spectroscopy (XPS; PHI-Quantro-SXM, Ulvac-PHI Corp., Japan). The elements composition and depth distribution were tested by Auger Electron Spectroscopy (AES; PHI-700). The binding energy of contaminated carbon (C1 s: 284.8 eV) was used as the reference.

## 3 RESULTS AND DISCUSSION

### 3.1 Oxidation weight gain

The oxidation weight gain was considered as a main method to evaluate whether chemical reactions take place and the chemical rate. In the following experiment, the qualitative changes in urea were treated as the experimental parameters. Urea has been traditionally considered to be decomposed into $NH_3$ and HNCO at high temperature. Alzueta et al. obtained the following rates and products channels for the decomposition of urea at high temperature when it is fed as an aqueous solution (Alzueta et al. 2000):

$$CO(NH_2)_2 \rightarrow NH_3 + HCNO \qquad (1)$$

$$CO(NH_2)_2 + H_2O \rightarrow 2NH_3 + CO_2 \qquad (2)$$

Figure 1 shows curve of zirconium hydride oxidation weight gains with the increasing urea concentration. It depicts that oxidation weight gain increases with the increasing urea. Some groups have investigated thermal decomposition characteristics of urea solution (Hong-Kun et al. 2010) and indicated the effective rate of decomposition is about 73.5% at 723 K, while up to 100% at 873 K. Under the experimental conditions, the carbon and nitrogen atom concentrations dissolved in the surface of the zirconium hydride samples increased in the atmosphere with the addition of urea, which enhances the atomic diffusion speed, leading to oxidation weight gained.

### 3.2 Phase structure (XRD)

Figure 2 presents X-Ray Diffraction spectrum for coatings grown on zirconium hydride. By comparing the substrate spectrum and samples with coatings, it can be observed that the main phase structure of coating is composed of monoclinic $ZrO_2$ (m-$ZrO_2$) and tetragonal $ZrO_x$ (t-$ZrO_x$). $ZrO_2$ belongs to stable monoclinic below 1000°C. Phase transition relates to the changed temperature. Monoclinic phase change to tetragonal phase begins from 1170°C (Kuznetzov 1968). In this study, factors' influence on the existence of tetragonal $ZrO_x$ (t-$ZrO_x$) was rooted in stress state or the grain size effect (Kee-Nam et al. 1987, Li et al. 2006). Meanwhile, the tetragonal phase presents black color, dense and strong adhesive with substrate. Coatings can be peeled off in the process of phase transition. Therefore, ZrN phase emerged at the position of $2\theta = 33.8°$, which indicated that the zirconium hydride capture N atom formed compounds.

From Figure 2(a), the strong-intensity peaks in the XRD spectrum correspond to sample $ZrH_{1.8}$, indicating that the obtained coatings are so thin that the X-ray penetrated the substrate reflecting

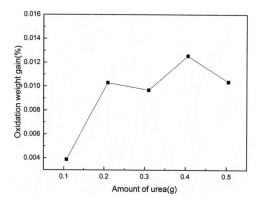

Figure 1. Curve of zirconium hydride oxidation weight gains with the increasing urea concentration.

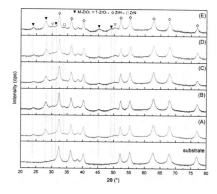

Figure 2. XRD patterns of coatings under different amounts of urea (substrate: non-coating, A:m = 0.1 g, B:m = 0.2 g, C:m = 0.3 g, D:m = 0.4 g, E:m = 0.5 g).

matrix signal. The purpose of this work is to fabricate composite coatings with carbon, nitrogen, and oxygen elements. Nevertheless, no carbon signal emerged in the XRD spectrum. The following techniques were adopted to identify the composition of the formed coating.

### 3.3 Morphology (SEM)

The oxidation weight gain is not obvious under the urea amount of m = 0.4, 0.5 g in Figure 2. Take the sample (m = 0.4 g), for example. Figure 3 shows SEM map of the surface of coatings under the amount of urea (m = 0.4 g). It can be observed that the surface coating is dense. Some particles were accumulated. Dense coatings may play an effective role in preventing hydrogen escaping.

The cross-section morphologies of the coating grown on zirconium hydride are shown in Figure 4. The thickness of the total coating is about 3.6 μm. It can be observed that the coating is composed of a dense layer. It is considered that the dense layer played an effective role in preventing hydrogen escaping. The holes were formed in the polishing process for microstructure observation. Meanwhile, Energy Dispersive Spectrometer (EDS) was brought to verify the existence of carbon, nitrogen, and oxygen elements distributed in formed coatings.

Figure 5 shows energy spectrum analysis of the cross-section of coating. The amount of oxygen element existing in the coating is higher than that in the substrate. It is ensured to be one zirconium oxidizing material by *in situ* oxidation reaction. Under experimental temperature, urea was decomposed into complicated chemical compound including carbon, nitrogen, and oxygen source. Furthermore, little amount of N element existed in the surface of $ZrH_{1.8}$ with thickness of nanometer scale. With the increase in the line scanning depth, distribution of an element is mainly made up of zirconium element which is up to substrate. It is concluded the coating is about 3.6 μm. Zirconium

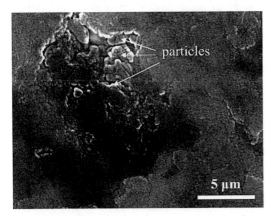

Figure 3. SEM map of the surface of coating under the amount of urea (m = 0.4 g).

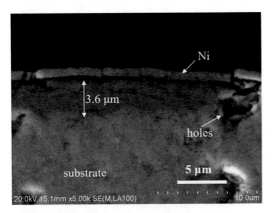

Figure 4. Cross-section morphologies of the coating grown on zirconium hydride.

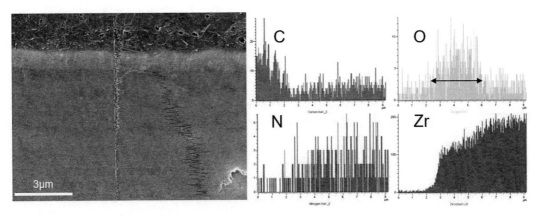

Figure 5. Energy spectrum analysis of cross-section of coating.

191

hydride can capture nitrogen, oxygen easily, and corresponding coatings were generated and the coating is made up of complex compound including Zr, N, and O elements.

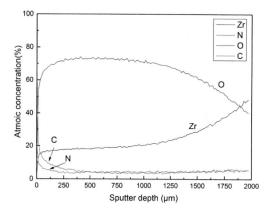

Figure 6. Relative atom concentration-depth profile from AES analysis of coating, where the sputter rate is 29 nm/min.

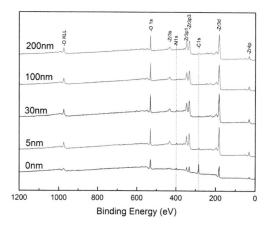

Figure 7. XPS full spectrum of coating under different sputter depths (m = 0.4 g).

### 3.4 Composition and depth distribution (AES)

Figure 6 depicts relative atom concentration-depth profile from AES analysis of coating, where the sputter rate is 29 nm/min. Four elemental species, zirconium, nitrogen, oxygen, and carbon, were identified in the coatings by assignment of corresponding signals observed in the AES spectra. It shows that the atomic concentration of carbon and nitrogen decreased continuously, while oxygen and zirconium increased. The depths of carbon and nitrogen are 100 and 130 nm, respectively. As seen, carbon-nitrogen-rich layers were distributed on the surface of coatings, the oxygen layers passed through the whole coatings. Zirconium alloy involved a number of different physical and chemical processes (Alzueta et al. 2000, Schaber et al. 1999). Urea was considered to be decomposed into $NH_3$ and HNCO at high temperature. Oxidation process is further complicated possibly by with various intermediate non-stoichiometric oxides and dissolution of oxygen into metal substrates.

### 3.5 Chemical binding states (XPS)

XPS was introduced to investigate the chemical binding states in the coatings, and the scanning range is 0~1200 eV. XPS full spectrum of coating under different sputter depths is shown in Figure 7. It is seen that coatings were composed of carbon, nitrogen, oxygen, and zirconium elements. With the increase in the sputter depth, oxygen and zirconium peaks become stronger, and carbon and nitrogen peaks become weaker, in accordance with AES or EDS results. XPS spectrum was collected at a depth of greater than 30 nm, excluding the effects of pollution elements (carbon, nitrogen).

Figure 8 depicts the Zr3d, N1 s, and O1 s XPS spectra. The Zr 3d spectra are characterized by doublet terms Zr 3d3/2 and Zr 3d5/2 due to spin-orbit coupling. The formation of ZrN is observed at binding energies of 179.4 eV (Matsuoka et al. 2008), and that of $ZrO_2$ at 182.9 eV. The O1 s spectra can be curve-fitted with two peaks: one peak at 531.3 eV is attributed to $Zr_2N_2O$

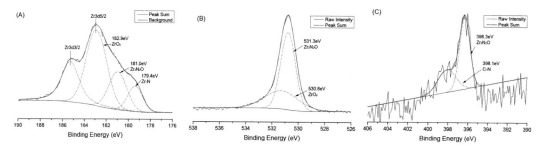

Figure 8. Elements of narrow spectrum of coating under the depth of 30 nm (A: Zr3d, B: O1s, C: N1s).

(Bazhanov et al. 2005), and the other peak at 530.8 eV to $ZrO_2$. The N 1s spectra were decomposed into three peaks, one peak at 397.9 eV to ZrN, one at 396.6 eV to $Zr_2N_2O$, and one around at 399.1 eV to absorb nitrogen (Re et al. 2003).

# 4 CONCLUSIONS

An *in situ* oxidation technique was used to fabricate composite coatings as hydrogen permeation barriers by utilizing urea as reactants with zirconium hydride at an elevated temperature. Oxidation weight gains increases with the increasing urea. The main phase structure of coating is composed of monoclinic $ZrO_2$ ($m$-$ZrO_2$) and tetragonal $ZrO_x$ ($t$-$ZrO_x$). There exists ZrN phase, indicating that the zirconium hydride captures compounds composed of nitrogen atom. SEM shows the coating composed of a dense layer. AES shows C, N, and O elements that existed in the coating. XPS depicts that the coating included $ZrO_2$, and ZrN, $Zr_2N_2O$ bonds. It has been concluded that the dense coatings and the special phase structure may play an effective role in preventing hydrogen escaping from zirconium hydride.

ACKNOWLEDGMENTS

This work was financially supported by the National Natural Science Foundation of China (No. 5140041034).

REFERENCES

Alzueta, M. U., Bilbao, R., Millera, A., M. Oliva, A. & Ibañez, J. C. (2000). Impact of New Findings Concerning Urea Thermal Decomposition on the Modeling of the Urea-SNCR Process. *Energy & Fuels,* 14, 509–510.
Aroutiounian, V. (2007). Metal oxide hydrogen, oxygen, and carbon monoxide sensors for hydrogen setups and cells. *International Journal of Hydrogen Energy,* 32, 1145–1158.
Bazhanov, D. I., Knizhnik, A. A., Safonov, A. A., Bagaturyants, A. A., Stoker, M. W. & Korkin, A. A. (2005). Structure and electronic properties of zirconium and hafnium nitrides and oxynitrides. *Journal of Applied Physics,* 97, 044108-044108-6.
Causey, R. A., Karnesky, R. A. & Marchi, C. S. (2012). 4.16–Tritium Barriers and Tritium Diffusion in Fusion Reactors. *Comprehensive Nuclear Materials,* 511–549.
Fan, X., Chen, W., Yan, S., Yan, G., Wang, Z. & Xu, Z. (2014). Oxide Films on Surface of ZrH_(1.8) Prepared by Sol-Gel Method with Different pH Values. *Chinese Journal of Rare Metals,* 38, 250–255.
Hollenberg, G. W., Simonen, E. P., Kalinin, G. & Terlain, A. (1995). Tritium/hydrogen barrier development. *Fusion Engineering & Design,* 28, 190–208.
Hong-Kun, L., Yang, W. J., Zhou, J. H., Liu, J. Z., Zhang, M. & Feng-Rui, L. I. (2010). Investigation on Thermal Decomposition Characteristics of Urea Solution Under High Temperature. *Zhongguo Dianji Gongcheng Xuebao/proceedings of the Chinese Society of Electrical Engineering,* 30, 35–40.
Kee-Nam, C., Su-Il, P. & Jong-Kyo, C. (1987). A study on the mechanism of iodine-induced stress-corrosion cracking of Zircaloy−4. *Journal of Nuclear Materials,* 149, 289–295.
Konashi, K., Ikeshoji, T., Kawazoe, Y. & Matsui, H. (2003). A molecular dynamics study of thermal conductivity of zirconium hydride. *Journal of Alloys & Compounds,* 356, 279–282.
Kuznetzov, V. A. (1968). Crystallisation of titanium, zirconium and hafnium oxides and some titanate and zirconate compounds under hydrothermal conditions. *Journal of Crystal Growth,* 3, 405–410.
Lee, R. W., Frank, R. C. & Swets, D. E. (1962). Diffusion of Hydrogen and Deuterium in Fused Quartz. *The Journal of Chemical Physics,* 36, 1062.
Li, J., Bai, X. & Zhang, D. (2006). Study on the Anodic Oxide Film and Autoclaved Oxide Film of Zircaloy-4. *Rare Metal Materials & Engineering,* 35, 1002–1005.
Matsuoka, M., Isotani, S., Sucasaire, W., Kuratani, N. & Ogata, K. (2008). X-ray photoelectron spectroscopy analysis of zirconium nitride-like films prepared on Si (100) substrates by ion beam assisted deposition. *Surface & Coatings Technology,* 202, 3129–3135.
Monev, M. (2012). Hydrogen permeation in steel during electroplating of Zn and Zn-Cr coatings. *Journal of the Electrochemical Society,* 159, D730-D736.
Re, M. D., Gouttebaron, R. & Dauchot, J. P. (2003). Study of ZrN layers deposited by reactive magnetron sputtering. *Surface & Coatings Technology,* 174, 240–245.
Schaber, P. M., Colson, J., Higgins, S., Dietz, E., Thielen, D., Anspach, B. & Brauer, J. (1999). Study of the urea thermal decomposition (pyrolysis) reaction and importance to cyanuric acid production. *American Laboratory,* 31, 13–21.
Serra, E., Calza, B. A., Cosoli, G. & Pilloni, L. (2005). Hydrogen Permeation Measurements on Alumina. *Journal of the American Ceramic Society,* 88, 15–18.
Shoji, T. & Inoue, A. (1999). Hydrogen absorption and desorption behavior of Zr-based amorphous alloys with a large structurally relaxed amorphous region. *Journal of Alloys & Compounds,* 292, 275–280.
Yan, G. Q., Chen, W. D., Zhong, X. K. & Yan, S. F. (2013). Properties of oxide coating on the surface of ZrH 1.8 prepared by microarc oxidation with different positive voltages. *Rare Metals,* 32, 169–173.
Ye, J., Nidegawa, Y. & Sakakibara, Y. (2013). An In-Situ Denitrification and Oxidation Process with Injection of Electrolytic Hydrogen and Oxygen. *Proceedings of Environmental & Sanitary Engineering Research,* 69, III_99-III_104.
Zhao, P., Kong, X. G. & Zou, C. P. (2005). Study on Hydrogen Barrier of Cr-C Alloy Fabricated by Electroplating upon Zirconium Hydride. *Nuclear Power Engineering.*

*Advances in Energy and Environment Research – Achour & Wu (Eds)*
© 2017 Taylor & Francis Group, London, ISBN 978-1-138-62682-9

# Glucose biosensor based on glucose oxidase immobilized on branched TiO₂ nanorod arrays

S. Liu & D. Fu

*State Key Laboratory of Bioelectronics, Southeast University, Nanjing, China*
*School of Biological Science and Medical Engineering, Southeast University, Nanjing, China*
*Suzhou Key Laboratory of Environment and Biosafety, Suzhou, China*

ABSTRACT: A glucose biosensor was fabricated by immobilizing Glucose Oxidase (GOx) on branched TiO₂ nanorod arrays (TiO₂ BNRs) using BSA-glutaraldehyde cross-linking technique. The structure and morphology of electrode material were characterized by scanning electron micrographs, X-ray diffraction. The electrochemical performances of the glucose biosensor were conducted by cyclic voltammetry and chronoamperometry measurements. The biosensor exhibited sensitivity of 139.69 $\mu A\,mM^{-1}\,cm^{-2}$, a linear response to glucose over a concentration range of 0.04 to 1 mM (R = 0.9888), detection limit of 28.67 $\mu M$ with the signal-to-noise ratio (S/N) of 3, and also a good reproducibility.

## 1 INTRODUCTION

The determination of blood glucose is significant for diagnosis and therapy of diabetics (Liu 2003, Mizutani F 1997). Amperometric enzymatic biosensors is widely used in the detection of glucose due to the high efficiency and specificity of enzyme. As an ideal mode enzyme, Glucose Oxidase (GOx) has been extensively used in the detection of glucose. However, the immobilization of enzyme on the electrode is a very important factor in fabricating glucose biosensors and various nanomaterials have been used for this purpose (Li 2012), such as gold nanoparticles (Galhardo 2012), platinum nanoparticles (Rakhi 2009), and carbon nanoparticles (Lu 2008). The nanostructured materials could remarkably improve the sensor activity because of larger surface areas which permit a higher enzyme loading.

Titanium dioxide (TiO₂), an important semiconductor with a wide bandgap of 3.2 eV, has attracted extensive attention in view of its outstanding physicochemical properties, such as suitable band position, non-toxicity, low cost and chemical stability. In the past few decades, TiO₂ has been extensively studied as an advanced material in various fields such as dye-sensitized solar cells, biomedical devices, photocatalysis and electrochemistry (Chen 2011). Various TiO₂ nanostructures such as nanoparticles, 1D nanowires, nanorods, and 2D nanosheets, etc., have been explored. Three-dimensional (3D) nanostructures based on hierarchically assembly have recently received significant attention due to unique characteristics such as large surface area, efficient charge separation and high carrier mobility (Liu 2014).

Single crystalline branched TiO₂ nanorod arrays (TiO₂ BNRs), a three-dimensional nanostructure with branches was grown along the backbone of the TiO₂ NRs, (Su 2013) can result in an increased surface area and enhanced electron conductivity, which is required for improved biosensor performance. In this study, the GOx enzyme is immobilized onto a TiO₂ BNRs modified FTO electrode by an optimized cross-linking technique though BSA-glutaraldehyde (Wang 2014). The obtained electrode showed a good performance for glucose detection.

## 2 EXPERIMENTAL SECTION

### 2.1 *Reagents and apparatus*

Glucose Oxidase (GOx, from Aspergillusniger, 136,300 units/g solid) was supplied by Sigma-Aldrich Co., hydrogen peroxide (H₂O₂, 30 wt.%), D-(+) glucose and other chemical reagents purchased from Sinopharm Chemical Reagent Co.. All electrochemical experiments and measurements were carried out in a 0.2 M Phosphate Buffer Solution (PBS) of pH 6.8 at room temperature. All other chemicals and reagents are of analytical grade and were prepared using distilled water.

### 2.2 *Characterization*

The crystalline structure of the prepared electrodes were investigated by a high resolution X-ray diffractometer (D8 Discover, Bruker AXS, Germany) at a voltage of 40 kV, a current of 40 mA with Cu Kα radiation source (λ = 1.5406 Å). The Morphology and composition of the samples were characterized by Field Emission Scanning Electron Micro-scope (FESEM; Zeiss Model ULTRA plus, Germa-ny). Transmission Electron Microscopic (TEM) images were obtained on a JEOL JEM-2100 electron microscope at an accelerating voltage of 200 kV. All the electrochemical measurements (except EIS) were performed on a CHI 760D electrochemical work-station (Shanghai Chenhua Instrument Co., China). Saturated Calomel Electrode (SCE) and platinum plate as reference electrode and counter electrode, respectively.

### 2.3 Growth of TiO₂ NRs and branched TiO₂ BNRs

Synthesis of $TiO_2$ BNRs on FTO substrates involves two steps. The FTO substrate (15*20 mm) ultrasonically cleaned in acetone, isopropyl alcohol, ethanol, deionized water for 15 minutes respectively. The $TiO_2$ nanorod arrays were obtained by hydrothermal methods: 40 mL of concentrated hydrochloric acid added into 50 mL of deionized water. After stirring for 5 min, the solution was mixed with 400 µL of titanium tetrachloride. Then 20 ml of the mixture prepared above was injected into a stainless steel autoclave with a Teflon lining with a volume of 50 ml. Put the cleaned FTO substrates into the Teflon lining with the conducting side facing up. The hydrothermal synthesis was conducted at 180°C for 2 h. After synthesis, the autoclave was cooled to room temperature, and FTO substrates were taken out, washed with deionized water, and dried in ambient air.

The FTO coated $TiO_2$ NRs was immersed in a 0.2 M $TiCl_4$ solution for 18 h at 25°C, then rinsing with pure ethanol, and dried at 80°C, finally the samples were annealed in air at 450°C for 30 min. After that $TiO_2$ BNRs covered FTO was obtained.

### 2.4 Immobilization of GOx

The GOx was immobilized onto the $TiO_2$ BNRs electrode by means of an optimized cross-linking technique. For the immobilization of GOx, the $TiO_2$ BNRs was immersed in PBS for 5 min. 25 mg BSA and 100 µL glutaraldehyde were dissolved in 1 mL PBS solution (pH = 6.8). The GOx solution was prepared by dissolving 2.5 mg GOx in 100 µL PBS solution. 40 µL mixture solution of BSA and glutaraldehyde was dropped onto the surface of $TiO_2$ BNRs electrode, then dried in air for 15 min. After that, 40 µL GOx solution was dropped on to the surface of the electrode. It allowed the amino groups of GOx to react with the free aldehyde group of glutaraldehyde. As a result, the GOx could be immobilized at the upper surface of the electrode. Consequently, the active sites of GOx were able to react with glucose efficiently when used for the detection of glucose. After drying at 4°C overnight, the electrode was immersed in PBS to remove unimmobilized enzymes, and then dried at room temperature.

## 3 RESULTS AND DISCUSSION

### 3.1 Morphology and structure

Figure 1a shows the typical FESEM images of $TiO_2$ NRs on FTO substrates. It can be observed that the FTO substrate was uniformly covered with ordered $TiO_2$ nanorods. After the treatment by $TiCl_4$ solution and anneal, nano-branches appeared along the trunks of the $TiO_2$ which were showed in Figure 1(b and c). It indicates that the $TiO_2$ NRs were decorated with secondary nanostructures and displayed a branched feature which greatly improves the surface area of $TiO_2$ layer. Figure 1d shows the TEM image of $TiO_2$ BNRs. As shown in Figure 1d, a hierarchically assembled 3D nanostructure includes branches grown along the backbone of the $TiO_2$ NRs.

To examine the phase structure of the materials, XRD was obtained, as shown in Figure 2. It indicates that FTO substrates have a tetragonal rutile phase of $SnO_2$ (JCPDS 41-1445). All the peaks of $TiO_2$ BNRs can be indexed to pure rutile phase of $TiO_2$ (JCPDS 21-1276).

### 3.2 Electrochemical experiment

Figure 3a shows the typical Cyclic Voltammetric (CV) sweep curve for $TiO_2$ BNRs and $TiO_2$ BNRs/

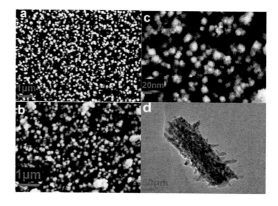

Figure 1. FESEM images of $TiO_2$ NRs (a), $TiO_2$ BNRs at low magnification (b) and high magnification (c). TEM image of $TiO_2$ BNRs (d).

196

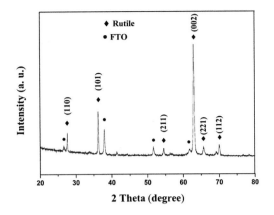

Figure 2. XRD of TiO$_2$ BNRs electrode.

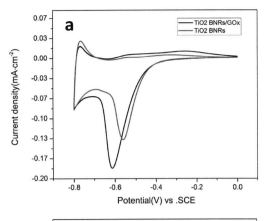

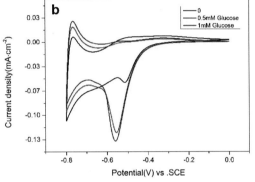

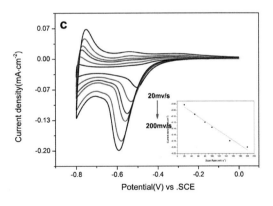

Figure 3. Detection of glucose in 0.2 M PBS (pH 6.8). (a) CV curves of TiO$_2$ BNRs, TiO$_2$ BNRs/GOx electrodes with 1 mM glucose at a scan rate of 100 mV s$^{-1}$. (b) CV curves of TiO$_2$ BNRs/GOx electrodes in different glucose concentrations (0, 0.5 and1 mM) at a scan rate of 100 mV s$^{-1}$. (c) CV curves of TiO$_2$ BNRs/GOx electrodes measured at different scan rate (20–200 mV s$^{-1}$) with 1 mM glucose.

GOx electrode respectively in 0.2M PBS (pH = 6.8) with 1 mM glucose at scan rate of 100 mv s$^{-1}$. TiO$_2$ BNRs electrode have a reduction peak at −0.56 V owing to the reduction of Ti$^{4+}$ to Ti$^{3+}$ when potential is lower than −0.4 V. These peaks are characteristic for nanostructured TiO$_2$ and have been interpreted as the reversible filling of surface states below the conduction band edge (Bisquert 2008). After the immobilization of GOx, the reduction peaks moved to more negative potential and current was also increased, that confirmed the loading of GOx at the electrode. The CV curves of TiO$_2$ BNRs/GOx electrode in different glucose concentration were displayed in Figure 3b. Obviously, the cathodic peak current exhibited an increase with glucose concentration which indicated TiO$_2$ BNRs/GOx electrode had an electrocatalytic activity for glucose oxidation. Figure 3c showed the CV curves of the TiO$_2$ BNRs/GOx electrode at scan rates from 20 mv s$^{-1}$ to 200 mv s$^{-1}$ with 1 mM glucose in electrolyte solution. The reduction peak current presented a linear relationship vs. the scan rate indicated that the electrochemical reaction was a surface-controlled process (inset of Figure 3c).

The typical amperometric responses of the enzyme electrodes in a stirred PBS (0.2 M, pH = 6.8) at potential of −0.5 V (vs. SCE) are shown in Figure 4. The corresponding calibration curve of the sensor is shown in inset of Figure 4. The TiO$_2$ BNRs/GOx based biosensors exhibited a linear dynamic range from 0.04 to 1 mM (R = 0.9888), sensitivity of 139.69 µA mM$^{-1}$ cm$^{-2}$ and detection limit of 28.67 µM with the signal-to-noise ratio (S/N) of 3.

The stability of TiO$_2$ BNRs/GOx biosensor was investigated over several days (Figure 5a). After being stored for 7 days, the response decreased about by 13%. This implied that the TiO$_2$ BNRs/GOx retain the bioactivity of enzyme and moderate stability.

In addition, the selectivity of the TiO$_2$ BNRs/GOx biosensor was evaluated in the presence of other interference species such as Uric Acid (UA), Ascorbic Acid (AA) and NaCl (Figure 5b).

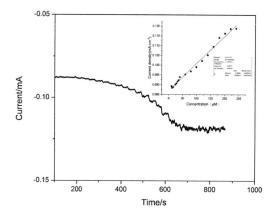

Figure 4. Amperometric response curve of TiO₂ BNRs/ GOx electrode at a potential of −0.5 V upon sequential addition of glucose.

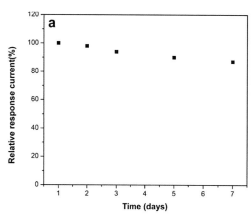

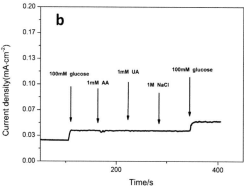

Figure 5. (a) Stability of the TiO₂ BNRs/GOx biosensor stored in dry condition at 4°C. (b) TiO₂ BNRs/GOx electrode amperometic response to UA, AA and NaCl.

It can be seen that the electrode had a remarkable response to glucose and was insensitive to interfering species.

## 4 CONCLUSIONS

In summary, We have successfully synthesized branched TiO₂ nanorods electrode and immobilizate glucose oxidase on it to use as glucose biosensor. This glucose biosensor shows a linear response to glucose over a concentration range of 0.04 to 1 mM (R = 0.9888), sensitivity of 139.69 $\mu A$ mM$^{-1}$ cm$^{-2}$ and detection limit of 28.67 $\mu M$ with the signal-to-noise ratio (S/N) of 3.

## REFERENCES

Bisquert J, Fabregat-Santiago F, and Mora-Sero I, et al. (2008). A review of recent results on electrochemical determination of the density of electronic states of nanostructured metal-oxide semiconductors and organic hole conductors[J]. Inorg Chim Acta, 361(3): 684–698.

Chen, X, Liu, L, Yu, PY, and Mao, SS. (2011). Science. 331, 746–750.

Galhardo KS, Torresi RM, and de Torresi SIC. (2012). Improving the performance of a glucose biosensor using an ionic liquid for enzyme immobilization. On the chemical interaction between the biomole cule, the ionic liquid and the cross-linking agent. Electrochim Acta 73: 123–128.

Li JH, Kuang DZ, Feng YL, Zhang FX, and Liu MQ. (2012). Glucose biosensor based on glucose oxidase immobilized on a nanofilm composed of mesoporous hydroxyapatite, titanium dioxide, and modified with multi-walled carbon nanotubes. Microchim Acta 176(1–2): 73–80.

Liu C, Li Y, and Wei L, et al. (2014). CdS quantum dot-sensitized solar cells based on nano-branched TiO₂ arrays. Nanoscale Res Lett 9(1): 1–8.

Liu S, and Ju H. (2003). Reagentless glucose biosensor based on direct electron transfer of glucose oxidase immobilized on colloidal gold modified carbon paste electrode[J]. Biosens. Bioelectron, 19(3): 177–183.

Lu J, Do I, Drzal LT, Worden RM, and Lee I. (2008). Nanometal-decorated exfoliated graphite nanoplatelet based glucose biosensors with high sensitivity and fast response. ACS Nano, 2(9): 1825–1832.

Mizutani F, and Yabuki S. (1997). Rapid determination of glucose and sucrose by an amperometric glucose-sensing electrode combined with an invertase/mutarotase-attached measuring cell [J]. Biosens. Bioelectron, 12(9): 1013–1020.

Rakhi RB, Sethupathi K, and Ramaprabhu S. (2009). A glucose biosensor based on deposition of glucose oxidase onto crystalline gold nanoparticle modified carbon nanotube electrode. J Phys Chem B 113(10): 3190–3194.

Su F, Wang T, and Lv R, et al. (2013). Dendritic Au/ TiO₂ nanorod arrays for visible-light driven photoelectrochemical water splitting. Nanoscale. 5(19): 9001–9009.

Wang W, Xie Y, and Wang Y, et al. (2014). Glucose biosensor based on glucose oxidase immobilized on unhybridized titanium dioxide nanotube arrays. Microchim Acta. 181(3–4): 381–387.

*Advances in Energy and Environment Research – Achour & Wu (Eds)*
© 2017 Taylor & Francis Group, London, ISBN 978-1-138-62682-9

# Hydrothermal fabrication of hierarchically boron- and lanthanum-modified TiO$_2$ nanowire arrays on FTO glasses

W. Zhang & D. Fu

*State Key Laboratory of Bioelectronics, Southeast University, Nanjing, China*
*College of Biological Science and Medical Engineering, Southeast University, Nanjing, China*
*Suzhou Key Laboratory of Environment and Biosafety, Suzhou, China*

ABSTRACT: We report the synthesis of boron and lanthanum co-modified TiO$_2$ with single crystalline phase by the one-pot hydrothermal method. A double-layered photocatalyst made of TiO$_2$ flowers (TFs) as the top layer and TiO$_2$ nanowire arrays (TNWs) as the bottom layer has been obtained. The hierarchical boron and lanthanum co-modified TiO$_2$ nanostructure exhibits excellent catalytic activity for photocatalytic degradation.

## 1 INTRODUCTION

Over the past 4 decades, combustion by products, industrial wastewater discharge, and pesticides abuse have resulted in serious environmental problems (Irie 2003). Until Fujishima and Honda (Fujishima 1972) reported that titanium dioxide has a photocatalytic activity in water in 1972; semiconductor photocatalysis has been one of the most promising technologies used for environmental cleaning to solve the problems of the current environmental pollution faced by human beings. Photocatalytic degradation of various hazardous organic compounds dissolved in water using photocatalysts has been widely studied (Chen 2010, Turchi 1990). Photocatalysis using semiconductors such as TiO$_2$ has been demonstrated to be an inexpensive and effective method for treating a wide range of pollutants, including alkanes, alcohols, carboxylic acids, alkenes, phenols, dyes, PCBs, aromatic hydrocarbons, halogenated alkanes, surfactant, and pesticides, in both water and air (Williams 2008, Nosaka 1998).

However, TiO$_2$ has a large band gap which makes it active only in the ultraviolet range. It seriously limits the practical application of the TiO$_2$ photocatalysis. Design and development of TiO$_2$ photocatalysts, which operate effectively under visible light or solar beam irradiation, are required for a large-scale utilization of TiO$_2$ photocatalysts (Tachikawa 2011). There have been many attempts to modify TiO$_2$ by tuning the crystallite size, depositing noble metals, compositing semiconductors, sensitizing with other semiconductor materials that have small band gaps such as CdSe, and doping with metal impurities such as Fe, Zn,

Sn, or nonmetal impurities such as B, C, N, and F (Yu 2002, Ohno 2003, Woan 2009, Zhang 2014). Among these approaches, doping impurities make it as the most effective method for enhancing the photocatalytic activity of TiO$_2$.

In this report, we present the study of preparing TiO$_2$ particles co-doped with boron and lanthanum (B-La-TiO$_2$) on FTO glasses by the hydrothermal method. The photocatalytic performance was investigated by photocatalytic degradation of methyl orange under visible light irradiation, which shows improved photocatalytic activity compared with undoped TiO$_2$.

## 2 EXPERIMENTAL SECTION

### 2.1 *Materials*

All solvents and chemicals were used without further purification. Lanthanum nitrate hexahydrate (>99.9%), Tetra-N-Butyl Titanate (TNBT, >97%), and boric acid (ACS reagent, >99.5%) were purchased from Sigma-Aldrich. Ethanol (>99.7%), acetylacetone (>99.5%), and hydrogen nitrate (≥65%) were purchased from Sinopharm Chemical Reagent Beijing Co, Ltd. FTO glasses were purchased from Nip-pon Sheet Glass, Japan, and the resistance is ~14 ohm per square.

### 2.2 *Characterization*

The crystalline structure of samples was analyzed by X-Ray Diffraction (XRD, Bruker AXS D8). The morphology was characterized by Scanning Electron Microscopy (SEM, Hitachi S-4800).

### 2.3 *Preparation of B-La-TiO₂ nanowire arrays*

Highly ordered B-La-TiO$_2$ nanowire arrays were pre-pared by the one-pot hydrothermal method. 1 mL Tetra-N-Butyl Titanate (TNBT) was added to the mixture solvent containing 30 mL deionized water and 30 mL HCl (37%). Then, 0.01 M lanthanum nitrate hexahydrate and 0.01 M boric acid were added into the mix solution subsequently. Before transfer-ring the mixture solution to a 100 mL Teflon-lined stainless steel autoclave, the solution should be stirred for 30 min. Then, two pieces of cleaned FTO glasses (15 mm × 20 mm, ultrasonic processing in acetone, 2-propanol, and deionized water for 10 min, respec-tively) were put into the solution and placed at an angle against the wall of the Teflon lined with the con-ducting side facing down. The hydrothermal process was conducted at 150 °C for 5 h in a vacuum oven and then cooled down to room temperature under flowing water. Then, the products were transferred to a muffle furnace and calcinated for 3 h at 700°C.

### 2.4 *Photoreactivity measurement*

The reactions were carried out in a multifunctional photochemical reactor (BL-GHX-V, Shanghai Bilon Experiment Equipment Co. Ltd., Shanghai, China) with freshly prepared pure TiO$_2$ and doped TiO$_2$. Photocatalysts were transferred into a beaker containing 100 mL methyl orange aqueous solution to a concentration of 10 mg L$^{-1}$. Before exposure under the xenon lamp, the mixture was reacted in the dark under strong stirring at room temperature for 30 min. The suspension liquid was then exposed to light under stirring at room temperature for 2 h. Aliquots (2 mL) of the solution were taken out every 30 min and centrifuged under 10000 rad min$^{-1}$ for 8 min, before transferring the supernatant to a cuvette and determining the absorbance under 464 nm.

## 3 RESULTS AND DISCUSSION

### 3.1 *XRD measurement*

The phase structures of the samples were analyzed by XRD spectra as shown in Figure 1. As can be seen in Figure 1, all peaks of the samples are in accordance with the standard pattern of TiO2 (JCPDS 21–1276) which present rutile phase. Fur-thermore, B and La elements doping did not affect the XRD pattern. This phenomenon may be due to the low concentration of those species that did not affect the crystal structure.

### 3.2 *SEM measurement*

Figure 2 shows the Scanning Electron Microscopic (SEM) images of the as-prepared B-La-TiO$_2$

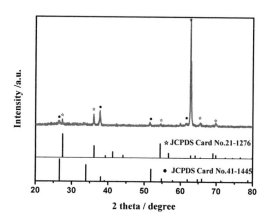

Figure 1. XRD pattern of the as-prepared B-La-TiO$_2$ sample.

samples on FTO glasses. It exhibited regular arrays for the undoped TiO$_2$ (Fig. 2a), while crater becomes forming when doping with B and La (Fig. 2b-2d). Moreover, there were a lot of flower-like structures appearing for the B-La-TiO$_2$ samples, which may be due to the doping species.

In order to verify the element in nanowire arrays and in nano-flowers, SEM-EDS measurement was performed. As shown in Figure 3, we can easily discover that main elements in both underlying nanowire array and top flowers were B, La, Ti, and O elements, which confirm the formation of TiO$_2$.

Other SEM-EDS images were further analyzed for the change when the doping ratio is increasing. Figure 4 exhibited the SEM image of B-La-TiO$_2$ samples with five times doping ratio. Interestingly, the morphology did not show the conventional nanowire arrays or nano-flowers, but displayed a newfangled nano-cluster morphology. EDS mapping images was carried out to confirm the distribution of elements, and obtained a clear distribution of all B (red), La (green), Ti (blue), and O (cyan) elements. This phenomenon illustrated that, if the doping ratio of B and La increased, a morphological change will appear. The interested novel morphology has a high surface area that is beneficial to the catalytic activity of photocatalytic degradation process.

### 3.3 *Photocatalysis experiment*

Figure 5 shows the degradation rate vs. reaction time images of the as-prepared B-La-TiO$_2$ sam-ples. We can know that when the reaction time is greater than 5 h, the rate of degradation remains at a high value. Therefore, we used the 5 h through our whole experiment.

Figure 6 shows the degradation rate vs. cal-cination temperature images of the as-prepared B-La-TiO$_2$ samples. As is easily seen that when the

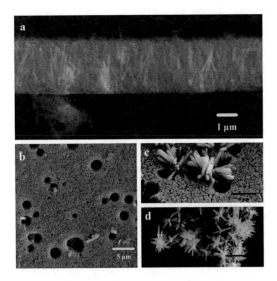

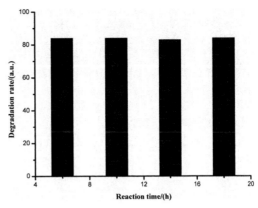

Figure 5. Degradation rate vs. Reaction time images of the as-prepared B-La-TiO$_2$ samples.

Figure 2. SEM images of the as-prepared TiO$_2$ (a) and B-La-TiO$_2$ (b-d) samples.

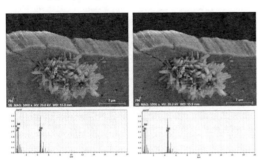

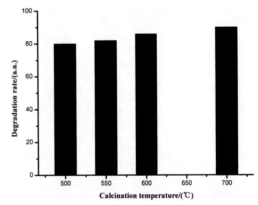

Figure 3. SEM-EDS pattern of the as-prepared B-La-TiO$_2$ sample.

Figure 6. Degradation rate vs. calcination temperature images of the as-prepared B-La-TiO$_2$ samples.

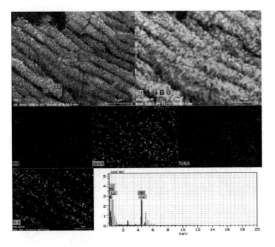

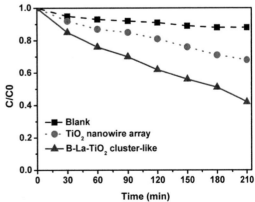

Figure 4. SEM-EDS images of the as-prepared B-La-TiO$_2$ samples with the increased doping ratio.

Figure 7. Degradation rate vs. irradiation time images of the as-prepared B-La-TiO$_2$ samples.

temperature is fixed at 700 °C, the catalytic activity becomes best (indicating a high degradation rate).

Figure 7 shows the degradation rate of blank, $TiO_2$ nanowire array, and B-La-TiO2 cluster-like samples.

When co-doped with B and La, the photocatalytic activities were enhanced compared with undoped $TiO_2$. The co-doped $TiO_2$ exhibited excellent photocatalytic activity, which could be assigned to the synergistic effect of doped B and La atoms. The crucial factor in the improvement of photocatalytic activity depends on prolonging the life of e- and H+ in the photocatalysts.

## 4  CONCLUSIONS

A novel photocatalyst, B-La-$TiO_2$ nanowires or cluster-like samples were synthesized by a simple one-step hydrothermal method. These B-La-$TiO_2$ nanostructured products exhibit excellent photocatalytic activity when serving as a photocatalyst for the degradation of methyl orange aqueous solution (organic dye). The efficiency of photocatalytic reached 60%, which is higher than the undoped $TiO_2$ sample (25%). This result provides a potential way for cleaning water pollution or air pollution.

## REFERENCES

Chen, X., Shen, S., & Guo, L. (2010). Semiconductor-based photocatalytic hydrogen generation. *Chem. Rev.*, 110, 6503–6570.

Fujishim,a A., & Honda, K. (1972). Photolysis-decomposition of water at the surface of an irradiated semiconductor. *Nature*, 238, 37–38.

Irie, H., Watanabe, Y., & Hashimoto, K. (2003). Nitrogen-concentration dependence on photocatalytic activity of TiO$_2$-x Nx powders. *J. Phys. Chem. B*, 107, 5483–5486.

Nosaka, Y., Kishimoto, M., & Nishino, J. (1998). Factors governing the initial process of TiO$_2$ photocatalysis studied by means of in-situ electron spin resonance measurements. *J. Phys. Chem. B*, 102, 10279–10283.

Ohno, T., Mitsui, T., & Matsumura, M. (2003). Photocatalytic Activity of S-doped TiO$_2$ Photocatalyst under Visible Light. *Chem. Let.*, 32, 364–365.

Turchi, C. S., & Ollis, D. F. (1990). Photocatalytic degradation of organic water contaminants: mechanisms involving hydroxyl radical attack. *J. Catal.*, 122, 178–192.

Tachikawa, T., Yamashita, S., & Majima, T. (2011). Evidence for crystal-face-dependent TiO$_2$ photocatalysis from single-molecule imaging and kinetic analysis. *J. Am. Chem. Society*, 133, 7197–7204.

Williams, G., Seger, B., & Kamat, P. V. (2008). TiO$_2$-graphene nanocomposites. UV-assisted photocatalytic reduction of graphene oxide. *ACS nano*, 2, 1487–1491.

Woan, K., Pyrgiotakis, G., & Sigmund, W. (2009). Photocatalytic carbon nanotube TiO$_2$ composites. *Adv. Mater.*, 21, 2233–2239.

Yu, J. C., Yu, J., & Ho, W. (2002). Effects of F-doping on the photocatalytic activity and microstructures of nanocrystalline TiO$_2$ powders. *Chem. Mater.*, 14, 3808–3816.

Zhang, W., Li, X., & Jia, G. (2014). Preparation, characterization, and photocatalytic activity of boron and lanthanum co-doped TiO$_2$. *Catal. Commun.*, 45, 144–147.

Advances in Energy and Environment Research – Achour & Wu (Eds)
© 2017 Taylor & Francis Group, London, ISBN 978-1-138-62682-9

# Spatio-temporal distribution of the heavy metals pollution in the sediments of Minjiang Estuary

Qinqin Sun, Meixue Luo & Houde Ji

*Fujian Provincial Key Laboratory of Coast and Island Management Technology Study, Fujian Institute of Oceanography, Xiamen, China*

ABSTRACT: Based on the Geographic Information System (GIS) and geochemical techniques, the heavy metal contents of the surface sediment and sedimentary cores in Minjiang Estuary were measured to study the spatial and temporal distribution rules, and the preliminary judgment of the main source of heavy metal pollution at the Minjiang Estuary area was made through the geo-statistical methods of GIS.

## 1 INTRODUCTION

The heavy metal elements in sediments have become one of the serious hazards to ecological environment since they are hard to degrade but easy to accumulate (Li et al., 2000; Bollhofer et al., 2001). Estuaries and coastal zones are not only dense areas for human activities, but also accumulation areas for various pollutants. The heavy metal pollutants from weathered ores, fire coals and automobile exhausts get into estuaries and coastal zones with the air and water flowing (Dao et al., 2010; Lee et al., 2007; Miller et al., 2007), and pose grave threats to the marine ecological environment accumulation and diffusion. At present, most of the researches on the heavy metal elements in sediments are focused primarily on the total amount characteristics and biological enrichments of the heavy metal elements, while very few aim to acquire comprehensive spatial distribution information by modeling the information of intra-zone discrete observation points as a continuous trend change (Hao et al., 2008). In addition, owing to the low availability of historical sampled data, there are not many studies on the temporal regularity of heavy metal pollution and the analysis of source.

## 2 STUDY AREA AND METHODOLOGY

### 2.1 Overview of the study area

The Minjiang River is the first largest river in Fujian Province, located in the economically developed southeastern coastal area of China. As a typical subtropical river in which there is sufficient water, it flows through topographic reliefs. With the rapid development of the Economic Zone on the West Side of the Straits, and the accelerating urbanization of the middle-lower Minjiang River reaches, the heavy metal pollutants from human living and enterprise production have increased obviously (Hou et al., 2009), and entered the Minjiang Estuary region along the tributaries of the Minjiang River posing threats to the quality of the sediment at the Minjiang Estuary, then affecting the marine environmental quality in the whole Taiwan Strait.

### 2.2 Research methodology

1. Geo-statistical analysis
The surface sediment at the Minjiang Estuary was sampled on the spot, and meanwhile data was collected with 44 stations set up there. A geo-statistical analysis was carried out on the discrete sampled data of the surface sediment at the Minjiang Estuary. With the content characteristics and spatial location of heavy metals combined together, an optimized semivariable function model was used to describe the spatial distribution pattern of the heavy metals in the surface sediment (Trangmar et al., 1985; López-Granados et al., 2002; Kravchenko, 2003). Kriging interpolation was used to generate a spatial distribution map for the content of heavy metals in the study area in ArcGIS, to reveal the regional distribution characteristics of the heavy metal content in the surface sediment at the Minjiang Estuary. The semivariable function model is shown in Formula (1):

$$r(h) = \frac{1}{2N(h)} \sum_{i=1}^{N(h)} [Z(x_i) - Z(x_i + h)]^2 \qquad (1)$$

where, $r(h)$ is a semivariable function; $h$ denotes the distance between sampling points; $N(h)$ spacing distance denotes the number of sampling points of h;

$Z(x_i)$ and $Z(x_i + h)$ are the observed value of regionalized variable $Z(x)$ on spatial locations $x_i$ and $x_i + h$.

2. Isotopic dating

One sedimentary core was collected from the Minjiang Estuary, and $^{209}$Po was taken as a tracer for the dating analysis by $^{210}$Pb. The radioactivity of $^{210}$Pb at different depths in the rock core was measured, and after the background value was deducted, $^{210}$Pb$_{ex}$ was obtained. The deposition rate of the sediment was estimated according to $^{210}$Pb$_{ex}$ (Chernyshev et al., 2007). In line with the vertical change process of $^{210}$Pb$_{ex}$, the formula below was used for deposition rate calculation: $S_R = -\lambda/2.303$ K, where, $S_R$ denotes deposition rate (cm/a), $\lambda$ denotes the decay constant (0.0311 yr$^{-1}$) of $^{210}$Pb, K denotes the gradient of the linear fitting straight line between $^{210}$Pb$_{ex}$ and depth after natural logarithm was taken of it. On the basis of accurate dating, isotopic dating method was used to determine the time for the content of heavy metals in the sediment columns, revealing the temporal distribution characteristics of the heavy metal content in the sediment in this region, and the historical variation trend of heavy metal pollution inside the basin.

## 3 RESULTS AND DISCUSSION

### 3.1 *The spatial distribution characteristics of the heavy metal content in the sediment at the Minjiang Estuary*

1. The content of Hg, Cu, Pb, Cd, Zn and As in the surface sediment in the Minjiang Estuary region does not exceed the standard for I class of marine sediment quality, suggesting that the heavy metal content in the sediment in this region is relatively low, and the sediment is relatively clean. From 2002 to 2014, in the surface sediment in the Minjiang Estuary region, the content of Hg basically didn't change, that of Pb first decreased then increased, that of Cd decreased gradually, and that of As first increased then decreased. The content of Cu and Zn was unknown due to data deficiencies.

Table 1. Statics of heavy metals content in surface sediments of Minjiang Estuary (ME).

| 1 | ME (this study, 2014) | ME (Wang et al., 2002) | ME (Ren, 2010) | I class of sediment quality (GAQS, 2002) |
|---|---|---|---|---|
| Hg | 0.08 | 0.079 | 0.08 | 0.2 |
| Cu | 11.38 | / | 20.4 | 35 |
| Pb | 25.3 | 29.7 | 20 | 60 |
| Cd | 0.08 | 0.244 | 0.13 | 1 |
| Zn | 62.21 | / | / | 150 |
| As | 7.8 | 2.44 | 9.3 | 20 |

It can be seen from the spatial distribution of the content of each heavy metal element that (Fig. 1), the content of Pb in the surface sediment in the tributary on the south side of the Minjiang Estuary is higher than that in the tributary on the north side of the Minjiang Estuary, and that the content of Pb is the highest in the region to the southeast of Langqi Island. This reflects that part of the Pb in this region comes from the Minjiang River, while the rest comes mainly from Tantou-Meihua Sea Area in Changle. This may have something to do with the land-based pollutant emission from the industrial parks in Jinfeng and Wenling Town, Changle City on the south bank of the Minjiang Estuary, and oil leakage from fishing vessels at the wharf. High-content Hg exists to the east of Langqi Island, high-content As exists to the north and southeast of Langqi Island, high-content Cu exists to the north and east of Langqi Island, and high-content Zn exists to the southeast of Langqi Island. The above several elements have basically the same distribution characteristics, suggesting that there are roughly similar influencing factors for them. It is possible that after pollutant source emission, under the influence of topographic and hydrodynamic conditions, pollutants were enriched in the sea area near the southeast side of Langqi Island, leading to a sharp increase in the content of Hg, As, Cu and Zn in the surface sediment in this region. For element Cd, its content in the surface sediment in the tributary on the south side of the Minjiang Estuary is higher than in the tributary on the north side of the Minjiang Estuary. High-content Cd exists to the west and east of Langqi Island far away from it, suggesting that it might come from land-based pollutant emissions and ocean dumping.

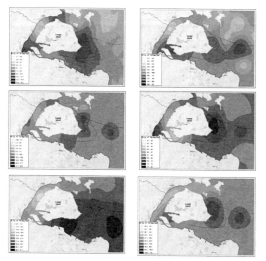

Figure 1. The heavy metals content distribution of surface sediments in Minjiang Estuary.

Overall, the surface sediment in the Minjiang Estuary region is relatively clean, and the content of Hg, Cu, Pb, Cd, Zn and As does not exceed the standard for I class of marine sediment quality. These heavy metal elements are mainly distributed at the Minjiang Estuary, as well as the areas to the east and south of Langqi Island. The reason, on the one hand, reflects the effect of river-carried land-based pollutants on this region, and on the other hand, reflects the impact of topographic and hydrodynamic conditions on the quality of the surface sediment in this region.

According to spatial interpolation of the discrete data, the continuous spatial information is also one of the major objectives of research on geo-statistics. In this study, Krig method was used to spatially interpolate the discrete sampled data to analyze the spatial distribution features of the heavy metals in the surface sediment in the Minjiang Estuary region. Krig method is an approach for linear unbiased optimal estimation of regionalized variable value in a finite region based on variable function theory and structural analysis. The following parameters are determined for the variation function model of heavy metal content (Robertson et al., 1997).

According to the structural features of variable function, range is a turning point via which a regionalized variable is turned from a state in the existing space into a state in a nonexistent space. The size of drill base value reflects the rangeability of regionalized variable. The ratio between nugget value and drill base value (nugget-base-ratio) is an important index that may reflect the spatial heterogeneity of regionalized variable. This ratio reflects which of regional factor (natural factor) and random factor (human factor), two spatial variance components, plays a dominant role.

As shown in Table 2, none of As, Cu, Pb, Hg, Zn and Cd has a large range, which indicates that these elements are correlated weakly to each other in space. In other words, these heavy metal elements have independence and instability. The nugget-base-ratio of Cd, As and Cu is less than 0.5, suggesting that these 3 elements have relatively significant spatial autocorrelation in this study, and regional factor (natural factor) has a greater

Table 2. The semi-variogram parameters of heavy metals content in Minjiang Estuary.

| Elements | $C_0$ | $C_1$ | $C_0/(C_0+C_1)$ | Range |
|---|---|---|---|---|
| As | 5.91 | 10.03 | 0.37 | 0.06 |
| Cu | 58.04 | 67.15 | 0.46 | 0.02 |
| Pb | 74.59 | 61.54 | 0.55 | 0.09 |
| Hg | 0.002 | 0.001 | 0.67 | 0.05 |
| Zn | 925.91 | 612.91 | 0.60 | 0.04 |
| Cd | 0.00 | 0.007 | 0.00 | 0.04 |

effect on their spatial distribution pattern. The nugget-base-ratio of Hg, Zn and Pb is greater than 0.5, suggesting that these 3 elements have relatively weak spatial autocorrelation in this study, and random factor (human factor) has a greater effect on their spatial distribution pattern.

### 3.2 The time change rule of the heavy metal content in the sediment columns from the Minjiang Estuary region

According to analytical testing of the sedimentary core sampled from the outside of the Minjiang Estuary (Fig. 2), the deposition rate of the

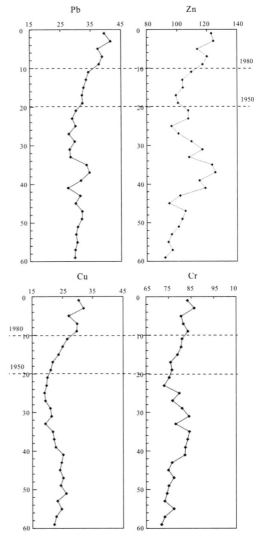

Figure 2. Heavy metals content of the sedimentary core in Minjiang Estuary.

205

sampling site at the Minjiang Estuary equals to around 0.31 cm per year, so the heavy metal content in the section of the sediment columns that is 10~20 cm high represents the heavy metal content in the sediment columns formed from 1950 to 1980. According to the analysis, the content of the heavy metal elements (Pb, Zn, Cu and Cr) in the sediment columns taken from the Minjiang Estuary increased slightly after 1950 and faster after 1980. The heavy metal content changed slightly in the lower part of the sediment columns.

## 4 CONCLUSIONS

The heavy metal content in the surface sediment in the Minjiang Estuary region does not exceed the standard for category-I marine sediment, so this sea area is relatively clean. A variable function parameter model was established, indicating that As, Cu, Pb, Hg, Zn and Cd have a small range. This suggests that these elements are weakly correlated to each other in space. Comparatively speaking, Cd, As and Cu have relatively significant spatial autocorrelation in this study, and regional factor (natural factor) has a greater effect on their spatial distribution pattern, while the nugget-base-ratio of Hg, Zn and Pb is greater than 0.5, reflecting that these 3 elements have relatively weak spatial autocorrelation in this study, and random factor (human factor) has a greater effect on their spatial distribution pattern.

The spatial distribution of the surface sediment in the Minjiang Estuary region shows that the heavy metal elements are mainly enriched at the Minjiang Estuary, as well as the sea areas to the east and south of Langqi Island, not only due to land-based pollutants brought by the river, but also due to the impact of the topographic and hydrodynamic conditions in this region. According to isotopic dating and heavy metal tests on the sediment columns taken from the Minjiang Estuary, the content of Cu, Pb, Zn and Cr increased obviously after 1980, reflecting that the rapid development of industrial modernization has affected the change of the heavy metal content in sediments in this region.

## ACKNOWLEDGMENTS

This study was financially supported by National Natural Science Foundation of China (grants No. 41106073, 40906047) and Natural Science Foundation of Fujian Province (2016 J01178).

## REFERENCES

Bollhofer A, Rosman KJR. 2001. The temporal stability in lead isotopic signatures at selected sites in the Southern and Northern Hemispheres. Geochimica et Cosmochimica Acta, 66(8): 1375–1386.

Chernyshev IV, Chugaev AV, Shatagin KN. 2007. High-precision Pb isotope analysis by multicollector-ICP-mass-spectrometry using 205Tl/203Tl normalization: optimization and calibration of the method for the studies of Pb isotope variations. Geochem Int, 45: 1065–76.

Dao L, Morrison L, Zhang C. 2010. Spatial variation of urban soil geochemistry in a roadside sports ground in Galway, Ireland. Science of the Total Environment, 408: 1076–1084.

General Administration of Quality Supervision (GAQS). 2002. Inspection and Quarantine of the People's Republic of China. GB 18668–2002 Marine sediment quality. Beijing: Standards Press of China.

Hao Zejiang, Zhou Feng, Guo Huaicheng. 2008. Spatial analysis of heavy metals in top marine sediment in Hong Kong. Research of Environmental Sciences, 21(6): 110–117.

Hou Xiaolong, Huang Jianguo, Liu Aiqin. 2009. Heavy Metals Pollution and It's Assessment in the Wetlands of Min River Estuary in Fujian Province. Journal of Agro-Environment Science, 28(11): 2302–2306.

Kravchenko AN. 2003. Influence of spatial structure on accuracy of interpolation methods. Soil Science Society America Journal, 67: 1564–1571.

Lee CS, Li XD, Zhang G, et al. 2007. Heavy metal and Pb isotopic composition of aerosol in urban and suburban areas of Hong Kong and Guangzhou, South China-Evidence of the long-range transport of air contaminants. Atmospheric Environment, 41: 432–447.

Li X, Wai O, Li YS, et al. 2000. Heavy metal distribution in sediment profiles of the Pearl River estuary, South China. Applied Geochemistry, 15: 567–581.

López-Granados F, Jurado-Expósito M, Atenciano S, et al. 2002. Spatial variability of agricultural soil parameters in southern Spain. Plant and Soil, 246(1): 97–105.

Miller JR, Lechler PJ, Mackin G, et al. 2007. Evaluation of particle dispersal from mining and milling operations using lead isotopic fingerprinting techniques, Rio Pilcomayo Basin, Bolivia. Science of the Total Environment, 384: 355–373.

Ren Baowei. 2010. Distribution and potentia l ecological risk assessment of heavy metals in surface sediments of Minjiang Estuary. Journal of Fujian Fisheries, 3: 46–50.

Robertson GP, et al. Soil resources, microbial activity, and primary production across an agricultural ecosystem. Ecological Applications, 1997, 7: 158–170.

Trangmar BB, Yost RS, Uehara G.1985. Application of geostatistics to spatial studies of soil properties. Advances in Agronomy, 38: 45–94.

Wang Xian, Li Wenquan, Zhang Fan. 2002. Assessment on present status and quality of sedimentin Fujian shore bay. Acta Oceanologica Sinica, 24 (4): 127–131.

*Advances in Energy and Environment Research – Achour & Wu (Eds)*
© *2017 Taylor & Francis Group, London, ISBN 978-1-138-62682-9*

# Carrying capacity and ecological footprint of Taiwan

Yung-Jaan Lee
*Chung-Hua Institution for Economic Research, Taiwan*

Chuan-Ming Tung
*Department of Urban Planning and Disaster Management, Ming Chuan University, Taipei City, Taiwan*

Shih-Chien Lin
*Department of Architecture and Urban Design, Chinese Culture University, Taipei City, Taiwan*

ABSTRACT:  Taiwan is located in the Western Pacific where typhoons and earthquakes frequently strike. The environment and geological conditions are sensitive and vulnerable to these phenomena. In recent years, rapid socio-economic growth has increasingly put pressure on land development on the island of Taiwan, causing frequent conflicts over land in mountain, river, and sea areas. Additionally, the increasing impact of climate change has made the maintenance of a stable equilibrium of the ecosystem, ensuring the survival of the population, and the sustainable development of the nation, important issues that must be actively faced, taking into account limitations on land and resources. "Doubly exposed" to globalization and global warming, Taiwan must actively ensure the balance between national land conservation for the future and economic development in its four regions. Taiwan must also coordinate regional development and effectively distribute resources among regions to achieve equal development and mitigate the challenges of resource shortages. Studies of carrying capacities have focused on environmental carrying capacity, recreational carrying capacity, and urban carrying capacity, mainly exploring the limits on population that are imposed by a single environmental system. Few studies of the amount of population that can be supplied by the land as part of the overall environmental system, the amount of available resources, the amount of consumption, the consumption differences and developmental issues, have been published. This study examines Taiwan's carrying capacity (resource supply side) from the perspective of "land resources", "food resources", and "carbon emissions". It determines the population size that can be supported by each resource to reveal the current levels of development of Taiwan's regions and future expected demands on each resource, and the carrying capacity of each resource in each region is compared with the ecological footprint (resource demand side) to support a discussion of regional adaptations. This study provides innovative thinking and methods to reduce resource shortages and the pressure that is imposed by climate change, to seek new opportunities for Taiwan's progress toward sustainable development.

## 1 INTRODUCTION

Taiwan has a steep topology, hills and mountainous woodlands throughout, and vulnerable geological conditions. On account of Taiwan's "double exposure" to economic globalization and climate change (global warming) (Kelman *et al.*, 2015; Silva *et al.*, 2010), typhoons, earthquakes, and other disasters cause severe damage to the land, soil erosion and environmental deterioration problems that have a severe impact on already vulnerable geological conditions and limit the area of useable land. Research on climate change in recent years has revealed that the current average global temperature is 4°C higher than it was in the pre-industrial era, and its rate of increase may reach a maximum by 2060 (Betts *et al.*, 2011). Warming of the climate may increase the temperature in 2016 by 1°C, forcing 60 million people, the largest number in modern history, to migrate (World Economic Forum, 2016).

The World Bank's 2015 annual report noted that climate change is the greatest threat to humanity and the Earth. It must be properly handled if extreme poverty is to be eradicated (World Bank, 2015). The setbacks of mitigation of and adaptation to climate change are regarded as posing the greatest risks for the future (World Economic Forum, 2016). The deterioration of ecological regions tends to cause poverty to persist. According to another World Bank report, all forms of pollution together take approximately nine million lives on average annually (World Bank, 2015).

Under globalization, the effective distances among countries are gradually decreasing. Interconnection that is associated with economic development, political policies, and social networks, increases the developmental potential of countries, but it may also cause environmental stress and generate risks of disaster. UN-Habitat (2011) noted that the world's urban population surpassed its rural population for the first time in 2007. In 1950, more than 70% of the global population lived in rural areas and less than 30% lived in urban areas. However, by 2014, 54% of the world's population lived in urban areas. Urban population is expected to grow continuously. By 2050, only 34% of the world's population will live in rural areas, according to the UN Habitat's estimates, and 66% will live in urban areas.

Overshooting in developed countries has caused environmental damage. Brown (2011) noted that in order to maintain current levels of consumption, 1.5 Earths would be required. For example, unsustainable overshooting of energy and natural capital has caused the world's Ecological Footprint (EF) to surpass the Earth's biocapacity by almost a half (UNISDR, 2015). The concept of EF was developed by Wackernagel & Rees (1996). It is the area of bioproductive land that is required to provide the consumed resources and absorb the waste of a population or economic system. Therefore, EF is an important metric of the consumption of the Earth's resources by humans (Borucke *et al.*, 2013; Lee, 2015; Lee & Peng, 2014; Wackernagel, 2014). EF is mostly the sum of six items, which are cropland footprint, grazing footprint, forest footprint, fishing ground footprint, carbon footprint, and built-up land. Of these, cropland footprint, carbon footprint, and built-up land are extremely useful in determining Taiwan's carrying capacity. They specify the supply of, and demand for, food resources, carbon emissions, and residential land for the current population, allowing consumer demand in Taiwan to be better understood.

Taiwan is an island country and is therefore severely threatened by climate change. Both active mitigating measures and adaptive measures must be taken to minimize the impact of climate change, and to support sustainable environmental, social, and economic development. In response to climate change and taking into account the conditions of the land and limitations on resources, the Taiwanese government must be more active in balancing the national land conservation and economic development. It must begin with the planning of sustainable national land use plan, and pursue economic development in a way that takes into account environmental carrying capacity. Population density and resource distribution must be controlled, and the development, effective use and allocation of resources among regions must be coordinated to ensure equitable development to reduce resource shortages, and to ensure sustainable development.

Taiwan is densely populated, has unevenly distributed resources, and faces overdevelopment. Development that does not consider carrying capacity has resulted in an imbalance of the nation's spatial order, damaging the land and contributing to environmental disasters, which make the land highly vulnerable to climate change impacts. Owing to the potential effects of unpredictable climate disasters on Taiwan, current land supply and consumption in Taiwan's four regions must be studied. Strategies for coping in extreme climate situations can then be developed. However, most studies of carrying capacity have investigated the limits that are imposed on a population by a single environmental system. This method neglects the amount of land that can be supplied and the amount of resources that can be consumed; it also neglects variations in resource consumption by the population among regions, so other regional variations and problems associated with regional development are difficult to understand.

This study examines carrying capacity (resource supply side) with respect to land resources, food resources, and carbon emissions in Taiwan. The population size can be sustained by each resource is used to show the current developmental situations of regions and to elucidate future demands. The carrying capacity of each resource in Taiwan's four regions is determined. The EF (resource demand side) is considered, and built-up land footprint, cropland footprint, and carbon footprint are calculated from the EF and compared among Taiwan's regions. Variations of supply and demand among regions are analyzed to elucidate living conditions, food consumption, and carbon emissions in each region. This aims to promote innovative thinking and methods to reduce resource shortages and the pressures that are imposed by climate change, in order to elucidate new opportunities to promote Taiwan's sustainable development.

## 2 LITERATURE REVIEW

According to a report by UNEP (2011), 30–50% of urban populations in developing countries live in environmentally vulnerable areas. The environmental risks in these areas include flooding, housing shortages, and poor infrastructure. Food consumption by the urban population is increasing annually. From 1992 to 2007, the consumption of fish and seafood increased by 32% and that of meat increased by 26%. Cities are becoming increasingly large centers of consumption. Globalization not only involves urbanization, but also leads to economic, social, and environmental issues, such as uneven income distribution, unbalanced housing supply, increased crime rates, and poor air quality. Urbanization centralizes risks in urban areas, resulting in the development there of the fundamental causes of environmental change (such as greenhouse gas emissions), and increasing their vulnerability to extreme weather (Tarhan et al., 2016). Urban populations consume 75% of global energy (United Nations, 2009) and are responsible for 80% of carbon emissions (World Bank, 2010).

According to the Fifth Assessment Report (AR5) of the Intergovernmental Panel on Climate Change (IPCC) in 2014, scientists now believe more than at any time previously that human activities are responsible for global warming. Since 1750, human activities have increased the concentration of carbon dioxide, a greenhouse gas, in the atmosphere (IPCC, 2014). According to statistical data that were provided by the International Energy Agency (OECD/IEA) in 2015 and other statistical data from the Taiwan Environmental Protection Agency, in 2013, Taiwan emitted a total of 2.487 million tons of carbon dioxide (Taiwan EPA, 2015), accounting for 0.77% of total global emissions, putting Taiwan in the 23rd position globally. Relative to its total population, Taiwan has high carbon emissions. Therefore, the carbon carrying capacity that is required to Taiwan's carbon dioxide emissions is an important topic.

With respect to terrestrial ecosystems, forests are important resources for absorbing carbon dioxide. They absorb approximately 1.85 billion tons of carbon dioxide annually, and have a unique role in the mitigation of carbon emissions (Lee, 2005). Forests can provide ecosystem services, such as providing raw materials for goods, regulating local and global climates, buffering climate events, regulating the hydrological cycle, and protecting watersheds (Martire et al., 2015).

Global population growth and the substantial increase in demand for food by the middle class are responsible for a high risk related to food security. Climate change has increasingly limited agricultural production (World Economic Forum, 2016). According to a report by the Food and Agriculture Organization (FAO), by 2050, the net urban population growth in developing countries will lead to high demands for water and food in urban areas. The report also identifies many problems that are caused by urbanization; for example, dense populations in metropolitan areas and high-density developments mostly live in very small spaces. Sixty percent of rapidly growing cities in many developing countries will face high risks of natural disasters (FAO, 2015). Overall, highly developed cities have a high demand for natural resources

and impose a heavy load on the carrying capacity. In face of extreme weather challenges, this fact is unfavorable for sustainable development. Therefore, controlling the consumer demand for resources and understanding carrying capacity (environmental resource supply) and EF (human demand) are important.

The land resource carrying capacity is the population that can be sustained by a given area. The population carrying capacity of land resources is used to analyze consumption and food production, and to address the balance between human demand and resource supply. Since environmental carrying capacity is non-recoverable, in situations with environmental limitation and environmental overuse, spatial adaptation strategies have become a primary consideration. This study analyzes resources that absorb carbon dioxide (forests), food, and residential areas, calculating their respective carrying capacities and EFs to determine whether a supply and demand imbalance exists given the current supply of, and demand for, resources in Taiwan. Furthermore, in the face of the climate change, strategies for adapting to the aforementioned supply and demand differences, and the issues that arise from them, must be developed.

## 2.1 Carrying capacity

The carrying capacity of a species refers to the maximum population of that species that the environment can sustain given their need for food, a habitat, water, and other necessities. From the perspective of population biology, carrying capacity can be defined as the maximum load on the environment (Hui, 2006). Odum (1971) defined carrying capacity as the maximum number of a single species that can be continuously supported by the environment. Rees (2000) defined carrying capacity as the maximum resource consumption rate and minimum waste emission rate that can be maintained without reducing the diversity of the ecosystem functions. Restated, the total resource consumption in a given area must not exceed the total amount of carrying capacity of that area, to prevent acceleration of the consumption of natural resources (Daly, 1990).

Carrying capacity should not only be considered with respect to resource supply in determining the population that can be accommodated. Resources required to deal with the waste that is produced during purification and consumption should also be considered. In an investigation of the complex system of humankind and the environment, carrying capacity should incorporate natural resource systems, human activity systems, and urban development. Carrying capacity should include environmental, economic and social systems. Carrying capacity is not fixed. It changes with lifestyle, technological development, and the accessibility of facilities (Nieswand & Pizor, 1977). Sustainable urban development is based on balanced population growth, economic development, availability of natural resources, and the ecosystem. Excessive growth of the population or industry affects the local environmental carrying capacity (Gilbert & Braat, 1991).

Human carrying capacity is defined as the population that can be accommodated without doing natural, social or environmental damage (Abernethy, 2001). The development of human activities should not exceed the carrying capacity of a given area, which depends on its resilience. If population or human activities exceed the carrying capacity, then negative effects will occur (Graymore *et al.*, 2010). Although technology can have positive effects, it sometimes causes problems (Martire *et al.*, 2015). In the past, carrying capacity has been mostly regarded as the population that the land can accommodate. This study uses the above concepts to determine the maximum amount of resources that the land can provide for human life.

## 2.2 Ecological footprint

Concerned with sustainable development, Wackernagel & Rees (1996) converted conventional carrying capacity in ecological economics into an EF concept that is easier to understand. EF differs from the environmental carrying capacity that was proposed in the 1960 s. Carrying capacity is the maximum population that a given area can accommodate given that development does not harm the ecosystem. Sustainable national land use should take into account land carrying capacity. Owing to industrialization, urban ecosystems must support extensive production and consumption activities on limited land (Tarhan *et al.*, 2016). The linear metabolic waste output has prevented the recycling use of resources, reducing the land's carrying capacity. Cities form artificial ecosystems that cause ecological black hole. Therefore, national land resource planning should consider the load that can be carried by the land.

Bioproductivity is used to determine the EF as an estimate of the land area for resource consumption and waste absorption by a specific population or economic system. As long as any material or resource is being consumed, land must be provided from one or several ecosystems. This land must provide the functions of related resources or decompose waste as required by the consumption (Wackernagel & Rees, 1996). Accordingly, the EF is proportional to the environmental footprint; a larger footprint indicates a greater environmental impact. It is inversely proportional to the surface area of the bioproductive land that is used by each person; a larger footprint indicates that each person can use less bioproductive land.

In Taiwan, the earliest study of EF was performed by Lee and Chen in 1997; their work referenced Wackernagel & Rees (1996), and divided ecological productivity of the land into six categories, which were fossil energy land, farmland, woodland, grassland, built-up land, and ocean. Since their calculations did not involve an "equivalence factor", their calculated EFs differed from the actual values. In 2003, the Global Footprint Network (GFN) constructed a global footprint database to help people to understand natural carrying capacity, how it is used, and who uses it; users could set up "footprint accounts" to determine the amount of natural resources they used.

The report by Taiwan's Council of Agriculture (COA) that was published in 2005 considered the ten-year period from 1994 to 2003 as the base period to construct a statistical database for determining Taiwan's EF. Equivalence factors were included in the analysis to elucidate the change in Taiwan's EF from 1994 to 2003. A follow-up study in 2006 was the first to reveal the pressure that was imposed by carbon dioxide emissions, and it estimated the EF of each county and city in Taiwan. The results indicated that consumption patterns and regional characteristics are closely related to EF (Lee, 2005).

Lee & Peng (2014) estimated the EF of Taiwan from 2008 to 2011. The annual increase in these years exceeded that from 1997 to 2007. Accordingly, policies to reduce carbon emissions and save energy must be strengthened to manage ecological resources in Taiwan. Mancini *et al.* (2015) noted that Average Forest Carbon sequestration (AFCs) is a key parameter in determining carbon footprint. Those authors were the first to include carbon emissions from wildfires, soil, and harvested wood products in their calculation. Forest carbon sequestration was considered in determining the amount of emitted carbon that is absorbed by the environment while providing for human consumption.

In summary, carrying capacity and EF are opposite concepts. Carrying capacity specifies the ability of the land to accommodate population as the

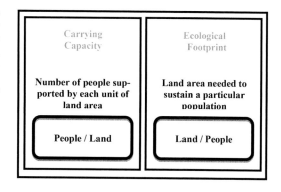

Figure 1. Concepts of carrying capacity and EF.

number of people that each unit of land area can support, whereas EF is the area of land (or water) that is needed to support a population. Figure 1 shows the two concepts. A carrying capacity analysis reveals the amount of resources that can be provided by the ecological environment, and is the supply side. EF analysis reveals the amount of land areas that an individual consumes, and is the demand side.

Only a few studies have examined the carrying capacity and EF of specific regions simultaneously. To investigate whether human activities in various regions of Taiwan are approaching their natural limits, or whether the total load of Taiwan is within the carrying capacity, this study examines the relationship between supply of and demand for resources. Carrying capacity is the amount of land that must be provided to support human activities. The EF specifies the amount of resources that must be consumed to ensure human survival and social operation. This study will examine whether over-consumption occurs in the regions of Taiwan, and will examine the progress of the country toward sustainability.

## 3  RESEARCH METHOD

This study analyzes the national land carrying capacity and EF of the northern, central, southern, and western regions of Taiwan to determine whether the current supply of and demand for environmental and economic resources are

Table 1.  National land carrying capacity and EF evaluation.

| Type | | | Calculation concept/Unit |
|---|---|---|---|
| Land carrying capacity | Land for production | Carrying capacity of food areas | 1. Using cultivated area to calculate the population that can be accommodated<br>2. Using arable area to calculate the population that can be accommodated<br>3. Agricultural land—Using farmed area to calculate the population that can be accommodated<br>[Formula] Population that can be accommodated (person) = Population that can be accommodated per unit of area (person/hectare) × Cultivated (or arable) area (hectare) |
| | Land for living | Carrying capacity of residential areas | The development status of land for residential—Population carrying capacity is estimated from the urban planning<br>Estimation for urban planning: Future population = (residential area × 120% + commercial area × 180%) ÷ Average residential area per person (50 m²/person)<br>Estimation for non-urban planning: Future population = Net population density (person hectare) × Habitable land area (hectare)<br>Future population (person): Planned population in urban area (person) + Population that can live on non-urban land (person) |
| Carbon Carrying Capacity | | | Sum of the areas of coniferous trees, mixed pine and hardwood forests, hardwood trees, and bamboo forest surface area (hectare) |
| EF | | | [Formula] EF (gha) = [ Productions + Imports—Exports (metric ton)] / Global biological productivity (metric ton/hectare) × equivalence factor (gha/hectare)<br>EF per capita (gha/person) = EF/Total population<br>Food footprint: Sum of productions of grains, tuber crops, sugar and honey, seeds and oilseeds, and fruit production (and will be converted to land areas)<br>Carbon footprint (gha) = CO2 emission (metric ton) × (1 – 1/3 or 1/4) / Rate of carbon absorption by forests (1.8 metric ton/hectare) × Forest equivalence factor (gha/hectare)<br>This study uses information on carbon dioxide emissions for each country form Unit 8: Energy Indicators from the "Key World Energy Statistics", published by the IEA. Chinese Taipei_CO2/POP denotes Taiwan's carbon dioxide emission per capita (unit: tons). |

balanced, and to develop future spatial adaptive strategies for each region. Based on grey literature (Schöpfel & Farace, 2010) (concerning national spatial planning policy, national regional plans, and related plans for counties and cities), 20 in-depth interviews with experts and professionals were conducted from December 2014 to May 2015, and experts, scholars, and government representatives from the northern, central, and southern regions were invited to participate in panel discussions. A total of 12 experts, scholars, and 30 government representatives participated. An international forum on carrying capacity and urban resilience was held on January 1, 2015 for the purposes of this study. Professors from the U.S. and Hong Kong were invited to share their experiences.

This study defines national land carrying capacity as "the population that can be accommodated by a nation's land given the supply of and demand for natural resources". This definition allows three components of carrying capacity—residential, food, and forestry resources. National land carrying capacity is divided into "productive land use" and "residential land use". The consumption of resources in four regions of Taiwan is analyzed. Table 1 presents the methods used.

The method of calculating the footprint for the land's carbon dioxide absorption uses the amount of bioproductive areas for forestation to absorb (sequester) carbon dioxide in the air. A third or a quarter of emissions are absorbed by the oceans and so must be subtracted from the calculated absorption by the land (IGBP et al., 2013; Lee, 2015). However, the 2011 version that was published by the GFN (2013) uses ocean absorption data from Khatiwala et al. (2009) and an ocean absorption rate of 2.3 Pg C yr−1. The formula uses the findings of IGBP et al. (2013) that a quarter of carbon dioxide emissions are absorbed by the ocean with ocean acidification. Therefore, this study uses 2009 as a cut-off point; the ocean absorption rate before 2009 was calculated using the factor of 1/3,

and that after and including 2009 was calculated using the factor of 1/4. The calculation method is:

Land required to absorb carbon dioxide (carbon footprint) (global hectare, gha) = CO2 emissions (metric tons) × (1 − 1/3 or 1/4)/Rate of absorption of carbon by forests (1.8 metric tons/hectare) × Forest equivalence factor (gha/hectare).

## 4 RESULTS

### 4.1 Carrying capacity

Residential carrying capacity is calculated from two factors, which are "estimated urban population" and "Floor Area Ratio" (FAR). Food carrying capacity is calculated from "cultivated area" and "arable area", whose values are taken from the 2014 Agricultural Statistical Report, published by Taiwan's COA. Carbon carrying capacity was calculated by summing the area of coniferous trees, mixture of pine and hardwood forests, hardwood trees, and bamboo forests, comprising the total forested area for carbon absorption, using data from the 2014 Agricultural Statistics Yearbook. Table 2 presents each carrying capacity.

### 4.2 Carbon footprint and food footprint

The main source of data for calculating carbon footprints is the "Key World Energy Statistics", published annually by the IEA website. This book (Unit 8: Energy Indicators) provides information on carbon dioxide emissions in countries worldwide. Chinese Taipei_CO2/POP denotes Taiwan's carbon dioxide emissions per capita.

An analysis reveals that of the regional food footprints in Taiwan, the northern region has the highest, which is 14,773,708 gha, and the eastern has the lowest, which is 463,736 gha. The carbon footprint is larger in the northern region than in all other regions. Table 3 summarizes these results.

Table 2. Carrying capacity in each region of Taiwan.

| Population that can be accommodated by carrying capacity Current population | | Northern region | Central region | Southern region | Eastern region | Unit |
|---|---|---|---|---|---|---|
| Population | | 10,477,807 | 5,788,242 | 6,415,872 | 558,718 | Person |
| 1. Residential carrying capacity | Planned urban population | 2,234,610 | 1,013,027 | 2,588,088 | 466,747 | Person |
| | FAR estimate | 844,995 | 2,585,611 | 3,986,916 | 594,097 | Person |
| 2. Food carrying capacity | Cultivated area | 61,986 | 299,357 | 248,422 | 58,184 | Hectare |
| | Arable area | 123,941 | 290,446 | 292,119 | 93,106 | Hectare |
| 3. Carbon carrying capacity | Forested area | 70,628,464 | 29,094,642 | 32,787,941 | 2,670,677 | Hectare |

## 4.3 *Analysis and discussion*

The carrying capacity analysis and EF analysis (Table 4) are combined to elucidate issues that are currently faced by each region of Taiwan, and the impact of the rise of metropolitan areas on national land planning and regional integration is summarized.

### 4.3.1

The residential carrying capacities of the regions in Taiwan appear to reveal that supply exceeds demand, and the extents to which supply exceeds demand follow the order, southern region > northern region > central region > eastern region (based on urban plan population).

The analysis of urban lands and non-urban lands yields an estimate of the population that can be accommodated in each region. Based on the planned urban population, the northern, central, southern, and eastern regions are estimated to accommodate 12,712,417 persons, 6,801,269 persons, 9,003,960 persons, and 1,025,465 persons, respectively. These numbers in these four regions exceed the actual numbers of residents, indicating the divergence between urban plans and the current situation. Based on the estimated FAR, the northern, central, southern, and eastern regions can accommodate 11,322,802 persons, 8,373,853 persons, 10,402,788 persons, and 1,152,815 persons, respectively. The planned urban population that can be accommodated in the northern region is 1,389,615 persons higher than the FAR estimation. The status of resources for living in urban planning areas and rural areas is examined. Resources for living are concentrated in the northern region— mainly in Taipei. Most of the areas in the central region with a large difference between residential supply and demand are in Taichung. The areas of greatest difference between residential supply and demand in the southern region are mostly

Table 3. Regional food and carbon footprints.

| EF | Northern region | Central region | Southern region | Eastern region | Taiwan | Unit |
|---|---|---|---|---|---|---|
| Food footprint | 14,773,708 | 5,325,183 | 6,736,666 | 463,736 | 27,299,293 | gha |
| Food footprint per capita | 1.41 | 0.92 | 1.05 | 0.83 | 1.05 | gha/person |
| Carbon footprint | 70,096,529 | 28,478,151 | 31,822,725 | 2,670,672 | 133,068,077 | gha |
| Carbon footprint per capita | 6.69 | 4.92 | 4.95 | 4.77 | 4.88 | gha/person |

Table 4. Results of analysis of EF of each region of Taiwan.

| | | Northern region | Central region | Southern region | Eastern region | Unit |
|---|---|---|---|---|---|---|
| Population in 2014 | | 10,477,807 | 5,788,242 | 6,415,872 | 588,718 | person |
| Residential carrying capacity | Planned urban population | 12,712,417 | 6,801,269 | 9,003,960 | 1,025,465 | person |
| | FAR estimation | 11,322,802 | 8,373,583 | 10,402,788 | 1,152,815 | person |
| Difference between supply and demand | Planned urban population | +2,234,610 | +1,013,027 | +2,588,088 | +466,747 | person |
| | FAR estimation | +844,995 | +2,585,341 | +3,986,916 | +594,097 | person |
| Food carrying capacity | | 14,773,708 | 5,325,183 | 6,736,666 | 463,736 | gha |
| Food carrying capacity | Agricultural area | 54,546.87 | 263,079.23 | 149,237.8 | 40,338.98 | hectare |
| | Arable area | 123,940.88 | 290,445.68 | 292,118.68 | 93,106.04 | hectare |
| Difference between supply and demand | Agricultural area | −14,719,161 | −5,062,103 | −6,587,428 | −423,397 | hectare |
| | Arable area | −14,649,767 | −5,034,737 | −6,444,547 | −370,629.9 | hectare |
| Carbon footprint | | 70,096,529 | 28,478,151 | 31,882,725 | 2,670,672 | gha |
| Carbon carrying capacity | | 70,628,464 | 29,094,642 | 32,787,941 | 2,670,677 | hectare |
| Difference between supply and demand | | +531,935 | +616,491 | +965,216 | +5 | hectare |

the coastal areas of Tainan Municipality and Kaohsiung Municipality; most of the supply is in Chiayi City. The areas of greatest difference between residential supply and demand in the eastern region are mostly in Hualien City and Ji'an Township in Hualien County, and Taitung City in Taitung County. The rest of the townships are located in smaller urban planning areas and rural areas.

The calculations reveal that, regardless of whether the planned urban population or FAR is used to estimate the relationship between current population and residential carrying capacity in each region of Taiwan, the residential supply exceeds residential demand in each region. Given Taiwan's recently slowing population growth, which is turning negative, the priorities of strategies to adapt to residential supply and demand should focus on adjusting populations in urban lands and non-urban lands in each region to prevent unnecessary urban sprawl and overexploitation of non-urban lands.

4.3.2

The food footprint in each region of Taiwan is higher than the food carrying capacity, indicating that food-producing land in Taiwan falls far short of being able to meet the current demand of the population. The difference between the supply of and demand for food is greatest in the northern region, where the efficiency of cultivated land is poor and food consumption is high, making the region vulnerable to extreme weather.

Table 4 shows that food footprint in the northern region exceeds those in other regions, and in regions, the food footprint exceeds the food carrying capacity (agricultural area or arable area). Food footprints are converted into land areas herein. In 2011, for example, the food footprint of the northern region was 14,773,708 gha, which is equivalent to 20 times the area of the northern region, which would have to be converted into arable land to provide food for the region's inhabitants. The food footprint in the central region is 5,325,183 gha, which is equivalent to five times the area of that region, converted into arable land. The food footprint of the southern region is 6,736,666 gha, which is equivalent to 6.7 times the area of that region, converted into arable land. The food footprint of the eastern region is 463,736 gha, which is equivalent to 0.6 times the area of the eastern region, if converted into arable land areas. Restated, food demand is less than supply only in the eastern region; food demand in other regions greatly exceeds supply, and the issue is most serious in the northern region.

Overall, 7.6 times the area of Taiwan was needed to produce the food that was consumed in Taiwan in 2011. This result reveals that food consumption in Taiwan far exceeds the extent to which the land can provide it, and the issue is most serious in the northern region. According to analytical results concerning food carrying capacity, based on estimates of either agricultural land area or arable land area, food carrying capacity is too low to produce food for the current population. Accordingly, unless some of food is imported, then the current food production in Taiwan does not suffice. The difference between supply of and demand for food in the northern region is far greater than that in the southern and central regions. Owing to a shortage of arable land and a lack of effective farming, the northern region, with huge food consumption, depends on the other regions (southern, central, and eastern regions) and looks overseas to meet its massive demand for food. Based on these results, with respect to arable land in Taiwan, the control of quality and total amount of development must be adopted to focus more on protecting arable land within each region to reduce the imbalance between Taiwan's food supply and demand.

Agricultural land use trends from 2010 to 2014 were analyzed (Fig. 2). The results reveal that the total agricultural land area in those five years exceeded the actual arable area, and farming efficiency was lowest in the northern region. Figure 2 reveals that both agricultural land and arable land were gradually lost in the northern region. From experience, once agricultural land has been converted into construction or industrial land, the probability of rehabilitation is extremely small, and the conversion may even affect the neighboring environment, destroying Taiwan's agricultural sustainable development. Under the threat of extreme weather, the problem of the conversion of agricultural land should be faced properly. To ensure food security and agricultural sustainable development, the protection and maintenance of agricultural land should be improved, and the conversion of agricultural land to other uses should be reduced.

Changes in people's eating habits have reduced the demand for rice, increased the demand for wheat, and gradually increased meat consumption. Accordingly, demands for both wheat and feed grain (such as corn) have increased. Furthermore, the demand for soybeans for processing is large, and the production of wheat, feed corn, and soybeans in Taiwan is small. The reliance is reflected in Taiwan's low food self-sufficiency (about 34%), which endangers the food security of Taiwan. To ensure food security, Taiwan's COA mandated a food self-sufficiency rate of 40% by 2020, in the hope of improving the efficiency of agricultural land use, and guiding the import of substitution crops to ensure food security in Taiwan. Therefore, increasing the rate of cultivation of farmland in each region should be the primary goal.

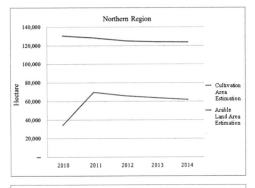

Northern Region

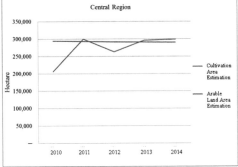

Central Region

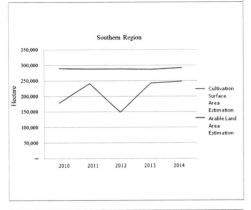

Southern Region

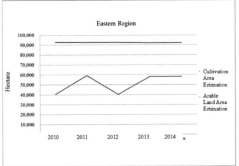

Eastern Region

Figure 2. Changes in areas of agricultural and arable land in four regions of Taiwan in 2010–2014.

Figure 2 shows that only in the central region is the arable land area similar to the current cultivated land area. The area of arable land in other regions exceeds the area of cultivated land, especially in northern and eastern regions. Therefore, effectively increasing the area of cultivated land in the northern and central regions to increase food supply within those regions, and reducing excessive food consumption in the northern region, are important issues in Taiwan.

4.3.3
Regions with a carbon carrying capacity that is almost equal to the carbon footprint have balanced supply and demand. However, the carbon footprint per capita is high throughout Taiwan, and especially so in the northern region.

The region of Taiwan with the highest carbon footprint is the northern region, whose footprint is 26 times that of the eastern region, which has the smallest footprint. The carbon footprint per capita is highest in the northern region (6.69 gha/person). However, a comparison of the carbon carrying capacity and the carbon footprint of each region in Taiwan reveals that the difference between supply and demand is positive everywhere, indicating balanced supply and demand. With respect to forest supply stability, this study examines changes in forested areas from 2010 to 2014. The forested area in each region is stable, neither increasing nor decreasing significantly. During these five years, Taiwan's forest coverage rate was 60.7%, which is more than double the global average, indicating that Taiwan's forest coverage and forest reserves are good. About 754,000,000 tons of forest carbon are stored in Taiwan, of which 36% are stored in broadleaf forest, 21% in pine forest, and 3.4% in bamboo forest, and mixed bamboo and wood forest (COA, 2015). Taiwan's *Greenhouse Gas Reduction and Management Act* of 2015 requires that the Forestry Bureau should regularly provide information on the carbon sink quantity; promote the management of forest resources and the enhancement of forest functions, and continuously monitor changes in the forest carbon sink quantity to monitor the balance between supply and demand. Although the carbon carrying capacity of Taiwan can still cope with the overall carbon footprint, effective carbon reduction strategies should still be used to reduce the high carbon footprint in metropolitan areas in the face of climate change.

In summary, the supply of and demand for residential, food, and forest resources in each region of Taiwan are as follows. The northern region has the highest residential carrying capacity, and the carrying capacity in each region indicates that supply exceeds demand. Therefore, Taiwan does not need continuously to develop land for construction in the future.

Research results concerning food resources reveal a shortage of food supply in each region, so food must be imported. Agricultural land use data from 2010 to 2014 demonstrate that not all arable land in each region of Taiwan is used; only in the central region has a higher usage of arable land than the other regions. The productivity or usage rate of agricultural land should be improved to deal with the issue of insufficient food supply. With respect to the balance between the supply of and demand for carbon resources from 2011 to 2014, data from Taiwan's COA indicate that forest conservation in Taiwan is relatively successful. However, as well as using forest land to absorb carbon emissions, Taiwan should develop incentives to educe emissions in order to achieve the goal of sustainable development, implement Taiwan's Greenhouse Gas Reduction and Management Act, and achieve annual carbon reduction targets.

## 5 CONCLUSIONS

The World Economic Forum (WEF) notes that the effect of climate change on food production and water resource supply will be greater than expected. WEF quotes the US National Intelligence Council that the relationship between food, water, energy, and climate change is one of the four major themes that will determine the state of the world in 2030 (WEF, 2015). The 11th edition of this report also notes that the East Asia and Pacific regions, where Taiwan is located, have the weakest protections against the global risk of "natural disasters", "extreme weather events", and "failure of governance" in the face of climate change (WEF, 2016).

The British risk management consulting agency, Verisk Maplecroft, listed Taiwan as one of the four countries with extreme risk of exposure to four types of natural disaster (earthquake, tsunami, storm, flood and landslide (mudslide)) in the Natural Hazards Risk Atlas, 2011 (Verisk Maplecroft, 2011). Since Taiwan is an island country, climate change could have a great impact on it. Accordingly, as well as mitigation measures, appropriate regional adaptive measures should be increased in response to climate change to minimize the impact of any related crisis, and socio-economic development should be guided toward sustainability.

The residential carrying capacity of each region of Taiwan is sufficient to meet, and even exceed, demand. Regardless of whether the planned urban population or FAR is used to estimate the current population, the residential carrying capacity exceeds demand. Distributing the population between urban land and non-urban land should be the main concern of future attempts to prevent unrestricted urban sprawl and over-exploitation of non-urban land.

The production of sufficient food to meet demand in Taiwan would require 7.6 times the area of the island. The food footprint of the northern region is equivalent to 20 northern regions; that of the central region is equivalent to five central regions; that of the southern region is equivalent to 6.7 southern regions; and that of the eastern region is equivalent to 0.6 eastern regions. These results show that the overall food footprint in Taiwan is quite high, and that Taiwan's food carrying capacity cannot meet the demand of the current population for food. Comparing the differences between arable land area and cultivated land area in the four regions clearly reveals that the food carrying capacity per capita in the northern region is excessive. The arable land in that region is clearly insufficient and apparently inefficient. The northern region has become a huge center of consumption that requires food from other regions and abroad, making the northern region vulnerable to direct and indirect impacts of disasters. Increasing the rate of cultivation of arable land in the northern and eastern regions; ensuring the total area of agricultural land and agricultural use, limiting urban sprawl, and reducing food waste in the northern region, are equally important and must be urgently promoted. The overall carbon carrying capacity and carbon footprint of Taiwan indicate balanced supply and demand for carbon emissions. Taiwan's successful conservation of forest has resulted in a forest coverage rate of 60.7%, which suffices to absorb the carbon footprint. In response to international rules concerning greenhouse gas reduction and management, annual, industrial, and personal carbon footprints should be monitored more closely to reduce the carbon footprint per capita in each region (and especially in the northern region). Databases on carbon emissions of Taiwan's energy and raw materials should be established in support of the ultimate goal of carbon neutrality.

As well as controlling the overall quantity of agricultural land resources in Taiwan, and protecting these land resources and designating them as sensitive areas for production, the agricultural land use rate should be increased. Since environmentally sensitive land in non-urban areas (such as important habitats, nature reserves, wildlife reserves, national forests, national parks, and protected areas along the coast) have "mitigation" characteristics that can reduce the impact of a disaster, the spread of urban areas into non-urban land should be prevented. Promoting brownfield redevelopment and government asset activation in recreation industrial areas should be the preferred strategy. Ensuring quality farmland for the production of food is a priority, and the protection of environmentally sensitive areas or those prone to disasters should be a guiding principle of land use in future economic development.

With respect to the northern region where most of the population of Taiwan's is located, follow-up research should investigate the compositions of its food footprint and its carbon footprint to develop effective strategies for reducing food consumption and carbon emissions, and to reduce the reliance of this huge center of consumption on productive land. Following the passing of the *National Land Use Act,* based on the above findings, climate adaptation strategies should be considered in national land use planning, and interactions among regions should be considered in formulating adaptation strategies for Taiwan in the face of climate change.

Faced with limited environmental resources, imbalances of resource supply and demand, and the multiple effects of climate change-induced disasters, the central government, county and city governments, and township district offices, should implement policies and practices to respond to associated challenges that account for geospatial and socio-economic development. Adaptation strategies in the future should exhibit cross-boundary and cross-sector integration attributes.

## ACKNOWLEDGMENTS

The authors would like to thank the Ministry of Science and Technology of the Republic of China, Taiwan, for financially supporting this research under Contract No. MOST 104-2410-H-170-003.

## REFERENCES

Abernethy, V.D. 2001. Carrying capacity: The tradition and policy implications of limits. *Ethics in Science and Environmental Politics* 23: 9–18.

Betts, R., Collins, M., Hemming, D., Jones, C., Lowe, J. & Sanderson, M. 2011. When could global warming reach 4°C? *Philosophical Transactions of the Royal Society* 369: 67–84.

Borucke, M., Moore, D. & Cranston, G. et al., 2013. Accounting for demand and supply of the biosphere's regenerative capacity: the National Footprint Accounts' underlying methodology and framework. *Ecological Indicators* 24: 518–533.

Brown, L.R. 2011. *World on the edge.* Norton: Earth Policy Institute.

COA 2015. *Status and Prospects of Forest Resources.* Executive Yuan Number 3465 Council Meeting Presentation. Taipei: COA. (in Chinese).

Daly, H.E. 1990. Toward some operational principles of sustainable development. *Ecological economics 2(1):* 1–6.

FAO 2015. *Towards a water and food secure future—critical perspectives for policy-makers.* Marseille: World Water Council.

GFN 2013. Accounting for demand and supply of the biosphere's regenerative capacity: The National

Footprint Accounts' underlying methodology and framework. *Ecological Indicators* 24: 518–533.

Gilbert, A.J. & Braat, L.C. (eds) 1991. *Modeling for Population and Sustainable Development.* New York: Routledge.

Graymore, M.L.M., Sipe, N.G. & Rickson, R.E. 2010. Sustaining human carrying capacity: a tool for regional sustainability assessment. *Ecological Economics* 69: 459–468.

Hui, C. 2006. Carrying capacity, population equilibrium, and environment's maximal load. *Ecological Modelling* 192: 317–320.

IGBP, IOC & SCOR 2013. *Ocean Acidification Summary for Policymakers—Third Symposium on the Ocean in a High-CO$^2$ World.* International Geosphere-Biosphere Programme. Sweden: Stockholm.

IPCC 2014. Climate Change 2013: The Physical Science Basis. Summary for Policymakers. Cambridge, United Kingdom and New York, NY, US: Cambridge University Press.

Kelman, I., Gaillard, J.C. & Mercer, J. 2015. Climate change's role in disaster risk reduction's future: beyond vulnerability and resilience. *International Journal of Disaster Risk Science* 6: 21–27.

Khatiwala, S., Primeau, F. & Hall, T., 2009. Reconstruction of the history of anthropogenic CO2 concentrations in the ocean. *Nature* 462: 346–350.

Lee, G.-Z. 2005. Natural air-conditioning system and human-mad forests. *Science Development.* 388: 28–35. (in Chinese).

Lee, Y.-J. 2005. *Analysis and comparison of the ecological footprint in Taiwan.* Taipei: Council of Agriculture. (in Chinese).

Lee, Y.-J. 2015. Land, carbon and water footprints in Taiwan. *Environmental Impact Assessment Review* 54: 1–8.

Lee, Y.-J. & Peng, L.-P. 2014. Taiwan's ecological footprint (1994–2011). *Sustainability* 6: 6170–6187.

Mancini, M.S., Galli, A., Niccolucci, V., Lin, D., Bastianoni, S., Wackernagel, M. & Marchettini, N. 2015. Ecological footprint: refining the carbon footprint calculation. *Ecological Indicators.* 61, doi: 10.1016/j.ecolind.

Martire, S., Castellani, V. & Sala, S. 2015. Carrying capacity assessment of forest resources: enhancing environmental sustainability in energy production at local scale. *Resources, Conservation and Recycling* 94:11–20.

Nieswand, G.H. & Pizor, P.J. 1977. How to apply carrying capacity analysis. Management and Control of growth, Washington, DC: Urban Land Institute Prex.

Odum, H.T. 1971. *Environment, Power, and Society.* New York: Wiley-Interscience.

OECD/IEA 2015. *Key World Energy STATISTICS.* Paris: OECD.

Rahmstorf, S., Cazenave, A., Church, J., Hansen, J., Keeling, R., Parker, D. & Somerville, R. 2007. Recent climate observations compared to projections, Science, 316, p.709 (http://www.sciencemag.org/content/316/5825/709.full).

Rees, W.E. 2000. Eco-footprint analysis: merits and brickbats. *Ecological Economics 32(3):* 371–374.

Schöpfel, J. & Farace, D.J. 2010. Grey literature. In M.J. Bates & M.N. Maack (eds), *Encyclopedia of Library and Information Sciences* (3rd Ed.): 2029–2039. Boca Raton: CRC Press.

Silva, J.A., Eriksen, S. & Ombe, Z.A. 2010. Double exposure in Mozambique's Limpopo River basin. *Geographical Journal*, 176(1), 6–24.

Taiwan EPA 2015. *2015 National Greenhouse Gas Inventory Report of Taiwan*. Taipei: Taiwan EPA. (in Chinese).

Tarhan, C., Aydin, C. & Tecim, V. 2016. How can be disaster resilience built with using sustainable development? *Procedia—Social and Behavioral Sciences* 216: 452–459.

UNEP 2011. Year Book 2011: Emerging Issues in our Global Environment.

UN-Habitat 2011. *Global Report on Human Settlements: Cities and climate Change: Policy Directions* (United Nations Human Settlements Programme) London: Earthscan.

UNISDR 2015. *Making development sustainable: the future of disaster risk management*. Global Assessment Report on Disaster Risk Reduction. Geneva, Switzerland: UNISDR.

United Nations 2009. The Millennium Development Goals Report. New York: United Nations.

United Nations 2015. United Nations Expert Group Meeting on the Post-2015 Era: Implications for the Global Research Agenda on Population and Development-Report of the meeting. New York: United Nations.

Verisk Maplecroft 2011. *Natural hazards risk atlas 2011*. Bath, UK: Verisk Maplecroft.

Wackernagel, M. 2014. Comment on "Ecological footprint policy? land use as an environmental indictor". *Journal of Industrial Ecology* 18(1): 20–23.

Wackernagel, M. & Rees, W.E. 1996. *Our ecological footprint: reducing human impact on the earth*. Gabriola Island, B.C.: New Society Publishers.

WEF 2015. *The global risks report 2015* (10th Edition). Geneva: World Economic Forum.

WEF 2016. *The global risks report 2016* (11th Edition). Geneva: World Economic Forum.

World Bank 2010. *The World Bank Annual Report*. Washington, D.C.: World Bank.

World Bank 2015. *World Bank Annual Report 2015*. Washington, DC: World Bank. doi: 10.1596/978-1-4648-0574-5.

*Advances in Energy and Environment Research – Achour & Wu (Eds)*
© *2017 Taylor & Francis Group, London, ISBN 978-1-138-62682-9*

# Water quality index of various Kuwaiti seas, and methods of purification

K. Chen
*American International School of Kuwait, Hawalli, Kuwait*

ABSTRACT: With an objective of measuring the water quality of Kuwaiti Seas and testing water purification methods, the water quality index of 3 Kuwaiti beach samples were taken before and after a 4-step water purification system using filtration, effective microbial consortium, electrochemical remediation, and distillation. The initial and final water quality indices were compared and each method's applicability, effectiveness, and overall successfulness were assessed.

## 1 INTRODUCTION

Kuwait borders the Persian Gulf and is home to numerous bodies of water that contribute largely towards Kuwaiti economy, recreation and transportation. However, many pollutants such as industrial discard, oil related chemicals, hospital waste (contaminated with bacteria and microbes after treating infectious diseases), and sewage effluent are frequently dumped or washed into the ocean (Environmental disasters at Kuwaiti beaches, hospital area). These pollutants damage Kuwait's water quality and pose detrimental consequences to both marine and human health.

The disposal of man-made chemicals into the ocean can cause liver disease, immunosuppression, endocrine system damage, reproductive malformations, growth and development issues, as well as nervous and digestive system problems in marine organisms (2016). This is an issue because it disturbs the aquatic food chain and will eventually impact humans through typhoid or cholera (Toumi, 2010). The rising levels of ammonia, phosphate, sulfur, salinity and bacteria content in oceans from industrial waste also strains Kuwait's water treatment systems by requiring additional money and effort to be invested into purification.

Kuwait's 5 water treatment plants: the Ardiya plant, Rigga station, East, Jahra station, Umm Al-Hayman station, and Sulaibiya station (2016) are actively supplying purified water to meet agricultural, industrial, commercial and domestic needs. In fact, the Sulaibiya station alone provides for more 26% of Kuwait's overall water demand (2016). The station also employs the most advanced water treatments technology in Kuwait, using reverse osmosis, and anaerobic, anoxic and aerobic treatment methods to purify secondary effluent.

Measured on a 100-point, the Water Quality Index (WQI) serves as an increment of water quality. A companion of WQI before and after receiving purification can indicate towards treatment effectiveness. To calculate the WQI, 9 factors comprising of: temperature, pH, dissolved oxygen, biochemical oxygen demand, fecal coliform bacteria, turbidity, total suspended solids, total phosphorus, and total nitrogen are taken into consideration. Each test result corresponds to a Q-value that is then multiplied by a weight specific to the factor being measured because parameters vary in importance when determining water quality. The results are then added towards the final WQI. Weighing factors and the WQI legend are as follows:

This study will compare the initial and final WQI of ocean water samples from Salwa Beach (A1), Sulaibikhat Beach (A2), and North West Beach (A3) after being treated with a complete 4-step water purification system using filtration, effective microbial consortium, electrochemical remediation, and distillation. This combination of physical, biological, and chemical processes is currently unemployed by any of Kuwait's treatment plants, but can all potentially be utilized either individually, or together as an add-on treatment to existing plants. Each method's applicability, effectiveness, and overall successfulness will also be assessed.

## 2 METHODOLOGY

### 2.1 *Sample collection*

250 ml of water samples were collected from each testing site and stored in Nalgene's 125 ml HDPE bottles. Rubber gloves were worn during the process.

Table 1. Water quality factor and weights.

| Factor | Weight |
|---|---|
| Temperature | 0.10 |
| pH | 0.11 |
| Dissolved Oxygen | 0.17 |
| Biochemical Oxygen Demand | 0.11 |
| Fecal Coliform | 0.16 |
| Turbidity | 0.08 |
| Total Suspended Solids | 0.08 |
| Total Phosphate | 0.10 |
| Total Nitrates | 0.10 |

Table 2. WQI legend.

| Range | Quality |
|---|---|
| 90–100 | Excellent |
| 70–90 | Good |
| 50–70 | Medium |
| 25–50 | Bad |
| 0–25 | Very Poor |

## 2.2 WQI testing

The Estuary and Marine monitoring kit was used for testing numerous WIQ factors. Utilized reagents include dissolved oxygen tables, biochemical oxygen demand tables, total phosphate tablets, and total nitrate tablets. Temperature was measured using a temperature probe, pH measured with pH strips, and total suspended solids were measured as the mass (in grams) difference between a coffee filter before and after having samples poured through. 25 ml of samples were used. The measured mass was multiplied by a factor of 40, and then converted into milligrams. Fecal coliform cultures were grown on fecal agar.

## 2.3 Filtration (Water treatment step-1)

*50 ml of water was used for the treatment of each sample, and purification results are measured through changes in the sample's salinity, pH, total hardness, total metals, phosphate, nitrate, and alkalinity levels during each step of the treatment process. Salinity was tested using a refractometer, and pH, total hardness, total metals, and alkalinity were measured using their corresponding test strips. The concentration of phosphate and nitrate were tested using tablets from the Estuary and Marine monitoring kit.

Contaminated water first influents through a preliminary screen that it removes any visible debris or suspended solids before entering a 5-inch filtration tank with a 4-inch diameter. The tank is filled with a 40 dm$^3$ layer of assorted pebbles on top, followed by 40 dm$^3$ of gravel, 290 dm$^3$ of sand and 30 dm$^3$ layer on the bottom. This arrangement ensures that polluting constituents are removed based on particle size. Large particles are trapped in the pebbles and gravel, while finer parties such as microorganisms, environmental persistent pharmaceutical, and organic or inorganic materials blocked by the sand. The arrangement also prevents the filter from clogging, and the layer of gravel under the sand ensures that no sand passes into the subsequent treatment tank.

## 2.4 Effective microbial consortium (Water treatment step-2)

In the second purification tank (same dimensions as first), 6 µL of Effective Microorganisms (EM) comprising of Lactobacillus, Pseudomonas, Aspergillus, Saccharomyces and Streptomyces are used to break down organic matter in samples (Monica, 2016). The microbial culture either converts the organic material found in water samples into oxidized products, or uses it as energy and carbon source for new microbial cell growth (Shalaby, 2016). Generated products then settle to the bottom of the tank as sludge over time. The purification runs for 90 minutes, during which air is also pumped into the tank through fusion air pump 200 to create an oxygen rich environment for microorganisms.

## 2.5 Electrochemical remediation (Water treatment step-3)

For this method, the samples are transferred from the secondary purification tank into a 100 ml beaker. Electrochemical remediation a chemical approach to the traditional coagulation method of water treatment. Coagulation occurs when ultrafine particles and pollutants in water are clumped together to form a floc in the presence of a flocculating agent. To initiate the electrocoagulation process, 2 aluminum electrodes are bent and submerged into the sample being tested, with 1.5 cm exposed. An alligator clip (red and black) is then attached to the exposed end of each electrode. The free ends of the alligator clips are clipped onto a 9-volt battery (red on positive end and black on negative) to conduct an electrical charge throughout the sample. As the current flows, the positive electrode (anode) undergoes oxidation and releases aluminumcations, while the negative electrode (cathode) reacts with water to form hydrogen gas and hydroxyl ions. The coagulation process starts as the metal cations and charged ultra-fine particles come together to be neutralized, and the hydroxyl ions precipitate with certain pollutants. The created electrical current also terminates bacteria or protozoan cysts. The electrocoagulation process is run for 120 minutes, then screened through a coffee filter where the floc is trapped, and ready to be distilled.

## 2.6 Distillation (Water treatment step-4)

To set up the apparatus for distillation, a ring clamp is first placed on a retort stand, and covered with a hard mesh sheet. A 500 ml Erlenmeyer flask is then placed up atop the mesh and help in place with another clamp. The flask is connected to a Liebig condenser with a 500 mm 3-way vigreux distillation head. The condenser is held in place by a second retort stand and connects to a 250 ml collection Erlenmeyer flask placed flat on a table. The sample is poured into the first (higher) flask, then heated over a hot plate until the water temperature reaches 100°C. After reaching its boiling point, the water sample will evaporate as steam and travel upwards through the distillation head. Once the gas comes in contact with the cold surface of the condenser, the gaseous volatile condensates back into its liquid state and flow downward into the collection flask. The distillation process separates industrial waste and hardening agents such as calcium, phosphorous, mercury, and arsenic from water, as the pollutants have a higher boiling point than water, and are left behind while the water evaporates. Any harmful bacteria and viruses are also killed off during the heating process. The only substances distillation is unable to remove are volatile organic chemicals. However, these were pollutants were already eliminated after the effective microorganisms and electrochemical remediation treatments.

## 3 RESULTS AND DISCUSSION

### 3.1 Initial WQI

The calculated WQI reveals that water samples from all 3 beaches are of medium quality with Sulaibikhat Beach, Salwa Beach and North West Beach having the best, middle and worse quality respectively. No results yielded for the cultivation of fecal coliform in any of the 3 beaches, despite testing with both fecal coliform tablets supplied in the monitoring kit and fecalagar. Therefore, fecal coliform results were taken from the report *Microbial water quality and sedimentary faecal sterols as markers of sewage contamination in Kuwait* [9]. Temperature change was measured as the temperature difference from sampling site to 1 mile upstream, and observed to be most drastic in Sulaibikhat and North West Beach. Turbidity was 0 in all samples, and

measurements of pH, dissolved oxygen, and biochemical oxygen demand were all generally healthy as well. The primarily concerning index is total suspended solids, with measurements from all 3 beaches drastically exceeding both the ideal values (50 mg/L) and index parameters at 500 mg/L. Total phosphate and nitrate measurements were also very high. Although, nitrogen and phosphate are essential nutrients to marine life, their overabundance causes adverse health and ecological effects. Firstly, excess nitrogen and phosphate are introduced into water through sewage and fertilizer runoffs, or from the atmosphere that carries nitrogen-containing compounds derived from automobiles and other sources (Perlman, 2016). Therefore, high levels of these elements in the water samples could be a result of industrial pollution. Consequently, excess nitrogen and phosphate causes overstimulation of growth in aquatic plants and algae that leads to hypoxic water conditions or eutrophication-in which case, marine animal and plant diversity will be impacted (Perlman, 2016). Additionally, the excess inputs can also cause imbalance to the nutrient and material cycling process as more phytoplankton/vegetation than can be consumed by the ecosystem is produced. Finally, humans are also impacted by excess nitrogen and phosphate in the ocean because we are susceptible to the escalated bacterial and toxic growth in presence of additional nutrients (Perlman, 2016).

### 3.2 Purification results

Filiation was successful in lowing salinity levels in all samples, as salt particles became trapped inside the sand. A decrease in total metals is also seen, but totally hardness levels remained 425 ppm for all samples. From this, it can be assumed that ultra-fine matter, such as calcium and magnesium particles are too small to be removed by sand. The alkalinity of each sample was also lowered dramatically. No qualitative changes were observed.

Total metals, alkalinity, and pH decrease for all samples. However, nitrite, phosphate and total hardness remain the same because they are inorganic compounds and therefore, could not be broken down and used by the effective microorganisms. The salinity level in A3 unexpectedly goes up by 10%. No qualitative changes were observed.

The salinity lowered slightly for each sample and total hardness remained the same. However, total metals decreased by half in samples A1 and A3 and 67% in A2. Phosphate was also removed from all samples. This suggests that dur-

Table 3. Initial WQI.

| Factor | A1 results | Q-value | A2 results | Q-value | A3 results | Q-value | Ideal results | Q-value |
|---|---|---|---|---|---|---|---|---|
| Temperature Change (°C) | 3 | 80 | 6 | 65 | 7 | 50 | 0 | 100 |
| pH | 7 | 70 | 7.5 | 90 | 7 | 70 | 8 | 100 |
| Dissolved Oxygen (% saturation) | 75 | 75 | 64 | 64 | 62 | 62 | 100 | 100 |
| Biochemical Oxygen Demand (mg/L) | 5 | 55 | 5 | 55 | 5 | 55 | 0 | 99 |
| Fecal Coliform (#/100 mL) | 2329 | 21 | 1 | 99 | 1 | 99 | 1 | 99 |
| Turbidity (NTU) | 0 | 99 | 0 | 99 | 0 | 99 | 0 | 99 |
| Total Suspended Solids (mg/L) | 4520 | 20 | 5640 | 20 | 3640 | 20 | 50 | 88 |
| Total Phosphate (ppm) | 1 | 40 | 2 | 28 | 4 | 18 | 0 | 100 |
| Total Nitrates (ppm) | 5 | 85 | 20 | 38 | 30 | 28 | 0 | 99 |
| WQI | 60 (Medium) | | 64 (Medium) | | 59 (Medium) | | 99 (Excellent) | |

Ideal Q-values gathered from pathfinderscience.net [8].

Table 4. Samples before treatment.

| Factor | A1 | A2 | A3 |
|---|---|---|---|
| Salinity (% sat) | 44 | 48 | 46 |
| pH | 7.5 | 8 | 7 |
| Total Hardness (ppm) | 425 | 425 | 425 |
| Total Metals (ppm) | 35 | 35 | 50 |
| Phosphate (ppm) | 1 | 2 | 4 |
| Nitrate (ppm) | 5 | 20 | 30 |
| Alkalinity (ppm) | 120 | 180 | 80 |

Table 7. Samples after electrochemical remediation.

| Factor | A1 | A2 | A3 |
|---|---|---|---|
| Salinity (% sat) | 15 | 20 | 20 |
| pH | 6 | 6 | 6 |
| Total Hardness (ppm) | 425 | 425 | 425 |
| Total Metals (ppm) | 10 | 10 | 5 |
| Phosphate (ppm) | 0 | 0 | 0 |
| Nitrate (ppm) | 3 | 5 | 0 |
| Alkalinity (ppm) | 0 | 0 | 0 |

Table 5. Samples after filtration.

| Factor | A1 | A2 | A3 |
|---|---|---|---|
| Salinity (% sat) | 18 | 21 | 10 |
| pH | 6.5 | 7 | 6.5 |
| Total Hardness (ppm) | 425 | 425 | 425 |
| Total Metals (ppm) | 35 | 20 | 20 |
| Phosphate (ppm) | 4 | 2 | 4 |
| Nitrate (ppm) | 5 | 5 | 20 |
| Alkalinity (ppm) | 80 | 80 | 40 |

Table 8. Samples after distillation.

| Factor | A1 | A2 | A3 |
|---|---|---|---|
| Salinity (% sat) | 0 | 0 | 0 |
| pH | 6 | 6 | 6.5 |
| Total Hardness (ppm) | 0 | 0 | 0 |
| Total Metals (ppm) | 5 | 5 | 5 |
| Phosphate (ppm) | 0 | 0 | 0 |
| Nitrate (ppm) | 0 | 0 | 0 |
| Alkalinity (ppm) | 0 | 0 | 0 |

Table 6. Samples after effective microorganisms.

| Factor | A1 | A2 | A3 |
|---|---|---|---|
| Salinity (% sat) | 18 | 21 | 20 |
| pH | 6.5 | 6.5 | 6 |
| Total Hardness (ppm) | 425 | 425 | 425 |
| Total Metals (ppm) | 20 | 15 | 10 |
| Phosphate (ppm) | 4 | 2 | 4 |
| Nitrate (ppm) | 5 | 5 | 5 |
| Alkalinity (ppm) | 40 | 40 | 0 |

ing coagulation, the hydroxyl ions coagulated with phosphate compounds and was removed as floc. During the remediation process, the water became very murky and gray as aluminum was rubbed off of the metal electrodes and floc filled the sample. The water became clear again after screening. Left over sediments had an initially guppy consistency that dried into powder.

Distillation was the most effective treatment method because salinity, total hardness, phosphate, nitrate, and alkalinity all lower to 0 after the process. This occurred because the boiling point of water is lower than all of the listed indices, and was therefore able to be separated into a different flask after condensation. The boiling points of sodium chloride, calcium and magnesium (total hardness), phosphate, and nitrate are 1,413°C, 1,484°C, 1,091°C, 215°C, and 380°C respectively. Total metals were reduced by half in A1 and A2,

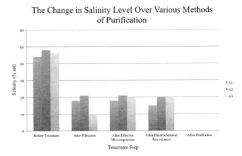

Figure 1.   The change in salinity levels over various methods of purification.

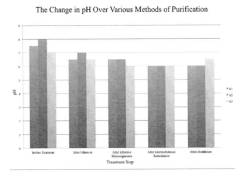

Figure 2.   The change in pH over various methods of purification.

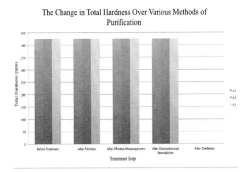

Figure 3.   The change in total hardness over various methods of purification.

and remained the same in A3. This suggests that the first 2 samples contained metals with a range of boiling points both higher and lower than that of water, while A3 only contained metals with a boiling point the same as, or lower than water. No qualitative changes were observed.

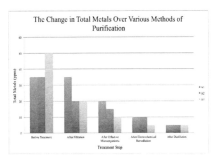

Figure 4.   The change in total metals over various methods of purification.

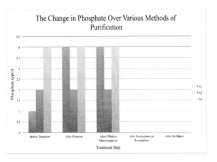

Figure 5.   The change in phosphate over various methods of purification.

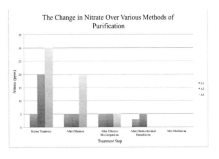

Figure 6.   The change in nitrate over various methods of purification.

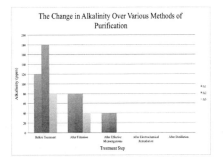

Figure 7.   The change in alkalinity over various methods of purification.

223

### 3.3 *Final WQ*

The newly calculated WQI reveals that the water quality from all 3 locations improved after purification. Namely, the improvement in temperature change, biochemical oxygen demand, fecal coliform, total suspended solids, total phosphate, and total nitrate were particularly high, as they all reached ideal values. While dissolved oxygen levels lowered drastically and significantly decreased the WQI from the value it could have been (because dissolved oxygen has the heaviest weight at 0.17), the purification system is still successful overall because high levels of this factor is only necessary for marine life. Therefore, having high dissolved oxygen levels is not a requirement concerning the agricultural, industrial, commercial and domestic applications of treated water. Also seen from the collected data, the final level of suspended solids decreased significantly from their initial measurement, with a reduction by 3670 mg/L, 4600 mg/L, and 2940 mg/L from Salwa, Sulaibikhat, and North West Beach samples respectively. However, these values are still much higher than the typical parameters of suspended solids at 500 mg/L.

As indicated from collected data, filtration was the most effect method of reducing salinity and alkalinity. Being highly viable and economical, this method could easily operate on an industrial scale, making it an ideal add on-treatment to any of Kuwait's existing purification plants.

Effective microbial consortium on the other hand, is a modern treatment method effective in reducing total metal and alkalinity. Water treated with effective organisms is non-toxic and safe to dispose because it only contains beneficial microorganisms. This process is also more environmentally friendly than conventional methods of water purification because the biological nature of the treatment leaves significantly less leftover brine/sediments. Effective microbial consortium is currently unemployed by any of Kuwait's treatment plants, but will make a potentially greatly successful and viable addition.

Next, electrochemical remediation was successful in lowering phosphate, nitrate and alkalinity levels. Although this method was effective in removing ultra-fine particles and pollutants the previous methods could not, the process required a great deal electricity to run and is not economical. However,

| Factor | A1 results | Q-value | A2 results | Q-value | A3 results | Q-value | Ideal results | Q-value |
|---|---|---|---|---|---|---|---|---|
| Temperature Change (°C) | 0 | 100 | 0 | 100 | 0 | 100 | 0 | 100 |
| pH | 6 | 45 | 6 | 45 | 6.5 | 55 | 8 | 100 |
| Dissolved Oxygen (% saturation) | 25 | 18 | 26 | 19 | 25 | 18 | 100 | 100 |
| Biochemical Oxygen Demand (mg/L) | 0 | 99 | 0 | 99 | 0 | 99 | 0 | 99 |
| Fecal Coliform (#/100 mL) | 1 | 99 | 1 | 99 | 1 | 99 | 1 | 99 |
| Turbidity (NTU) | 0 | 99 | 0 | 99 | 0 | 99 | 0 | 99 |
| Total Suspended Solids (mg/L) | 850 | 20 | 1040 | 20 | 2700 | 20 | 50 | 88 |
| Total Phosphate (ppm) | 0 | 100 | 0 | 100 | 0 | 100 | 0 | 100 |
| Total Nitrates (ppm) | 0 | 99 | 0 | 99 | 0 | 99 | 0 | 99 |
| WQI | 74 (Good) | | 74 (Good) | | 75 (Good) | | 99 (Excellent) | |

## 4 CONCLUSION

The data analysis revealed that the 4-step water purification system using filtration, effective microbial consortium, electrochemical remediation, and distillation water purification was successful, as WQI of all water samples improved after the treatment. Initially, the highest and lowest scoring water quality had a difference of 5 points (North West Beach had a WQI of 59 and Sulaibikhat Beach had a WQI of 64). However, after completing the purification cycle, this difference was reduced to 1 point. This suggests that the designed treatment system is able to consistently purify water-no matter the contamination level, to a high extent.

the left over sediments form this method dries to powder and is easier to dispose of than regular brine.

Lastly, distillation was the most effective purification method tested as it reduced salinity, total hardness, total metals, phosphate, nitrate, and alkalinity all to 0. Although the process used in this study cannot be imitated industrially, solar distillation is a modern method of water treatment operating on the same principle and would be highly applicable in Kuwait's weather.

In summary, the purpose of this study was to compare the initial and final WQI of ocean water samples from Salwa Beach, Sulaibikhat Beach, and North West Beach after being treated with a complete 4-step water purification system using filtration, effective microbial consortium, electro-chemical remediation, and distillation. The findings indicate towards the success of the tested treatment methods, and each process's ability to remove certain pollutants allows for target specific purification. For example, because filtration is effective in reducing salinity and alkalinity, and electrochemical remediation in lowering phosphate and nitrate levels, a combination of these two methods can be utilized to remove all major contaminants. To conclude, applying target specific purification allows water to be treated more efficiently and economically, and allows for these methods to be applied to existing treatment plants as their function is needed.

## ACKNOWLEDGMENTS

This work was supported by Alpha Bio-Laboratory, special thanks Dr. I-Chen for making this report possible.

## REFERENCES

Lyons, B. P., et al. "Microbial Water Quality and Sedimentary Faecal Sterols as Markers of Sewage Contamination in Kuwait." *Marine Pollution Bulletin* 100.2 (2015): 689–698. Web. 27 July 2016.

Monica, S, et al. *Formulation of Effective Microbial Consortia and Its Application for Sewage Treatment.* School of Biosciences and Technology, VIT University, Vellore, India: Journal of Microbial & Biochemical Technology, Jan. 2011. Web. 23 July 2016.

Perlman, Howard. "Nitrogen and water: USGS water science school." *USGS.* USGS—U.S. Geological Survey, n.d. Web. 23 July 2016.

Shalaby, Emad A. "Prospects of Effective Microorganisms Technology in Wastes Treatment in Egypt." 1.3 (2011): n.pag. Web. 6 July 2016.

"Sulaibiya wastewater treatment." Water Technology. Water Technology, n.d. Web. 5 July 2016.

"The Effects of Ocean Pollution on Marine Mammals." *Ocean Pollution and Marine Mammals—BlueVoice. org.* N.p., n.d. Web. 04 July 2016.

Toumi, Habib, and Bahrain Bureau Chief. "Kuwait's polluted seawaters a threat to marine life." Environment,. Gulf News, 14 Mar. 2010. Web. 5 July 2016.

"Water Quality Index Protocol." *Path Finder Science.* n.d. Web. 27 July 2016.

"Water resources in Kuwait." Beatona Kuwait Offical Water Portal. 5 July 2016. Web. 5 July 2016.

"Water treatment alternatives—distillation." All About Water. 2004. Web. 6 July 2016.

*Advances in Energy and Environment Research – Achour & Wu (Eds)*
*© 2017 Taylor & Francis Group, London, ISBN 978-1-138-62682-9*

# Identification of bacterial species in Kuwaiti waters through DNA sequencing

K. Chen
*American International School of Kuwait, Hawalli, Kuwait*

ABSTRACT: With an objective of identifying the bacterial diversity associated with ecosystem of various Kuwaiti Seas, bacteria were cultured and isolated from 3 water samples. Due to the difficulties for cultured and isolated fecal coliforms on the selective agar plates, bacterial isolates from marine agar plates were selected for molecular identification. 16S rRNA genes were successfully amplified from the genome of the selected isolates using Universal Eubacterial 16S rRNA primers. The resulted amplification products were subjected to automated DNA sequencing. Partial 16S rDNA sequences obtained were compared directly with sequences in the NCBI database using BLAST as well as with the sequences available with Ribosomal Database Project (RDP).

## 1 INTRODUCTION

Bacteria are key microbes in marine microbiology that play fundamental roles in the ocean environment. Apart form being the base of the food chain, supplying more than half of the world's oxygen, controlling the role of marine nutrients, and being major processors of greenhouses gases, they are also rapidly responsive to changes in their ecosystem, making bacteria an ideal indictor of marine change. One such case is the microbialization of bacteria in contaminated waters over the past decade. The growing pathogenicity of marine microbes can cause disease in both marine and human life. Upon contact or consumption of specific bacteria, humans can experience cramps, nausea, diarrhea, tuberculosis, legionella pneumonia, typhoid, and cholera. Bacterial diseases of fish include-but are not limited to: hemorrhagic septicemia, vibriosis, and columnaris (Sharp, 2009). The susceptibility of infection from pathogenic bacteria reinforces the importance of water purification before industrial, agricultural, or domestic applications. It is also ideal that water treatment processes use purification methods that limit leftover waste, as brine sediments often contain pathogens not killed off during treatment, and disposing them back into the ocean reintroduces the harmful bacteria into the aquatic environment in abundance.

There is growing interest in marine bacteria in the scientific world because pathogenic, or otherwise, their and identification serves many purposes. Firstly, extremophiles can be used as a measure of water pollution levels or aquatic condition because these species only grow in extreme environments. For example, halophilic bacteria only thrive in waters of high salt concentration, and are an indicator of overly high salinity. Next, being aware of pathogenic bacteria and locations were specific genera reside allows researchers, scientists and doctors to better prevent disease that could be contracted in the area-some of which can be very dangerous. Treatment plants can also better allocate purification methods for more effect water treatment with an understanding of the bacterial genera residing in specific marine regions. For example, bacteria killing methods of purification-such as electrochemical remediation, will be more efficient in treating water samples containing pathogens than other methods that do not eliminate bacteria. Employing target specific ways of water purification is more effective, efficient, and economical. Finally, bacterial identification is also important because many bacterial species such as: Marinobacter hydrocarbonoclasticus, Carboxydothermus hydrogenoformans, and Sphingomonas harbor properties that can be industrially useful in pharmaceutical, biotechnological, agricultural, chemical, and environmental fields.

The purpose of this study is to investigate the diversity and properties of marine bacteria species from Salwa Beach (N1), North West (N2) and Sulaibikhat Beach (N3) by directly culturing seawater samples, PCR amplification of 16S rRNA gene, and DNA sequencing.

## 2 METHODOLOGY

### 2.1 *DNA growth and preparation*

Bacteria from each beach was first grown by spreading 3 µL of each sample onto separate marine agar plates, then incubated at 28°C in an bacteriological incubator for 24 hours. Grown bacteria cultures were then collected and mixed into cell pellets of 10 µL of sterile distilled water. 48 samples of bacteria were collected from each beach.

*Cultivation was also attempted on MacConkey, and nutritional agar plates. However, results indicated that bacteria could not be cultured on said plates, which made the identification of fecal coliforms based on culturing conditions and microscopical morphologies impossible.

## 2.2 PCR

Bacterial 16S rDNA was amplified from the extracted genomic DNA by using the following universal eubacterial 16S rRNA primers: forward primer 5' AGAGTTTGATCCTGGCTCAG 3' and reverse primer 5' ACGGCTACCT TGTTAC-GACTT 3'. An ABI PCR System 9700 was used to perform the PCR protocol in a volume of 24 µL of bacteria strain identification solution mix containing $10 \times 2.5$ µL buffer, 1.25 µL magnesium chloride, 1 µL dNTP, 0.05 µL primer 1, 0.05 µL primer 2, 0.1 µL enzyme taq, and 19.05 µL of distilled water per 1 µL of template bacteria. Thermal cycling consisted of a 10 minute initial denaturation at 95°C, 30 cycles of denaturation at 94°C for 30 seconds, 30 seconds of annealing at 60°C, and extension for 2 minutes at 72°C with a 10 minute final maintenance at 4°C (Jeshveen, 2012).

## 2.3 Agarose gel electrophoresis

From each PCR product, an aliquot of 15 µL was subject to 1.0% agarose gel electrophoresis containing 0.5 × TBE buffer (pH 8.0), and 5 µL of ethidium bromide was used to stain the gel. Electrophoresis was conducted at 100 Volt, 400 mA for 20–30 minutes with 0.8 µL of 100 bp DNA marker. After, the DNA bands were observed under ultraviolet light using gel documentation system (Jeshveen, 2012).

## 2.4 Sequencing of 16S rRNA genes

Sequencing reaction was carried out using ABI PRISM 310 Genetic Analyzer (PE applied Biosystems). For the sequencing reaction, Big Dye Ready Reaction DyeDeoxy Terminator Cycle Sequencing kit (Perkin-Elmer) was employed.

## 2.5 Data analysis

Sequence analysis was performed with sequences in the NCBI database using BLAST as well as with sequences available with Ribosomal Database Project (RDP)

# 3 RESULTS AND DISCUSSION

## 3.1 Salwa Beach

Total bacteria counts: 85

It can be inferred that Salwa beach is home to an abundance of tapeworms, as 6 out of 8 pathogenic bacteria identified were of such. Tapeworms, which generally reside in areas of inadequate sanitation is an indication of the area's contamination levels. Salwa beach is also

Table 1. Identified bacteria isolates in Salwa Beach after data analysis.

| Isolate | Identified bacterial strain |
|---------|----------------------------|
| N1-001 | Uncultured bacterium clone ncd981d10c1 |
| N1-002 | Exiguobacterium acetylicum |
| N1-003 | Cronobacter sakazakii |
| N1-004 | Uncultured bacterium isolate KWSUB08_76 16S |
| N1-005 | Marinobacter sp. MMRF630 16S |
| N1-006 | Alteromonas macleodii |
| N1-007 | Marinobacter litoralis |
| N1-009 | Marinobacter sp. HOT2G5 16S |
| N1-010 | Alteromonas sp. NBF17 16S |
| N1-011 | Alteromonas macleodii |
| N1-012 | Bacterium SCSIO13055 16S |
| N1-013 | Bacterium SCSIO13055 16S |
| N1-014 | Pseudoalteromonas sp. Bac192 16S |
| N1-015 | Alteromonas macleodii |
| N1-016 | Alteromonas sp. SS12.5 16S |
| N1-017 | Phaeobacter caeruleus |
| N1-018 | Enterobacter cloacae |
| N1-019 | Alteromonasmacleodii strain CSB14KR 16S |
| N1-020 | Sphingomonas |
| N1-021 | Spirometra erinaceieuropaei |
| N1-022 | Alteromonas sp. OCN004 16S |
| N1-023 | Hydatigera taeniaeformis |
| N1-024 | Gamma proteobacterium |
| N1-025 | Schistosoma rodhaini |
| N1-026 | Erythrobacter citreus |
| N1-027 | Exophiala mesophila 3-isopropylmalate dehydrogenase |
| N1-029 | Tropicibacter sp. MCCC1 A07686 16S |
| N1-030 | Xanthomonas oryzaepv. Oryzae |
| N1-031 | Pseudoalteromonas sp. F497 16S |
| N1-032 | Pseudomonas sp. X3–2 |
| N1-033 | Pseudomonas psychrophila |
| N1-034 | Colacium sp. Ungok033107C |
| N1-035 | Citromicrobium bathyomarinum |
| N1-036 | Candida catenulate |
| N1-038 | Diphyllobothrium latum |
| N1-041 | Schistosoma rodhaini |
| N1-045 | Spirometra erinaceieuropaei |
| N1-046 | Tetrapisispora phaffii |
| N1-047 | Plasmodium falciparum |

Table 2. Pathogenic bacteria found in Salwa Beach isolates after data analysis.

| Isolate | Description |
|---------|-------------|
| N1-003 | Cronobacter sakazakii: causes cronobacter disease with fatality rates of 50–80%. Infants and immunocompromised adults are most susceptible (Healy, 2009). |
| N1-018 | Enterobacter cloacae: Enterobacter cloacae infection has the highest mortality rate amongst all Enterobacter infections. It causes bacteremia, skin and soft-tissue infections, urinary tract infections, endocarditis, intra-abdominal infections, septic arthritis, osteomyelitis, CNS infections, ophthalmic infections, and lowers respiratory tract infections. |
| N1-021 | Spirometra erinaceieuropaei: is a tapeworm that causes a parasitic infection called sparganosis (Bennett, 2014). |
| N1-023 | Hydatigera taeniaeformis: a parasitic tapeworm with cats as their primary definitive hosts. |
| N1-025 | Schistosoma rodhaini: commonly known as blood-flukes, they are parasitic flatworms that cause schistosomiasis in humans-an infection considered by the World Health Organization to be the second most socioeconomically devastating parasitic disease infecting hundreds of millions worldwide. |
| N1-030 | Xanthomonas oryzaepv. Oryzae: causes bacterial blight, a serious disease for rice, grass, and sedges worldwide (Zhang, 2013). |
| N1-038 | Diphyllobothrium latum: is the largest human tapeworm and the principle species of Diphyllobothrium causing diphyllobothriosis. It infects humans through the somsuptiono of raw or undercooked seafood, but can also affect fish and other mammals. |
| N1-041 | Schistosoma rodhaini: refer to N1-025 |
| N1-045 | Spirometraerinaceieuropaei: is a tapeworm that infects humans and domestic animals. In humans, is causes sparganosis (Bennett, 2014). |
| N1-047 | Plasmodium falciparum: is a protozoan parasite that is the deadliest of five human malaria species, being responsible for more than 85% of all malaria cases, and has the highest mortality rate amongst all malaria species (Bowman, 1999). |

Table 3. Bacteria with industrial uses found in Salwa Beach isolates after data analysis.

| Isolate | Description |
|---------|-------------|
| N1-020 | Sphingomonas: typically found in environments polluted with toxic compounds, they have the ability to utilize the contaminants as nutrients. Their biodegradative and biosynthetic capabilities have been utilized for numerous biotechnological applications including bioremediation of environmental contaminants for the production of extracellular polymers such as sphingans. It is also effective in degrading degrade alkylated polyaromatic hydrocarbon (2-methylphenanthrene), commonly known as polystyrene plastic (Matuzahroh, 1999). |
| N1-031 | Pseudoalteromonas sp. F497: capable of synthesizing biologically active molecules that produce compounds beneficial to eukaryotic organisms. |
| N1-046 | Tetrapisispora phaffii: killer yeast that secretes a glycoprotein known as Kpkt that has antimicrobial activity with spoilage yeasts under winemaking conditions. It can be used to combat contaminating wild yeasts in food, to control pathogenic fungi in plants, and to develop novel antimycotics for the treatment of human and animal fungal infections in the medical field (Comitini, 2009). |

a harbor. Many tapeworms could dwell inside the intestines of fish, so extra persuasions should be taken when eating undercooked seafood to prevent infections.

There is no pattern to the useful bacterial species identified in Salwa beach, but the particularly useful Sphingomonas was cultured and identified. Its unique ability to synthesize and degrade toxic compounds in water could be applied as an effective organism in water purification-a function that can be utilized in Kuwait's many water treatment plants.

## 3.2 North West Beach

Total bacteria counts: 112

Schistosoma mattheei was the singular pathogenic bacteria culture and identified in North West Beach, and it has no direct threat to either humans or marine life. The area of the collected sample should be taken into consideration, as it is the closest to a direct polluting source (industrial power plant) with notably less fish than in popular fishing regions such a Salwa Beach. This could possibly explain the sample's lack of pathogenic bacteria, as many species of pathogens reside in marine life.

Table 4. Identified bacteria isolates in North West Beach after data analysis.

| Isolate | Identified bacterial strain |
|---------|------------------------------|
| N2-001 | Muricauda sp. SA1 16S |
| N2-002 | Alteromonas macleodii strain G7C_40 m_M_02 16S |
| N2-003 | Myxococcus fulvus |
| N2-004 | Marinobacterium stanieri |
| N2-005 | Idiomarina aquimaris |
| N2-006 | Marinobacter hydrocarbonoclasticus |
| N2-007 | Spongiibacte rmarinus |
| N2-010 | Flavobacteriales bacterium |
| N2-011 | Marinobacter sp. NNA5 16S |
| N2-012 | Alteromonas macleodii |
| N2-013 | Corynebacterium humireducens |
| N2-014 | Rhodobacteraceae bacterium |
| N2-015 | Bacillus pumilus |
| N2-016 | Schistosoma mattheei |
| N2-017 | Muricauda sp. mur1 16S |
| N2-018 | Flavobacteriales bacterium |
| N2-019 | Marinobacter hydrocarbonoclasticus |
| N2-020 | Marinobacter flavimaris |
| N2-021 | Rhodobacteraceae bacterium |
| N2-022 | Marinobacter Pelagius |
| N2-023 | Rhodobacteraceae bacterium |
| N2-024 | Loxodontaafricana ALG12, alpha-1,6-mannosyltransferase |
| N2-025 | Pectobacterium sp. SCC3193 |
| N2-026 | Brassica napus |
| N2-027 | Schistosoma curassoni |
| N2-028 | Marinobacter flavimaris |
| N2-029 | Erythrobacter citreus |
| N2-030 | Marinobacter hydrocarbonoclasticus |
| N2-036 | Novosphingobium sp. SG8916S |
| N2-037 | Marinobacter sp. G5-4a |
| N2-038 | Rhodobacteraceae bacterium |
| N2-039 | Mus musculus |
| N2-040 | Solitalea canadensis |
| N2-041 | Pseudoalteromonas sp. ECSMB85 16S |
| N2-042 | Marinobacter flavimaris |
| N2-043 | Spongiibacter marinus |
| N2-044 | Schistosoma curassoni |
| N2-045 | Cyanobium sp. NIES-981 |
| N2-046 | Marinobacter sp. DY57-1 16S |
| N2-047 | Uncultured organism clone ctg_CGOF252 16S |
| N2-048 | Roseovarius pacificus |

Table 5. Pathogenic bacteria found in North West Beach isolates after data analysis.

| Isolate | Description |
|---------|-------------|
| N2-016 | Schistosoma mattheei: a parasite of sheep, goats, and cattle in regions of West Africa (Pfukenyi, 2006). |
| N2-027 | Schistosoma mattheei: refer to N2-016 |
| N2-044 | Schistosoma mattheei: refer to N2-016 |

Table 6. Bacteria with industrial uses found in North West Beach isolates after data analysis.

| Isolate | Description |
|---------|-------------|
| N2-006 | Marinobacter hydrocarbonoclasticus: are important to bioremediation because they can effectively degrade hydrocarbons-especially those found in alkanes and major components of oil as they are tolerant of high salinity and can grow aerobically and anaerobically (Hamdan, 2011). The marinobacter hydrocarbonoclasticus' distinctive ability makes it ideal for treating recalcitrant pollutants. |
| N2-015 | Bacillus pumilus: is used as an active ingredient in agricultural fungicides because its growth on plant roots prevents rhizoctonia and fusarium spores from germinating. Bacillus pumilus proteases are also used in chemical, detergent, food, and leather industries as antimicrobials and antifungals. |
| N2-019 | Marinobacter hydrocarbonoclasticus: refer to N2-006 |
| N2-029 | Erythrobacter citreus: is a potential candidate for removing poisons because it can resist and reduce the toxic compound tellurite (Yurkov, 1996). |
| N2-030 | Marinobacter hydrocarbonoclasticuss: refer to N2-006 |

3 isolates of Marinobacterhydrocarbonoclasticus was found out of the 48 bacteria samples taken, suggesting that the species is abundant in North West Beach waters. Its properties are very useful in purifying Kuwaiti water, as it is able to degrade hydrocarbons found in oil. Oil is an abundant resource in Kuwait that frequently pollutes the county's water sources, but is difficult to remove once contaminated. Therefore, the potential of using bacteria to purify this recalcitrant pollutant is exciting and could perhaps be more cost efficient than currently employed methods.

### 3.3 Sulaibikhat Beach

Total bacteria counts: 126

Species of bacteria from the Vibrio genus was especially prevalent in Sulaibikhat Beach, with 6 out of 10 identified isolates fitting into the category. Both Vibrio alginolyticus and Vibrio parahaemolyticus, along with Providenciarettgeri, are all foodborne infections. Therefore, precautions should be taken against eating undercooked food fished from this area.

A wide diversity of environmentally useful bacterial species was identified in Sulaibikhat Beach. The most abundant was Exiguobacterium undae. Its property of reverting mealworms

Table 7. Identified bacteria isolates in Sulaibikhat Beach after data analysis.

| Isolate | Identified bacterial strain |
| --- | --- |
| N3-001 | Carboxydothermus hydrogenoformans |
| N3-002 | Lactococcus garvieae |
| N3-003 | Vibrio alginolyticus |
| N3-004 | Vibrio parahaemolyticus |
| N3-005 | Vibrio sp. H-188 16S |
| N3-006 | Virgibacillus sp. SK33 16S |
| N3-007 | Vibrio alginolyticus |
| N3-008 | Vibrio alginolyticus |
| N3-010 | Exiguobacterium undae |
| N3-012 | Uncultured gamma proteobacterium |
| N3-013 | Exiguobacterium undae |
| N3-014 | Exiguobacterium undae |
| N3-015 | Exiguobacterium undae |
| N3-017 | Vibrio sp. 3677 |
| N3-018 | Vibrio alginolyticus |
| N3-019 | Vibrio sp. B2-15-2 16S |
| N3-020 | Vibrio sp. H-219 16S |
| N3-021 | Vibrio sp. 3683 |
| N3-022 | Vibrio orientalis |
| N3-023 | Vibrio natriegens |
| N3-024 | Providencia rettgeri |
| N3-025 | Clupeaharengus translocase |
| N3-026 | Anas platyrhynchos |
| N3-027 | Drosophila takahashii |
| N3-028 | Vibrio sp. sw9 16S |
| N3-029 | Uncultured bacterium clone I3Q1XXJ01 APUOI 16S |
| N3-030 | Nippostrongylus brasiliensis |
| N3-031 | Vibrio sp. 3677 |
| N3-033 | Vibrio alginolyticus |
| N3-034 | Vibrio sp. B2-15-2 16S |
| N3-035 | Vibrio sp. H-219 16S |
| N3-036 | Vibrio sp. 3683 |
| N3-038 | Vibrio orientalis |
| N3-039 | Vibrio natriegens |
| N3-040 | Providencia rettgeri |
| N3-041 | Clupea harengus |
| N3-042 | Anas platyrhynchos |
| N3-043 | Drosophila takahashii |
| N3-044 | Vibrio sp. sw9 16S |
| N3-045 | Uncultured bacterium clone I3Q1XXJ01 APUOI 16S |
| N3-046 | Nippostrongylus brasiliensis |

to feed solely on polystyrene has enormous potential as an ecofriendly way of degrading plastic. Next, Carboxydothermus hydrogeno-formans's diet of only toxic carbon monoxide and water, which later produces hydrogen gas, could provide an alterative for oil. Lastly, a useful species of Vibrio, sp. B2-15-2, was identified amongst its pathogenic counterparts and demonstrates the impressive biodiversity of marine bacteria and their respective properties.

Table 8. Pathogenic bacteria found in Sulaibikhat Beach isolates after data analysis.

| Isolate | Description |
| --- | --- |
| N3-002 | Lactococcus garvieae: is a zoonotic pathogen that causes infective endocarditis, spondylitis, osteomyelitis, and liver abscesses in humans. But mostly affects the Japanese yellowtail, rainbow trout, and grey mullet (Chao, 103). |
| N3-003 | Vibrio alginolyticus: like most Vibrio infections, this is a foodborne disease that causes vibriosis, otitis and wound infection. |
| N3-004 | Vibrio parahaemolyticus: also cause vibriosis, along with gastrointestinal illness when ingested |
| N3-007 | Vibrio alginolyticus: refer to N3-003 |
| N3-008 | Vibrio alginolyticus: refer to N3-003 |
| N3-018 | Vibrio alginolyticus: refer to N3-003 |
| N3-030 | Nippostrongylus brasiliensis: is a gastrointestinal roundworm that primarily infects rats (Maizels, 2016). |
| N3-033 | Vibrio alginolyticus: refer to N3-003 |
| N3-040 | Providencia rettgeri: causes infections involving the urinary tract with symptoms of polyuria, hematuria, fevers, and dysuria. It has also recently been implicated as etiologic agents in traveler's diarrhea. |
| N3-046 | Nippostrongylus brasiliensis: refer to N3-030 |

Table 9. Bacteria with industrial uses found in Sulaibikhat Beach isolates after data analysis.

| Isolate | Description |
| --- | --- |
| N3-001 | Carboxydothermus hydrogenoformans: is an extremely thermophilic bacteria that feeds almost entirely on toxic carbon monoxide and water, and produces hydrogen gas as part of its metabolism (Wu, 2005). Its ability to rapidly and efficiently conduct this conversion makes it highly valuable for the future where hydrogen gas is being studied as potential biofuel. |
| N3-013 | Exiguobacterium undae: According senior research engineer Wei-Min Wu in his report "Biodegradation and Mineralization of Polystyrene by Plastic-Eating Mealworms. 2. Role of Gut Microorganisms", injections of exiguobacterium undae into the gut of mealworms allows them to survive solely on a diet of polystyrene, and thus, can significantly help in the biodegradation and mineralization of plastic (Yang, 2015). |
| N3-014 | Exiguobacterium undae: refer to N3-013 |
| N3-015 | Exiguobacterium undae: refer to N3-013 |
| N3-034 | Vibrio sp. B2-15-2 16S: is capable of producing biosurfactant that can be applied into environmental, technical, and potentially pharmaceutical industries (Hu, 2015). |

## 4  CONCLUSION

The data analysis revealed that Kuwaiti water samples harbor extremely diverse bacteria belonging to over 53 bacterial genera. 24 genera were isolated in Salwa Beach, 17 in North West Beach, and 12 in Sulaibikhat Beach. In Salwa Beach samples, identified genera consisted primarily of Alteromonas (17.5% out of 40), North West Beach samples consisted of prominently Marinobacter (26.8% out of 41 sequences), and Sulaibikhat Beach samples consisted of mostly Vibrio (46.3% out of 41 sequences). Alteromonas and Marinobacter strains were found to be abundant and distributed in both Salwa and North West Beach, while none were cultured and identified in Sulaibikhat Beach.

Each sampled location exhibited specific patterns of cultivated pathogen bacterial species that reveals information regarding each beach's unique marine condition and life. In Salwa Beach, most pathogens were tapeworms, only Schistosoma mattheei were isolated from North West Beach, and 3 pathogenic species of Vibrio was found in Sulaibikhat Beach. Since tapeworms and Vibrio species are ubiquitous in aquatic environments copious in marine life, the intense mariculture in Salwa and Sulaibikhat Beach may be partially accountable for the high occurrence of these pathogens.

On the other hand, identified industrially useful bacteria demonstrated no pattern in accordance to sampled location, but similarities in valuable property. While a wide range of diverse, unrelated bacterial cultures were isolated in each beach, the majority of all had biodegradative and biosynthetic capabilities. For example, Sphingomonas found in Salwa Beach can utilize toxic compounds and contaminants as nutrients, Marinobacter hydrocarbonoclasticus found in North West Beach can effectively degrade hydrocarbons, and Carboxydothermus hydrogenoformans found in Sulaibikhat Beach can feed almost entirely on toxic carbon monoxide and water, while producing hydrogen gas. Upon further study, these species-along with the others identified, potentially offer incredible scientific breakthroughs in the usage of microorganisms in pharmaceutical, biotechnological, agricultural, chemical, and environmental fields.

In summary, the purpose of this study was to cultivate, isolate, identity, and determine properties of marine bacterial species in Kuwait from Salwa Beach (N1), North West (N2) and Sulaibikhat Beach (N3) through PCR amplification of 16S rRNA gene, and DNA sequencing of bacteria grown from collected water samples. The findings indicate towards a high level of macrobiotic diversity, and shed light on the ecological functions of both pathogenic and useful bacterial isolates-information that can be used towards precautionary causes or further inquiry and useful application.

## ACKNOWLEDGMENTS

This work was supported by Alpha BioLaboratory, special thanks Dr. I-Chen for making this report possible.

## REFERENCES

"Bacillus pumilus." *Microbe Wiki*. n.d. Web. 12 July 2016.

"Bacteria and virus contamination of water." *Aqua Bright*. 2013. Web. 11 July 2016.

Bennett, HM, et al. "The Genome of the Sparganosis Tapeworm Spirometra Erinaceieuropaei Isolated from the Biopsy of a Migrating Brain Lesion." *Genome biology*. 15.11 (2014): n.pag. Web. 12 July 2016.

Bowman, S, et al. "Plasmodium falciparum Sanger." *Sanger*. 1999. Web. 16 July 2016

"Carboxydothermus hydrogenoformans." *Microbe Wiki*. n.d. Web. 12 July 2016.

"Causes and symptoms of Vibrio parahaemolyticus - Minnesota Dept. Of health." *Minnesota Department of Health*. 10 Mar. 2015. Web. 12 July 2016.

CDC. "Diphyllobothrium—biology." *Parasites—Diphyllobothrium Infection*. CDC, 15 Aug. 2010. Web. 16 July 2016.

Chao, C. T., C. F. Lai, and J. W. Huang. "Lactococcus Garvieae Peritoneal Dialysis Peritonitis." 33.1 (2013): n.pag. Web. 12 July 2016.

Comitini, F, and M Ciani. "The Zymocidial Activity of Tetrapisispora Phaffii in the Control of Hanseniaspora Uvarum During the Early Stages of Winemaking." *Letters in applied microbiology*. 50.1 (2009): 50–6. Web. 16 July 2016.

"Enterobacter infections: Practice essentials, background, Pathophysiology." *Medscape*. 2 June 2016. Web. 12 July 2016.

Hamdan, Leila J., and Preston A. Fulmer. *Effects of COREXIT® EC9500A on Bacteria from a Beach Oiled by the Deepwater Horizon Spill*. Washington, DC 20375, USA: Inter-Research, 31 Mar. 2011. Web. 12 July 2016.

Healy, B, et al. "Cronobacter (Enterobacter Sakazakii): An Opportunistic Foodborne Pathogen." *Foodborne pathogens and disease*. 7.4 (2009): 339–50. Web. 12 July 2016.

Hu, Xiaoke, Caixia Wang, and Peng Wang. "Optimization and Characterization of Biosurfactant Production from Marine Vibrio Sp. Strain 3B-2." 6. (2015): n.pag. Web. 16 July 2016.

"Hydatigera taeniaeformis—WormBase paraSite." *Wormbase Parasite*. 2016. Web. 12 July 2016.

Jeshveen, S. S., et al. *Optimization of Multiplex PCR Conditions for Rapid Detection of Escherichia Coli O157:H7 Virulence Genes*. Malaysia: nternational Food Research Journal, 2012. Web. 12 July 2016.

Maizels, Rick. "Nippostrongylus brasiliensis." *Rick Maizels' Group*. n.d. Web. 12 July 2016.

"Marine Microbes." *Australian Institute of Marine Science.* jurisdiction = Commonwealth of Australia; corporate Name = Australian Institute of Marine Science, 24 July 2000. Web. 18 July 2016.

"Marine Microbes: Did You Know?" *National Oceanic and Atmospheric Administration.* n.d. Pdf. 18 July 2016.

"Marinobacter hydrocarbonoclasticus." *Microbe Wiki.* n.d. Web. 12 July 2016.

Matuzahroh, Ni, et al. "In-Vitro Study of Interaction Between Photooxidation and Biodegradation of 2-Methylphenanthrene by Sphingomonas Sp. 2MPII." *Chemosphere* 38.11 (1999): 2501–2507. Web. 12 July 2016. Pan, J, Q Huang, and Y Zhang. "Gene Cloning and Expression of an Alkaline Serine Protease with Dehairing Function from Bacillus Pumilus." *Current microbiology.* 49.3 (2004): 165–9. Web. 12 July 2016.

Pfukenyi, DM, et al. "Epidemiological Studies of Schistosoma Mattheei Infections in Cattle in the Highveld and Lowveld Communal Grazing Areas of Zimbabwe." *The Onderstepoort journal of veterinary research.* 73.3 (2006): 179–91. Web. 12 July 2016.

"Providencia infections: Background, Pathophysiology, Epidemiology." *Medscape.* 10 Apr. 2016. Web. 16 July 2016.

"Pseudoalteromonas." *Microbe Wiki.* n.d. Web. 12 July 2016.

Sharp, Merck, and Dohme Corp. "Bacterial diseases of fish: Aquarium fishes: Merck veterinary manual." *Merck Manuals.* 2009. Web. 11 July 2016.

"Vibrio infections: Background, Pathophysiology, Epidemiology." *Medscape.* 2 June 2016. Web. 12 July 2016.

WHO. "Schistosomiasis." *World Health Organization.* World Health Organization, 22 Feb. 2016. Web. 12 July 2016.

Wu, Martin, et al. "Life in Hot Carbon Monoxide: The Complete Genome Sequence of Carboxydothermus Hydrogenoformans Z-2901." *PLoS Genetics* 1.5 (2005): e65. Web.

Yang, Yu, et al. "Biodegradation and Mineralization of Polystyrene by Plastic-Eating Mealworms: Part 2. Role of Gut Microorganisms." *Environmental Science & Technology* 49.20 (2015): 12087–12093. Web.

Yurkov, V., J. Jappe, and A. Vermeglio. "Tellurite Resistance and Reduction by Obligately Aerobic Photosynthetic Bacteria." 62.11 (1996): n.pag. Web. 12 July 2016.

Zhang, Haitao, and Shiping Wang. "Rice Versus Pv.: A Unique Pathosystem." *Current Opinion in Plant Biology* 16.2 (2013): 188–195. Web. 12 July 2016.

*Advances in Energy and Environment Research – Achour & Wu (Eds)*
*© 2017 Taylor & Francis Group, London, ISBN 978-1-138-62682-9*

# Topographic feature line extraction from point cloud based on SSV and HC-Laplacian smoothing

Wei Zhou, Rencan Peng & Lihua Zhang
*Department of Hydrography and Cartography, Dalian Naval Academy, Dalian, China*
*PLA Key Laboratory, Dalian, China*

Wanjin Wang
*The PLA 91550 Unit, Liaoning, Dalian, China*

ABSTRACT:   In this paper, a new method for extracting topographic feature lines from point cloud based on SSV and HC-Laplacian smoothing method is proposed. As the calculation of the curvature based on surface fitting is quite complex and less efficient, Signed Surface Variation (SSV) is introduced first to extract the potential feature points. Then, the potential feature points are segmented into different clusters by region growing based on the Euclidean distance and SSV. By using the Ball Pivoting algorithm, the mesh surface of each cluster is reconstructed and the boundary points are recognized. The HC-Laplacian smoothing method can prevent the shrinkage of the feature lines, and the SSV weight can effectively accelerate the convergence. Combining the two advantages, the potential feature points in each cluster are iteratively thinned using an SSV weighted HC-Laplacian smoothing method, in which the boundary points and non-boundary points have different surrounding neighbors and weight. Finally, the feature lines are achieved by connecting these extracted points based on the Minimum Spanning Tree (MST) algorithm. The comparison with D8 flow direction method shows that the method proposed in this paper can extract the ridge lines and valley lines from point cloud in various terrains with high efficiency.

## 1   INTRODUCTION

The topographic feature lines are the skeleton of the geomorphology, which can describe the topographic relief and terrain complexity. Extraction of topographic feature lines is a common issue in geoinformatics research, and it is of great significance in point cloud registration, data reduction, and three-dimensional (3D) surface reconstruction (Bisheng Yang et al., 2016; Lin Yang et al., 2014; Jingzhong Xu et al., 2014). Most works of the topographic feature line extraction are conducted based on the raster model (Frans Persendt et al., 2015; Fatih Gulgen et al., 2010; G. Mandlburger et al., 2011). The raster-based algorithm has several advantages, such as simple data structure, easy saving, and convenient management. However, this algorithm could operate only in two-dimensional (2D) space, and just interpolate the height from a more or less smoothed Digital Elevation Model (DEM), which will cause the precision loss of the geographic coordinate during the re-sampling or the interpolation process. Therefore, pure point-based method may be a good option to extract feature lines, which could avoid the precision loss of the geographic coordinate caused by re-sampling or interpolation.

However, to the best of the authors' knowledge, only limited work on the topographic feature line extraction based on the pure point method has been published, and the 2D approximation of the feature line increased the complexity (Christian Briese, 2004).

A number of effective algorithms of extracting feature lines from the original point cloud have been proposed by many researchers in the fields of reverse engineering and object recognition. For instance, Signed Surface Variation (SSV) was proposed by Nie et al. (2015). Except for the ability to represent the local surface variation, SSV can also distinguish concave surfaces from convex ones, so it is considered as a good approximation to the surface curvature. However, in this method, feature points were obtained by iteratively thinning the boundary points by the bilateral filtering algorithm, and it was difficult to obtain accurate results when the regions surrounded by the boundary points are complex. Altantsetseg et al. (2013) used a curvature-weighted Laplacian-like thinning method to extract feature points from detected potential feature points; however, this method would shrink the feature lines.

By taking the advantages of the above two methods into account, an optimized method for extracting topographic feature lines from terrain point cloud is proposed in this paper. This method will make full use of the information in the point cloud, and the topographic feature lines are connected by the original points, which could effectively avoid the precision loss of the geographic coordinate due to re-sampling or interpolation. Besides, the SSV is introduced for detecting the potential feature points, because it is easy to calculate. In addition, the SSV-weighted HC-Laplacian smoothing method is used for preventing the shrinkage of the feature lines and accelerating the convergence in the thinning process of potential feature points. The results indicate that this method can efficiently extract the ridge lines and valley lines from the complicated terrain point cloud data.

## 2 FRAMEWORK

Figure 1 depicts the flowchart of this method. The input is the distributed terrain point cloud with X, Y, and Z coordinates, and the outlier and the off-terrain points have been filtered. Our algorithm starts by calculating the SSV for each point, and then the potential feature points are extracted if the absolute value of SSV of this point is greater than the user-defined threshold value (Section 3.1). After that, the region growing algorithm based on the Euclidean distance and SSV is used to segment the potential feature points, in order to sort the feature lines, which are close to each other (Section 3.2). Because the boundary points and non-boundary points are processed differently in the HC-Laplacian smoothing algorithm, the boundary points in each cluster should be recognized in advance. Therefore, the classical and efficient Ball Pivoting Algorithm (BPA, F. Bernardini et al., 1999) is used to reconstruct the mesh surface of each cluster for recognizing the boundary points. Subsequently, the potential feature points in each cluster are iteratively thinned by a HC-Laplacian smoothing method, and the SSV at each point is selected as the weight function for accelerating this thinning process (Section 3.3). Then, the feature lines are constructed by connecting these extracted feature points based on the Minimum Spanning Tree (MST) algorithm (Section 3.4).

## 3 METHODOLOGY

### 3.1 *Potential feature points detection*

Except for the ability to represent the local surface variation, SSV can also distinguish concave surfaces from convex ones. Moreover, the calculation of SSV is easier and more efficient than that of the curvature based on the surface fitting. In addition, the SSV has the strong ability to suppress the noise. Therefore, the SSV is introduced in this paper, and the potential feature point is detected if the absolute value of SSV is higher than the user-defined threshold value.

In order to compute the SSV of each point $p$ of point cloud $P$, the $3 \times 3$ covariance matrix $C$ for a sample point $p$ is defined as:

$$C = \frac{1}{k} \begin{bmatrix} p_1 - \bar{p} \\ \cdots \\ p_k - \bar{p} \end{bmatrix}^T \cdot \begin{bmatrix} p_1 - \bar{p} \\ \cdots \\ p_k - \bar{p} \end{bmatrix}^T \quad (1)$$

where $\bar{p} = \sum_{i=1}^{k} p_i / k$ is the centroid of the $k$ nearest neighbors $p_i$ of $p$, and $k$ is set to 30 in this paper.

As has been demonstrated in M. Pauly et al. (2002), principal component analysis of local point neighborhoods can be used to calculate the surface variation $\sigma_n(p)$:

$$\sigma_n(p) = \frac{\lambda_0}{\lambda_0 + \lambda_1 + \lambda_2} \quad (2)$$

where $\lambda_i$ $(i = 0,1,2)$ is the eigenvalues of $C$, with $\lambda_0 < \lambda_1 < \lambda_2$.

On the basis of the improvement of the theory mentioned above, Nie et al. (2015) proposed SSV, which is given by:

$$SSV(p) = \begin{cases} \sigma_n(p), & \boldsymbol{n_p}\boldsymbol{n_p} \cdot p\bar{p} \geq 0 \\ -\sigma_n(p), & \boldsymbol{n_p}\boldsymbol{n_p} \cdot p\bar{p} < 0 \end{cases} \quad (3)$$

where $n_p$ is the normal vector of $p$.

Then, the set of potential feature points $P_f \subset P$ can be created by the following formula:

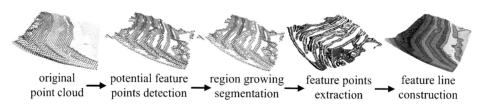

original point cloud → potential feature points detection → region growing segmentation → feature points extraction → feature line construction

Figure 1. Flowchart of the feature line extraction algorithm framework.

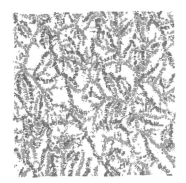

Figure 2.    Detected potential feature points.

$$P_f = \{p | p \in P, \ \|SSV(p)\| > H_{SSV}\} \quad (4)$$

where $H_{SSV}$ is the user-defined threshold value. In this method, $H_{SSV}$ is selected in the range of 10–15% of the averaged surface variation of the point cloud. The potential feature points detected using SSV in the test area are shown in Figure 2. The red and green dots represent the potential ridge points and valley points, respectively.

### 3.2    Region growing segmentation

In order to avoid confusing or incorrectly sorting the feature lines, which are close to each other, the potential feature points are segmented into different clusters by region growing algorithm, which has been proposed in T Rabbani et al. (2006). In this paper, the process of segmentation is to merge the points those are close enough in terms of the SSV constraint. The details of the steps are given in Algorithm 1.

---

**Algorithm 1** Region growing segmentation

---

**Inputs:** Potential feature points $P_f$, Normal vectors $N$, signed surface variations $SSV$, and angle threshold $\theta_{th}$

---

**Initialize:** Region List $R$, Current Seeds $S$, Current region $R_c$

while $P_f$ is not empty

   find point $P_{min}$ with minimum absolute value of $SSV$ in $P_f$

   insert point $P_{min}$ to $S$ and $R_c$

   remove point $P_{min}$ from $P_f$

   for each point $s_n \in S$

     find $k$ nearest neighbors $B_s$ of $s_n$

     **for** each point $p_i \in B_s$

       **if** $P_f$ contains $p_i$ and $\cos^{-1}(|N(p_i), N(p_j)|) < \theta_{th}$

       && $SSV(p_i) \cdot SSV(p_j) > 0$

       insert point $p_i$ to $R_c$ and $S$

       remove point $p_i$ from $P_f$

     end if

    end for

  end for

  Insert current region $R_c$ to global segment list $R$

end while

---

Since the SSV constraint is added, the angle threshold $\theta_{th}$ can be selected in a larger range. As shown in Fig. 3, the region growing segmentation can effectively sort the ridge feature points and the valley feature points.

### 3.3    Feature points extraction

In order to extract the feature points, the pseudo-feature points, which are obtained through the region growing segmentation, should be eliminated in each cluster. Many algorithms have been introduced for thinning point cloud (B. Ramamurthy et al., 2015; J. Cao, 2010; V. Sam, 2013), e.g. the Laplacian smoothing method is a classical and simple one (L. Freitag, 1999). However, this method will cause the deformation and shrinkage of the feature lines, which are undesirable. Vollmer et. al. (1999) showed that this problem can be resolved effectively by the HC-Laplacian smoothing method, which is given by:

$$p_i' = p_i - \left( \beta b_i + \frac{1-\beta}{|Adj(i)|} \sum_{j \in Adj(i)} b_j \right) \quad (5)$$

$$p_i = \frac{1}{|Adj(i)|} \sum_{j \in Adj(i)} q_j \quad (6)$$

$$b_i = p_i - q_i \quad (7)$$

where $q_i$ is the current point before the application of the Laplacian smoothing algorithm, $p_i$ is the new position of vertex $i$, which is produced by the Laplacian smoothing algorithm, $p_i'$ is the new position of vertex $i$ after the application of the HC-Laplacian smoothing algorithm, $q_j$ is the adjacent vertices of vertex $i$, $Adj(i)$ is the set of

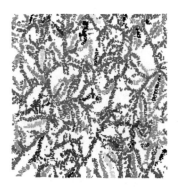

Figure 3.    The result of region growing segmentation.

adjacent vertices of one vertex $i$, $|Adj(i)|$ is the number of the adjacent vertices, $b_i$ is the difference between the new position $p_i$ and the old position $q_i$ (In (L. Freitag, 1999), $b_i = p_i - (\alpha o_i + (1 - \alpha) q_i)$, and to simplify the calculation in our letter, set $\alpha = 0$).

In order to accelerate the thinning process, the SSV weight function is introduced in this work. The specific method is as follows:

$$p_i' = p_i - \left( \beta b_i + \frac{1-\beta}{\sum\limits_{j \in Adj(i)} \omega(q_j)} \sum\limits_{j \in Adj(i)} \omega(q_j) b_j \right) \quad (8)$$

$$p_i = \frac{1}{\sum\limits_{j \in Adj(i)} \omega(q_j)} \sum\limits_{j \in Adj(i)} \omega(q_j) q_j \quad (9)$$

$$b_i = p_i - q_i \quad (10)$$

where $\omega(q_j)$ is the weight function, which is selected as $\omega(q_j) = SSV(q_j)^2$.

As the input point cloud is irregularly distributed, the point density could be very low in some region. Thus, the $Adj(i)$ should be selected as the $k$ nearest neighbors instead of the neighbors in the fixed distance. In addition, the $Adj(i)$ should be treated differently between the non-boundary points and boundary points. When vertex $i$ is on the boundary of the detected potential feature point set, the point cannot be completely surrounded by its neighbors, as shown in Fig. 4. Therefore, the boundary points of each cluster should be extracted before the process of the HC-Laplacian smoothing. In our method, $k$ is fixed at 6 and 4 for the non-boundary points and boundary points, respectively.

In order to extract the boundary points, the BPA is introduced. The BPA is a classical algorithm of mesh surface reconstruction, which can effectively recognize the boundary points from each segmented cluster, without changing the position of the original points. The result of the surface reconstruction is illustrated in Figure 5.

Figure 5. Surface reconstruction of each segmented region.

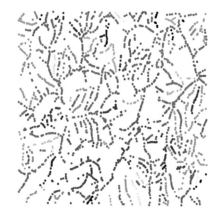

Figure 6. Extracted feature points.

The thinning process for the boundary points and non-boundary points are handled differently. After each iteration of the thinning process, the absolute value of SSV of every point in the fixed distance is compared (the fixed distance is chosen as the mean point distance of the original point cloud), and the point with the smallest absolute value of SSV is eliminated. Then, the point in the original segmented cluster with the same index number is removed. After that, the potential feature point cloud without these points is used as the input in the next iteration. The process continues until no point meets the "removing condition". The feature point extraction process is summarized in Algorithm 2. Fig. 6 shows an example of the feature points extracted by the iterative thinning process.

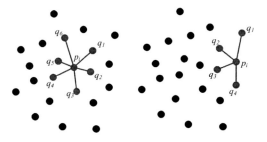

Figure 4. $k$ nearest neighbors of the non-boundary point and boundary point.

**Algorithm 2** Feature points extraction

---

**Inputs:** Segmented clusters $C$ i.e. $P_f$, Average point distance of the original point cloud $\bar{d}$, signed surface variations SSV

---

**Initialize:** The number of the removed points $n_r$, a temporary point set $P_t$

do
  $n_r = 0$
  **for** each cluster $C_i \in C$
    $P_t = C_i$
    **for** each point $p_j \in P_t$
      apply SSV weighted HC-Laplacian smoothing to $P_t$ in Eq.(8)
    **end for**
    **for** each point $p_j \in P_t$
      find all nearest neighbors $B_{\bar{d}}$ in a sphere of a radius $\bar{d}$, record the indices of neighbors in $C_i$
      **for** each $q_k \in B_d$
        **if** $\left| SSV_{q_k} \right|$ is minimum
          remove point with the index $i_{q_k}$ from $C_i$
          $n_r$++
        **end if**
      **end for**
    **end for**
  **end for**
while $n_r \neq 0$

---

### 3.4 Feature line construction

After extracting the feature points from each cluster, the feature points can be divided into ridge or valley points, depending on the sign of the SSV. Then, the ridge feature lines and the valley feature lines are respectively constructed by MST algorithm. As the feature points are relatively sparse, Kruskal's MST algorithm is chosen for increasing the operating speed. First, the connected graph is constructed by the feature points, and the edge weight is given as:

$$C(p, p_i) = \frac{|SSV(p) - SSV(p_i)|}{|SSV(p) + SSV(p_i)|} + \gamma \cdot \boldsymbol{p} - \boldsymbol{p}_i \qquad (11)$$

where $p_i$ is a neighbor point of $p$ with a fixed distance (the fixed distance is set as $10\bar{d}$), and $\gamma = 0.1$ is an additional parameter that can balance SSV weights against Euclidean distance. The first term on the right-hand side of the formula indicates the difference between the SSV of the two feature points. The higher difference implies that these two points are less likely connected. The MST is the tree connecting all feature points in each cluster so that the sum of the edge weight is minimum. Figure 7 shows the MST of a test area generated with this algorithm.

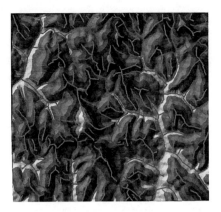

Figure 7.   Final result of the feature line extraction.

## 4   RESULTS AND DISCUSSION

The proposed method has been implemented using the C++ programming language. In order to demonstrate the capability of the derived algorithm, two point cloud data sets were used in this section, which have filtered the non-terrain points and the outliers. The first data set is provided by ISPRS Commission III, Working Group III (http://www.itc.nl/isprswgIII-3/filtertest/index.html), which contains complex terrains such as terrace, bluff, depression, and flat. In addition, the density of the points changes significantly throughout the data (c.f. detail of Figure 8a). The other data set is produced by random sampling from a tidal flat DEM, which has small elevation changes (see Fig. 8b). The detailed parameters of the two point cloud data sets are shown in Table 1.

The results of the topographic feature line extraction are presented in Figures 9 and 10. The lines are respectively overlaid on hillshaded DEM and TIN surface composed of the two point clouds. It is important to note that only the ridge lines are shown in Figure 9, for visualizing more clearly, and also in Figure 10, only the valley lines are plotted. Visual inspection shows that our method successfully extracted most of the topographic feature lines. However, many short branches are formed, and some feature lines are misconnected. The reasons for these omissions may include the following: (1) in the bluff region, the point density is too low to accurately calculate the SSV (c.f. detail of Figure 9) and (2) the feature lines constructed by the MST are not pruned and optimized (c.f. detail of Figure 10a).

In order to evaluate our method, the classic D8 flow direction algorithm (K. Jenson et al., 1988) is used for comparison. The D8 method is based on hydrology using the raster model, and it is

Table 1. Parameters of the point cloud data sets.

| Cloudpoints | Points | Point density ($pts/m^2$) | Size (m × m) | Elevation (m) |
|---|---|---|---|---|
| 1 | 32989 | 0.18 | 430.41 × 473 | 0–79.05 |
| 2 | 202160 | 0.06 | 3837.5 × 2956.25 | 0–313.32 |

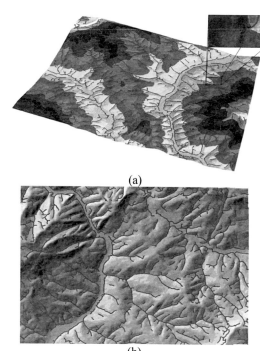

(a)

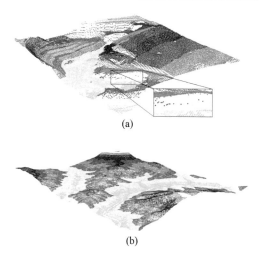

(a)

(b)

Figure 8. Original point clouds of the test data sets.

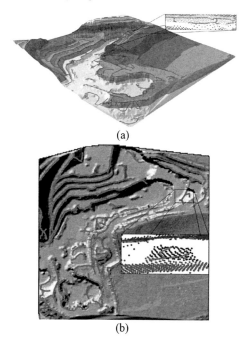

(a)

(b)

Figure 9. The extracted ridge feature line of the first point cloud.

(b)

Figure 10. The extracted valley feature line of the second point cloud.

available in the ArcHydro software package. By comparing the results obtained by our method (red ridge lines in Figure 11a and blue valley lines in Figure 11b) and D8 method (yellow lines), it is clear that our method can extract two boundary lines in the very plain tidal flats, whereas the feature lines detected by D8 method are not concord to the topographical feature in these areas (Fig. 11b). Moreover, visual interpretations show that our method performed better than the D8 method in the area of complex terrain (Fig. 11a). Therefore, we conclude that our method can not only effectively extract feature lines in the terrace area, but satisfactory results can also be achieved in some micro-landforms. However, the D8 method does not perform appreciably, and sometimes the detected feature lines mismatch the topographical feature.

The above experiments indicate the effectiveness and certain universality of the proposed method for extracting topographic feature lines in various terrains. More experiments are required and will be performed in our future work to assess the feasibility and robustness of our method on point clouds with different point density over different types of terrains.

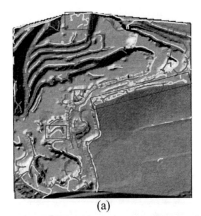

(a)

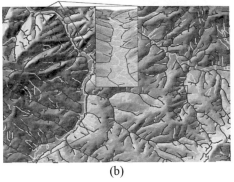

(b)

Figure 11. Comparison of the results obtained by our method (red ridge lines in Figure 11a, and blue valley lines in Figure 11b) and D8 method (yellow lines).

## 5 CONCLUSIONS AND OUTLOOK

In this paper, we present a pure point-based novel method, which could avoid the precision loss of the geographic coordinate during the re-sampling or the interpolation process, for extracting topographic feature lines from unorganized point cloud data sets. The SSV instead of the surface curvature is introduced to distinguish the concave–convex quality of the surface. In addition, the HC-Laplacian operator is used for preventing the shrinkage of the feature lines in the smoothing process of potential feature points. In order to accelerate the smoothing process, an SSV-based weight function is applied.

The experimental results in this study show that the topographic feature lines extracted by the proposed method are superior to those extracted by the DEM-based D8 method, particularly in the plain tidal flat area and the terrace area. However, the feature lines are misconnected in some areas, which may be because the point density is too low or the MST is not pruned and optimized.

In future work, we will improve the pruning and optimization processes and enhance the adaptive ability of the algorithm for various point densities. Moreover, in addition to the visual inspection, statistical analysis is needed to validate the effectiveness of the method.

## ACKNOWLEDGMENT

This paper was supported by the National Natural Science Foundation of China (41171349, 41471380) and the National High Technology Research and Development Program of China (2012 AA12 A406).

First Author: Zhou Wei, PhD, majors in the theories and methods of charting, point cloud data processing. E-mail: fangzhouwei@163.com.

Corresponding Author: PENG Rencan, PhD, Professor. E-mail: pengrencan63@163.com.

## REFERENCES

Altantsetseg, E., Y. Muraki, K. Matsuyama, et al. Feature Line Extraction from Unorganized Noisy Point Clouds using Truncated Fourier Series [J]. Visual Computer, 2013, 29(6–8): 617–626.

Bernardini, F., J. Mittleman, H. Rushneier, et al. The Ball-pivoting Algorithm for Surface Reconstruction [J]. IEEE Transactions on Visualization and Computer Graphics, 1999, 5(4): 349–359.

Bisheng Yang, Zhen Dong, Fuxun Liang, et al. Automatic Registration of Large-scale Urban Scene Point Clouds Based on Semantic Feature Points [J]. ISPRS Journal of Photogrammetry and Remote Sensing, 2016, 113: 43–58.

Cao, J., A. Tagliasacchi, M. Olson, et al. Point Cloud Skeletons via Laplacian-Based Contraction[C]. Smi. Shape Modeling International Conference, Aix En Provence, France, June, 2010: 187–197.

Christian Briese. Three-dimensional Modelling of Breaklines from Airborne Laser Scanner Data[C]. Interational Archives of Photogrammetry, Remote Sensing and Spatial Information Sciences, 2004, 35: 1097–1102.

Fatih Gulgen, Turkay Gokgoz. A New Algorithm for Extraction of Continuous Channel Networks without Problematic Parallels from Hydrologically Corrected DEMs [J]. Boletim de Ciencias Geodesicas, 2010, 16(16): 20–38.

Freitag. L., On combining Laplacian and optimization based mesh smoothing techniques[C]. Symposium on Joint Asme, Asce, 1999, 220: 37–44.

Jenson, K., O. Domingue. Extracting topographic structure from digital elevation data for geographic information system analysis [J]. Photogrammetric Engineering and Remote Sensing, 1988, 54(11): 1593–1600.

Jingzhong Xu, Yuan Kou, Jun Wang. High-precision DEM Reconstruction Based on Airborne LiDAR Point Clouds [C]. Proc. of SPIE, 2014, 9158(3): 1187–1190.

Lin Yang, Yehua Sheng, Bo Wang. LiDAR Data Reduction Assisted by Optical Image for 3D Building Reconstruction [J]. Optik, 2014, 125(20): 6282–6286.

Mandlburger, G., M. Vetter, N. Pfeifer, et al. Derivation of a Countrywide River Network Nased on Airborne Laser Scanning DEMs-results of a Pilot Study[C]. International Congress on Modelling and Simulation, 2011, 8(1): 2423–2429.

Nie Jianhui, Liu Ye, Gao Hao, et al. Feature Line Detection from Point Cloud Based on Signed Surface Variation and Region Segmentation [J]. European Journal of Mineralogy, 2015, 5(4): 755–762.

Pauly, M., M. Gross, LP. Kobbelt. Efficient simplification of point-sampled surfaces [J]. Visualization, Vis IEEE, 2002, 1(4): 163–170.

Persendt, F.C., C. Gomez. Assessment of Drainage Network Extractions in a Low-relief Area of the Cuvelai Basin (Namibia) from Multiple Sources: LiDAR, Topographic Maps, and Digital Aerial Orthophotographs [J]. Geomorphology, 2015, 494(s 1–2): 1–8.

Rabbani, T, FAVD. Heuvel, G. Vosselman. Segmentation of Point Clouds using Smoothness Constraint[C]. Interational Archives of Photogrammetry, Remote Sensing and Spatial Information Sciences, 2006, 36: 248–253.

Ramamurthy, B., J. Doonan, J. Zhou. Skeletonization of 3D Plant Point Cloud using a Voxel Based Thinning Algorithm[C]. 23rd European Signal Processing Conference, Nice, France, 2015: 2686–2690.

Sam, V., H. Kawata, T. Kanai. A Robust and Centered Curve Skeleton Extraction from 3D Point Cloud [J]. Computer-Aided Design and Applications, 2013, 9(6): 869–879.

Vollmer, J., R. Mencl, H. Muller. Improved Laplacian Smoothing of Noisy Surface Meshes [J]. Computer Graphics Forum, 199, 18(3): 131–138.

*Advances in Energy and Environment Research – Achour & Wu (Eds)*

# Multi-scale representation of a digital depth model based on the positive direction rolling ball transform algorithm

Jian Dong, Rencan Peng & Lihua Zhang
*Department of Hydrography and Cartography, Dalian Naval Academy, Dalian, China*
*PLA Key Laboratory, Dalian, China*

Jiansheng Yuan
*The Navigation Guarantee Department of the Chinese Navy Staff, Beijing, China*

ABSTRACT: The multi-scale representation of spatial data is one of the most important and difficult problems in the field of Geographic Information System (GIS). Digital Depth Model (DDM) is the digitized model reflecting the depth change of ocean. As an important representation model of seafloor reliefs, DDM is not only the main source of information guarantying safety navigation, but also the information platform for marine geoscience research, maritime engineering, subaqueous archaeology, and so on. With the development of digital ocean industry and ocean environment projects, the application fields of DDM are expanding continuously, which leads to tremendous requirements of DDM multi-scale representation. In fact, in the same area, DDM of different scales are the different digitized representation forms of identical marine topography. Therefore, it is an effective approach to DDM multi-scale application by studying a multi-scale representation algorithm based on original DDM. As an important visualization expression format of DDM, bathymetric contour shares the same restriction rules as DDM in multi-scale representation. Existing multi-scale representation algorithms of DDM are mostly bi-dimensional extension of generalization algorithms for two-dimensional bathymetric contour graphics. These algorithms mainly focus on the geometry characteristics of DDM and simplify DDM by preserving the feature points and filtering ordinary grid points. While DDM generalization is not a simple process of accepting or rejecting grid points of DDM, some factors including geopraphic and scale characteristics of DDM should be considered to maintain the consistency of spatial cognition and abstract grade. Mainly focused on the geographic and scale characteristics of DDM, based on the analysis of essential principles of planar rolling circle transform, this paper has brought forward a multi-scale representation of DDM based on the positive direction rolling ball transform algorithm, by using different sizes of three-dimensional balls instead of planar circles rolling over the upper surface of DDM, which will preserve the positive, flat reliefs of DDM and reduce or fill up the negative reliefs of DDM contrarily, and realizes multi-scale representation of DDM from the viewpoint of guarantying safety navigation. Besides, the paper also expatiates the keystone and solution steps of the algorithm. The experiments show that the algorithm can preserve the basic terrain characteristics of the DDM according to the safety navigation principles, meanwhile, with high computing efficiency.

## 1 INTRODUCTION

Digital Depth Model (DDM for short) is the digitized model reflecting the depth change of ocean (Thomas Porathe, 2006). It is also the common three-dimensional digitized model expressing the basic relief characteristics of seafloor. According to different data structures, DDM can be divided into grid DDM and irregular DDM. DDM herein means grid DDM. As an important model for the representation of seafloor reliefs, DDM is not only the main source of information guarantying safe navigation, but also the information platform for marine geoscience research, maritime engineering, subaqueous archaeology, and so on (Thomas Porathe, 2006; HU Jinxing et al., 2004). With the development of digital ocean industry and ocean environment projects, DDM is applied in an increasing number of fields. The growing application of DDM results in the demand for multi-scale representation of DDM. For a long time, given the technical constraints, DDM of different scales was built by repetitive digitizing. The building process was expensive and inefficient (Ai Tinghua et al., 2005; Jia Juntao et al., 2008). To solve this problem, some scholars tried to obtain different-resolution DDM from the same data source.

As an important visualization expression format of DDM, bathymetric contour shares the same restriction principles as DDM in multi-scale representation (Thomas Porathe, 2006; Christensen A.H.J, 2003; Christensen A.H.J, 1999). Therefore, DDM multi-scale representation algorithms currently are always constructed by dimensional extension of planar generalization algorithms for bathymetric contour. Existing algorithms mainly focus on the geometry characteristics of DDM, and simplify DDM by preserving the feature points and filtering ordinary grid points (Wan Gang et al., 1999; Yang Zuqiao et al., 2005; Li Jingzhong et al., 2009; Fei Lifan et al., 2006); however, DDM generalization is not a simple process of accepting or rejecting DDM grid points, and some factors including geographic and scale characteristics of DDM should be considered to maintain the consistency of spatial cognition and abstract grade (Hu Peng et al., 2009; Li Zhilin, 2005). Christensen (1999, 2003) put forward the concept of waterlining and medial-axis transformation (rolling circle transform) based on planar buffer function and applied it in cartographic line generalization. Smith (2003) analyzed the possibility of applying rolling circle transform to the process of bathymetric contour generalization, and showed that the new generalization algorithm fulfills the DDM generalization principles of expanding shallow water area and reducing deepwater area. By preserving convex feature points of navigation significance, and displacing concave feature points according to threshold value (circle radius) reasonably, the algorithm realizes the multi-scale representation and automatic generalization of bathymetric contour according to the safety navigation principles (Smith Shepard M, 2003; Gao Wangjun, 2009).

By means of rolling circle transform dimensional extension, this paper has brought forward a multi-scale representation of DDM based on positive direction rolling ball transform algorithm, by using different sizes of three-dimensional balls instead of planar circles rolling over the upper surface of DDM, which preserves the positive, flat reliefs of DDM and reduces or fills up the negative reliefs of DDM contrarily, and realizes the multi-scale representation of DDM from the viewpoint of safety navigation.

## 2   ANALYSIS OF ROLLING CIRCLE TRANSFORM

Rolling circle transform can be described as the process of a planar circle rolling over one side of cartographic line and forming an inner tracing line,

and can be divided into left and right side rolling circle transform (Smith Shepard M, 2003; Hu Peng et al., 2002; Hu Peng et al., 2005). Mathematically, rolling circle transform is equal to the combination of left and right side buffer boundary transform. For a cartographic line $T$ with point collection $\{Q_1 \cdots Q_{N+1}\}$ the left (right) side buffer transform $K_L(r)(K_R(r))$ can be defined as follows (Guo Renzhong et al., 2000; Peng Rencan et al., 2002):

$$\begin{cases} T \cdot K_L(r) = \left\{ P_l \,\middle|\, \left\{ d_e(P_l, T_i) \middle\| \overline{T_i} \times \overline{P_l Q_i} \middle| \le 0 \middle| T_i \in T \right\} = r \right\} \\ T \cdot K_R(r) = \left\{ P_r \,\middle|\, \left\{ d_e(P_r, T_i) \middle\| \overline{T_i} \times \overline{P_r Q_i} \middle| \ge 0 \middle| T_i \in T \right\} = r \right\} \end{cases} \quad (1)$$

where $r$ represents buffer radius; $T \cdot K_L(r)$ represents the left side buffer boundary of cartographic line $T$; $T \cdot K_R(r)$ represents the right side buffer boundary of cartographic line $T$; $P_l$, $P_r$ represents arbitrary point of buffer boundary $T \cdot K_L(r)$ and $T \cdot K_R(r)$, respectively; $d_e$ represents planar Euclidean distance calculation; $\times$ represents vector product calculation; and $\|$ represents module calculation. Left side buffer boundary transform $K_L(r)$ preserves the convex, plain parts and reduces or flattens the concave parts of the left side of cartographic line $T$ and right side buffer boundary transform $K_R(r)$ preserves the concave, plain parts and reduces or flattens the convex parts of right side of cartographic line $T$. Therefore, rolling circle transform can be defined as follows (Smith Shepard M, 2003; Hu Peng et al., 2002):

$$\begin{cases} T \cdot V_L(r) = T \cdot K_L(r) \cdot K_R(r) \\ T \cdot V_R(r) = T \cdot K_R(r) \cdot K_L(r) \end{cases} \quad (2)$$

where $V_L(r)$ represents the left side rolling circle transform; $V_R(r)$ represents the right side rolling circle transform; $T \cdot V_L(r)$ represents the cartographic line formed by left side buffer boundary transform $K_L(r)$ and right side buffer boundary transform $K_R(r)$, namely, an inner tracing line shaped by rolling circle rolling over the left side of cartographic line $T$; $T \cdot V_R(r)$ represents the cartographic line formed by right side buffer boundary transform $K_R(r)$ and left side buffer boundary transform $K_L(r)$, namely, an inner tracing line shaped by rolling circle rolling over the right side of cartographic line $T$.

As shown in Figure 1, for a bathymetric contour $T$ (where the left side is deepwater area and the right side is shallow water area), inner tracing line $T \cdot V_L(r)$ formed by left side rolling circle transform $V_L(r)$ preserves convex, plain parts and reduces or flattens the concave parts of bathymetric contour $T$; inner tracing line $T \cdot V_R(r)$ formed by right side rolling circle transform $V_R(r)$ preserves the concave,

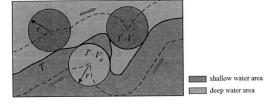

| shallow water area |
| deep water area |

Figure 1. Sketch map of rolling circle transform.

plain parts and reduces or flattens the convex parts of bathymetric contour $T$. However, considering the safety navigation principles of multi-scale representation of bathymetric contour, rolling circle transform is always applied to the side of deepwater area of bathymetric contour (corresponding to the left side rolling circle transform $V_L(r)$ as shown in Figure 1). Therefore, convex feature points of navigation significance are preserved, and concave feature points are displaced according to threshold value (buffer radius $r$) reasonably, which realizes the multi-scale representation and automatic generalization of bathymetric contour.

## 3 MULTI-SCALE REPRESENTATION OF DDM BASED ON POSITIVE DIRECTION ROLLING BALL TRANSFORM

The characteristics of planar rolling circle transform are of great significance in multi-scale representation of cartographic line (such as costal line, bathymetric contour, and island boundary) when considering the safety navigation principles (Christensen A.H.J, 2003; Christensen A.H.J, 1999; Smith Shepard M, 2003). This inspires us whether we could design a new multi-scale representation algorithm by using a three-dimensional ball rolling over specific side of DDM to realize the automatic generalization of DDM. If that is feasible, positive reliefs affecting navigation safety can be preserved, and negative reliefs of a certain scale can be reduced or filled up; therefore, DDM multi-scale representation can be simplified as the threshold value (buffer radius) regulation of three-dimensional rolling ball.

### 3.1 Three-dimensional buffer analysis

Three-dimensional buffer can be defined as follows (Zhou Peide, 2000; Zhu Changqing et al., 2006):

$$B(T,r) = \{P(x,y,z) \mid \{d_e(P,Q_T) \mid \\ Q_T(x_T, y_T, z_T) \in T\} \le r\} \tag{3}$$

where $B$ represents the buffer of spatial entity $T$; $P(x, y, z)$ represents an arbitrary point of buffer

$B$; $Q_T(x_T, y_T, z_T)$ represents a sampling point of spatial entity $T$; $d_e(P, Q_T)$ represents the three-dimensional Euclidean distance from point $P$ to sampling point $Q_T$. Buffer $B$ with radius $r$ can be described as the collection of point $P$ whose distance to sampling point $Q_T$ is less than or equal to buffer radius $r$.

### 3.2 Positive and negative direction buffer boundary transform

The key procedure of planar rolling circle transform is to generate the left (right) buffer boundary of cartographic line. As a special morphology type of three-dimensional single-value surface, one direction coordinate value of DDM can be calculated by single-value function of the other two direction coordinate values (Christensen A.H.J, 2003; Yang Zuqiao et al., 2005). That is, coordinate $(x_T, y_T, z_T)$ of grid point (sampling point) $Q_T$ fulfills the single-value function $z_T = f(x_T, y_T)$. According to formula 3, coordinate $(x, y, z)$ of an arbitrary point $P$ of DDM buffer boundary fulfills the single value function as follows:

$$\begin{cases} z' = f'_{Q_T}(x,y) = f(x_T, y_T) \\ \quad + [r^2 - (x - x_T)^2 - (y - y_T)^2]^{1/2} \\ z'' = f''_{Q_T}(x,y) = f(x_T, y_T) \\ \quad - [r^2 - (x - x_T)^2 - (y - y_T)^2]^{1/2} \end{cases} \tag{4}$$

where supposing that $z' \ge f(x_T, y_T)$, it means coordinate $(x, y, z')$ is above the DDM surface, and can be defined as upper buffer boundary point $P_u$. Otherwise, supposing $z'' \le f(x_T, y_T)$ indicates coordinate $(x, y, z'')$ is under the DDM surface, and can be defined as nether buffer boundary point $P_d$. Therefore, positive (negative) buffer boundary transform of DDM can be defined as follows:

$$\begin{cases} T \cdot K_U(r) = B'(T,r) = \{P_u(x,y,z') \mid z' \\ \quad = f'_{Q_T}(x,y) \mid Q_T(x_T, y_T, z_T) \in T\} \\ T \cdot K_D(r) = B''(T,r) = \{P_d(x,y,z'') \mid z'' \\ \quad = f''_{Q_T}(x,y) \mid Q_T(x_T, y_T, z_T) \in T\} \end{cases} \tag{5}$$

where $B'(T,r)$ represents the upper boundary of DDM buffer; $B''(T,r)$ represents the nether boundary of DDM buffer; positive direction buffer boundary transform $K_U(r)$ is characterized of preserving the convex, plain parts and reducing or flattening the concave parts of upper side of DDM; on the contrary, negative direction buffer boundary transform $K_D(r)$ is characterized of preserving the concave, plain parts and reducing or flattening the convex parts of the nether side of DDM.

As shown in Table 1, both positive relief $\alpha$ and negative relief $\beta$ consisted of DDM grid cell

*GFPH*, *FEDP*, *PDCB*, and *PBAH*. In the table, $P$ represents the terrain feature point and the other grid points share the same depth. Positive direction buffer boundary transform $K_U(r)$ makes the terrain feature point $P_u$ approach to the other grid points in $\beta \cdot K_U(r)$ (reducing or flattening the concave parts), while the distance from feature point $P_u$ to the other grid points in $\alpha \cdot K_U(r)$ remains unchanged (preserving the convex, plain parts); negative direction buffer boundary transform $K_D(r)$ makes the terrain feature point $P_d$ approach to the other grid points in $\alpha \cdot K_D(r)$ (reducing or flattening the convex parts), while the distance from feature point $P_d$ to the other grid points in $\beta \cdot K_D(r)$ remains unchanged (preserving the concave, plain parts).

### 3.3 Rolling ball transform

Rolling ball transform can be described as the process of a three-dimensional ball rolling over one side of DDM and forming an inner tracing surface, and can be divided into positive direction and negative direction rolling ball transform. Mathematically, rolling ball transform equals the combination of positive direction and negative direction buffer boundary transforms, namely:

$$\begin{cases} T \cdot V_U(r) = T \cdot K_U(r) \cdot K_D(r) \\ T \cdot V_D(r) = T \cdot K_D(r) \cdot K_U(r) \end{cases} \tag{6}$$

where $V_U(r)$ represents the positive direction rolling ball transform; $V_D(r)$ represents the negative direction rolling ball transform; $T \cdot V_U(r)$ represents the surface formed by positive direction buffer boundary transform $K_U(r)$ and negative direction buffer boundary transform $K_D(r)$, namely an inner tracing surface shaped by three-dimensional ball rolling over the upper side of DDM; and $T \cdot V_D(r)$ represents the surface formed by negative direction buffer boundary transform $K_D(r)$ and positive direction buffer boundary transform $K_U(r)$, namely an inner tracing surface shaped by three-dimensional ball rolling over the nether side of DDM.

As shown in Table 2, positive direction ball transform $V_U(r)$ makes the terrain feature point $P_{ud}$ approach to the other grid points in $\beta \cdot V_U(r)$ (reducing or filling up the negative reliefs), while the distance from feature point $P_{ud}$ to the other grid points in $\alpha \cdot V_U(r)$ remains unchanged (preserving the positive and flat reliefs); negative direction rolling ball transform $V_D(r)$ makes the

Table 1. Characteristic analysis of three-dimensional buffer boundary transform.

| Transform type | Positive relief $\alpha$ | Negative relief $\beta$ |
|---|---|---|
| Positive direction buffer boundary transform | | |
| Negative direction buffer boundary transform | | |

Table 2. Characteristic analysis of rolling ball transform.

| Transform type | Positive relief $\alpha$ | Negative relief $\beta$ |
|---|---|---|
| Positive direction rolling ball transform | | |
| Negative direction rolling ball transform | | |

terrain feature point $P_{du}$ approach to the other grid points in $\alpha \cdot V_D(r)$ (reducing or slicing off the positive reliefs), while the distance from feature point $P_{du}$ to the other grid points in $\beta \cdot V_D(r)$ remains unchanged (preserving the negative and flat reliefs).

### 3.4 Scale dependence analysis

On the basis of the characteristic analysis of rolling ball transform in section 3.3, we can infer that the positive and negative direction rolling ball transforms share the same construction principles, and the threshold value (buffer radius $r$) directly influences the change extent of DDM spatial geometry morphology. The characteristics of preserving positive, flat reliefs and reducing or filling up negative reliefs owned by positive direction rolling ball transform $V_U(r)$ correspond to the multi-scale representation principles of DDM (safety navigation principles). The extent of reducing or filling up negative reliefs and the threshold value (buffer radius $r$) satisfy the following formula:

$$r - r' = r - \sqrt{r^2 - \lambda^2 \xi^2} \leq \Phi \qquad (7)$$

where $r'$ represents the distance from the center point of rolling ball to the negative reliefs after filled up by positive direction rolling ball transform; $\xi$ represents the original grid cell size (resolution) of DDM; $\Phi(\Phi \geq 0)$ represents the filling up extent of negative reliefs, which directly affects the precision of DDM multi-scale representation. In practical application, $\Phi$ can be determined by the limit error (confidence level, 95%) of bathymetric survey specified in hydrographic specification (Liu Yanchun et al., 2006). That is, $\Phi = 2\sigma$ ($\sigma$ represents the mean square error of bathymetric survey). Therefore, the threshold value (buffer radius $r$) can be calculated as $r \geq \left( \sigma + \frac{\lambda^2 \xi^2}{4\sigma} \right)$. In other words, by way of positive direction rolling ball transform $V_U(r)$ with buffer radius $r \geq \left( \sigma + \frac{\lambda^2 \xi^2}{4\sigma} \right)$, negative reliefs with length (width), which less than or equal to $2\lambda\xi$, will be filled up thoroughly.

In Figure 2, $P_u$ represents the center point of rolling ball; $P_{ud}$ represents the terrain feature point of negative relief $\beta$ shaped by positive direction rolling ball transform $V_U(r)$; line $P_u P_{ud}$ intersects the plane $ACEG$ at point $P'_{ud}$; and the length of line $P_{ud} P'_{ud}$ reflects the filling up extent of negative relief $\beta$. Supposing that $P_{ud}$ coincides with $P'_{ud}$ (namely, the length of line $P_{ud} P'_{ud}$ is less than the limit error of bathymetric survey $2\sigma$) indicates negative relief $\beta$ is completely filled up. It should be noted that the change of threshold value (buffer radius $r$) will affect the position change of terrain feature point

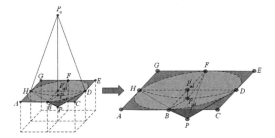

Figure 2. Scale-dependent analysis of rolling ball transform.

$P$, and as the threshold value increases, the length of line $PP_{ud}$ will continue to increase until an invariant value is obtained (where $r \geq \left( \sigma + \frac{\lambda^2 \xi^2}{4\sigma} \right)$). However, the length of line $PP_{ud}$ should not exceed the grid cell size of new scale ($2\lambda\xi$) when considering preserving the special morphology of seafloor concave reliefs (such as bluff, volcano, and trough valley) and the influence of scale factor $\lambda$.

## 4 ALGORITHM DESIGN AND REALIZATION

The key procedure of three-dimensional rolling ball transform is to generate the upper (nether) buffer boundary of DDM as DDM can be described as the finitude point collection of grid points in DDM, namely $T = \bigcup_{T=1}^{N} Q_T$, where $N$ represents the count of grid point $Q_T$ in DDM. Therefore, DDM upper (nether) buffer boundary $B'(T,r)(B''(T,r))$ can be represented by the finitude union of the upper (nether) buffer boundary $B'(Q_T,r)$ ($B''(Q_T,r)$) of grid point $Q_T$, namely $B'(T,r) = \bigcup_{T=1}^{N} B'(Q_T,r)$ $(B''(T,r) = \bigcup_{T=1}^{N} B''(Q_T,r))$. On the basis of this hypothesis, this paper establishes the DDM upper (nether) buffer boundary generation algorithm as follows.

### 4.1 Algorithm initialization

Essentially, DDM is the ordered numerical value array describing the spatial distribution morphology of seafloor. Therefore, DDM can be represented by ordered collection of two-dimensional matrix (Li Jingzhong et al., 2009; Guo Renzhong, 2000; Zhu Changqing et al., 2006), namely

$$T = \{ z_{ij} \mid i \in [0, n-1], j \in [0, n-1] \} \qquad (8)$$

where $i$, $j$ represents the row and column index of DDM separately; $z_{ij}$ represents the depth of grid point $(i, j)$; and $n$ represents the total number of rows and columns of DDM. On the basis of the

247

construction theory of DDM upper (nether) buffer boundary in section 3.2, $B'(T,r)$ and $B''(T,r)$ are both single value surface. Recur to formula 8, DDM upper and nether buffer boundary can be signed as $B'(T,r) = \{z'_{ij} = -v \mid i \in [0,n-1], j \in [0,n-1]\}$ and $B''(T,r) = \{z''_{ij} = v \mid i \in [0,n-1], j \in [0,n-1]\}$ separately, where $v$ represents an infinite positive integer.

For DDM upper (nether) buffer boundary $B'(T,r)$ $(B''(T,r))$ can be represented by the finitude union of the upper (nether) buffer boundary $B'(z_{ij},r)$ $(B''(z_{ij},r))$ of grid point $(i, j)$, therefore, $B'(z_{ij},r)$ and $B''(z_{ij},r)$ can be defined as the grid point collection of $B'(T,r)$ and $B''(T,r)$ in the spatial neighborhood $\Omega(i,j)$ of grid point $(i,j)$ separately, namely, $B'(z_{ij},r) = \{z'_{xy} = -v \mid (x,y) \in \Omega(i,j)\}$ and $B''(z_{ij},r) = \{z''_{xy} = v \mid (x,y) \in \Omega(i,j)\}$. Spatial neighborhood $\Omega(i,j)$ is determined by the row and column index of grid point $(i,j)$, namely,

$$\Omega(i,j) = \{(x,y) \mid x \in [\max(i-r,0),\min(i+r,n-1)], \\ y \in [\max(j-r,0),\min(j+r,n-1)]\} \tag{9}$$

According to formula 9, the upper (nether) buffer boundary $B'(z_{ij},r)$ $(B''(z_{ij},r))$ of grid point $(i,j)$ in spatial neighborhood $\Omega(i,j)$ can be calculated as follows:

$$\begin{cases} B'(z_{ij},r) = \{z'_{xy} = \max(z'_{xy}, z_{ij} \\ \quad + \sqrt{r^2 - \xi^2[(x-i)^2 + (y-j)^2]}) \mid (x,y) \in \Omega(i,j)\} \\ B''(z_{ij},r) = \{z''_{xy} = \min(z''_{xy}, z_{ij} \\ \quad - \sqrt{r^2 - \xi^2[(x-i)^2 + (y-j)^2]}) \mid (x,y) \in \Omega(i,j)\} \end{cases} \tag{10}$$

As shown in Fig. 3, spatial neighborhood $\Omega(i,j)$ projects as the grid cell collection in the circle range with radius $r$ and center point $(i,j)$. For an arbitrarily grid cell $x, y$ in spatial neighborhood $\Omega(i,j)$,the upper (nether) buffer boundary point $z_{xy}$ $(z_{xy})$ of grid pint $(i,j)$ can be calculated according to formula 10. Finally, by traversing each grid cell in spatial neighborhood $\Omega(i,j)$, the corresponding upper (nether) buffer boundary $B'(z_{ij},r)$ $(B''(z_{ij},r))$ of grid point $(i,j)$ can be calculated.

## 4.2 Union of grid point buffer boundary

On the basis of the construction theory of grid point buffer boundary in section 4.1, for an arbitrarily grid point $(a, b)$ distinguished from grid point $(i, j)$ $((a-i)(b-j) \neq 0)$, supposing that $\Omega(i,j) \cap \Omega(a,b) \neq \varphi$, union operation should be done to the upper (nether) buffer boundary of grid point $(a, b)$ and $(i, j)$ separately.

According to the construction theory of DDM upper (nether) buffer boundary in section 3.2, the

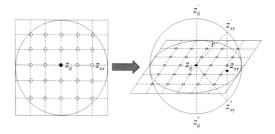

Figure 3. Construction of grid point upper (nether) buffer boundary.

upper buffer boundary point $z'_{xy}$ and the nether buffer boundary point $z''_{xy}$ are above and under the DDM surface, respectively, that is, $z'_{xy} \geq z_{ij}$ and $z''_{xy} \leq z_{ij}$. Therefore, for an arbitrary grid cell $x, y$ in spatial neighborhood $\Omega(i,j) \cap \Omega(a,b)$, the upper (nether) buffer boundary point $z'_{xy}$ $(z''_{xy})$ should be corrected to the corresponding maximum (minimum) value of upper (nether) buffer boundary point in spatial neighborhood $\Omega(i,j) \cap \Omega(a,b)$:

$$\begin{cases} z'_{xy} = \max(z'_{xy}, z_{ab} + \sqrt{r^2 - \xi^2[(x-a)^2 + (y-b)^2]}), \\ \quad (x,y) \in \Omega(i,j) \cap \Omega(a,b) \\ z''_{xy} = \min(z''_{xy}, z_{ab} + \sqrt{r^2 - \xi^2[(x-a)^2 + (y-b)^2]}), \\ \quad (x,y) \in \Omega(i,j) \cap \Omega(a,b) \end{cases} \tag{11}$$

Combining formulas 10 and 11, union operation of upper buffer boundary $B'(z_{ij},r)$ and $B'(z_{ab},r)$ (nether buffer boundary $B''(z_{ij},r)$ and $B''(z_{ab},r)$) of different grid points $(i, j)$ and $(a, b)$ can be defined as follows:

$$\begin{cases} B'(z_{ij},r) \cup B'(z_{ab},r) = \{z'_{xy} = \max(z'_{xy}, z_{ab} \\ \quad + \sqrt{r^2 - \xi^2[(x-a)^2 + (y-b)^2]}) \mid (x,y) \in (i,j) \cup \Omega(a,b)\} \\ B''(z_{ij},r) \cup B''(z_{ab},r) = \{z''_{xy} = \min(z''_{xy}, z_{ab} \\ \quad - \sqrt{r^2 - \xi^2[(x-a)^2 + (y-b)^2]}) \mid (x,y) \in (i,j) \cup \Omega(a,b)\} \end{cases} \tag{12}$$

## 4.3 Generation of DDM upper (nether) buffer boundary

By traversing each grid point $(i, j)$ of DDM and generating corresponding upper buffer boundary $B'(z_{ij},r)$ and nether buffer boundary $B''(z_{ij},r)$, and according to the union operation principles of grid point buffer boundary in section 4.2, DDM upper buffer boundary $B'(T,r)$ and nether buffer boundary $B''(T, r)$ can be generated by the union operation of upper buffer boundary $B'(z_{ij},r)$ and nether buffer boundary $B''(z_{ij},r)$ of grid point $(i, j)$, respectively, as

$$\begin{cases} B'(T,r) = \bigcup_{i=0}^{n-1}\bigcup_{j=0}^{n-1} B'(z_{ij},r) \\ B''(T,r) = \bigcup_{i=0}^{n-1}\bigcup_{j=0}^{n-1} B''(z_{ij},r) \end{cases} \qquad (13)$$

Algorithm evaluation should take stability, precision, and operation speed into account. As the data volume in GIS application is extremely large, the operation speed of an algorithm is of paramount importance. According to formula 13, the key to generate DDM upper buffer boundary $B'(T,r)$ (nether buffer boundary $B''(T,r)$) is the union operation of upper buffer boundary $B'(z_{ij},r)$ and $B'(z_{ab},r)$ (nether buffer boundary $B''(z_{ij},r)$ and $B''(z_{ab},r)$) of different grid points $(i, j)$ and $(a, b)$. On the basis of the construction theory of grid point buffer boundary in section 4.1, generation of upper buffer boundary $B'(z_{ij},r)$ (nether buffer boundary $B''(z_{ij},r)$) will take $r \times r$ times operation in spatial neighborhood $\Omega(i,j)$. According to the union operation principles of grid point buffer boundary in section 4.2, generation of upper buffer boundary $B'(z_{ab},r)$ (nether buffer boundary $B''(z_{ab},r)$) and it's union operation with $B'(z_{ij},r)$ ($B''(z_{ij},r)$) will take $r \times r$ times operation too. Therefore, by traversing each grid point $(i, j)$ of DDM, the generation and union operation of corresponding upper buffer boundary $B'(z_{ij},r)$ (nether buffer boundary $B''(z_{ij},r)$) will take $n \times n \times r \times r$ times operation. That is, the time complexity of the algorithm is $O(n^2 r^2)$.

## 5 EXPERIMENT RESULTS AND ANALYSIS

To validate the algorithm, this paper realizes the multi-scale representation algorithm of DDM based on positive direction rolling ball transform using VC++ program, and visualizes and analyzes the experiment results using Surfer8.0 software.

The experiment data are from the grid DDM data of a coastal region along East China Sea. The mean square error of bathymetric survey is 0.5 m, size of grid DDM is $100 \times 87$, and the original grid size (resolution) is 45 m.

Experiments are divided into five groups according to different scale factors. Figures 4 and 5 represent the contrast maps of bathymetric contour and marine topography surface, respectively, based on different scale factors ($\lambda = 3$ and $\lambda = 5$). It can be concluded from these figures that the positive direction rolling ball transform can distinguish the seafloor reliefs preferably. As the scale factor grows, the concave reliefs become reduced and filled up gradually, while the positive reliefs remain all the time. Quantitative analysis has been done to these five groups using the slope analysis function of Surfer8.0 software. Table 3 shows the experiment results.

Annotation: $\lambda = 1$ represents $2 \times 2$ grid cells of original DDM incorporate to one grid cell of new scale DDM, $\lambda = 2$ represents $4 \times 4$ grid cells of original DDM incorporate to one grid cell of new scale DDM, the rest can be done in the same manner.

The experiment results show that: 1) positive direction rolling ball transform preserves the basic characteristics of original data; skeleton reliefs of mountain ridge and mountain valley change little; and only part of detail reliefs smooth properly; 2) maximum depth remains unchanged, suggesting that positive reliefs are preserved; 3) as the scale factor increases, gradual increase is observed in the minimum and mean depth, suggesting that the proportion of shallow water points increases, which demonstrates that positive direction rolling ball transform exhibits the characteristics of reducing or filling up negative reliefs gradually; 4) slope value fluctuates in a certain range and the slope range decreases gradually. By factoring into the change of mean slope value, it can be seen that the detail relief feature points of adjacent scales

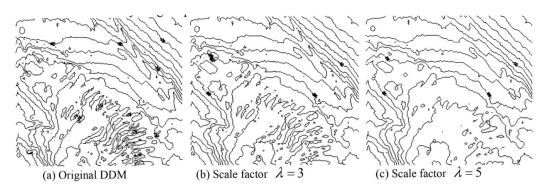

| (a) Original DDM | (b) Scale factor $\lambda = 3$ | (c) Scale factor $\lambda = 5$ |

Figure 4. Contrast map of bathymetric contour based on different scale factors.

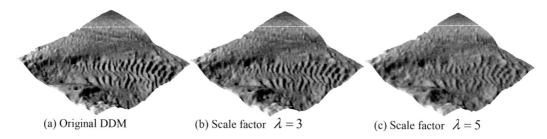

| (a) Original DDM | (b) Scale factor $\lambda = 3$ | (c) Scale factor $\lambda = 5$ |

Figure 5.   Contrast map of marine topography surface based on different scale factors.

Table 3.   Statistics of DDM multi-scale representation experiments.

| Statistic parameter | Original DDM | Scale factor | | | | |
|---|---|---|---|---|---|---|
| | | $\lambda = 1$ | $\lambda = 2$ | $\lambda = 3$ | $\lambda = 4$ | $\lambda = 5$ |
| Maximum depth | −13.528 | −13.528 | −13.528 | −13.528 | −13.528 | −13.528 |
| Minimum depth | −94.925 | −93.596 | −91.232 | −89.338 | −88.145 | −87.333 |
| Mean depth | −61.447 | −61.411 | −61.353 | −61.235 | −61.019 | −60.836 |
| Slope range | 0–10.304 | 0–9.265 | 0–8.014 | 0–7.572 | 0–7.066 | 0–6.347 |
| Mean slope value | 1.914 | 1.891 | 1.812 | 1.704 | 1.525 | 1.407 |
| Time consumed (s) | | 0.008 | 0.049 | 0.217 | 0.464 | 1.173 |

change more than other feature points of different scales. As the scale decreases and compression ratio increases, the entire DDM becomes smooth gradually and gently in line with the characteristics of DDM multi-scale representation.

## 6   CONCLUSIONS

The core of spatial data multi-scale representation is to establish an abstract model of scale dependence. By means of dimensional extension of planar rolling circle transform, this paper has brought forward a multi-scale representation of DDM based on positive direction rolling ball transform algorithm, and analyzes the spatial morphology influence of threshold value (buffer radius) to DDM. Theoretical analysis and experimental comparisons show that rolling ball transform sustains the geography and scale characteristics of DDM, and confirm the consistency of spatial cognition and abstract extent. While since the limit error of bathymetric survey adopts experiential parameter in the process of the algorithm, it remains to be seen if it fully meets the theoretical and universal requirements. Furthermore, research on the problem of self-adapting estimation for actual distribution of DDM grid points is needed. Besides, given the large size of the original DDM, the algorithm needs to be optimized to a higher degree.

## ACKNOWLEDGMENT

This paper was supported by the National Natural Science Foundation of China (41171349, 41471380) and the National High Technology Research and Development Program of China (2012 AA12 A406).

First Author: DONG Jian, Ph.D, lecturer, majors in the theories and methods of charting; the development and application of military oceanic geographical information system; and models and methods of marine delimitation. E-mail: navydj@163.com

Corresponding Author: PENG Rencan, Ph D, Professor. E-mail: pengrencan63@163.com

## REFERENCES

Ai Tinghua, Cheng Jianguo. Key Issues of Multi-Scale Representation of Spatial Data [J]. Geomatics and Information Science of Wuhan University, 2005, 30(5): 377–382.

Christensen A.H.J. Cartographic Line Generalization with Waterlines and Medial-Axes [J]. Cartography and Geographic Information Science, 1999, 26(1): 19–32.

Christensen A.H.J. Two Experiments on Stream Network Generalization [A]. Proceedings of the 21st International Cartographic Conference[C]. Durban: [s. n.], 2003.

Fei Lifan, He Jin, Ma Chenyan, et al. Three Dimensional Dauglas-Peucker Algorithm and the Study of Its Application to Automated Generalization of DEM [J]. Acta Geodaetica et Cartographica Sinica, 2006, 35(8): 278–284.

Gao Wangjun. Research of the Double Direction Buffering Model and Its Application on Chart Generalization [D]. Dalian: Dalian Naval Academy, 2009.

Guo Renzhong. Spatial Analysis [M]. Wuhan: Wuhan Technical University of Surveying and Mapping Press, 2000.

Hu Jinxing, MA Zhaoting, WU Huanping, et al. Massive Data Delaunay Triangulation Based on Grid Partition Method [J]. Acta Geodaetica et Cartographica Sinica, 2004, 33 (2): 163–167.

Hu Peng, Gao Jun. The Digital Generalization Principle of Digital Elevation Model [J]. Geomatics and Information Science of Wuhan University, 2009, 34(8): 940–942.

Hu Peng, Geng Xiepeng, Cao Feng. Morpha Transformation and Universal Convex Hull Algorithm [J]. Geomatics and Information Science of Wuhan University, 2005, 30(11): 1003–1007.

Hu Peng, You Lian, Yang Chuan-yong, et al. Map Algebra [M]. Wuhan: Wuhan University Press, 2002.

Jia Juntao, Zhai Jingsheng, Meng Chanyuan, et al. Construction and Visualization of Submarine DEM Based on Large Number of Multibeam Data [J]. Journal of Geomatics Science and Technology, 2008, 25(4): 255–259.

Li Jingzhong, Ai Tinghua, Wang Hong. The DEM Generalization Based on the Filling Valley Coverage [J]. Acta Geodaetica et Cartographica Sinica, 2009, 38(3): 272–275.

Li Zhilin. A Theoretical Discussion on the Scale Issue in Geospatial Data Handling [J]. Geomatics World, 2005, 3(2): 1–5.

Liu Yanchun, Xiao Fumin, Bao Jingyang, et al. Introduction to Hydrogrophy [M]. Beijing: Surveying and Mapping Press, 2006.

Peng Rencan, Wang Jiayao. A Research for Creating Buffer on the Earth Ellipsoid [J]. Acta Geodaetica et Cartographica Sinica, 2002, 31(3): 270–273.

Smith Shepard M. The Navigation Surface: A Multipurpose Bathymetric Database [D]. Durham, New Hampshire: University of New Hampshire, 2003.

Thomas Porathe. 3-D Nautical Charts and Safe Navigation [D]. Eskilstuna: Malardalen University Press Dissertation, 2006.

Wan Gang, Zhu Changqing. Application of Multi-Band Wavelet on Simplifying DEM With Lose of Feature Information [J]. Acta Geodaetica et Cartographica Sinica, 1999, 28(1): 36–40.

Yang Zuqiao, Guo Qingsheng, Niu Jiping, et al. A Study on Multi-Scale DEM Representation and Topographic Feature Line Extraction [J]. Acta Geodaetica et Cartographica Sinica, 2005, 34(2): 134–137.

Zhou Peide. Computational Geometry: Algorithms Analysis and Design [M]. Beijing: Tsinghua University Press, 2000.

Zhu Changqing, Shi Wenzhong. Spatial Analysis Modeling and Theory [M]. Beijing: Science Press, 2006.

*Advances in Energy and Environment Research – Achour & Wu (Eds)*
*© 2017 Taylor & Francis Group, London, ISBN 978-1-138-62682-9*

# Tectono-stratigraphy evolution of XuanGuang Basin (Anhui, East China) and its implication for shale gas exploration

Jianghua Chen

*College of Geosciences, China University of Petroleum, Beijing, China*

ABSTRACT: XuanGuang Basin has experienced three major phases of deformation since Mesozoic: NW-SE transpression (Middle Triassic-Middle Jurassic), NE-SW transtension (Late Jurassic-Early Cretaceous), and NW-SE extension (Middle-Late Cretaceous). Through the integration of all available information, the study clarified the tectono-stratigraphy evolution of XuanGuang Basin consisting of six major phases since Paleozoic: rifting basin development (Cambrian-Silurian), orogeny (Early-Middle Devonian), foreland basin filling (Carboniferous-Middle Triassic), overthrust (Late Triassic-Early Jurassic), thermal uplift (Late Jurassic-Early Cretaceous), and rift and basin forming (Cretaceous-Present). Subject to the tectono-stratigraphy evolution, several series of shale well developed in Paleozoic in the basin. However, the tectonic complexity resulted in poor shale gas potential.

## 1 INTRODUCTION

### 1.1 General context

XuanGuang Basin ["study area"] is located in the southeast of Anhui province of East China. It is a down-faulted basin mainly dominated by Cretaceous deposits with an area of 6,000 km² (Figure 1).

Because of its location close to the developed area of China (Yangtze delta economic zone including Shanghai) and its associated increasing needs for energy, significant efforts have been made for oil and gas exploration in XuanGuang Basin and its adjacent area since the 1950s. However, no breakthrough has ever been made up to now and the exploration status is still at a low level.

Since 2010, following the shale gas's boom in the United States, XuanGuang Basin attracts the attention again because of its well-developed Paleozoic shale underneath (Lower Cambrian Hetang formation, Lower Permian Gufeng formation, and Upper Permian Dalong formation) and the huge estimated shale gas potential.

### 1.2 Tectonic background

In the Lower Yangtze Basin, the tectonic environment during Caledonian and Hercynian periods (from Sinian to Early Triassic) was calm to form simple and stable structures, mainly characterized by rifting and depression structural framework. Since Middle Triassic, the study area has experienced three major phases of deformation.

– Phase 1: NW-SE transpression (Middle Triassic-Middle Jurassic). It corresponds to the collision of Yangtze plate and North China plate (Indosinian movement), resulting in strong uplifting and folding. This was the strongest tectonic movement since Paleozoic. Following it, the Upper Paleozoic strata were widely eroded in the study area. Two decollement levels are recognized, one at the base of Cambrian and the other at the base of Silurian.

– Phase 2: NE-SW transtension (Late Jurassic-Early Cretaceous). It corresponds to the NNW subduction of pacific plate to Euro-Asian plate (Middle Yanshanian) resulting in series of NNE directional sinistral strike-slip faults and WNW-ESE normal faults.

– Phase 3: NW-SE extension (Middle-Late Cretaceous to Eocene). It corresponds to late Yanshanian movement resulting in NE/NNE directional Cretaceous basin in the area.

At the end of the rift development, a phase of structural inversion reshaped the rift and deformed the syn-rift sediments, substantially uplifting the basin and creating large erosion.

Finally, the basin underwent a sag phase with slow subsidence and stabilization.

## 2 TECTONO-STRATIGRAPHY EVOLUTION

### 2.1 Stratigraphic framework

During the Paleozoic, the study area is characterized by a stable marine basin development with very broad carbonate depositional systems. However, during maximum sea level rise, widely spread organic-rich shale systems were deposited.

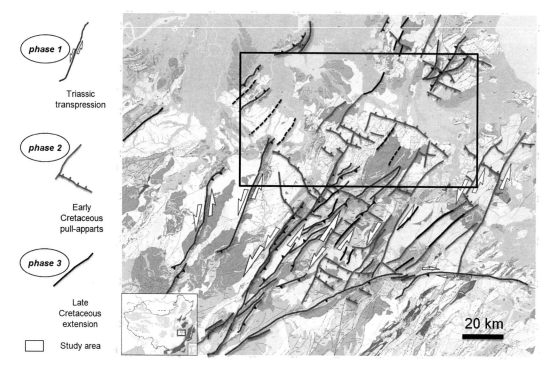

Figure 1. Location of the study area and its major tectonic events.

On the basis of the sedimentary field work and core observations of existing exploration wells in and around the study area, the Paleozoic main stratigraphy conclusions can be summarized as follows:

– Cambrian to Middle Ordovician depositional environments can be considered as deepwater environments (below the storm wave base) into the basin domain. Middle Ordovician to Middle Silurian depositional environments seems to be turbiditic with important tractive currents onto a ramp. From Middle Silurian to Devonian, series display sub-deltaic to deltaic deposits. Carbonate deposits of Carboniferous age are likely corresponding to platform environments. Permian deposits are mainly developed in coastal facies and platform facies.
– General coarsening and shallowing upward trend from Cambrian to Devonian: The basin history for this time interval seems relatively continuous (at the first and second orders). The progressive filling displays an obvious prograding system, particularly for the Silurian-Devonian age.
– No main lateral change (at least for Cambrian series): basinal marine shales seem to be widely and uniformly deposited at regional scale. Facies and thicknesses are equivalent over 50 km.

### 2.2  Tectono-stratigraphy evolution

Regardless of the scarcity of data and structural complexity of the area, the integration of all available information allowed the clarification of the tectono-stratigraphy evolution of the study area, supporting the selection of the most favorable area for new exploration campaign. Details are listed as below:

In the geological history, the area experienced series of tectono-stratigraphic evolutions.

Before Cambrian, in the early stage of Neo-Proterozoic, because of the closure of old Huanan Sea, Yangtze block and Cathaysian blocks merged, forming NE-SW oriented orogenic belt running through the study area. In the late stage of Neo-Proterozoic, rifting happened along the suture zone between Cathaysian and Yangtze blocks, starting the development of rifted basin. Since then, the area mainly experienced six major phases (Figure 2):

Phase 1: From Cambrian to Silurian ($\epsilon$-S) – Rifting Rifting took place in a transtensional setting dominated by NE-SW dextral faults associated with E-W normal faults. Progressive filling of the basin resulted from the deposition of basinal mudstones (Cambrian-Ordovician) to clastic turbiditic fans (Early Silurian) capped by shallow deltaic facies (Upper Silurian).

254

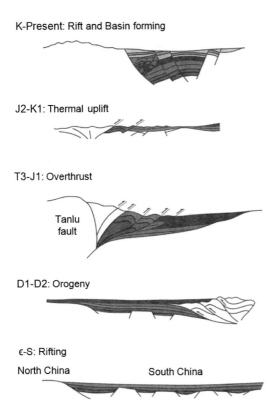

**K-Present: Rift and Basin forming**

**J2-K1: Thermal uplift**

**T3-J1: Overthrust**

Tanlu fault

**D1-D2: Orogeny**

**ε-S: Rifting**

North China          South China

Figure 2.  Tectono-stratigraphy evolution of the study area.

Paleozoic source rock deeply buried, resulting in high maturity (oil and dry gas window).

As collision progressed during Late Triassic times, the thrust and fold belts advanced into the inland in the Lower Yangtze Basin, where Xuan-Guang Basin belongs to and most of the earlier deposits were severely deformed, especially in the north part of the basin close to the suture zone along the Dabie orogenic belt.

Former NE-SW early Paleozoic faults were inverted into a right-lateral transpressional setting.

Phase 5: Late Jurassic—Early Cretaceous (J2-K1): Thermal Uplift & Erosion
During the late Jurassic, regional erosion without folding and thrusting can be interpreted as a thermal uplift.

Several igneous events during the Yanshan period (Jurassic & Early Cretaceous) controlled by the large faults oriented in NE-SW and near E-W directions.

Phase 6: Cretaceous-Present (K-Present): Rifting basin development across the entire western Pacific
During the early Cretaceous, NE-SW trans-tension is responsible for the development of pull-apart basins filled by continental facies. This event is interpreted to be related to the initiation of NW-SE oblique subduction of the Pacific plate.

During the late Cretaceous-Paleogene, a regional NW-SE extension is associated with the development of NE-SW normal faults giving the final shape of the Wuhu and XuanGuang basins.

This event is attributed to the slowing down of the Pacific subduction and subsequent back-arc extension. No folding/faulting is observed in the Neogene to Quaternary sediments.

At the end of the rift development, a phase of structural inversion reshaped the rift and deformed the syn-rift sediments, substantially uplifting the basin and creating large erosion.

Finally, the basin underwent a sag phase with slow subsidence and stabilization.

## Phase 2: Devonian Orogeny (D1-D2)

The Devonian is marked by a regional scale compression that led to the inversion of previous rift structures and further deposition of coastal sediments. Sedimentation hiatus was observed and there was no deposition of lower-middle Devonian.

## Phase 3: Carboniferous-Middle Triassic (C-T2)

Carboniferous shallow dolomitic breccias were observed in the outcrops, which may correspond to the end of the foreland basin filling.

During the Permian to early Triassic, a deepening of the basin is characterized by progradation and retrogradation cycles of the carbonate platform.

Basal transgressive black mudstones also correspond to shale gas objectives (Gufeng and Dalong Fm.).

## Phase 4: Late Triassic—Early Jurassic (T3-J1): Collision between North China and Yangtze Plate

Onset marked by a clear increase in subsidence as it entered a foreland basin as the Tethys Sea closed along the north and west margins of the Yangtze Plate.

Huge amounts of Triassic clastics sediments were developed in the Lower Yangtze Basin with

## 3  IMPLICATION FOR SHALE GAS EXPLORATION

### 3.1  Summary of structural configuration

Subject to the overall tectono-stratigraphic background of the study area, stable marine deposits are developed in Paleozoic. From late Triassic to Jurassic, the study area suffered a major compressive event, forming series of NE-SW trending folds. Folds are narrow in the northwest part while more wide in the southeast.

The youngest strata involved in the folding are Triassic. Faults developed in the study area are mainly NE-SW trending thrusts and NW-SE oriented transfer faults. Between Jurassic and Cretaceous, there is an obvious angular unconformity. Coarse Cretaceous continental clastics filled the local basinal structures. Igneous intrusions and extrusions are apparent. In accordance with tectonic background, there is a hiatus in Tertiary. Quaternary strata unconformably covered the Jurassic and Cretaceous sequences in the topographic lows, forming alluvial plains.

## 3.2 *Igneous rock development and distribution*

Igneous rocks mainly developed in the central south and north of the study area. The granites reflect different ages, which correspond to different phases of tectonic movements (Indosinian, Middle, and Late Yanshanian). Basalts are sparsely developed in the Cretaceous basin. The igneous rocks are mainly developed along the major faults and controlled by them, in the types of dyke and bodies. Inversion of Magnetotelluric (MT) data confirmed that the two large igneous rocks developing in the two sides of Cretaceous basin are not connected underneath.

## 3.3 *Potential implication for shale gas potential*

Within the Paleozoic series, different series have the shale gas potential and particularly within the Lower Cambrian Hetang Formation. However, the structural complexities identified in the study area are very significant and the following features may negatively affect the shale gas potential:

- narrow folds and thrusts, making the Paleozoic series uplifting and suffering severe erosion;
- closely spaced faults, affecting the gas retention conditions and implying limits on drilling and potential future development if any;
- natural fractures widely developed in the study area in different phases, also affecting the gas retention conditions and rock frackability;
- steep dips of the formations also affecting fracking and the ability to drill horizontally;
- igneous intrusions probably affecting the maturity;
- late phase of extension, which controlled the final shape of Cretaceous depression, making the Cambrian objective locally extremely deep (up to 6000 m). In addition, this late extensional phase had its large and destructive impact on the whole gas retention conditions of Paleozoic series.

## 4 CONCLUSIONS

XuanGuang Basin is in a complex tectonic background and has experienced three major phases of deformation since Mesozoic: NW-SE transpression (Middle Triassic-Middle Jurassic), NE-SW transtension (Late Jurassic-Early Cretaceous), and NW-SE extension (Middle-Late Cretaceous).

Through the integration of all available information, the tectono-stratigraphy evolution of Xuan-Guang Basin is clarified. It mainly experienced six major phases since Paleozoic: rifting basin development ($\epsilon$-S), orogeny (D1-D2), foreland basin filling (C-T2), overthrust (T3-J1), thermal uplift (J2-K1), and rift and basin forming (K2-present).

Subject to the tectono-stratigraphy evolution, several series of shale well developed in Paleozoic in the basin. However, the tectonic complexity resulted in poor shale gas potential.

## ACKNOWLEDGMENT

The authors thank the other members of the project team who contributed to this work. They also thank the experts for reviewing the article and their useful comments.

## REFERENCES

Chen, A.D., Liu, D.Y., Liu, Z.M., 1999. Mesozoic and Paleozoic Source rock study and resources evaluation of Jiangsu lower yangtze area [R].

Du, X.B., Zhang, M.Q., Lu, Y.C., Cheng, P., Lu, Y.B., 2014. Lithofacies and depositional characteristics of gas shales in the western area of the Lower Yangtze, China. Geological Journal, DOI: 10.1002/gj.2587.

He, Y.B., Luo, J.X., 2010. Lithofacies palaeogeography of the late Permian Changxing age in middle and upper Yangtze region. Journal of Palaeogeography 12, 497–514.

Xia, B.D., Lu, H.B., 1988. The sedimentologic and tectonic evolution in the Zhejiang-Anhui-Jiangsu region. Acta Geologica Sinica 62, 301–310 (in Chinese with English abstract).

Xia, B.D., Zhong, L.R., Fang, Z., Lu, H.B., 1995. The origin of bedded cherts of the early Permian Gufeng Formation in the Lower Yangtze area, Eastern China. Acta Geologica Sinica 8, 372–397.

Zhang, Y.H., 1996. The practical significance of overthrust structure in the oil and gas exploration of lower paleozoic in Yangtze basin. East China Petroleum Geology, China Sciences (Volume D), 26(6), 537–543.

*Advances in Energy and Environment Research – Achour & Wu (Eds)*
*© 2017 Taylor & Francis Group, London, ISBN 978-1-138-62682-9*

# The main factors analyzing the effect of microstructure cell on the mechanical properties of L-CFRP

Yiyong Yao & Xu Wang
*School of Mechanical Engineering, Xi'an Jiaotong University, Xi'an, ShanXi, China*
*Key Laboratory of Education Ministry for Modern Design and Rotor-Bearing System,*
*Xi'an Jiaotong University, Xi'an, ShanXi, China*

Rushan Dou
*School of Mechanical Engineering, Xi'an Jiaotong University, Xi'an, ShanXi, China*

ABSTRACT: This paper studied the effect of the microstructure cell on the mechanical properties of L-CFRP. Combined with finite element technology, and through a set of orthogonal tests, the main variables in the microstructure cell that affect the mechanical properties of L-CFRP were found, and the representative volume of interlayer-enhanced L-CFRP was analyzed. The results of simulation analysis and the experiments showed that the main factors affecting the mechanical properties of L-CFRP were the width-to-thickness ratio of the fiber cross-section and the ratio of fiber bundle clearance to thickness of the fiber cross-section. The simulation results were consistent with the experiment results.

## 1 INTRODUCTION

Because of their strong structure designability and strength designability, L-CFRPs are widely used in aviation, aerospace, shipbuilding, and other fields. However, according to the practical uses, the different precast structures and molding processes result in different structures of L-CFRP component with different mechanical properties. Therefore, the mechanical properties of L-CFRPs are difficult to predict.

Many scholars studied the mechanical properties of L-CFRPs with different microstructures (Sziroczak & Smith 2016, Yashiro & Ogi 2016) and found a relationship between the parameters of representative volume unit cell in L-CFRP and its mechanical properties. M'Membe et al. (2016) studied the effect of the fiber bundle maximum angle on the mechanical properties of composites, and the results showed that the increase of the fiber bundle maximum angle could bring certain damage to the mechanical properties of composites in the main direction. Stig & Hallström (2013) studied the effect of the fiber bundle buckling rate on the mechanical properties of composites, and the results showed that the higher the fiber bundle buckling rate, the poorer the mechanical properties of composites in the main direction. Scholars studied the effect of carbon fiber volume fraction on the mechanical properties of composites, and the results showed that the larger the carbon fiber volume fraction, the better the mechanical proper-

ties of composites (Yan et al. 2016 & Brunbauer et al. 2015). Dalaq and Abueidda (2016) studied the mechanical properties of three-dimensional (3D) braided composites, and the results showed that when the carbon fiber volume fraction was up to 75%, its elastic modulus in the main direction was 173.92 GPa.

In most of the existing studies, the mechanical properties of a composite are related to its fiber volume fraction. However, in practice, the same volume fraction could be corresponding to several different microstructures and different performances of L-CFRPs. Therefore, it is important to study the effect of the microstructure of representative volume unit cell on the mechanical properties of L-CFRP.

## 2 PARAMETER ANALYSIS OF UNIT CELL MICROSTRUCTURE

Because the structure of L-CFRP laminates has certain periodicity, as shown in Figure 1(a), the macro 3D representative volume unit cell could be used to represent the overall structure of the material, as shown in Figure 1(b). In L-CFRP laminates, the aeolotropic macroscopic properties can be determined by macro parameters such as carbon fiber volume fraction of macro cell, main fiber of carbon fiber bundle, braiding angle, and the cross-section shape of fiber bundle. The macro mechanical properties of composite could be obtained by the study of representative volume unit cell.

In the representative volume unit cell of interlayer-enhanced L-CFRP, the fiber bundles are crisscross and corrugated due to weave form. The cross-section of fiber bundle is oval and most of the space is filled, and there is very small amount of pure resin area. The parameters of unit cell microstructure that affect the mechanical properties of interlayer-enhanced L-CFRPs and their relationship are shown in Figure 2.

In order to verify the effect of microstructure parameters on mechanical properties of laminates, a set of orthogonal tests need to be designed, and the five variables should be turned into four dimensionless parameters: $P_1=w_f/t_f$, $P_2=s_f/t_f$, $P_3=d_f/t_f$, and $P_4=t_c/t_f$. In actual material preparation process, the control of four-group dimensionless parameters is implemented as follows: P1 is controlled through changing the containing amount of carbon fiber yarn in 2D braided carbon cloth fiber bundle; $P_2$ is controlled through changing the longitude and latitude structures of 2D braided carbon cloth; $P_3$ is controlled in the preparation process of carbon fiber laminate precast, which is implemented by

laying different length and volume interlayer fibers in water jet process; $P_4$ is controlled through laying 2D braided carbon cloth with different thickness in the laminate with same thickness.

## 3 MAIN FACTORS AFFECTING MECHANICAL PROPERTIES OF L-CFRP

### 3.1 ABAQUS finite element simulation orthogonal experiment design

The effects of different microstructures of unit cell on the mechanical properties of interlayer-enhanced L-CFRP are studied through finite element numerical simulation experiments. In order to further verify the accuracy of the simulation test, and combined with actual material condition, a set of orthogonal experiments need to be designed. There are three factor variables: the length of interlayer fiber, the type of carbon cloth, and the number of layers (Table 1).

Table 1. Factors analysis of composite orthogonal experiment.

| Level | Factor A<br>Type of fiber bundle | Factor B<br>Fabric structure/<br>(branch/cm²) | Factor C<br>Length of interlayer fibers/<br>(mm) | Factor D<br>Number of layers |
|---|---|---|---|---|
| 1 | 3k | $4 \times 4/3 \times 3$ | 0 | 3 |
| 2 | 6k | $5 \times 5/3.5 \times 3.5$ | 5 | 4 |
| 3 | 12k | $6 \times 6/4 \times 4$ | 10 | 5 |

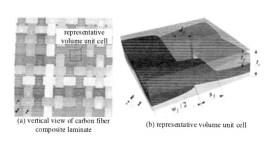

(a) vertical view of carbon fiber composite laminate

(b) representative volume unit cell

Figure 1. Partition of representative volume unit cell.

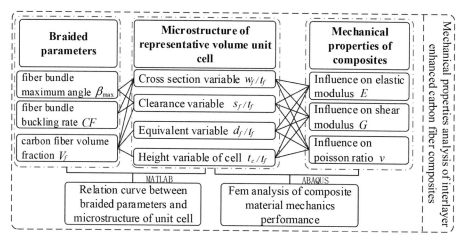

Figure 2. Factors affecting representative volume unit cell.

In Table 1, fiber bundle types of 3k, 6k, and 12k represent 3000, 6000, and 12000 branches of carbon fibers in a fiber bundle, respectively. Fabric structure represents the amount of carbon fiber bundle of carbon cloth in 1 cm$^2$. The length of interlayer fibers represents the length of added interlayer fibers in water jet process. The corresponding parameters of factor A and factor B in 2D braided carbon cloth are shown in Table 2.

In Table 1, the fiber bundle types 3k, 6k, and 12k represent 3000, 6000, and 12000 branches of carbon fibers in a fiber bundle, respectively. Fabric structure represents the amount of carbon fiber bundle of carbon cloth in 1 cm$^2$. The length of interlayer fibers represents the length of added interlayer fibers in water jet process. The corresponding parameters of factor A and factor B in 2D braided carbon cloth are shown in Table 2.

The testing program is shown in Table 3; the micron X-ray 3D imaging system is used to detect the carbon fiber volume fraction $V_f$ of test specimens. The carved different material test specimens are clamped on universal material testing machine. The standard to stop the test is the unloading of testing machine over 70%. There are five effective specimens in each test, and the displacement–load curves in the testing process are recorded. The damage types of specimen are according to the standard ASTM-D3039.

For the elastic mechanical properties of interlayer-enhanced L-CFRP, the level that the above four factors have no interaction is chosen as 5. Orthogonal table L25 (56) is used to perform the finite element orthogonal factor analysis, as shown in Table 4. The finite element orthogonal experi-

Table 2. Different specifications of 2D braided carbon cloth and its basic parameters.

| Specifications | Tensile strength/ (MPa) | Tensile modulus/ (GPa) | Density/ (g/m$^2$) | Fabric structure/ (branch/ cm$^2$) | Thickness/ (mm) |
|---|---|---|---|---|---|
| 3k | >3300 | >220 | 160 | 4 × 4 | 0.26 |
| 12k | >3300 | >220 | 560 | 4 × 4 | 0.56 |

Table 3. Orthogonal experiment scheme of L-CFRP microstructure.

| Fiber bundle type | Fabric structure/ (branch/ cm$^2$) | Length of interlayer fibers/ (mm) | Number of layers | $V_f$ |
|---|---|---|---|---|
| EX-1 | 3k | 4 × 4 | 0 | 3 | 0.19 |
| EX-2 | 3k | 5 × 5 | 5 | 4 | 0.29 |
| EX-9 | 12k | 4 × 4 | 5 | 3 | 0.50 |

Table 4. Finite element orthogonal factor analysis.

| Level | Factor A $P_1 = w_f/t_f$ | Factor B $P_2 = s_f/t_f$ | Factor C $P_3 = d_f/t_f$ | Factor D $P_4 = t_c/t_f$ |
|---|---|---|---|---|
| 1 | 2 | 0.1 | 0 | 2.2 |
| 2 | 4 | 0.5 | 0.1 | 2.4 |
| 3 | 10 | 10 | 0.4 | 3 |

Table 5. Finite element orthogonal experiment scheme.

| Factors | $P_1 = w_f/t_f$ | $P_2 = s_f/t_f$ | $P_3 = d_f/t_f$ | $P_4 = t_c/t_f$ |
|---|---|---|---|---|
| FE-1 | 2 | 0.1 | 0 | 2.2 |
| FE-2 | 2 | 0.5 | 0.1 | 2.4 |
| FE-25 | 10 | 10 | 0.3 | 2.6 |

Table 6. Material component performance parameters of composite microstructure.

| Material | E$_1$/GPa | E$_2$/GPa | G$_{12}$/GPa | G$_{23}$/GPa | v$_{12}$ |
|---|---|---|---|---|---|
| Impregnated carbon fiber bundle | 200 | 10 | 5 | 5 | 0.2 |
| Epoxy matrix | 3 | 3 | – | – | 0.35 |

ment scheme is shown in Table 5, and $t_f = 0.5$ mm is assumed.

Through ABAQUS finite element simulation, the relationship between microstructure of composites and its elastic mechanics could be obtained. The material component performance parameters of composite microstructure are shown in Table 6.

### 3.2 Process of ABAQUS finite element simulation test

To analyze the influence rule of two factors A and B on the mechanical properties of interlayer-enhanced L-CFRPs, the finite element simulation test is shown in Figure 3.

### 3.3 Results analysis of ABAQUS finite element orthogonal experiment

Establish finite element models of unit cells for different microstructures in the finite element orthogonal tests. Using ABAQUS software and through the analysis of the simulation results, the range of elastic mechanical properties of L-CFRP under the influence of each factor is shown in Table 7.

It is evident from the table that for elastic mechanical properties of composites such as E$_1$ & E$_2$, V$_{12}$ & V$_{21}$, and V$_{31}$ & V$_{32}$, A is the most significant factor, whereas for E$_3$, G$_{12}$, G$_{13}$ & G$_{23}$, and V$_{13}$ & V$_{23}$, B is the most significant factor.

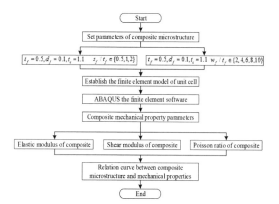

Figure 3. Process of ABAQUS finite element simulation test.

Table 7. Important sequences of factors affecting elastic mechanical properties of materials.

| | Factor A | Factor B | Factor C | Factor D | Importance sequences |
|---|---|---|---|---|---|
| $E_1$ & $E_2$ | 17.21 | 6.766 | 6.526 | 2.094 | A > B > D > C |
| $E_3$ | 0.254 | 1.376 | 0.96 | 0.302 | B > D > C > A |
| $G_{12}$ | 0.47 | 1.124 | 0.38 | 0.102 | B > A > D > C |
| $G_{13}$ & $G_{23}$ | 0.252 | 0.826 | 0.692 | 0.118 | B > D > A > C |
| $V_{12}$ & $V_{21}$ | 0.079 | 0.03 | 0.038 | 0.032 | A > D > C > B |
| $V_{13}$ & $V_{23}$ | 0.05 | 0.103 | 0.093 | 0.04 | B > D > A > C |
| $V_{31}$ & $V_{32}$ | 0.087 | 0.035 | 0.023 | 0.018 | A > B > D > C |

## 4 ANALYSIS OF THE MECHANICAL PROPERTIES OF MICROSTRUCTURE CELL

Four variables of composite microstructure: $P_3 = 0.2$ and $P_4 = 2.2$. Let $t_f = 0.5$ mm, $P_1 = 2, 4, 6, 8, 10$ and $P_2 = 0.5, 1, 2$. Establish 15 different finite element models of composite with different microstructures, and use ABAQUS software to calculate the corresponding mechanical properties parameters of each unit cell. There are three types of mechanical properties parameters: elastic modulus, shear modulus, and Poisson's ratio. The relative curves between mechanical properties parameters and composite microstructure are drawn accordingly.

### 4.1 *The influence rule of unit cell microstructure on elastic modulus of L-CFRP*

The influence rules of P1 and P2 on elastic modulus of L-CFRP are shown in Figure 4. Because the interlayer-enhanced L-CFRP is transversely isotropic, $E_1 = E_2$.

From the above, it can be seen that, first, the elastic moduli $E_1$, $E_2$, and $E_3$ in the directions 1,

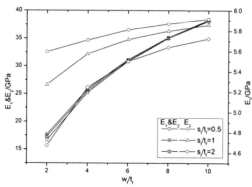

Figure 4. Influence rule of unit cell microstructure on elastic modulus of L-CFRP.

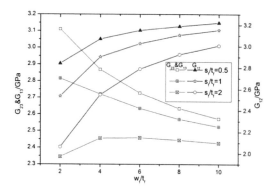

Figure 5. The influence rule of unit cell microstructure on shear modulus of L-CFRP.

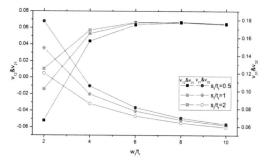

Figure 6. The influence rule of unit cell microstructure on Poisson's ratio of L-CFRP.

2, and 3, respectively, are gradually increased. Second, the elastic moduli of directions 1 and 2 are higher than direction 3. This is because 1 and 2 are the main directions of material, and the carbon fiber bundle of 2D braided carbon cloth bears the

Table 8. Calculation results of strength and modulus in tensile test.

| Factors | Specimen 1 $\sigma_t$ | $E_t$ | Specimen 2 $\sigma_t$ | $E_t$ | Specimen 3 $\sigma_t$ | $E_t$ | Specimen 4 $\sigma_t$ | $E_t$ | Specimen 5 $\sigma_t$ | $E_t$ | Mean $\sigma_t$ | $E_t$ |
|---|---|---|---|---|---|---|---|---|---|---|---|---|
| EX-1 | 252 | 17.1 | 234 | 16.8 | 227 | 15.5 | 262 | 16.3 | 242 | 15.0 | 243 | 16.1 |
| EX-2 | 305 | 20.8 | 299 | 18.5 | 296 | 19.6 | 330 | 19.1 | 366 | 24.2 | 319 | 20.4 |
| EX-9 | 432 | 27.1 | 421 | 26.2 | 428 | 25.8 | 467 | 31.3 | 472 | 30.0 | 444 | 28.1 |

main load. In direction 3, only the scattered inter-layer mixing fibers and matrix bear the load, so its elastic modulus is far less than that of directions 1 and 2. Third, the higher the carbon fiber volume fraction of composite, the higher the elastic modulus.

### 4.2 The influence rule of unit cell microstructure on shear modulus of L-CFRP

The influence rule of unit cell microstructure on shear modulus of L-CFRP is shown in Figure 5. Because the interlayer-enhanced L-CFRP is transversely isotropic, $G_{23} = G_{13}$.

As the aperture of 2D braided cloth decreases, the shear modulus in plane 12 of L-CFRP gradually increases. When the composite is under the stress of direction 3 in the normal plane of direction 1, L-CFRPs are more susceptible to shear failure, and hence the shear modulus in the normal plane of direction 1 gradually decreases.

### 4.3 The influence rule of unit cell microstructure on Poisson's ratio of L-CFRP

The influence rule of unit cell microstructure on Poisson's ratio of L-CFRP is shown in Fig. 6. Because the interlayer-enhanced L-CFRP is transversely isotropic, $V_{12} = V_{21}$ and $V_{31} = V_{32}$.

From the above, it could be seen that the change of Poisson's ratio in plane 12 of L-CFRP gradually increases and then begins to flatten. When the composite is under the stress of direction 3, Poisson's ratio of L-CFRP gradually decreases and then begins to flatten.

## 5 ANALYSIS OF EXPERIMENTAL RESULTS

Tensile strength and tensile elastic modulus of all tensile tests are shown in Table 8.

From the average value of tensile elastic modulus, the tensile test results could be analyzed, as shown in Table 9.

Results of orthogonal tests show that the elastic modulus of composite laminates in direction 1 are

Table 9. Elastic modulus analysis of the tensile experiment specimens (unit: GPa).

| Levels | Factor A Type of fiber bundle | Factor B Fabric structure/ (branch/ cm²) | Factor C Length of interlayer fibers/ (mm) | Factor D Number of layers |
|---|---|---|---|---|
| Mean 1 | 21.167 | 31.833 | 26.167 | 26.667 |
| Mean 2 | 25.067 | 27 | 26.5 | 25.733 |
| Mean 3 | 33.767 | 21.167 | 27.333 | 27.6 |
| Range | 12.6 | 10.666 | 1.166 | 1.867 |
| Levels | A3 | B1 | C3 | D3 |

mainly affected by factor A (fiber bundle type) and factor B (fabric structure). The experiment results are consistent with the simulation analysis. And factor C (length of interlayer fiber) has minimal effect on the elastic modulus of composite laminates in direction 1.

## 6 CONCLUSIONS

In this paper, the microstructure of representative volume unit cell in interlayer enhanced L-CFRP is taken as the study object. The five parameters of the microstructure are redefined as four dimensionless parameters. Combined with the finite element technology, a set of orthogonal tests are conducted to find the main variables affecting the mechanical properties of composites. Furthermore, the effect of width-to-thickness ratio of the fiber cross-section on the mechanical properties of L-CFRP is analyzed. The effect of the ratio of fiber bundle clearance to thickness of the fiber cross-section on its mechanical properties is also studied. Finally, in accordance with the finite element simulation tests, a set of experiments with different representative volume unit cell microstructures are conducted. The results show that the main factors that affect the mechanical properties of composite are the width-to-thickness ratio of the fiber cross-section and the ratio of fiber bundle clearance to thickness of the fiber cross-section. The simulation results are consistent with the experiment results.

## ACKNOWLEDGMENTS

The research work was supported by the National Natural Science Foundation of China under Grant No. 51275394.

## REFERENCES

Brunbauer, J., Stadler, H. & Pinter, G. (2015). Mechanical properties, fatigue damage and microstructure of carbon/epoxy laminates depending on fibre volume content. *International Journal of Fatigue, 70,* 85–92.

Dalaq, A. S., Abueidda, D. W., & Al-Rub, R. K. A. (2016). Mechanical properties of 3d printed interpenetrating phase composites with novel architectured 3d solid-sheet reinforcements. *Composites Part A Applied Science & Manufacturing, 84,* 266–280.

M'Membe, B., Gannon, S., Yasaee, M., Hallett, S. R., & Partridge, I. K. (2016). Mode ii delamination resistance of composites reinforced with inclined z-pins. *Materials & Design, 94,* 565–572.

Stig, F. & Hallström, S. (2013). Influence of crimp on 3d-woven fibre reinforced composites. *Composite Structures, 95*(1), 114–122.

Sziroczak, D. & Smith, H. (2016). A review of design issues specific to hypersonic flight vehicles. *J. Progress in Aerospace Sciences.*

Yan, S., He, P., Jia, D., Yang, Z., & Duan, X., et al. (2016). Effect of fiber content on the microstructure and mechanical properties of carbon fiber felt reinforced geopolymer composites. *Ceramics International, 42*(6), 7837–7843.

Yashiro, S. & Ogi, K. (2016). 7—high-velocity impact damage in cfrp laminates. *Dynamic Deformation Damage & Fracture in Composite Materials & Structures,* 169–191.

*Advances in Energy and Environment Research – Achour & Wu (Eds)*
*© 2017 Taylor & Francis Group, London, ISBN 978-1-138-62682-9*

# Structural characteristics of genomic islands associated with *RlmH* genes in *Bacillus*

Lingwei Su, Xuan Peng, Qingze Zha, Shidan He, Mengjie Yang & Lei Song
*College of Life and Environmental Sciences, Shanghai Normal University, Shanghai, China*

ABSTRACT: In this paper, 12 genomic islands flanked by *RlmH* genes were determined in the genus *Bacillus*, such as in *Bacillus cereus*, *Bacillus thuringiensis*, *Bacillus cytotoxicus*, and *Bacillus cellulosilyticus*. Multiple genomic islands that are the tandem genomic islands and the nest genomic islands were recognized in *B. cytotoxicus* NVH 391–98 and three *B. thuringiensis* chromosomes, respectively. Bce10987GI$^{RlmH}$ and three BtGIs$^{RlmHb}$ play an important role in type III restriction system. 5'GAAC3' in flanking direct repeats was predicted as a core region and a cutting site of the serine recombinases in some GIs$^{RlmH}$. Another sequence, namely 5'GTATCA3' was speculated as a binding site of the serine recombinases that can become the potential tools for engineering mammalian genomes.

## 1 INTRODUCTION

Genomic Islands (GIs) that contain at least one mobile gene and flanking Direct Repeats (DRs) are the products of Horizontal Gene Transfer (HGT) (Dobrindt et al. 2004). The mobile gene proteins in the GIs often are the tyrosine-type integrases or recombinases, but are rarely studied as the serine-type integrases or recombinases. The first found insertion sites of the GIs are the tRNA genes, soon afterward, can be the structural genes, sRNA genes, tmRNA genes, intergenic spacers, and so on (Bellanger et al. 2014). *Bacillus* includes some pathogenic species, for example, *B. anthracis*, *B. cereus*, and *B. thuringiensis*. We first found Bce10987GI$^{RlmH}$ flanked by 3' end of *RlmH* gene, which is a structural gene functioning as a 23SRNA methylase (Ero et al. 2008) through alignment of *RlmH* gene region among intraspecific chromosomes. *RlmH* gene first became an integration site in the Staphylococcal Cassette Chromosome *mec* (SCC*mec*) (Boundy et al. 2013). SCC*mec* is a large DNA cassette that contains a methicillin-resistant gene *mecA* in *Staphylococcus aureus* and was named as a new antibiotic-resistant GI (Katayama et al. 2000). Methicillin-Resistant *S. Aureus* (MRSA) can promote hospital-acquired infections and result in a high mortality rate. Two cassette chromosome recombinases A and B (*ccrA* and *ccrB*) delete the SCC*mec* from and reintegrate it into the chromosome. ccrA and ccrB proteins are partially homologous to TP901-1 integrase, which is a Large Serine Recombinase (LSR) in the temperate Lactococcal bacteriophage TP901-1 (Christiansen et al. 1996; Yuan et al. 2008). Furthermore, interaction between ccrA and ccrB can complete site-specific excision

and insertion of SCC*mec* (Wang et al. 2010). *RlmH* gene is highly conserved among all *Bacillus* chromosomes. We suspected that more GIs$^{RlmH}$ will be found in *Bacillus* chromosomes. Thus, 12 GIs$^{RlmH}$, namely Bce10987GI$^{RlmH}$, BtHD-789GI$^{RlmH}$, BtHD-789GI$^{RlmHs}$, Bcy391–98GI$^{RlmH-1}$, Bcy391-98GI$^{RlmH-2}$, BtHD1002GI$^{RlmHb}$, BtHD1002GI$^{RlmHs}$, BcFRI-35GI$^{RlmH}$, BcAH820GI$^{RlmH}$, BtAM65-52GI$^{RlmHb}$, BtAM65-52GI$^{RlmHs}$, and BcelDSM2522GI$^{RlmH}$ were determined in *Bacillus* chromosomes through multiple methods. Meanwhile, multiple GIs were found, namely Bcy391-98GI$^{RlmH-1}$ and Bcy391-98GI$^{RlmH-2}$ are the tandem GIs, BtHD-789GI$^{RlmHb}$ and BtHD-789GI$^{RlmHs}$, or BtHD1002GI$^{RlmHb}$ and BtHD1002GI$^{RlmHs}$, or BtAM65-52GI$^{RlmHb}$ and BtAM65-52GI$^{RlmHs}$ are the nest GIs. A part of DRs (5'GAACCGTATCA3') was predicted as the action region of the mobile gene proteins that are structurally similar to TP901-1 integrase in some GIs$^{RlmH}$ of *Bacillus*.

## 2 MATERIALS AND METHODS

All nucleotide and protein sequences of *Bacillus* genomes were downloaded from NCBI (ftp://ftp.ncbi.nlm.nih.gov/genomes/). The alignment results of the nucleotide and protein sequences were obtained by Blastn or Tblastn from NCBI (https://blast.ncbi.nlm.nih.gov/Blast.cgi). The intraspecific chromosomes were compared by WebACT (http://www.webact.org/WebACT/generate). The flanking DRs were aligned by ClustalW (http://www.genome.jp/tools/clustalw/). Molecular evolutionary genetics analysis of the mobile gene proteins in GIs$^{RlmH}$ was performed using MEGA7 software (Kumar et al. 2016). A mobile gene protein sequence (BCE_RS26800) in Bce10987GI$^{RlmH}$ was aligned with all complete genomes. Protein sequences highly similar to BCE_RS26800 were determined in some *Bacillus* chromosomes. The left and right flanking sequences of Bce-10987GI$^{RlmH}$ were aligned with the *Bacillus* chromosomes that contain the highly similar protein (BCE_RS26800). Thus, some GIs flanked by *RlmH* genes and contain homologous mobile gene proteins were determined in some *Bacillus* chromosomes. Every newly determined GI$^{RlmH}$ was aligned with the *Bacillus* chromosomes that contain the GI$^{RlmH}$. All of mobile gene proteins in GIs$^{RlmH}$ were structurally predicted by PDB (http://www.rcsb.org/pdb/home/home.do).

## 3 RESULTS AND DISCUSSION

### 3.1 *General characteristics of GIs flanked by RlmH genes in Bacillus*

Initially, eight GIs integrated into *RlmH* genes, Bce10987GI$^{RlmH}$, BtHD-789GI$^{RlmHb}$, Bcy391-98GI$^{RlmH-1}$, BtHD1002GI$^{RlmHb}$, BcFRI-35GI$^{RlmH}$, BcAH820GI$^{RlmH}$, BtAM65-52GI$^{RlmHb}$, and BcelDSM2522GI$^{RlmH}$, were determined in *B. cereus* ATCC 10987, *B. thuringiensis* HD-789, *B. cytotoxicus* NVH 391-98, *B. thuringiensis* HD1002, *B. cereus* FRI-35, *B. cereus* AH820, *B. thuringiensis* serovar israelensis strain AM65-52, and *B. cellulosilyticus* DSM 2522 (see Table 1), respectively. Among them, Bcy391-98GI$^{RlmH-1}$ was not flanked by *RlmH* gene that is localized upstream of the *RlmH* gene. Bcy391-98GI$^{RlmH-2}$ that is integrated into *RlmH* gene was recognized through interspecific alignment of the *RlmH* gene region between *B. cytotoxicus* NVH 391-98 and *B. thuringiensis* HD-789. Bcy391-98GI$^{RlmH-1}$ and Bcy391-98GI$^{RlmH-2}$ were named as the tandem GIs and collectively own one flanking sequence. We also found that BtHD-789GI$^{RlmHb}$, BtHD1002G-I$^{RlmHb}$, and BtAM65-52GI$^{RlmHb}$ contain two mobile gene clusters. One gene cluster includes one recombinase, and one resolvase is located near one flanking DR sequence, which is far away from insert site (*RlmH* gene) and another resolvase is located in the middle of three GIs$^{RlmHb}$ (see Table 1). We speculated whether one new GI$^{RlmH}$ is located inside BtGI$^{RlmHb}$. Thus, three BtGIs$^{RlmHs}$, BtHD-789GI$^{RlmHs}$, BtHD1002GI$^{RlmHs}$, and BtAM65-52GI$^{RlmHs}$, were determined through alignment of three *B. thuringiensis* chromosomes with Bcy391-98GI$^{RlmH-1}$. Three BtGI$^{RlmHb}$ and BtGI$^{RlmHs}$ are named as the nest GIs, because useful flanking DRs of BtGI$^{RlmHs}$ are shorter than those of BtGI$^{RlmHb}$. GIs$^{RlmH}$ are heterogenous

fragments, because GC% of GIs$^{RlmH}$ (i.e. about 30%) is obviously lower than that of Bacillus genomes (i.e. about 35%). Finally, 12 GIs$^{RlmH}$ were determined and some GIs$^{RlmH}$ have complex structure, named as the tandem or the nest GIs, in *Bacillus*.

*RlmH* (rRNA large subunit methyltransferase H) gene (or named *YbeA*) has the function of methylation modification at nucleotide m$^3\Psi$1915 in 23S rRNA of *Escherichia coli* (Ero et al. 2008). RlmH and homologous gene proteins are conserved in most bacterial chromosomes, such as *Haemophilus influenzae*, *S. aureus*, *Streptococcus pneumoniae*, and *Enterococcus faecalis*, but are not found in the Archaea and Eukaryota. Absence of RlmH protein only causes slight decrease in the growth rate of *E. coli* (Ero et al. 2008). A SCC*mec* element that contains a β-lactam antibiotic-resistant gene (*mecA*) was found and flanked by 3' end of *RlmH* gene in *S. aureus* that also belongs to *Bacillales* (Boundy et al. 2013). A novel SCC*fusC* element containing a fusidic acid-resistant gene (*fusC*) and a polyamine-resistant gene (*speG*) was also integrated into 3' end of *RlmH* gene and located upstream from a SCC*mec* element in *S. aureus* (Lin et al. 2014). We speculated that the SCC*fusC* and its downstream SCC*mec* element are the multiple GIs, tandem GIs, or nest GIs.

The nucleotide sequences of BtHD-789GI$^{RlmHb}$, BtHD1002GI$^{RlmHb}$, and BtAM65-52GI$^{RlmHb}$ are the same and belong to *B. thuringiensis* genomes. About 5.3kb (5190228-5195531) of Bce10987GI$^{RlmH}$ is high similar (97%) to a part of nucleotide sequences of BtGI$^{RlmHb}$ except BtGI$^{RlmHs}$ and contains about type III restriction system genes. The nucleotide sequence of Bcy391-98GI$^{RlmH-1}$ is similar to that of BcFRI-35GI$^{RlmH}$. The nucleotide sequence of BcAH820GI$^{RlmH}$ is highly similar to that of three BtGI$^{RlmHs}$. The nucleotide sequences of Bcy391-98GI$^{RlmH-2}$ and BcelDSM2522GI$^{RlmH}$ are unique through alignment with all of the complete genomes in bacteria.

### 3.2 *Mobile genes and DR analysis*

Mobile gene proteins in GIs$^{RlmH}$ annotated as the resolvase family recombinases by CDD (http://www.ncbi.nlm.nih.gov/Structure/cdd/wrpsb.cgi) were mainly classified into four types (I–IV) by MEGA7 (see Fig. 1). Types I and III or Types II and IV mobile genes are closely adjacent (see Table 1). Thus, we speculated that Type I and Type III mobile genes, as *ccrA* and *ccrB* in SCC*mec*, constitute Group-I mobile genes, and Type II and Type IV mobile genes can constitute Group-II mobile genes. Group-I and Group-II mobile genes locate near another flanking DR and far away from *RlmH* genes. This result showed that the GIs$^{RlmH}$ should be deleted from the original chromosome and

Table 1. General characteristics of GIs$^{RlmH}$ in *Bacillus*.

| Strain | GI | Range | Size | Insertion site | DR(UP\\\DOWN) | Mobile genes | GC% of GI (GC% of genome) |
|---|---|---|---|---|---|---|---|
| *B. cereus* AT CC 10987 | Bce10987 GI$^{RlmH}$ | 5185816–5195809 | 9994 | BCE_RS26845 complement (5195787..5196266) | ATTTATGATACGGTTCTCCC\\\ ACTTGTGATACGGTTCCCCAC | BCE_RS26800 5185818..5187386 BCE_RS26805 complement (5187574..5188350) | 29.87 (35.52) |
| *B. thuringiensis* HD-789 | BtHD-789 GI$^{RlmHb}$ | 4987525–5009137 | 21613 | BTF1_25905 complement (5009115..5009594) | ATTTATGATACAGTTCTCCC\\\ ACTTGTGATACGGTTCCCCAC | BTF1_25795 4987527..498095 a new gene 4989286..4990059 | 29.93 (35.18) |
| | BtHD-789 GI$^{RlmHs}$ | 4997486–5009132 | 11647 | | TTTACTATCACTTGTGATACGGTTC\\\ TTTACCATCACTTGTGATACGGTTC | BTF1_25850 4997455..4999044 BTF1_25860 complement (4999672..5000550) | 30.08 (35.18) |
| *B.cytotoxicus* NVH 391-98 | Bcy391-98 GI$^{RlmH-1}$ | 4036530–4045502 | 8973 | | TTGTGATACGGTTCTCC\\\ TTGTGATACGGTTCACC | BCER98_RS20005 complement(4036494..4038097) BCER98_RS20010 complement (4038285..4039061) | 30.13 (35.89) |
| | Bcy391-98 GI$^{RlmH-2}$ | 4045484–4058128 | 12645 | BCER98_RS20115 complement (4058108..4058635) | ACTTGTGATACGGTTCACC\\\ ACTTATGGTAAGGTTCCCC | BCER98_RS20080 complement (4051760..4052347) BCER98_RS20110 (4056571..4058136) | 30.90 (35.89) |
| *B. thuringiensis* HD1002 | BtHD1002 GI$^{RlmHs}$ | 3896936–3908582 | 11647 | AS86_RS25740 3896474..3896953 | GAACCGTATCACAAGTGATGG TAAA\\\GAACCGTATCACAAGTGA TAGTAAA | AS86_RS25785 3905494..3906396 AS86_RS25795 3907024..3908613 | 30.08 (35.07) |
| | BtHD1002 GI$^{RlmHb}$ | 3896931–3918543 | 21613 | | GTGGGGAACCGTATCACAAGT\\\ GGGGAGAACTGTATCATAAAT | AS86_RS25835 3916009..3916785 AS86_RS25840 3916973..3918541 | 29.93 (35.07) |
| *B. cereus* FRI-35 | BcFRI-35 GI$^{RlmH}$ | 1609324–1619073 | 9750 | BCK_07995 1608867..1609346 | GTGGGGAACCGTATCACAAGT GATA\\\GGGGAGAACCGTATCAT AAATAGTA | BCK_08035 (1616537..1617313) BCK_08045 (1617499..1619067) | 30.09 (35.45) |

*(Continued)*

265

Table 1. (*Continued*)

| | | | | | | | |
|---|---|---|---|---|---|---|---|
| *B. cereus* AH820 | BcAH820 GI^RlmH | 11542 | 5262651–5274192 | BCAH820_5565 complement (5274175..5274678) | TTACTTATGATACGGTTC\\\ TCATTTGTGATACGGTTC | BCAH820_RS27595 complement (5262655...5264202) BCAH820_RS27605 complement (5264889..5265707) | 30.19 (35.31) |
| *B. thuringiensis* serovar israelensis strain AM65-52 | BtAM65-52 GI^RlmHb | 21613 | 5449640–5471252 | ATN07_28320 complement (5471230..5471709) | ATTTATGATACAGTTCTCCC\\\ ACTTGTGATACGGTTCCCCAC | ATN07_28220 complement (5449642..5451210) ATN07_28225 complement (5451398..5452174) | 29.93 (35.27) |
| | BtAM65-52 GI^RlmHs | 11647 | 5459601–5471247 | | TTTACTATCACTTGTGATACGGTTC\\\ TTTACCATCACTTGTGATACGGTTC | ATN07_28265 complement (5459570..5461159) ATN07_28275 complement (5461787..5462689) | 30.08 (35.27) |
| *B. cellulosilyticus* DSM 2522 | BcelDSM25 22GI^RlmH | 31338 | 4614898–4646235 | BCELL_RS21200 complement (4646215..4646694) | CATCACTTATGATACGGTTCTCC\\\ CATCACTTGTGATACGGTTCACC | BCELL_RS21045 complement (4614904..4616481) BCELL_RS21165 4637409.. 4638284 | 30.37 (36.52) |

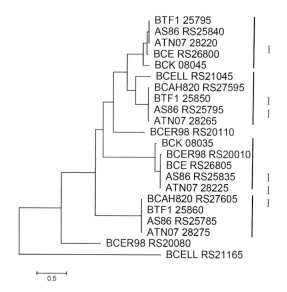

Figure 1. Phylogenetic analysis of mobile gene proteins in GIs^RlmH of *Bacillus* using MEGA7 software.

reintegrated into related chromosomes by Group I or II mobile genes. However, 5' end of nucleotide sequence of Type I and Type II mobile genes and one flanking sequence of DRs are overlapped. Type I and Type II mobile genes need be verified whether they are active. BTF1_25795 (Type I mobile gene) is not matched with one Type III mobile gene and BCER98_RS20010 (Type III mobile gene) is not matched with one Type I mobile gene. A new gene (Type III mobile gene), which is highly similar to AS86_RS25835, was found in 4989286–4990059 bp of *B. thuringiensis* HD-789 chromosome, and its initiation code is GUG (representing valine) through alignment with AS86_RS25835. BCER98_RS20005 is highly similar to BTF1_25795, but the absence of one base (A) (near 4036892 bp) of BCER98_RS20005 promotes it as a pseudogene. BCELL_RS21045 is highly similar to Type II mobile genes, but its matched gene, which should be highly similar to IV-type mobile genes, was not found. BCELL_RS21165 is far away from all of mobile gene proteins through MEGA analysis. This showed that BCELL_RS21165 could not play a role in mobility of BcelDSM2522GI^RlmH. BCER98_RS20080 and BCER98_RS20110, which are also far away from mobile gene, proteins need to be verified whether they can together work and finish deletion and reintegration of Bcy391-98GI^RlmH-2.

The DRs of Bce10987GI^RlmH, BcFRI-35GI^RlmH, Bcy391–98GI^RlmH-1, BtHD-789GI^RlmHb, BtHD-1002GI^RlmHb, and BtAM65–52GI^RlmHb, which contain Group I mobile genes, were aligned by ClustalW (see Fig. 2a) and were named as Group I

DRs. Meanwhile, the DRs of BcAH820GI[RlmH], BcelDSM2522GI[RlmH], BtHD-789GI[RlmHs], BtH-D1002GI[RlmHs], and BtAM65-52GI[RlmHs], which contain Group II mobile genes, were aligned by ClustalW (See Fig. 2b) and were named as Group II DRs. Results showed that Group I and II DRs are highly similar. We speculated that if Type I and II mobile genes participate in recognizing the binding site in the DRs; meanwhile, Type III and IV mobile genes participate in recognizing the cutting site in the DRs, the cutting sites of Group I and II mobile genes will be GG dinucleotide and 5'GAAC3', respectively, and their binding site is 5'GTATCA3'. Another hypothesis is that if Type I and II mobile genes participate in recognizing the cutting site in the DRs, their cutting site and binding site will be 5'GAAC3' and 5'GTATCA3', respectively.

Type I–IV mobile gene proteins were structurally similar to TP901-1 integrase, which was first found as a resolvase-like protein in Bacteriophage TP901-1 and furthermore was called as an LSR (Yuan et al. 2008). The DRs of Bacteriophage TP901-1 include a core (5'TCAAT3') and an identical region (5'AAGGTAA') (Stoll et al. 2002). TC dinucleotide plays an important role in the deletion and reintegration of Bacteriophage TP901-1 (Breüner et al. 2001). TP901-1 integrase is a potential tool for directed rebuilding eukaryotic genomes (Stoll et al. 2002).

```
HD1002b-up      GTGGGGAACCGTATCACAAGT----
FRI-35-up       GTGGGGAACCGTATCACAAGTGATA
HD-789b-down    GTGGGGAACCGTATCACAAGT----
10987-down      GTGGGGAACCGTATCACAAGT----
AM65-52b-down   GTGGGGAACCGTATCACAAGT----
NVH391-98-1-down  --GGTGAACCGTATCACAA------
NVH391-98-1-up    --GGAGAACCGTATCACAA------
HD-789b-up      GGGGAGAACTGTATCATAAAT----
HD1002b-down    GGGGAGAACTGTATCATAAAT----
AM65-52b-up     GGGGAGAACTGTATCATAAAT----
FRI-35-down     GGGGAGAACCGTATCATAAATAGTA
10987-up        GGGGAGAACCGTATCATAAAT----
                ** **** ****** **
```

Figure 2a. Alignment of DRs of the GIs[RlmH] containing Group I mobile genes.

```
HD-789s-down    ---GAACCGTATCACAAGTGATGGTAAA
HD1002s-up      ---GAACCGTATCACAAGTGATGGTAAA
AM65-52s-down   ---GAACCGTATCACAAGTGATGGTAAA
AM65-52s-up     ---GAACCGTATCACAAGTGATAGTAAA
HD-789s-up      ---GAACCGTATCACAAGTGATAGTAAA
HD1002s-down    ---GAACCGTATCACAAGTGATAGTAAA
AH820-down      ---GAACCGTATCACAAATGA-------
DSM2522-up      GGAGAACCGTATCATAAGTGATG-----
AH820-up        ---GAACCGTATCATAAGTAA-------
DSM2522-down    GGTGAACCGTATCACAAGTGATG-----
                *********** ** * *
```

Figure 2b. Alignment of DRs of the GIs[RlmH] containing Group II mobile genes.

We found that the DRs of some GIs[RlmH] also include a core (5'GAAC3') and an identical region (5'GTATCA3'). We speculated that 5'GAAC3' contains the cutting site of the mobile gene proteins in some GIs[RlmH]. Meanwhile, the identical region (5'GTATCA3') of flanking DRs should be suggested as a binding region of the mobile gene proteins in some GIs[RlmH]. Type I–IV mobile gene proteins, as the TP901-1 integrase homologs, are also possible to complete the rearrangement of eukaryotic genomes.

ACKNOWLEDGMENT

This study was funded by grants from Shanghai university students' innovation program (No. 201610270162), China.

REFERENCES

Bellanger, X. et al. Conjugative and mobilizable genomic islands in bacteria: evolution and diversity. FEMS Microbiol Rev. 2014, 38(4):720–760.

Boundy, S. et al. Characterization of the *Staphylococcus aureus* rRNA methyltransferase encoded by *orfX*, the gene containingthe staphylococcal chromosome Cassette *mec* (SCC*mec*) insertion site. J Biol Chem. 2013, 288(1): 132–140.

Breüner, A. et al. Resolvase-like recombination performed by the TP901–1 integrase. Microbiology. 2001, 147(Pt 8): 2051–2063.

Christiansen, B. et al. A resolvase-like protein is required for the site-specific integration of the temperate lactococcal bacteriophage TP901–1. J Bacteriol. 1996, 178(17): 5164–5173.

Dobrindt, U. et al. Genomic islands in pathogenic and environmental microorganisms. Nat Rev Microbiol. 2004, 2(5):414–424.

Ero, R. et al. Identification of pseudouridine methyltransferase in *Escherichia coli*. RNA. 2008, 14(10):2223–2233.

Katayama, Y. et al. A new class of genetic element, staphylococcus cassette chromosome *mec*, encodes methicillin resistance in *Staphylococcus aureus*. Antimicrob Agents Chemother. 2000, 44(6): 1549–1555.

Kumar, S. et al. MEGA7: Molecular Evolutionary Genetics Analysis Version 7.0 for Bigger Datasets. Mol Biol Evol. 2016, 33(7): 1870–1874.

Lin, Y. T. et al. A novel staphylococcal cassette chromosomal element, SCC*fusC*, carrying *fusC* and *speG* in fusidic acid-resistant methicillin-resistant *Staphylococcus aureus*. Antimicrob Agents Chemother. 2014, 58(2): 1224–1227.

Stoll, S.M. et al. Phage TP901-1 site-specific integrase functions in human cells. J Bacteriol. 2002, 184(13): 3657–3663.

Yuan, P. et al. Tetrameric structure of a serine integrase catalytic domain. Structure. 2008, 16(8): 1275–1286.

Wang, L. et al. Roles of CcrA and CcrB in excision and integration of staphylococcal cassette chromosome *mec*, a *Staphylococcus aureus* genomic island. J Bacteriol. 2010, 192(12): 3204–3212.

*Advances in Energy and Environment Research – Achour & Wu (Eds)*
© *2017 Taylor & Francis Group, London, ISBN 978-1-138-62682-9*

# Study of the preparation and application of modified luffa sponges in the pollutant removal process of the eco-pond-ditch system

L.W. Kong & R.W. Mei
*Environmental Research and Design Institute of Zhejiang Province, Hangzhou, China*

S.P. Cheng
*Key Laboratory of Yangtze River Water Environment, Ministry of Education, Tongji Universty, China*

Y. Zhang, D. Xu & L. Wang
*Wuhan Zexi Pro-Creation Environmental Protection Technology Co. Ltd., Wuhan, China*

X. Cui
*Geographical Environment and Science Institute of Zhejiang Normal University, Jinhua, China*

ABSTRACT: In this paper, we study the scientific preparation and application of the modified Luffa Sponge (LS) as carrier in wastewater treatment process, and the effects of surface modification and removal of pollutant were investigated. Results showed that the hydrophobic modified LS fiber after $Fe^{3+}$ and ozone (in the air) treatment could achieve a contact angle of 113.4°, and it was a highly suitable filler due to its high microbial biofilm-forming ability and prolonged service life. Scanning Electron Microscope (SEM) images also showed that the modified LS fiber sacrifices surface area to some extent for a stable and not easily degraded structure. The modified LS fiber showed high performance in the removal of $COD_{Cr}$, but low performance in the removal of TN and TP. Irrespective of the modification state, the LS fiber in a ditch system has a significant effect on its treatment.

## 1 INTRODUCTION

At present, water treatment filler, which is the carrier material for biochemical treatment, has been playing an important role in the modern water treatment process. The preparation and application of different fillers improved the water treatment technology to meet various needs (Wu et al. 2015, Feng et al. 2015). Many types of fillers, including elastic filler, biological filler, fiber packing, and new types of packing, have been invented according to the different needs of wastewater treatment worldwide since the 1980s (Wu et al. 2016, Wang et al. 2015). However, disadvantages due to the characteristics of various fillers, such as easy plugging of the rigid filler, short operation cycle time of the soft packing filler, high cost, and difficulty in the formation of biofilm for semi-soft packing also exist. Therefore, regular filler is insufficient to meet the current water treatment technology. Because of the limited availability of resources and environmental pollution, development of fillers with characteristics of environmental friendliness, high performance, and low price is of paramount importance.

In this research, LS fiber is chosen as the carrier filler because of its unique porous physical structure, high mechanical strength, conducive to biofilm

formation, and most importantly, low cost, thereby widening its range of water treatment application prospects. Many studies (Shen et al. 2012, An et al. 2016) show that the LS material outperforms a variety of traditional engineering materials. However, the surface of natural LS fiber is rich in hydroxyl, which increases its water absorption rate. Liu et al. (2016) enhanced the hydrophilic property of LS fiber to obtain high water absorption ability. They prepared LS by grafting polymerization of acrylamide (AM) on luffa cylindrica and subsequent partial hydrolysis under alkaline conditions. Results reveal that the hydrolyzed LS particularly exhibit high water absorption capacities of 75 g/g. On the contrary, the hydrophilic idea limits its development and application in another field like wastewater treatment.

Therefore, a variety of modification methods should be used to reduce the hydroxyl on surface of LS, improve fiber resistance to microbial corrosion, and prolong the service life.

The ecological pond-ditch system (eco-pond-ditch system) has been proved effective in intercepting nitrogen and phosphorus pollutants of non-point source pollution, thereby increasing its research attention and promotion. However, there are many problems in the application and promotion of eco-pond-ditch system. One of these problems is the submerged and emergent aquatic plant system, which is easily affected by seasonal factors, and its efficiency is very different between winter and plant-growing seasons. In summary, modified LS fiber, which less affected by seasonal factors, could be an alternative to traditional aquatic plants, especially in strengthening the decontamination effect of the system.

Therefore, this paper reveals a method to modify LS to prolong its service life, conduct surface cationization treatment and ozone treatment to enhance microbial biofilm formation, and test the modified LS in the eco-pond-ditch system treating micro-polluted water.

## 2 MATERIALS AND METHODS

### 2.1 Setup of the bench scale experiment

The direction of the wastewater flow is shown in Fig. 1. All the tank, pond, and ditches systems are plastic materials, with dimensions of 0.5 m*0.4 m *0.25 m, 0.6 m*0.45 m*0.3 m, and 0.4 m*0.3 m *0.2 m, respectively. The collecting tank is filled with wastewaters from enterprise park life sewage; eco-pond and three ditches are loaded with mud from farmland and have 1/5 tank capacity; eco-pond plants: submerged plant *Myriophyllum* and emergent aquatic plant *Thalia*; Ditch 1# system: only plants *Calamus*; Ditch 2#: only fixed modified LS (suspension installation); and Ditch 3#: plants *Calamus* coupled with modified LS. The density of planting and modified LS installation is nine plants for per square meter. The influent characteristics of $COD_{Cr}$, TN, and TP are shown in Table 1.

### 2.2 Modification of LS fiber process

The raw LS fiber is cut into several pieces. As the key part of modification, the method of preparation of hydrophobic LS is in accordance with the patent application, thus the subsequent of this paper is based on the finished products. As shown in Table 2, six blocks of raw LS fiber (1#) (blank group),

Table 1. Influent characteristics of COD.

| Item | Mean (mg/L) |
| --- | --- |
| $COD_{Cr}$ | $29.35 \pm 24.15$ |
| TN | $2.19 \pm 0.83$ |
| TP | $0.22 \pm 0.15$ |

Table 2. Material modification process and test arrangement.

| Items | Ozone (Air) | $Fe^{3+}$ + Ozone (Air) | Ozone + $Fe^{3+}$ (Solution) | Tests items | Tips |
| --- | --- | --- | --- | --- | --- |
| 1#-Raw material (6 blocks) | 0# | 0# | 0# | FTIR, CA, SEM | Blank |
| 2#-Raw material (18 blocks) | 1#-a | 1#-b | 1#-c | FTIR, CA, SEM | Ozone method |
| 3#-Hydrophobic (18 blocks) | 2# | 3# | 4# | FTIR, CA, SEM | Ozone and $Fe^{3+}$ |

*Ozone concentration = 16 mg/L, while gas output = 3 SCCM, disposal time = 20 min.

18 blocks of raw LS fiber (2#), and 18 blocks of hydrophobic modified LS fiber (3#) are arranged to carry out three different series of experiments: ozone treatment (in the air), $Fe^{3+}$+ozone treatment (in the air), and ozone+ $Fe^{3+}$ treatment (in solution). Temperature of $Fe^{3+}$ bath treatment is set to 45°C in 1 h, and the parameters for ozone treatment are: ozone concentration = 16 mg/L, gas output = 3 SCCM, and treatment time = 20 min.

### 2.3 Experiment of biofilm formation on the materials

Nine different types of materials are accordingly arranged to conduct microbial biofilm test, including the raw LS fiber (0#), 1#-a, 1#-b, 1#-c, 2#, 3#, 4#, 5#, and 6#. The 5# material represents suspended filler, while the 6# material is composite packing, and both of them are very commonly used filler in wastewater treatment field. These fillers are immersed in an aerobic tank of a pilot scale sewage treatment plant for 5 and 16 days, each time they are placed in an 80 °C oven for drying to a constant weight at regular intervals. The material performance is determined by increasing the weight of the microbial biofilm. Specific working conditions and parameters of aerobic tank of the pilot scale plant are as follows: settling performance of activated sludge value, SV < 10%, pH = 7.69, and DO = 4.57 mg/L. The aerobic tank is shown in Fig. 2.

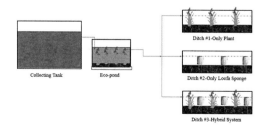

Figure 1. Bench scale experiment of three eco-pond-ditch coupling systems.

270

Figure 2. Aerobic tank of pilot scale wastewater treatment plant.

## 2.4 Method and equipment for material test

The parameters of the method are: $COD_{Cr}$, microwave digestion titration; $NH_4^+$-N, Nessler's reagent colorimetric method (GB7479–1987); TN, alkaline potassium persulfate digestion UV spectrophotometric method (GB11894–1989); $NO_3^-$-N, ultraviolet spectrophotometry (HJ/T346–2007); $NO_2^-$-N, spectrophotometry (GB7493–1987); and TP, ammonium molybdate spectrophotometric method (GB11893–1989). Software origin 9.0 is used for the gram and data analysis, and the main methods of measurement were the industry standard method and the national standard method. The Fourier Transform Infrared (FTIR) spectrometer, Scanning Electron Microscope (SEM), and Contact Angle (CA) tests are conducted in Key Laboratory of Organosilicon Chemistry and Material Technology, Hangzhou Normal University. Changes on the surface of the material were observed by FTIR (Nexus Fourier Transform Infrared Spectrometer, Thermo Nicolet Co, USA), while changes in the material morphology were observed by SEM (Field Emission Scanning Electron Microscope S-4800, Hitachi Co, Japan).

## 3 RESULTS AND DISCUSSION

### 3.1 Characteristic of the prepared hydrophobic LS fiber

Most of the studies on LS are mainly focused on structural and mechanical properties (Shen et al. 2013, Chen et al. 2014, An et al. 2016), cell immobilization (Hideno et al. 2007, Behera et al. 2011, Saratale et al. 2011), absorption properties, and so on, while only few studies have been conducted on the surface hydrophobic modification of LS fiber. Therefore, this paper is an important report for revealing one of the hydrophobic methods and modification effect of the LS fiber.

Fig. 3 (a) shows the image of a raw LS fiber, Fig. 3 (b)–Fig. 3 (e) show raw LS fibers before different hydrophobic modification treatment, and Fig. 3 (f)–Fig. 3 (k) show the dripping test effects with or without hydrophobic modification materials. It can be seen from Fig. 3 (h)–Fig. 3 (k) that the hydrophobic effect of LS is better than that of Fig. 3 (f), as the raw LS material has lower contact angle and higher surface tension. Meanwhile, Fig. 3 also shows that different methods have different hydrophobic effects with the same dose of water drops.

According to the experimental setup, the contact angles of different materials are found to obtain precise data. Fig. 4 (a) shows raw LS (0#), Fig. 4 (b) shows raw LS with $Fe^{3+}$ + ozone treatment (1#-b), and Fig. 4 (c) shows hydrophobic LS with $Fe^{3+}$ + ozone treatment (3#). In this paper, the contact angles of 1#-b and 3# material are 92.05° and 113.4°, respectively, while the 0# material has no result because of the lack of measuring method. Chen (Chen et al. 2011) found that LS fiber has a contact angle of 51.2°, thus 3# material accordingly has the best hydrophobicity in this paper. It is very strange that 1#-b material also has a contact angle above 90°, as this material has not been hydrophobic modified. The only explanation could be the $Fe^{3+}$ + ozone treatment process has improved the roughness of the LS fiber surface, which, as one of the important conditions, can increase the hydrophobicity of the solid material (Wenzel 1949). The specific mechanism on $Fe^{3+}$ and ozone to increase the surface hydrophobicity needs further study.

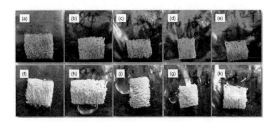

Figure 3. Dripping test of LS fibers before and after hydrophobic modification.

Figure 4. Contact angles of raw materials and modified materials.

### 3.2 Preparation of modified LS fibers and biofilm test

Fig. 5 (a) shows the preparation of $Fe^{3+}$ solution, while Fig. 5 (d) indicates raw LS fibers in the bath treatment process. Fig. 5 (b) and Fig. 5 (e) show that LS fibers are being ozone-treated in the air and in solution environment, respectively. Fig. 5 (c) and Fig. 5 (f) show the nine materials before and after being immersed in aerobic tank for testing biofilm formation, which clearly shows the differences in color. Table 3 shows the contrast test between LS fiber before and after modified and other fillers in terms of microbial biofilm formation. It can be found from Table 3 that the ability of biofilm formation on the surface of raw LS fiber decreases after ozone treatment. The weight decreases from 7.63 g/g (0#) to 5.85 g/g for 1#-a material and to 5.82 g/g for 2# material. It may be related to the increase of surface roughness of the LS fiber, caused by the high-energy physical method, which can enhance hydrophobicity.

Table 3 also shows that raw LS fiber and composite packing exhibit better microbial biofilm forming performance. However, the raw LS fiber is easier to be degraded and has a short service life, while the composite packing can be easily hardened, heavy loaded, and needs to be installed scaffold artificially. In the meantime, 3# modified LS fibers showed higher increase in the weight of biofilm than that of 2# and 4#, and has merits of simple installation and no secondary pollution. Although 1#-b and 1#-c modified LS fibers show better performance in the formation of microbial biofilm, they are not good at long-term operation as engineering filler in wastewater treatment field, because of their enhanced hydrophilic properties. Therefore, 3# modified LS fiber is a highly suitable filler due to its high microbial biofilm-forming ability and prolonged service life.

### 3.3 SEM characteristic of the modified LS fibers

Fig. 6 shows the SEM images of 0#, 1#-b, and 3# materials, and it is evident that changes are obvious. The roughness of fiber surface shown in Fig. 6(a) is very similar to that observed in Fig. 6(b). However, the difference between them is that the former is dark, while the latter is bright. This can be explained by the preparation process. According to Table 3, 0# is raw LS fiber, while 1#-b is raw LS fiber modified by $Fe^{3+}$ and ozone process material. The brightness could be due to the physical and chemical reactions taking place on the surface of the LS fiber.

Fig. 6(c) shows significant difference from Fig. 6(a) and Fig. 6(b). It seems very smooth than that of 0# and 1#-b for a hydrophobic "membrane" forming on surface of 3# modified LS fiber caused by formerly modification. With the enhancement of ozone treatment, the hydrophobic property improves, its surface energy decreases much, and micro- or nanometer-leveled roughness becomes significant. Besides, the gaps between

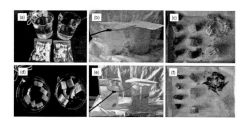

Figure 5. Preparation of modified LS fibers and biofilm test of nine materials.

Figure 6. SEM images of 0# (a), 1#-b (b), and 3# (c) material (20.0 kv, X 30, 1 mm).

Table 3. Biofilm contrast test of between LS fiber before and after modification and other fillers (g/g).

| | Blank | Raw LS fibers | | | LS fibers after hydrophobic modification | | | Raw other materials | |
| | | Ozone | $Fe^{3+}$ & Ozone (Air) | Ozone & $Fe^{3+}$ (Solution) | Ozone | $Fe^{3+}$ & Ozone (Air) | Ozone & $Fe^{3+}$ (Solution) | Suspended filler | Composite packing |
| | — | | | | | | | | |
| Item | 0# | 1-a# | 1-b# | 1-c# | 2# | 3# | 4# | 5# | 6# |
|---|---|---|---|---|---|---|---|---|---|
| Original weight | 0.19 | 0.25 | 0.32 | 0.17 | 2.73 | 3.09 | 2.17 | 3.66 | 2.57 |
| After 5 days | 0.20 | 0.26 | 0.33 | 0.18 | 2.75 | 3.19 | 2.22 | 3.69 | 2.82 |
| After 16 days | 0.22 | 0.28 | 0.37 | 0.21 | 3.05 | 3.48 | 2.31 | 3.69 | 3.30 |
| Average weight gain | 7.63 | 5.85 | 7.62 | 11.27 | 5.82 | 6.16 | 3.18 | 0.41 | 13.37 |

the tiny fiber structures decrease, and surface area reduce consequently. Surface area data tested by BJH method show that the cumulative desorption surface areas of 0#, 1#-b, and 3# fibers are 2.885E+00, 3.012E+00, and 3.737E-01 $m^2/g$, respectively. Therefore, 3# modified LS fiber sacrifices the surface area to some extent for a stable and not easily degraded structure. This is the reason for the degraded performance of 3# modified LS fiber compared with 0# and 1#-b in forming biofilm.

### 3.4  FTIR characteristic of the modified LS fibers

Because of ozone modification, the surface of the LS fiber becomes rougher, and the specific surface area increases, but the specific changes in the surface have a direct effect on the water absorption properties of the materials. Therefore, infrared spectrum analysis is needed to study the reaction mechanisms and changes in the surface groups.

Fig. 7 shows the infrared spectra of 0#, 1#-b, and 3#. As seen in the graph, the presence of absorption peaks at 3301 and 3333 $cm^{-1}$ indicates the occurrence of O–H bond stretching. The emergence of strong absorption peaks at 2918 and 2962 $cm^{-1}$ accounts for the asymmetric stretching of the C–H bond. The peak absorption at 1651 $cm^{-1}$ indicates C = C bond stretching vibration, and

the peak absorption at 1258 $cm^{-1}$ indicates C–H bond bending. However, some of the characteristic peaks disappear or weak between 690 and 843 $cm^{-1}$, and the emergence of a new absorption peak at 789 $cm^{-1}$ evidences the bending of the = C–H bond. Changes in the peaks discussed above prove that $Fe^{3+}$, ozone, and hydrophobic modification methods have produced significant changes on the surface of the LS fiber.

### 3.5  Bench scale application of modified LS fibers in eco-pond-ditch system

Figs 8–10 show the changes of $COD_{Cr}$, TN, and TP concentrations after being treated by the eco-pond-ditch coupling system, respectively. Sets 1–5 represent collecting tank, eco-pond, ditch 1#, ditch 2#, and ditch 3#, respectively. It is evident from these figures that the coupling system has obvious effect on the removal of organic pollutant and N, P nutrients. The eco-pond system accounts for most of the removal rate, while the eco-ditch system contributes to achieve further purification effect.

The modified LS fiber shows great performance in the removal of $COD_{Cr}$. For the modified LS fibers applied in ditch 2# and ditch 3#, the removal effect is very clear when in contrast to ditch 1#, and there appears a descending trend from ditch 1# to ditch 3#. However, it does not perform remarkably

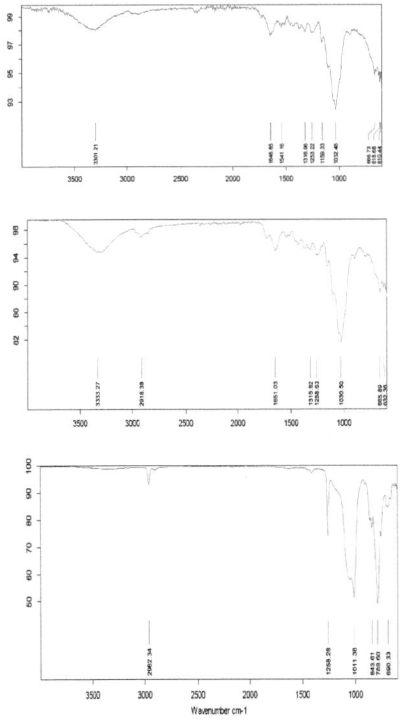

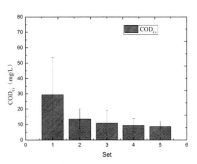

Figure 8.  Removal of $COD_{Cr}$ by the eco-pond-ditch system.

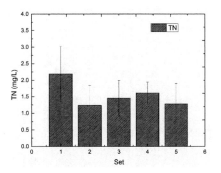

Figure 9.  Removal of TN by the eco-pond-ditch system.

Figure 7.  FTIR images of 0# (a), 1#-b (b), and 3# (c) material.

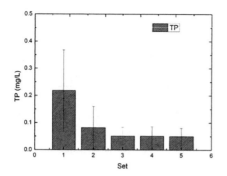

Figure 10. Removal of TP by the eco-pond-ditch system.

Table 4. Concentration of $COD_{Cr}$, TN, and TP after treatment by ditch systems (mg/L).

| Item | Ditch 1# | Ditch 2# | Ditch 3# |
|------|----------|----------|----------|
| $COD_{Cr}$ | $10.94 \pm 8.30$ | $9.47 \pm 4.25$ | $8.69 \pm 3.68$ |
| TN | $1.45 \pm 0.54$ | $1.61 \pm 0.34$ | $1.28 \pm 0.62$ |
| TP | $0.051 \pm 0.033$ | $0.051 \pm 0.036$ | $0.050 \pm 0.031$ |

in removal of TN and TP. Overall, with or without the modified LS fiber in ditch system has great influence on its treatment effect.

The mechanisms of the eco-ditch system mainly depend on two aspects: one is for the growth and metabolism of submerged and emergent plant, which has significant effect on the removal of pollutant; another is the effect of modified LS fiber as carrier for microbial growth and decontamination. However, which plays a more important role in terms of pollutant removal is uncertain, and remains to be further studied.

## 4 CONCLUSIONS

1. The hydrophobic modified LS fiber after $Fe^{3+}$ and ozone (in the air) treatment achieves a contact angle of 113.4°, and could prolong service life. Thus, it is suitable as a carrier for microbial biofilm growth.
2. SEM graphs show that the modified LS fiber has less surface area but a stable structure to effectively resist degradation.
3. The modified LS fiber performs well in $COD_{Cr}$ removal, but not in the removal of TN and TP. Irrespective of modification state, the LS fiber in ditch system has significant effect on its treatment.

## ACKNOWLEDGMENTS

This work was supported by Major Science and Technology Program for Water Pollution Control and Treatment of China 12th Five-Year Plan (No.2014ZX07101–012–03), Science and technology project of Zhejiang Province (2014F50025) and Environmental Protection Program of Hubei Province (2015HB14). We thank Mr. Xin Yao, Ms. Jie Gao and Lu Wang for their assistance during the work.

## REFERENCES

An X.Y., Fan H.L. 2016. Hybrid design and energy absorption of luffa-sponge-like hierarchical cellular structures. *Materials & Design*, 106:247–257.

Behera S., Mohanty R.C., Ray R.C. 2011. Ethanol production from mahula (Madhuca latifolia L.) flowers with immobilized cells of Saccharomyces cerevisiae in Luffa cylindrica L. sponge discs. *Applied Energy*, 88 (1):212–215.

Chen Q., Shi Q., Gorb S.N., Li Z.Y. 2014. A multiscale study on the structural and mechanical properties of the luffa sponge from Luffa cylindrica plant. *Journal of Biomechanics*, 47(6):1332–1339.

Chen M., Wei L., Wang Y., Zhao J.X. 2011. Performance Study about Film of Mercerizing Reguar and Microcrystalline Cellulose in the Ionic Liquids. *Journal of Anhui Agricultural Sciences*, 39(16):9486–9487. (In Chinese).

Feng L.J., Yang G.F., Zhu L., Xiangyang Xu, Gao F., Mu J., Xu Y.M. 2015. Enhancement removal of endocrine-disrupting pesticides and nitrogen removal in a biofilm reactor coupling of biodegradable Phragmites communis and elastic filler for polluted source water treatment. *Bioresource Technology*, 187:331–337.

Hideno A., Ogbonna J.C., Aoyagi H., Tanaka H. 2007. Acetylation of loofa (Luffa cylindrica) sponge as immobilization carrier for bioprocesses involving cellulose. *Journal of Bioscience and Bioengineering*, 103(4):311–317.

Liu Z., Pan Y.X., Shi K., Wang W.C., Peng C., Li W., Sha D., Wang Z., Ji X.L. 2016. Preparation of hydrophilic luffa sponges and their water absorption performance. *Carbohydrate Polymers*, 147:178–187.

Shen J.H., Xie Y.M., Huang X.D., Zhou S.W., Ruan D. 2012. Mechanical properties of luffa sponge. *Journal of the Mechanical Behavior of Biomedical Materials*, 15:141–152.

Shen J.H., Xie, Y.M. Huang X.D., Zhou S.W., Ruan D. 2013. Behaviour of luffa sponge material under dynamic loading. *International Journal of Impact Engineering*, 57:17–26.

Saratale G.D., Saratale R.G., Chang J.S., Govindwar S.P. 2011. Fixed-bed decolorization of Reactive Blue 172 by Proteus vulgaris NCIM-2027 immobilized on Luffa cylindrica sponge. *International Biodeterioration & Biodegradation*, 65(3):494–503.

Wang J.H., Ji C.T., Yan Y.T., Zhao D., Shi L.Y. 2015. Mechanical and ceramifiable properties of silicone rubber filled with different inorganic fillers. *Polymer Degradation and Stability*, 121:149–156.

Wenzel R.N. 1949. Surface roughness and contact angle. *J Phys Colloid Chem*, 53:1466–1467.

Wu S.Q., Qi Y.F., Fan C.Z., He S.B., Dai B.B., Huang J.C., Zhou W.L., Gao L. 2016. Application of novel catalytic-ceramic-filler in a coupled system for long-chain dicarboxylic acids manufacturing wastewater treatment. *Chemosphere*, 144:2454–2461.

Wu S.Q., Qi Y.F., Yue Q.Y., Gao B.Y, Gao Y., Fan C.Z., He S.B. 2015. Preparation of ceramic filler from reusing sewage sludge and application in biological aerated filter for soy protein secondary wastewater treatment. *Journal of Hazardous Materials*, 283(11):608–616.

*Advances in Energy and Environment Research – Achour & Wu (Eds)*
© *2017 Taylor & Francis Group, London, ISBN 978-1-138-62682-9*

# Spatio-temporal distribution of PM$_{2.5}$ concentration and its correlation with environmental factors in different activity spaces of an urban park

Huan Xu, Jianhui Hao, Hong Li & Jingjing Zhao

*Department of Landscape Architecture, City and Environment College, Jiangsu Normal University, China*

ABSTRACT: This study aims to investigate spatio-temporal distribution of PM$_{2.5}$ concentration in different activity spaces of urban park and analyze the correlation between PM$_{2.5}$ concentration and main environmental factors. Here, six typical activity spaces of Yunlong Park are taken as the research object, and PM$_{2.5}$ concentration and environmental factors of each space were monitored simultaneously under mild-to-moderate air pollutions. The results show that the difference between the maximum and minimum values in the same time is 1.35–1.65 times, and the maximum difference between the different time periods is 1.77 times in one day. Moreover, the PM$_{2.5}$ concentration of different activity spaces has different tendency of average daily variation. Open Plaza and Great Lawn have waveform trends, the minimum value appears at 16:00; Avenue and Wisteria Pergola have stepwise rising trends, the minimum value appears at 11:00; and Waterfront Platform and Waterfront Pavilion have M-shaped trends, the minimum value appears at 13:00. The highest values of PM$_{2.5}$ concentration at all the monitoring sites appear around 19:00. The PM$_{2.5}$ concentration has negative correlation with temperature and wind velocity, and positive correlation with relative humidity. Experimental results can be used for guiding citizens to select reasonable time and place for activities, as well as provide reference and basis for layout planning of activity space in urban parks.

## 1 INTRODUCTION

With the increase of urbanization, cities in Middle and Eastern China have been severely polluted. The haze weather characterized by fine particulate matter (PM$_{2.5}$) have accounted for more than one-third of the total number of days in each year, and becomes the main factor to endanger human health and atmospheric environmental quality. PM$_{2.5}$ can suspend in the air for a long time, which has small particle size, contain plenty of poisonous and harmful substances and can be transmitted over a long distance. The long residence time in the atmosphere with PM$_{2.5}$ pollution can lead to a variety of respiratory disease and cardiovascular disease, and even increase people's risk of premature death (Pope et al. 2002, Brook et al. 2004, Johnson & Graham 2005). Urban park, consisting of many recreational activity spaces, is the most frequently used green space of urban residents, and plays an important role in mitigating PM$_{2.5}$ pollution and meeting urban residents' leisure and fitness. Among PM$_{2.5}$-based pollution weather, mild-to-moderately polluted weather (150 μg/m$^3$ > PM2.5 > 75 μg/m$^3$) accounts for ≥ 70%. Besides, urban parks do not play significant role in reducing PM$_{2.5}$ under heavy air pollutions (PM$_{2.5}$ > 150 μg/m$^3$) (Wang et al. 2014a, Xiao et al. 2015), and urban residents should minimize outdoor activities. Therefore, it is necessary to investigate which activity space has lower PM$_{2.5}$ concentration, and determine what type of activity space is most suitable at what time for activities under mild-to-moderate air pollutions.

This paper takes Yunlong Park in Xuzhou City as an example and selects six typical activity spaces from the park. PM$_{2.5}$ concentration and environmental factors of each space were monitored simultaneously under mild-to-moderate air pollutions, and the uneven distribution characteristic of PM$_{2.5}$ concentration in urban park was analyzed. Experimental results can be used for guiding citizens to select reasonable time and place for activities, as well as provide reference and basis for layout planning of activity space in urban parks.

## 2 MATERIALS AND METHODS

### 2.1 *Study area*

Xuzhou City is located in the southeast of North China Plain (116°22'~118°40'E, 33°43'~34°58'N). Its average temperature is 14°C, sunshine hours is 2221h, the average annual rainfall is 857 mm (rainfall mainly concentrates in summer), and the annual average frequency of static wind reaches 20%. Yunlong Park is located in Quanshan District of Xuzhou City, which connects Yunlong Mountain in the east and covers an area of about 25 hectares, with water area of 8 hectares. It is a large urban comprehensive park, and attracts 1.7 million tourists each year.

## 2.2 *Study methods*

Six typical activity spaces of Yunlong Park were selected: Open Plaza (A1), Avenue (A2), Waterfront Platform (A3), Waterfront Pavilion (A4), Wisteria Pergola (A5), and Great Lawn (A6). According to the ambient air quality index, technical requirements (HJ633-2012) released by the China Ministry of environmental protection, 3 days with mild-to-moderately polluted air (150 $\mu g/m^3 > PM_{2.5}$ >75 $\mu g/m^3$) were selected from April 2016. The high-precision $PM_{2.5}$ detector (CW-HAT200S) was used to measure the $PM_{2.5}$ mass concentration of each space, and monitoring data were collected simultaneously by six groups. The measurement time interval was 7:00–20:00, with readings taken once each hour. Three sequential data were collected at a time, and the mean value was used as the concentration value of $PM_{2.5}$. The sampling time of each data is 60 s, and the sampling height is 1.5 m above the ground. The temperature, relative humidity, and wind velocity of the monitoring site were measured by using the multi-functional environment measuring instrument (HT-8500) at the same time. Finally, collected data were analyzed by Microsoft Excel 2010 and IBM SPSS Statistics 22.0.

## 3 RESULTS AND DISCUSSION

### 3.1 *General distribution characteristics of $PM_{2.5}$ concentration*

The mean value of $PM_{2.5}$ concentration in six activity spaces of Yunlong Park is shown in Fig. 1. The mean concentration value of Open Plaza is up to 130.6 $\mu g/m^3$, followed by that of Great Lawn (128.4 $\mu g/m^3$), Waterfront Platform (119.4 $\mu g/m^3$), Waterfront Pavilion (107.5 $\mu g/m^3$),

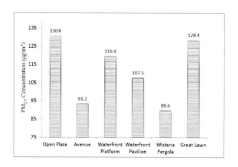

Figure 1. Mean value of $PM_{2.5}$ concentration in six activity spaces.

Avenue (93.2 $\mu g/m^3$), and Wisteria Pergola (89.6 $\mu g/m^3$). The maximum difference between mean concentration values of six activity spaces is 1.46 times. The difference between the maximum and minimum values of $PM_{2.5}$ concentration at different activity spaces in the same time is 1.35–1.65 times (Fig. 2). The highest value of $PM_{2.5}$ concentration (148.6 $\mu g/m^3$) appears in the Great Lawn at 19:00, the lowest value (83.8 $\mu g/m^3$) appears in the Wisteria Pergola at 7:00, and the difference is 1.77 times.

### 3.2 *Daily variation characteristics of $PM_{2.5}$ concentration*

The daily variation characteristics and regression simulation curve of $PM_{2.5}$ concentration in six activity spaces of Yunlong Park are shown in Fig. 3. As can be seen from the variation trends of $PM_{2.5}$ concentration, Open Plaza and Great Lawn show waveform change trends, and have three peak values and two valley values. The peak values of Open Plaza and Great Lawn in the morning and evening are higher than that at noon. Because the surrounding of Open Plaza and Great Lawn is relatively open, and affected by the automobile exhaust at rush hours, the $PM_{2.5}$ concentration value is higher at 9:00, 14:00, and 19:00. Moreover, the temperature is low and air humidity is higher in the morning and evening. Because of the increasing temperature, dropping humidity and illumina-

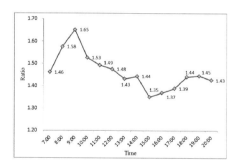

Figure 2. Ratio of maximum value to minimum value in the same time.

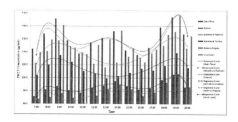

Figure 3. Daily variation characteristics and regression simulation curve of $PM_{2.5}$ concentration.

tion at noon, the peak value at noon is lower than that in the morning and evening.

The PM$_{2.5}$ concentration of Avenue and Wisteria Pergola has stepwise increasing trends, that is, peak value in the morning < peak value at noon < peak value in the evening. Because the Avenue and Wisteria Pergola have higher canopy density, windless environment leads to poor dispersion conditions, PM$_{2.5}$ tends to accumulate in the space (Wu et al. 2016). Moreover, it is less affected by ambient temperature, humidity, and illumination; PM$_{2.5}$ concentration increases stepwise, and finally reaches a high level.

The PM$_{2.5}$ concentration values of Waterfront Platform and Waterfront Pavilion show M-shaped trends, present two peak values and one valley value. The peak value in the evening is higher than that in the morning. Because these two monitoring sites are close to water, and the sunlight intensity and temperature gradually increase from 10:00 to 13:00, the air humidity near ground rapidly increases at noon. When the humidity is too high, fine particulate matters with water-soluble ions as the main components are more easy to absorb large amounts of water to increase the particle size, thus gravity settling occurs (Jin et al. 2012, Lang et al. 2013), so that the PM$_{2.5}$ concentration reaches the minimum value at 13:00.

The results indicate that under mild-to-moderate air pollution (150 μg/m³ > PM$_{2.5}$ > 75 μg/m³), the best activity time in Open Plaza and Great Lawn is around 16:00; the best activity time in Avenue and Wisteria Pergola is around 11:00; and the best activity time in Waterfront Platform and Waterfront Pavilion is around 13:00.

### 3.3 *Correlation between PM$_{2.5}$ concentration and environmental factors*

The correlation between PM$_{2.5}$ concentration of six activity spaces and environmental factors (temperature, relative humidity, and wind velocity) is shown in Fig. 4–Fig. 6. PM$_{2.5}$ concentration has negative correlation with temperature and wind velocity, and the corresponding correlation coefficients are R = −0.358 and R = −0.717 ($p < 0.01$). PM$_{2.5}$ concentration is positively correlated with relative humidity, whose correlation coefficient R = 0.340 ($p < 0.01$). The results show that PM$_{2.5}$ concentration gradually decreases when temperature and wind velocity increases. By contrast, PM$_{2.5}$ concentration gradually increases with the increase of relative humidity.

### 3.4 *Discussion*

In this study, the difference of PM$_{2.5}$ concentration between the maximum and minimum values in different activity spaces in the same time is 1.35–1.65 times and the maximum difference in a day is 1.77

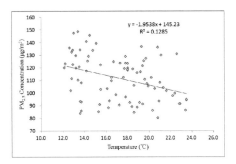

Figure 4.  Correlation between PM$_{2.5}$ concentration and temperature.

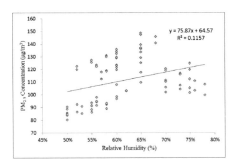

Figure 5.  Correlation between PM$_{2.5}$ concentration and relative humidity.

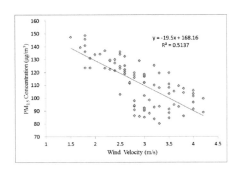

Figure 6.  Correlation between PM$_{2.5}$ concentration and wind velocity.

times, which are consistent with the results of research on the uneven distribution of PM$_{2.5}$ carried out by Wu et al. (2015) and Wen et al. (2012) in micro scale (1.1–2.1 times). Because the space barrier of urban park is mainly plants and terrain, the vertical height is lower than that of residential buildings and buildings along the street, and the difference of PM$_{2.5}$ concentration is slightly lower than that of the residential areas and urban streets.

The PM$_{2.5}$ concentration of different activity spaces has different tendency of average daily variation. Open Plaza and Great Lawn have waveform trends, Avenue and Wisteria Pergola have stepwise rising trends, and Waterfront Platform and Waterfront Pavilion have M-shaped trends. The results are consistent with the "Twin-peak Single-valley", "N-shaped" and "S-shaped" curves described in Li et al. (2014), Wang et al. (2013), Wang et al. (2014b), and Chen et al. (2016).

## 4 CONCLUSIONS

This paper takes six typical activity spaces of Yunlong Park as the research object, and $PM_{2.5}$ concentration and environmental factors of each space were monitored simultaneously under mild-to-moderate air pollutions. The main conclusions are drawn as follows:

In Yunlong Park, the average concentration of $PM_{2.5}$ in Open Plaza is 130.6 µg/m³, followed by that of Great Lawn (128.4 µg/m³), Waterfront Platform (119.4 µg/m³), Waterfront Pavilion (107.5 µg/m³), Avenue (93.2 µg/m³), and Wisteria Pergola (89.6 µg/m³). The maximum difference between six activity spaces is 1.46 times. The difference between the maximum and minimum values at different activity spaces in the same time is 1.35–1.65 times. And the maximum difference between the different time periods is 1.77 times in one day. From the results, under mild-to-moderate air pollutions, the overall mean values of Avenue and Wisteria Pergola are lower than that of other activity spaces.

The $PM_{2.5}$ concentration of different activity spaces has different tendency of average daily variation. Open Plaza and Great Lawn have waveform trends, the minimum value appears at 16:00; Avenue and Wisteria Pergola have stepwise rising trends, the minimum value appears at 11:00; and Waterfront Platform and Waterfront Pavilion have M-shaped trends, the minimum value appears at 13:00. The highest values of $PM_{2.5}$ concentration at all the monitoring sites appear around 19:00 pm.

In the six activity spaces of Yunlong Park, the $PM_{2.5}$ concentration has negative correlation with temperature and wind velocity, and positive correlation with relative humidity. Results show that PM2.5 concentration gradually decreases increase of temperature and wind velocity. By contrast, PM2.5 concentration gradually increases with the increase of relative humidity.

## ACKNOWLEDGMENTS

This study was supported by the National Natural Science Foundation of China (Grants No. 31500575, 31500579) and the Natural Science Foundation of Jiangsu Province (Grants No. BK20150231).

## REFERENCES

Brook, R. D., Franklin, B., Cascio, W., Hong, Y., Howard, G. and Lipsett, M. et al. 2004. Air pollution and cardiovascular disease: a statement for healthcare professionals from the expert panel on population and prevention science of the american heart association. Circulation, 109(21), 2655–2671.

Chen, F. T., Zhao, W. J., Yan, X. and Xiong, Q. L. 2016. Vertical distribution characteristics and origin of $PM_{2.5}$ concentration in asymmetrical street canyon. Chinese Journal of Environmental Engineering, 10(3): 1333–1339.

Jin, X., Cheng, M. T., Wen, T. X., Tang, G.Q., Wang, H. and Wang, Y. S. 2012. The variation of water-soluble inorganic ions during a heavy pollution episode in winter, Beijing. Environmental Chemistry, 31(6): 783–790.

Johnson, P. R. S. and Graham, J. J. 2005. Fine particulate matter National Ambient Air Quality Standards: public health impact on populations in the Northeastern United States. Environmental Health Perspectives. 113(9): 1140–1147.

Lang, F. L., Yan, W. Q., Quan, Z. and Cao, J. 2013. Size distribution of atmospheric particle number in Beijing and association with meteorological conditions. China Environmental Science, 33(7): 1153–1159.

Li, X. Y., Zhao, S. T., Li, Y. M., Guo, J. and Li W. 2014. Subduction effect of urban arteries green space on atmospheric concentration of $PM_{2.5}$ in Beijing. Ecology and Environmental Sciences, 23(4): 615–621.

Pope, III C. A., Burnett, R. T. and Thun, M. J., Calle, E. E., Krewski, D., Ito, K. and Thurston, G. D. 2002. Lung cancer, cardiopulmonary mortality, and long-term exposure to fine particulate air pollution. Jama, 287(9): 1132–1141.

Wang, G. Y., Bai, W. L., Li, X. Y. and Zhao, S. T. 2014a. Research of greenbelt design technology on $PM_{2.5}$ pollution reduction in Beijing. Chinese Landscape Architecture, 30(7): 71–76.

Wang, X. L., Wang, C., Gu, L., Wang, Q. and Wang, Y. Y. 2014b. Distribution characteristics of $PM_{2.5}$ concentration in different weather conditions in city street greenbelt in spring. Ecology and Environmental Sciences, 23(6): 972–978.

Wang, Y., Li, Y., Li, X., Zhao, S. and Guo, J. 2013. The effects of road green space on concentration distribution and subduction of $PM_{2.5}$ in Beijing City. Hubei Forestry Science & Technology, 42(6): 4–9.

Wen, Y. Z., Sun, Z., Deng, C. and Fu, Z. M. 2012. $PM_{2.5}$ concentration monitoring and analysis in Hangzhou Xiasha region. Journal of China University of Metrology, 23(3): 279–283.

Wu, B., Shen, X., Cao, X., Yao, Z. and Wu, Y. 2016. Characterization of the chemical composition of PM2.5 emitted from on-road China III and China IV diesel trucks in Beijing, China. Science of the Total Environment, 551–552: 579–589.

Wu, Z. W., Han, Y. T., Wang, Y. H. and Shan, H. N. 2015. The PM2.5 uneven distribution phenomenon of Beijing City in tiny scale. Environmental Science & Technology, 38(6P): 52–55.

Xiao, Y., Wang, S., Li, N., Xie, G. D., Lu, C. X., Zhang, B. and Zhang, C. S. 2015. Atmospheric PM2.5 removal by green spaces in Beijing. Resources Science, 37(6): 1149–1155.

*Advances in Energy and Environment Research – Achour & Wu (Eds)*
© *2017 Taylor & Francis Group, London, ISBN 978-1-138-62682-9*

# Effect of leachate concentration on pollutant disposal ratio and electricity production characteristics by a bio-cathode MFC

Jinfeng Hu, Longjun Xu, Qi Jing, Duowen Qing & Miao Xie
*State Key Laboratory of Coal Mine Disaster Dynamics and Control, Chongqing University, Chongqing, China*

ABSTRACT: In this paper, the effects of leachate concentration on the pollutant removal ability and electricity production characteristics by bio-cathode Microbial Fuel Cell (MFC) were studied. The results showed that the voltage of electricity produced changed periodically, and the average maximum voltage of MFCs decreased in the following order: chemical cathode MFC (CMFC) > aeration bio-cathode MFC (ABMFC) > no-aeration bio-cathode MFC (NBMFC), when different volume ratios of leachate were used as anode substrate. COD, $NH_4^+$-N, $NO_3^-$-N and $NO_2^-$-N in leachate could be effectively removed by three groups of MFC. With the increase of the volume ratio of leachate, the removal rate of COD increased and then decreased, while the total removal rate of N increased monotonically.

## 1 INTRODUCTION

Microbial Fuel Cell (MFC) is a type of reactor that can deal with sewage while producing electric energy. Biological cathode type MFC is a new type of microbial fuel cell using microorganisms as catalyst. It has characteristics of low operating cost, environment-friendliness, and nitrogen removal. Thus, scientists are more focused on this research direction (Zhao et al. 2006, Mansoorian et al. 2013, Kim et al. 2014). The Aged Landfill Leachate (ALL) is one of the globally recognized treatments of refractory organic wastewater containing high levels of organic pollutants, nitrogen, and phosphorus. At present, the chemical method, biological method, and combination technology have been adopted to deal with ALL. However, the process is complex, involves high processing cost, and is difficult to realize the resources (Xiong & Zheng 2015). The treatment of landfill leachate by MFC could produce electrical energy and reduce the energy consumption of sewage treatment. This method has opened up a new idea for landfill leachate treatment technology.

You et al. (2006) were the first to find that landfill leachate MFC can produce electrical energy. Maximum power densities of single-chamber and dual-chamber MFCs were 6817.4 and 2060.19 mW/m³, respectively. High salinity is conducive to reduce the internal resistance of MFC and improve the performance of electric power (Puig et al. 2010). Electricity production and pollutant removal efficiency of MFC were significantly improved when the refractory landfill leachate was pretreated by fermentation (Mahmoud et al. 2014, Tugtas et al. 2013). The COD removal rate is 70.4% and part

of ammonia nitrogen is transferred from anode to cathode (Yan et al. 2013). The effect of the ratio of anode to cathode area on electricity production and treatment efficiency of the ALL by biological cathode type MFC was studied by Cheng & Xu (2015). However, the non-biological cathode-type MFCs were used in these reports. For better tracking of electricity production and pollutant degradation, a dual-chamber aeration bio-cathode MFC (ABMFC) and a dual-chamber no-aeration MFC (NBMFC) were constructed to treat the ALL. For comparison, a dual-chamber chemical cathode MFC was also built by using potassium ferricyanide as electron acceptor (CMFC).

## 2 EXPERIMENTS

### 2.1 Experimental materials and composing of MFCs

The three groups of H-form MFC were constructed with organic glass, and carbon felt (4 cm × 4.5 cm × 1 cm) was selected as the electrode. Volumes of cathode and anode chamber were both 800 ml (d 153 mm, φ80 mm) and were separated by proton exchange membrane (Nafion 117, effective area: 7.07 cm²). Two electrodes were connected by a copper line and external resistance (1500Ω) to form a loop. The saturated calomel electrode was set as reference electrode in anode. MFC systems were operated at atmospheric pressure and room temperature. For ABMFC, dissolved oxygen in the cathode chamber was supplied with air. A small amount of air resulted from mechanical agitation in the cathode chamber for NBMFC. In

CMFC, potassium ferricyanide was used as electron acceptor.

## 2.2 Vaccination and operation of MFC

The ALL in the anode chamber was taken from a landfill in Chongqing, and inoculated microorganisms obtained from the MFC anode solution ran over 1 year in this laboratory. The inoculation proportion was 1:1 in both the cathode chamber and anode chamber. Experiments were performed on the three groups of MFC under the same conditions.

Taking 5d as the period in the first stage, the anode chamber was maintained with anaerobic condition, and the gradual domestications were done by using 20%, 40%, 60%, 80%, and 100% (volume fraction) of the landfill leachate as substrate. For ABMFC and NBMFC, and cathode solution consisted of glucose (2.0 g/L), $Na_2HPO_4 \cdot 12H_2O$ (8.95 g/L), $KH_2PO4$ (3.40 g/L), $NH_4Cl$ (0.2 g/L), KCl (0.13 g/L), trace metal ions (12.5 mg/L), and vitamin E (5 mg/L) ( Liu & Logan 2004), while the cathode solution of CMFC was only $K_3Fe(CN)_6$(25 mmol/L). In the second stage, CMFC and ABMFC with 100% leachate were consecutively run for 45 days, with regular addition of ferricyanide or carbon source in cathode.

All agents used in the experiment were of analytical grade, and obtained from Kelong Chemical Reagent Factory, Chengdu. The following are the characteristics of the original leachate: pH 8.84, COD 6842.1 mg/L, NH3-N 3520.9 mg/L, and conductivity 12.0 mS/cm.

## 2.3 Test parameters and methods

COD, $NH_3$-N, $NO_2^-$-N, and $NO_2^-$-N were detected by standard method (SEPA 2012). Output voltage (U) and anodic potential (EA) were automatically acquired and stored by Agilent 34970 A. The current (I) was calculated as I = U/R. The current density (j) and output power (p) were obtained as follows: $j(A/m^3) = I/V$ and $p(W/m^3) = U^2/R_{ex}V$,

where V is the anode effective volume ($m^3$) and $R_{ex}$ is the external resistance ($\Omega$). The internal resistance was tested by using polarization curve method (You et al. 2009). Coulombic efficiency was calculated using the reference method (Li & Zhang 2012).

## 3 RESULTS AND DISCUSSION

### 3.1 Effect of leachate concentration on the generation voltage of MFCs

With a cycle of 5 days, the microbial community of electrogenesis was gradually domesticated and enriched by using different leachate concentrations (20%, 40%, 60%, 80%, and 100%) in turn as anolyte of MFCs, and the electric voltage is shown in Fig. 1. It is evident from the figure fluctuation exists in the production of voltage after the anode solution of three groups of MFCs was replaced by 20% landfill leachate. This indicated that in electricity production, the microorganism has experienced the adaptation process of re-cultivation. The following laws were obtained.

The voltage of CMFC is the highest as 40% of leachate was used as initial anode substrate, and the average voltage is less than 630 mV. In each period, the voltage of CMFC was significantly increased when potassium ferricyanide was added. The output voltage significantly decreased with the fading of cathode ferricyanide solution after 2 days. This indicated that the output voltage of the CMFC is mainly affected by the amount of potassium ferricyanide in cathode chamber.

The voltage of ABMFC was the highest (381.9 mV) when the landfill leachate concentration of anode substrate was 60%, and the average voltage was about 380 mV. In each period, the voltage of ABMFC was decreased from the last highest voltage after the cathode or anode solution was changed. After 1 day, the output voltages were increased with the cathode, and suspended microorganisms and biological membrane electrode were

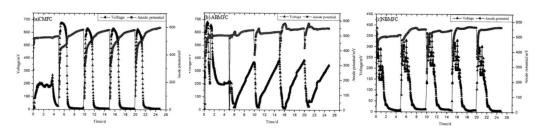

Figure 1. Current generation by MFCs using different concentrations of landfill leach in anode chamber.

gradually stable. This indicated that the output voltage of ABMFC was affected by the amount of dissolved oxygen and the stability of aerobic microorganism community.

The voltage of NBMFC was the highest (353.1 mV) when the anode liquid was 60% aging landfill leachate, and its average voltage was more than 350 mV. In each cycle, the voltage decreased when the cathode or anode solution was changed, then the voltage decreased irregularly, and finally, the voltage became less than 50 mV after 2 days. This indicated that the output voltage of NBMFC was mainly affected by the degrading activity of microorganism community due to the lack of nutrient contents in cathode.

The above phenomenon can be attributed to the following causes. At the beginning of electricity production, the voltages of three MFCs were fluctuated because of the competitive action between the electrogenesis microorganisms and degradation microorganisms. With the increasing of leachate volume ratio, the density of organisms increased in the anode substrate, which provided enough nutrients for their growth, and the output voltage also increased. Thus, it can be speculated that the output voltage increased with the increase of leachate volume ratio when the leachate of MFCs was at a low density level. However, when the leachate volume increased to a certain extent, it can not only bring in nutrient, but can also increase the release of harmful substances for microbial growth. Thus, it can be speculated that the output voltage of MFCs was decreased with the increase of leachate volume ratio (Lu et al. 2009).

### 3.2 Effect of leachate concentration on COD removal ratio

The COD concentrations and removal efficiencies after 5-day treatment for MFCs at initial concentrations of 20%, 40%, 60%, 80%, and 100% are shown in Fig. 2. It is evident from the figure that the removal amounts of COD in the anode of

ABMFC and NBMFC increased with the increase of leachate volume ratio. The COD was higher and more complex in anode solution that in the cathode solution, and was unfavorable for the growth of microorganisms. Thus, removal rates of COD in the cathode ABMFC and NBMFC were higher than those in the anode. Furthermore, the COD removal rate of NBMFC was higher than that of ABMFC in the anode, which indicated that it was easier to remove COD in NBMFC.

Comparing with the corresponding output voltage change trend, the removal rate of COD increased initially and then decreased with the increase of leachate volume ratio. When the anode substrate was 40% landfill leachate in CMFC, the concentration of COD decreased from initial 2716.8 mg/L to 834.0 mg/L, and the highest removal rate in anode was 30.7%. The concentration of COD decreased from initial 4025.3 mg/L to 2896.1 mg/L when the anode substrate was 60% landfill leachate in ABMFC, and the highest removal rate in anode was 28.0%. However, the removal rate of COD in cathode was the least (76.3%) in five ABMFCs. As the anode substrate was 60% landfill leachate in NBMFC, the concentration of COD decreased from initial 4025.3 mg/L to 2654.9 mg/L, and the highest removal rate in anode was 34.0%, while the removal rate of COD in cathode was the least (70.2%) in five NBMFCs.

### 3.3 Coulombic efficiency

The effect of leachate concentration on Coulombic efficiencies of MFCs is shown in Fig. 3. It is evident from the figure that the Coulombic efficiencies decreased with the decrease of leachate concentration. For ABMFC, NBMFC, and CMFC, the Coulombic efficiencies were 10.26, 4.3, and 1.46, respectively. This indicated that a small part of organic matter in leachate was converted into electrical energy, and the COD in leachate of ABMFC anode was more efficiently converted to electrical energy than in other cells of the anode.

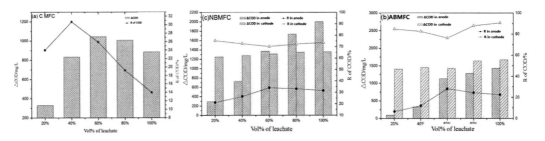

Figure 2. COD concentration and removal efficiencies after 5 day treatment for MFCs with different initial concentrations.

COD removal rate and Kulun efficiency were relatively low in this research. The possible reasons are as follows. In the anode, oxidation reaction produced electrons, and not only COD, but also ammonia nitrogen and nitrate occurred. Oxidation is not sufficient in 5 days, because the microorganisms have adaptation period in the new environment after the gradual change of the anode liquid.

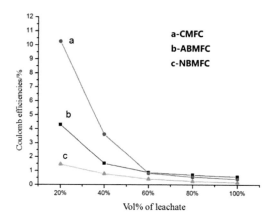

Figure 3. MFCs processing Coulomb different concentrations of landfill leachate.

## 3.4 Removal efficiencies on $NH_4^+$-N, $NO_3^-$-N, and $NO_2^-$-N

$NH_4^+$-N, $NO_3^-$-N, and $NO_2^-$-N concentrations and removal efficiencies after 5 day treatment for MFCs with different initial concentrations (20%, 40%, 60%, 80%, and 100%) are shown in Fig. 4, and the following results could be concluded. The $NH_4^+$-N removal amount increased with the increase of leachate volume ratio, but the of $NH_4^+$-N removal rate decreased and then increased in anode chambers of MFCs. The $NH_4^+$-N removal rate was the highest with 20% landfill. The removal rate of $NO_3^-$-N increased and then decreased, and it was the highest with 40% landfill leachate. It decreased in the following order: ABMFC (32.8%) > CMFC (32.2%) > CMFC (31.4%). It did not accumulate $NO_2^-$-N in all anode solution of MFCs. For chambers in MFCs, the effluent concentrations of $NH_4^+$-N and $NO_3^-$-N were higher than influent concentrations. The removal rate was negative, which indicated that a part of $NH_4^+$-N accumulated the cathode chamber through proton exchange membrane and some $NH_4^+$-N groups were oxidized into $NO_3^-$-N groups.

There was no relationship between effluent concentration of $NH_4^+$-N and initial concentration in

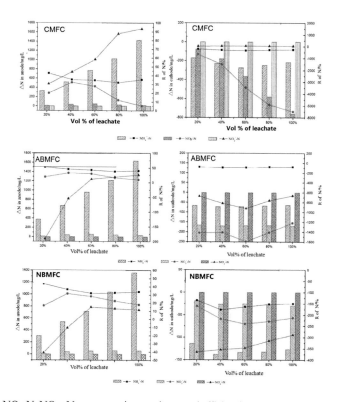

Figure 4. $NH_4^+$-N, $NO_3^-$-N, $NO_2^-$-N concentrations and removal efficiencies.

anode leachate, and the order of effluent concentration was CMFC (about 310 mg/L) > NBMFC (about 215 mg/L) > NBMFC (about 150 mg/L). This indicated that the initial leachate concentration of $NH_4^+$-N had a small impact on the degradation of cathode $NH_4^+$-N in MFCs. The $NO_3^-$-N concentrations for the cathode effluent of ABMFC and NBMFC also had no relationship with initial leachate concentration, which were 167 and 40 mg/L, respectively. However, $NO_3^-$-N concentrations of the cathode effluent for CMFC was positively correlated with the initial leachate concentration.

Total Nitrogen (TN) removal efficiencies after 5 day treatment for MFCs with different initial concentrations are shown in Fig. 5. It is evident from the figure that MFCs had successively removed N in leachate, and its removal increased with the increase of leachate proportion. The order of removal amount of TN was ABMFC > NBMFC > CMFC.

## 4 CONCLUSIONS

There was a periodic variation in the power voltage and dilution ratio of anode leachate in different cathode-type MFCs. In each period, the average maximum power generation voltages of the battery were about 630 mV (CMFC), 380 mV (ABMFC), and 330 mV (NBMFC).

The COD removal rates in the anode chamber first increased and then decreased with the volume ratio of the leachate for the three groups of MFCs. For ABMFC and NBMFC, the removal amount of COD in the anode chamber increased with the increase of the volume ratio of the leachate, and the COD removal rates in the cathode chamber were higher than those of the anode chamber. Three groups of MFCs were achieved for the removal of $NH_4^+$-N, $NO_3^-$-N, and $NO_2^-$-N, and all were $NH_4^+$-N diffusion phenomenon from the anode chamber to the cathode chamber. The removal rate of $NH_4^+$-N in the anode chamber was the largest, i.e. 20% of leachate, and the removal rate of $NO_3^-$-N in the anode chamber was the largest, i.e. 40% of leachate. No accumulation of $NO_2^-$-N was found in the anode chambers. The total removal amount for the three types of N increased with the increase of the volume ratio of the leachate, and when the volume ratio were the same, the order was ABMFC > NBMFC > CMFC.

## ACKNOWLEDGMENT

The authors would like to acknowledge the financial support from the Fundamental and advanced research projects of Chongqing Science and Technology Commission (2013 jjB20001).

## REFERENCES

Cheng, L.Y. & L.J. Xu. Effects of electrode surface area on electrical properties of microbial fuel cells used the aging landfill leachate as substrate. *J. Fuel Chem. Technol.*, 2015, 43: 1011–1017.

Li, J.T. & S.H. Zhang. Basic research on denitrifying microbial fuel cell. *China Environmental Science*, 2012, 32(4): 617–622.

Liu, H. & B.E. Logan. Electricity generation using an air-cathode single chamber microbial fuel cell in the presence and absence of a proton exchange membrane. *Environmental Science Technology*, 2004, 38(14): 4040–4046.

Lu, N., B. Zhou, L.F. Deng, S.G. Zhou & J.R. Ni. Starch processing wastewater treatment using a continuous microbial fuel cell with $MnO_2$ cathodic catalyst. *Journal of basic science and engineering*, 2009, 17: 65–73.

Kim, J., B. Kim, H. Kim, & Z. Yun. Effects of ammonium ions from the anolyte within bio-cathode microbial fuel cells on nitrate reduction and current density. *Int Biodeterior Biodegrad*, 2014, 95 (Part A): 122–126.

Mahmoud, M., P. Parameswaran, C.I. Torres, & B. Rittmann. Fermentation pre-treatment of landfill leachate for enhanced electron recovery in a microbial electrolysis cell. *Bioresource Technology*, 2014, 151: 151–158.

Mansoorian, H.J., A.H. Mahvi, A.J. Jafari, M.M. Amin, A. Rajabizadeh, & N. Khanjani. Bioelectricity generation using two chamber microbial fuel cell treating wastewater from food processing. *Enzyme Microb Technol*, 2013, 52 (6/7): 352–357.

Puig, S., M. Serra, M. Coma, M. Cabré, & M. Balaguer. Microbial fuel cell application in landfill leachate treatment. *Journal of Hazardous Materials*, 2010, 185, 763–767.

Tugtas, A.E., C. Pelin, & C. Baris. Bio-electrochemical post-treatment of anaerobically treated landfill leachate. *Bioresource Technology*. 2013, 128, 266–272.

Xiong, J. & Z. Zheng.Characteristics of the dissolved organic matter in landfill leachate and their removal technology: A review. *Environmental Chemistry*, 2015: 34(1): 44–53.

Yan, F., S.B. Huang, M.Y. Xu, & Y.G. Yang, A.J. Wang, G.P. Sun. Treatment of mature landfill leachate by microbial fuel cell. *Industrial Water Treatment*, 2013, 33(1): 52–55.

You, S.J., N.Q. Ren, Q.L. Zhao, P.D. Kiely, J.Y. Wang, F.L. Yang, L. Fu & L. Peng. Improving phosphate buffer-free cathode performance of microbial fuel cell based on biological nitrification. *Biosensors & Bioelectronics*, 2009, 24(12): 3698–3701.

You, S.J., Q.L. Zhao, J.Q. Jiang, J.N. Zhang, & S.Q. Zhao. Sustainable approach for leachate treatment: electricity generation in microbial fuel cell. *Environ Sci Health A*, 2006, 41(12): 2721–2734.

Zhao, F., F. Harnisch, U. Schroder, F. Scholz, P. Bogdanoff, & I. Herrmann. Challenges and constraints of using oxygen cathodes in microbial fuel cells. *Environment Science Technology*, 2006, 40: 5193–5199.

*Advances in Energy and Environment Research – Achour & Wu (Eds)*
© *2017 Taylor & Francis Group, London, ISBN 978-1-138-62682-9*

# Policy orientation and pattern path of sustainable development of modern agriculture in Suzhou City

Guosheng Ma & Juan Chen
*Suzhou Polytechnic Institute of Agriculture, Suzhou, Jiangsu, P.R. China*

ABSTRACT: Based on the field research and combined with the recent development tendency and regional characteristics of modern agriculture in Suzhou, this paper carries out a comprehensive analysis of the four principles of sustainable development of urban agriculture in Suzhou, including the ecology and brand-oriented development principle, the scale and characteristic construction principle, the refined and standardized production principle and the socialized and specialized service principle. It also presents four policy directions, the red-line, yellow-line, green-line and blue-line policies, as well as ten development patterns, agriculture demonstration garden, cooperative farm, technical demonstration base, family farm, tourism and leisure agriculture, agro-ecological circulation, ecological culture, production and marketing of agricultural products with brand features, the third-party specialized service.

## 1 INTRODUCTION

The integration of urban and rural development of Suzhou requires multi-functional and diversified agricultural development. According to statistics of Suzhou Statistics Bureau (2011, 2015), the cultivated area was 208,100 hectares in 2015, while in 1980 it was 373,300 hectares. The proportion of agricultural output in GDP of Suzhou dropped from 28.1% in the early era of reform and opening-up to about 1.6% in 2015.

Due to the increasingly rapid urbanization, gradually decreasing cultivated area, low agriculture's comparative benefit, how to achieve the sustainable development of urban modern agriculture has become a big issue.

## 2 THE SCALE DEVELOPMENT OF MODERN AGRICULTURE IN SUZHOU CITY

The flourishing modern agriculture in Suzhou has the regional, standardized and co-operative development. By the end of 2015, more than 90% of the contract land is the land circulation. The ratio of agricultural moderate scale operation area is to 92%.

The total area of agricultural park reaches 70,600 hectare. There are 181 leading agricultural enterprises of municipal level and above, and 4535 rural "three-cooperation" organizations. The rural per capita net income reached 25,580 yuan and the agriculture has undergone scale and industrial development.

The details are shown in Table 1.

Table 1. The scale and industrial development of agriculture in Suzhou City in the past five years.

| Industrial scale | 2015 | 2010 | Growth rate |
|---|---|---|---|
| Moderate Scale Management proportion | 92% | 70% | 31.43% |
| Cooperation organizations | 4535 | 3043 | 49.03% |
| Area of facility agriculture and fishery (ten thousand hectare) | 4.68 | 2.33 | 100.57% |
| Total area of the agricultural park (ten thousand hectare) | 7.06 | 3.30 | 113.94% |
| Leading enterprises of municipal level and above | 181 | 112 | 61.61% |
| Rural per capita net income (thousand yuan) | 25.58 | 14.46 | 76.90% |

## 3 SUSTAINABLE DEVELOPMENT PRINCIPLES OF MODERN AGRICULTURE IN SUZHOU

### 3.1 *The ecology and brand-oriented development*

Ecological civilization has become a national strategy and Suzhou has implemented the ecological compensation policy. According to the pollution sources census jointly issued by the Ministry of Environmental Protection, Ministry of Agriculture and other departments, agriculture has become an important source of pollution. As three quarters of the water in Taihu Lake and the longest coastline are in Suzhou, the development of modern agriculture should highlight the ecological development pattern. In the future, the development of modern agriculture in Suzhou should be based on the "Ecology First" concept and brand-development strategy. We should enhance the ecological functions of agriculture, improve the brand effect of agriculture and develop ecological farming with ecological cycle, ecological animal husbandry, ecological aquaculture and recreational fishery, sustainable ecological forestry with limited land resources in Suzhou. Meanwhile, the quality of cultivated land should be improved. The agricultural space will serve the public and become the garden and paradise of the city.

### 3.2 The scale and characteristic construction principle

To achieve sustainable development of modern agriculture in Suzhou, we should adopt the construction mode of park and farm proposed by the Twelfth Five-year Plan. The "fragmentation" problem that constrains its development should be solved by establishing cross-village agricultural park. The administrative organization of the agricultural park should be combined with the government, and its staffing level should be implemented. The cooperative farms of moderate scale, family farms and other farms should be supported and standardized. The "four-million-acre" contiguous space should be optimized. The distinguishing features could promote the scale and characteristic development of modern agriculture in Suzhou and make it better serve the modern city.

### 3.3 The refined and standardized production principle

Intensive cultivation is a fine tradition of farming culture in Suzhou. We should combine it with the modern agricultural technology means to develop the agriculture products with superior quality and local characteristics, as well as the facility agriculture and precision agriculture, and try to apply the Internet of Things to agriculture. The introduction of modern agricultural technology, and the establishment of local standard of agricultural production, and the construction of standardized agricultural park could promote the standardized production of agriculture in Suzhou. It is necessary to explore and implement the fine management mode of cultivated land resources which involves citizens in it, improve the technological level of agricultural production and management, thus increasing the labor-output ratio per unit area and bringing income growth for farmers. Besides, the idle land, roof of buildings, the agricultural use of some green land all make agriculture the primary industry of the city, as well as the modern and fashionable industry which can be fully integrated into the urban spatial layout.

### 3.4 The socialized and specialized service principle

During the "Twelfth Five-year Plan" period, a socialized service system has been basically established to serve the modern agriculture of the city. In the future, we should continue to promote its development, improve its level of service and make it more standardized. Agricultural sci-tech staff and professional teachers from the scientific research institutes and universities should be supported and encouraged to create or participate in the socialized service organizations and enterprises, aiming to improve the professional level of them. Vocational training should be carried out for farmers, and the grass-roots agro-technical stations and cooperatives should be used to improve the professional level of service of the socialized service organizations. Technical training, on-site meetings, demonstrations, seminars, study tours and other means are also needed.

## 4 POLICY SUPPORT OF THE SUSTAINABLE DEVELOPMENT OF MODERN AGRICULTURE IN SUZHOU

The policy support plays an important role in the sustainable development of modern agriculture and the policies mainly include protection policy, re-feeding policy, compensatory policy, and science and technology policy, as shown in Figure 1.

### 4.1 The red-line policy—protection by planning red line

Apply the protection measures for the agricultural land within the planned red line of the whole city. Strict protection policies should be implemented in

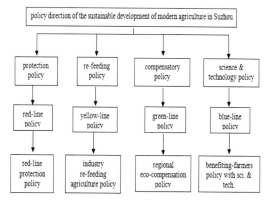

Figure 1. Policy direction of the sustainable development of modern agriculture in Suzhou.

the agriculture planning area and land occupancy is forbidden. If the occupancy is reasonable, all the projects must submitted for the approval of Municipal Planning Bureau, Bureau of Land and Resources and Council of Agriculture, and the a new base with the same area of the occupied land should be established in this administrative area as a compensation. We should adhere to the development strategy of "four million Mu (1 mu = 666.7 m$^2$)" and legalize the policy of project and develop this strategy.

### 4.2 *The yellow-line policy—industry re-feeding agriculture*

The benefiting-farmers policies should be developed and much favorable fiscal expenditure should go to agriculture and rural areas to ensure its increment and total quantity. Endowment insurance, unemployment insurance and other benefits should be made to work for all the rural people. People could receive payment on the basis of their property, shares of the land, collective assets, cooperatives and village-level entities. Policies should be developed to encourage the farmer entrepreneurship, such as rural home inns, online stores for agricultural products. Besides, the on-the-job training could also help famers broaden employment channels.

### 4.3 *The green-line policy—regional ecological compensation*

Uphold and improve the ecological compensation policy, increase and improve ecological compensation and expand its scope. The expenses of ecological compensation are covered by municipal and district-level public finance but in the future they will be covered by the municipal public finance. The amount will be adjusted according to the total GDP of the city.

### 4.4 *The blue line policy—benefiting-farmers policy with technology*

Promote the development of agricultural technology, support the innovation of agricultural technology and encourage university-industry collaboration. Establish the agricultural science and technology innovation union, and promote the construction of demonstration bases of agricultural technology in countries and towns. Explore the cloud platform of agricultural science and technology service. And try to construct a number of agricultural parks, demonstration bases, science-based agricultural leading enterprises and high-tech family farms.

## 5 SUSTAINABLE DEVELOPMENT PATTERNS OF MODERN AGRICULTURE IN SUZHOU

According to development status and trends of agriculture in Suzhou, ten patterns are presented in Figure 2, which can be divided into two types based on industrial patterns and business entities.

### 5.1 *Classification by agricultural business entities*

#### 5.1.1 *Agricultural demonstration garden*
It is an important carrier of agricultural science and technology research and demonstration. Therefore, we should construct a number of cross-village, cross-town and cross-country agricultural demonstration gardens with the market as an orientation, and promote the institutional innovation and mechanism innovation. The demonstration of new technologies, products and facilities will contribute to the modern construction of agriculture and promote the regional structure adjustment and industrial upgrading, which is an effective development pattern and can solve the "fragmentation" problem that constrains its development.

This development pattern is mainly applied to the large-scale agricultural projects (more than thousand acres).

#### 5.1.2 *Cooperative farm*
It is a new business pattern in rural areas which integrates the contracting system and cooperative system and this new agricultural development pattern is based on collective operation after land circulation. The cooperative farm is sponsored by the collective economic organizations and farmers and enterprises which are engaged in agricultural product processing and marketing can achieve cooperative operation by sharing and holding equities.

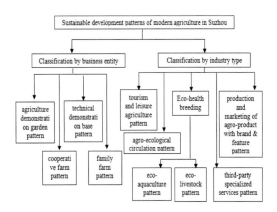

Figure 2. Sustainable development patterns of modern agriculture in Suzhou.

They can become shareholder and share profit on the basis of their land, property, or receive payment by their labor service.

This mode is suitable for medium-sized agricultural projects (about thousand acres).

### 5.1.3 *Technical demonstration base*

It is an important carrier of modern agricultural technological innovation and promotion determined by Provincial Council of Agriculture. One technical demonstration base of approximately 300 acres should be established in each country and some important villages, so that farmers will be exposed to the advanced agriculture technology. This will also contribute to the technological research of scientific and technical personnel and new agricultural technology transformation.

This pattern is mainly suitable for the agricultural projects which meet the government's requirement for agricultural technology demonstration bases.

### 5.1.4 *Family farm*

It is one of the new development patterns of the modern agriculture in recent years. It is mainly operated by family and the laborers are family members. The mechanized and technological production is adopted rather than the traditional agricultural production ways after the transfer of land.

This pattern is suitable for small-scale agricultural projects (about 50–150 acres).

### 5.2 *Classification by agricultural business entities*

### 5.2.1 *Tourism and leisure agriculture*

Tourism and leisure agriculture is a new mode of agricultural production and operation which appears in the agricultural structure adjustment process and it is also the main part of modern urban agriculture. Rural tourism gradually becomes a consumption hotspot and it has great market potential. On the one hand, visitors can learn about the lives of farmers through tourism and enjoy the local customs. Meanwhile, they can also be involved in the labor process. On the other hand, a variety of cultural resources like local rural heritage, historical sites are on view to visitors. This kind of edutainment will make the participants cherish the natural and cultural resources in rural areas.

This pattern is suitable for agricultural project with certain tourism resources.

### 5.2.2 *Agro-ecological circulation*

It is established based on the principles of ecology and economics and it keeps the integrality, coordination, circulation and regeneration. Besides, the modern scientific and technological achievements, modern management means, effective experience of traditional agriculture, biological measures, engineering measures, cultivation practices are all combined, so it can be used to get higher ecological and economic benefits, social benefits. As the main production mode to achieve agricultural transformation and upgrading, it focuses on resource recycling and could play agro-ecological function, protect agricultural ecological environment, improve the quality and safety of agricultural products and output pollution-free, green and organic agricultural products.

This mode is applicable to any type of agriculture project.

### 5.2.3 *Ecological culture*

It can be divided into ecological aquaculture and ecological animal husbandry. The former one is to make rational use of the relationship between the aquatic products and environment, combined with the biological measures, technical measures and engineering measures to achieve best economic benefit, social benefit and environmental benefit. It mainly includes recirculation aquaculture, fully-enclosed recirculation industrial aquaculture, rice-fish agriculture, fish-vegetable agriculture, three-dimensional vertical circulating aquaculture, etc. The ecological aquaculture is the modernization trend of fishery and is suitable for different aquaculture industries and ecological recreational fisheries. Ecological livestock breeding is one of the important ways to transform aquaculture production, promote the transformation and upgrading of animal husbandry and achieve agricultural modernization. As an eco-cycle way of animal husbandry, it mainly includes healthy aquaculture, agriculture and animal husbandry, three-dimensional planting and breeding, fermentation bed farming, etc., thus achieving safe farming, harmless manure disposal, ecological wastewater treatment and recycling and utilization of resources.

This mode is suitable for animal husbandry and poultry farming of different scales.

### 5.2.4 *Branding of characteristic agricultural production and marketing*

The characteristic development is the lifeblood of modern agricultural development in Suzhou, so it should be considered as an important means to promote the development of agricultural characteristics and actively adapt to the urban requirements. Promote branding of agricultural products of Suzhou, including crabs, tea (Bi luochun), Dongting loquat, "Three Whites" of Taihu, Eight Immortals (eight edible aquatic plants), sheep, pig, rice and some other famous agricultural products.

This mode is mainly suitable to the agricultural project with characteristic agricultural products.

The specialized service plays an important part in the modern agricultural development. The development of large-scale agriculture brings fine social division of labor and higher level of agricultural specialization. Therefore, we should continue to promote the social service system and improve its professional level.

The third party could provide specialized service for the modern agriculture, including agricultural production, processing, marketing, finance.

## 6 CONCLUSION

Modern agriculture is a general review of the past and an outlook on the future and it is changing and evolving every day. According to Z.Q. Fang, F.W. Wu, & W. Wang (2008), the urban agriculture refers to the new agriculture in the city or near the city. In H.X. Qin (2003), it is shown that, the urban modern agriculture is to combine these two concepts, an extension of urban agriculture and refinement of modern agriculture. The development of urban agriculture is a dynamic evolving process to meet the objective requirements of urban residents by adjusting the structure, accompanied by the urban-rural integration. It is to balance the supply and demand by M. Zhang et al. (2014). With the rapid progress of the urban-rural integration, the development of agriculture in Suzhou increasingly shows a trend of urban agriculture. To choose the right path to promote the rapid, healthy and sustainable development of agriculture combined with its own features is very important. At present, there are still some problems in development planning, management, market positioning and brand building in China, and the similarity of themes and the low-level blind expansion are widespread (B. Zhang & M.K. Liu 2013).

This paper argues that to achieve sustainable development of modern agriculture in Suzhou, we must first choose a good policy path to guide and support its sustainable development. Second, it is necessary to choose a good development pattern and provide a successful model based on the local agricultural development pattern. Finally, do a good job in selecting the technology which can promote the sustainable development of modern agriculture in Suzhou. In this way, the urban agriculture in Suzhou will get fast, healthy and sustainable development and take the lead to realize modernization.

## ACKNOWLEDGMENT

This research was financially sponsored by Qing Lan Project of Jiangsu Colleges & Universities, Jiangsu Agriculture San Xin Project (SXGC[2016]302), Jiangsu 333 High-level Talent Training Project, Six-group Peak Talent Project of Jiangsu, Suzhou City Association for Soft Science Research Project (SZKXKT2016-B06), and Suzhou City Association of Environmental Science for Soft Science Research Project (2016-03).

## REFERENCES

Fang, Z.Q., F.W. Wu, & W. Wang (2008). Review of some issues on the theoretical study of urban agriculture in China, *Chinese Agricultural Bulletin*, vol. 24 no. 8, pp. 521–525.
Qin, H.X. (2003). Discussion on Beijing urban modern agriculture development. In: Urban modern agriculture theory and practice. *Peking: China Agriculture Press*.
Suzhou Statistics Bureau (2011), Suzhou statistical yearbook 2011, *Peking: China Statistics Press*.
Suzhou Statistics Bureau (2015), Suzhou statistical yearbook 2015, *Peking: China Statistics Press*.
Zhang, B. & M.K. Liu (2013). Evolution and development mechanism of urban agricultural tourism in China. *Jiangsu Agricultural Science*, vol. 41 no. 2, pp. 419–422.
Zhang, M., F.J. Lu, R.L. & Miao, et al. (2014). Systematic analysis of the stage of urban agriculture development in large cities in China, *Jiangsu Agricultural Science*, vol. 42 no. 3, pp. 360–363.

*Advances in Energy and Environment Research – Achour & Wu (Eds)*
*© 2017 Taylor & Francis Group, London, ISBN 978-1-138-62682-9*

# A study on marketization of land element allocation in Beijing

Y.F. Yang & J.Q. Xie
*School of Land Science and Technology, China University of Geosciences, Beijing, China*

ABSTRACT: This paper analyzes the current situation of land element allocation and its difficulty in reforming Beijing land market. Literature analysis and empirical research method are used for analysis. Suggestions for promoting the marketization of land element allocation are to: 1) improve related policies and measures of the land market; 2) accelerate the ownership certification and illegal land resolution to speed up the establishment of a unified urban and rural construction land market; and 3) follow the rules of market economy and decentralize to let the market works. These are based on four principles: "uphold the public ownership", "uphold legislation first", "uphold the gradual reform", and "uphold the integration of urban and rural". When the government gives full play to the role of land elements in the market economy through the reform of land element market and establishes a unified land market in urban and rural areas, the economy can be increased quickly and smoothly.

## 1 INTRODUCTION

The land element is an important study point in the relationship between human economic activities and the natural environment. As early as the period of classical economics, William Petty and David Ricardo, representatives and scholars started discussing about the optimal allocation of the land. In the Neoclassical period, Marshall (1980) and Solow (1956) believed that technological advances can promote economic growth instead of land scarcity. Nichols (1970) explored the land value growth model (only containing the production model). Metezmakesr & Louw (2005) believed that the land and labor are the elements of national economic growth, shortage of industrial land will limit economic growth, and increasing the intensity of land use can increase social welfare.

The land investment contribution is more than capital and labor, to secondary and tertiary industry development of China (Mao Zhenqiang & ZuoYuqiang, 2007, Mao zhenqiang et al. 2009). Li Mingfeng (2010) showed the land element has contributed significantly to the current economic growth of China. The contribution rate of land element to China's economic growth is between 20% and 30% in 1997–2008. Feng Lei et al. (2008) constructed an economic growth model of land element and analyzed their contribution to economic growth. It is concluded that land element is important for the economic growth of China. Especially in the east area, economic development relies more on the contribution of the land element. Some studies on the relationship between the elements of land and economic growth, based on new classical economic theory, found that increase of land use technology will help promote economic growth (Wu Kangping & Yang Wanli 2009, Xia Fangzhou et al. 2014, Jiang Hai & Qu Futian 2009, ZhangYouxiang & Jin Zhaohuai 2012). Di Jianguang et al. (2013) showed that the contribution of construction land to non-agricultural economic growth was 11.81%. Zhang Leqin et al. (2014) studied the contribution of land elements to the economic growth of Anhui Province. Du Guanyin et al. (2010) analyzed the land total element productivity including construction investment. Zhao Ke et al. (2014) studied the effect of the quality of economic growth on urban land expansion.

The Third Plenary Session of the 18th Central Committee of the Communist Party of China offered, "focus on the market plays a decisive role in the allocation of resources, to deepen economic system reform and establish a unified urban and rural construction land market". To establish a unified urban and rural construction land market through the reform, the land resource benefits should be fully utilized in economic and social development. This market containing state-owned land and collective construction land is in line with the planning and use control of the premise. The price of land should be defined by this free market, to reflect the status of supply and demand flexibly, as well as scarcity of resources level. Land element in this paper is only limited to state-owned construction land and collective managed construction land, and does not include the land for public welfare, rural infrastructure construction, and other aspects of land resources.

## 2 CURRENT STATUS OF THE BEIJING LAND MARKET ELEMENTS ALLOCATION AND DIFFICULTIES OF REFORMATION

Beijing land market was formally established in April 2002, and has made significant contributions to the capital city construction and economic development. However, the land transaction is limited to state-owned construction land market, and the "decisive role" in the land element allocation has not been fully reflected. This is caused by not only the existing land laws and regulations, but also the practical problems that exist in the city's land management. Many difficulties in promoting marketization of land elements allocation must be tackled.

### 2.1 The dual urban–rural division of land management mechanism, leading to no formation of unified urban and rural construction land market in the city

In the use of state-owned land, commercial construction and industrial land have been taken into account, with the basic market allocation of land element. However, according to the land management law: "Any units and individuals which need to use land, must apply for the use of state-owned land, except setting up township enterprises, village residential construction, the township (town) village public facilities and public construction which have been approved by law, farmers collective owned land can be used." Collective land can only enter the market through the national expropriation. The rigid constraints of the law led to the state-owned construction land, and collective construction land management models are different and show dual structure. Exclusion of collective construction land from land market caused a failure of unified urban and rural construction land market setup, and it eventually cannot achieve "the same place, the same rights, equal price". Approved by the Central Government, Daxing District in Beijing is considered as pilot city to trade collective construction land in land market, from 2015.

### 2.2 State-owned construction land needs to further play a decisive role in the land market

Industrial, commercial, tourism, entertainment, commercial residential land, and land with more than two investors are required to take the form of auction to sell. Market players can have the land usage rights only through fair competition. Many lands are obtained based on a wide variety of "special purpose" allocated by government for no cost, such as government usage, infrastructure

Table 1. Comparison of land transaction supply ways of state-owned constructions in Beijing from 2006 to 2014.

| Year | Allocated land ha. | Selling land ha. |
|---|---|---|
| 2006 | 2086.16 | 2515.51 |
| 2007 | 1402.53 | 2438.75 |
| 2008 | 1409.29 | 1240.64 |
| 2009 | 721.62 | 1854.02 |
| 2010 | 264.22 | 2129.00 |
| 2011 | 585.88 | 2085.71 |
| 2012 | 449.84 | 1077.25 |
| 2013 | 350.42 | 1739.39 |
| 2014 | 554.91 | 1256.47 |
| 2015 | 694.72 | 808.31 |

Resources: Beijing land natural resources yearly check.

construction, and public welfare. However, with the development of economics and the changing of time, these lands are also engaged in business activities. For example, survey conducted on hundreds of typical lands of Xicheng District in 2012 found that the government-granted lands in this area are used for commerce, catering, and other activities, changing the usage of land area by more than 20% of the total area of the parcel. The proportion of land allocation is higher every year from 2006 to 2014 (Table 1). This clearly shows the intervention of the administrative measures in the land element allocation. This is not only conducive to maintain the fairness of the market, but also decreases the land revenue of the government.

### 2.3 Management of collective construction land is lagging behind: property rights & incomplete authority, percentage of collective construction land meeting the registration policy requirement is lower, collective construction of land market is facing many problems

Beijing launched a rural collective construction land use rights special activity in 2013, focusing on the right cadastral survey and registration, which was completed by the end of 2014. The statistics of the survey shows that collective construction land with clear property source to prove the ownership accounted for 26.8% and meeting the registration certification criteria accounted for 20.4%. Through investigation, we found that collective construction land is very complicated in Beijing. Most of them lack approval materials, ownership is not clear; furthermore, there are still a large number of rural collective construction lands in illegal sale, lease, and transfer of the contract. According to current policies, they do not meet registration criteria.

Clear property rights is a prerequisite for market transactions, and collective construction land without registration does not meet the market entry conditions following current law. Moreover, according to the relevant provisions of the current land management law, property law, and security law, the collective construction land cannot be mortgaged by financing or leasing under normal operating circumstances. Marketing of collective construction land is facing many problems, especially lack of support from related policies.

### 2.4 Government-led land supply and activity in the secondary land market is not high

In recent years, the government has reserved a large number of land resources by land development. The major land supply channel is managed by the government. Objectively, this is destined according to our social system, laws, regulations, and government functions, and supposed to strengthen the macro-control of the government, urban planning, and dominant socio-economic development. However, because of the rhythm, the total land supply, land, and even prices of the supplied land, many cases are determined by the needs of macro-control, rather than the market's spontaneous behavior. Market is difficult to timely, truly, and flexibly reflect the relationship between land supply and demand and the degree of scarcity of resources. Furthermore, there is power rent-seeking and other loopholes in the process of land transactions, which suppress the vitality of the element productivity. This is actually in contrast to the market, which plays a decisive role.

In addition, the city's state-owned construction land is not active, except mortgage, land lease, transfer, and other secondary market. In the last

Table 2. Transaction of the primary land market and secondary land market in Beijing from 2006 to 2014.

| Year | Primary land market ha. | Secondary land market ha. |
|---|---|---|
| 2006 | 4601.67 | 694.05 |
| 2007 | 3841.28 | 371.03 |
| 2008 | 2649.93 | 326.23 |
| 2009 | 2575.62 | 202.46 |
| 2010 | 2393.04 | 226.76 |
| 2011 | 2671.59 | 253.22 |
| 2012 | 1527.09 | 165.63 |
| 2013 | 2089.81 | 70.80 |
| 2014 | 1811.38 | 37.34 |
| 2015 | 1503.02 | 40.41 |

Resources: Beijing land natural resources yearly check.

3 years, for example, both the number of transactions or trading area and land transfer are far less than the same period in the primary market bidding to sell. On the contrary, many of the land transfer behaviors did not fulfill the legal formalities. They were processed by the two parties privately.

## 3 PRINCIPLES TO BE FOLLOWED TO PROMOTE THE REFORMATION

Understanding the purpose and significance of this reform from the strategic perspective of the capital development, full competition in the field of land elements is beneficial to reduce the element cost, increase the supply ability of the elements, and improve the international competitiveness of the industry. The following steps should be performed in the reform:

First, adhere to the nature of land ownership unchanged.

Second, adhere to the legislation. Before the reform, amend land management law and other laws and regulations in favor of the implementation of the reform.

Third, adhere to the gradual reform. On the basis of the overall framework of the reform, in accordance with the procedure steadily: pilot—summary—improve—the promotion of the steps.

Finally, adhere to the principle of urban and rural integration. Simultaneously develop the collective construction land market pilot and improve the existing state-owned construction land use, trading, and management policy system, in accordance with the requirements of the market.

## 4 RECOMMENDATIONS TO PROMOTE MARKETIZATION OF LAND ELEMENT ALLOCATION

Any reform involving land system is highly challenging, especially marketization reforms of land elements. Therefore, we cannot hope it to happen overnight. Land market reform can be processed following the below three steps.

### 4.1 Improve the land market supporting policies and systems gradually

Expand the scope of compensation for the use of land, refine non-profit land directory, and strictly distinguish uses and the nature of education facilities, research institutions, sports facilities, and medical health projects. All land shall be used in the production areas through open market transactions.

Explore other land supply methods, like land elasticity sell and sell after rent; study and formulate land secondary market management rules; and standardize the behavior of the transfer to activate the secondary market to create an open and transparent market environment for all market players. A fair and efficient, unified urban and rural construction land market is finally built.

## 4.2 Promote the ownership certification, planning revision, illegal land processing to establish a unified urban and rural construction land market

First, continue to actively and steadily promote the collective construction land use rights registration and ownership certification work, to provide a clear property rights system for the market-oriented reforms. Second, construct an appreciable overall land use plan, for the organic convergence of urban and rural planning and national economic and social development and planning. Third, increase the intensity of collective construction land use violation cases in treatment and clean up "historical account", for the collective management of construction land market in the city "Wrecker".

## 4.3 Respect the rules of market economy and decentralize, to let the market emit its effects

Reduce the burden on the land market, let the market perform in accordance with its internal rule, and let the market fully develop in a favorable environment. Other social issues and government functions should be solved in other effective and reasonable ways, such as "market do market businesses, government acts government".

The city land and resources departments should be solely dependent on the land market management functions, including the establishment of a unified land market trading platform, to develop trading rules, protect all types of property rights equally, and formulate a list of developed markets, with market regulators as "reference". Affordable housing and facilities for the elderly, educational facilities, and urban infrastructure construction, which need sponsor from government, can be resolved through redistribution of land market returns.

## 5 CONCLUSION

With an active market economy, the land element plays an important role in the economic development of Beijing. We need to break the dual structure of the urban–rural land system to give full play to the role of land elements in the market economy through the reform of land element market, establish a unified land market in urban and rural areas, and promote economic development of the capital quickly and smoothly.

## REFERENCES

Di, J.G. & Wu, K.P. (2013). The Research on the Contribution of Construction land expansion to China's Non-agricultural economic Growth. *J. Journal of Applied Statistics and Management.* 414–424.

Du, G.Y & Cai, Y.L. (2007). Analysis on Total Factor Productivity Including Construction Land Input in China from 1997 to 2007. *J. China Land Science.* 7, 59–65.

Feng, L. & Wei, L. (2008). Study on Contribution of Land Element to Economic Growth in China. *J. China Land Science.* 4–10.

Jiang, H. & Qu, F.T. (2009). Contribution and Response of Constructed Land Expansion to Economic Growth at Different Development Stages: A Case Study for Jiangsu. *J. China Population, Resources and Environment.* 19 (1): 70–75.

Li, M.F. (2010). Research on the Contribution of Land Element to China's Economic Growth. *J. Journal of China University of Geosciences: Social Sciences Edition.* (1): 60–64.

Mao, Z.Q. & Zuo, Y.Q. (2007). Study on Contribution Rate of Land to the Second & Service Industry Growth. *J. China Land Science.* (3): 59–63.

Mao, Z.Q. & Zuo, Y.Q. (2009). Re-discussion on Land Contribution to the Secondary and Tertiary Industries in China. *J. China Land Science.* (1): 19–24.

Nichols, D.A. (1970). Land and Economic Growth. *J. The American Economic Review.* 60(3): 332–340.

Solow W.A. (1956). Contribution to the Theory of Economic Growth. *J. Quarterly. J. Journal of Economics.* 70(1): 65–94.

Wu, K.P. & Yang, W.L. (2009). Economic Growth With Land Factors in New classical Theory. *J. System Engineering-Theory & Practice.* 50–55.

Xia, F.Z. & Yan, J.M. (2014). Study on the Effect of Land Elements on Technical Efficiency of Economic Growth in China Based upon Stochastic Frontier Analysis. *J. China Land Science.* (7): 4–10.

Zhang, L.Q. & Chen, S.P. (2014). the Contribution of Land to the Economic Growth and Inflection Point of Its Logistic Curve in Anhui Province in Recent 15 Years. *J. Sciatic Geographical Science.* (1): 40–46.

Zhang, Y.X. & Jin, Z.H. (2012). The Output Elasticity Of Urban Land and Its Contribution to Economic Growth. *J. Economic Theory and Business Management.* (9): 49–54.

Zhao, K. & Zhang, B.X. (2014). Theory of Economic Growth Quality Effect on Urban Land Expansion: An Empirical Study of 14 Cities in Liaoning Province. *J. China Population, Resources And Environment.* 24(10): 76–84.

*Motivation, automation and electrical engineering*

*Advances in Energy and Environment Research – Achour & Wu (Eds)*
© 2017 Taylor & Francis Group, London, ISBN 978-1-138-62682-9

# Design and implementation of a $PM_{2.5}$ remote sensing monitoring system based on Hadoop

L. Wang
*State Key Laboratory of Remote Sensing Science, Institute of Remote Sensing and Digital Earth, Chinese Academy of Sciences, Beijing, China*
*Computer and Information Engineering College, Henan University, Kaifeng, China*

S. Xu
*School of Mathematics and Computer Science, Xinyang Vocational and Technical College, Xinyang, China*

F.B. Zheng
*Computer and Information Engineering College, Henan University, Kaifeng, China*

Y.D. Si
*State Key Laboratory of Remote Sensing Science, Institute of Remote Sensing and Digital Earth, Chinese Academy of Sciences, Beijing, China*

Q. Ge
*Computer and Information Engineering College, Henan University, Kaifeng, China*

ABSTRACT: With the development of space technology of China, remote sensing data are growing exponentially, which has put forward a high demand of the storage and computation capabilities of related application systems. In this paper, a $PM_{2.5}$ remote sensing monitoring system based on the framework of cloud computation and Hadoop platform is designed. By using HDFS and MapReduce technologies, the redundant storage and parallel processing of massive data are achieved. Through the speedup analysis of the $PM_{2.5}$ product from the GF-1 satellite, it validates the high processing efficiency and availability of our system.

## 1 INTRODUCTION

$PM_{2.5}$, also known as fine particulate matter, is the general term for air-suspended particles whose diameter is less than or equal to 2.5 micron. It can be suspended in air for a long time. The higher the concentration is in the air, the severer air pollution will become. With the help of ground monitoring stations, one can be kept informed of the $PM_{2.5}$ levels around the monitoring stations, but these conventional monitoring methods lack the ability of an effective large-scale monitor. Remote sensing data have the characteristics of large-scale covered, quasi real-time accessing, dynamical updating, and low cost and therefore, the use of the satellite remote sensing technology to monitor $PM_{2.5}$ embraces the features of large-scale monitoring, faster data accessing, shorter accessing cycle, less restricted conditions, etc. With the continuous development of China's aerospace industry, series of satellites, such as the FY, the HJ, and the GF, have been successfully launched. Among

which, the GF-1, which was successfully launched by using a Long March 2C launch vehicle at our Jiuquan Satellite Launch Center on 26 April 2014, is marked as the first star of the national science and technology project under the high-resolution Earth observation system. After the successful launch of the GF-1 satellite, it is able to provide highly precise and wide-range space observation services for the Ministry of Land and Resources, Ministry of Agriculture, environmental protection department, geographic mapping, marine, water conservancy, forestry resources monitoring, and city and public health emergency area. With the advantages of high spatial resolution imaging for earth observation and significant combination, the GF-1 satellite also shows strong potential in the field of atmospheric remote sensing.

The ultrahigh spatial resolution (16 m) of the GF-1 satellite results in an exponential growth of data. Therefore, in the course of remote sensing applications, it is no doubt that computing tasks are bound to multiply. For example, based on the

NASA data collected by using a Moderate Resolution Imaging Spectroradiometer (MODIS), it often takes several hours or even longer to conduct the inversion of parameters of atmospheric pollution in China at 1 km resolution with a single machine. For sub-meter GF-1 data, even resampled to 100 m, its calculation time is still theoretically 100 times longer than the time taken by MODIS 1 km data. Consequently, the requirements of GF-1 on computer data processing are also on the rise. It is the emergency of parallel processing that solves the problem that a single-machine environment cannot deal with two or more tasks simultaneously. Graham E. discussed that the MPI parallel processing could achieve shared storage, by virtue of being logic-simple and easily deployed. However, MPI is weak at fault tolerance and unbalanced loading, which means that once the node fails, all the related tasks must be recalculated.

Being an open-source distributed computing framework of Apache open-source organizations, Hadoop is the general term that is used to denote HDFS (Hadoop Distributed File System, the Distributed File System), MapReduce, HBase, Hive, ZooKeeper, etc. At present, a cloud computing platform, with a distributed computing open source framework—Hadoop—as a core has proposed a new solution to large-scale data computation. Hadoop has been widely used in the field of Web searching, e-commerce advertising analysis, and recommendation system on the Internet. There are many international and domestic scholars who have tried to combine Hadoop and remote sensing to develop, for example, Fang Y studied the distributed processing of vector spatial data based on an open source Hadoop; Yue Z, by using Hadoop implements, built an urban transport carbon emissions data mining prototype system. On the basis of above-mentioned studies, in this paper, an attempt is made to design a system to monitor $PM_{2.5}$ distribution of certain regions or the whole country based on the Hadoop platform by using GF-1 data. In this paper, the first part introduces the architectural design of the overall system; the second part describes the core algorithm of this system in detail; the third part shows performance analysis and examples test, and the last part presents the summary and prospects of the system.

## 2 SYSTEM ARCHITECTURE DESIGN

The overall framework of the system is similar to the cloud computing architecture, which could be respectively divided into infrastructure layer, platform layer, application layer, and clients. These four layers are arranged from the top layer to bottom. Based on the massive remote sensing

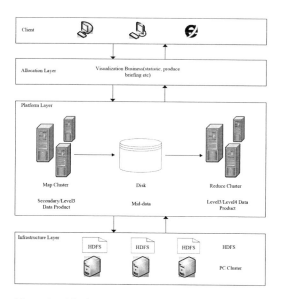

Figure 1. The framework of the system.

data mainly provided by the HDFS, MapReduce provides efficient parallel processing on these data (as shown in Figure 1).

### 2.1 Infrastructure layer

An infrastructure layer is a collection of virtualized hardware resources (compute, storage, and network) and related management functions. Through abstracting physical resources with virtualization technology and hiding unnecessary details from the users, developers can implement some or all functions in a virtual environment that users seriously require in reality. Sub-meter remote sensing data can be identified as the "golden data" in the international field of remote sensing. Due to the importance of satellite data, the method of multi-copy storage of HDFS is exploited to prevent GF-1 data from being lost in the case of a hardware failure, and improve the response speed, truly realizing high-reliability and high-availability storage of a tremendous amount of information. Therefore, an infrastructure layer is mainly composed of PC clusters and a distributed file system that is deployed on PC clusters. The internal storage of satellite data is usually realized in the form of blocks, by default under HDFS, the size of each block is 64 M. According to the total amount of GF-1 data and cluster conditions, an infrastructure layer can adjust the block size and divide the data gained from the satellite based on this changed size. A HDFS consists of a NameNode node and numerous DataNodes nodes. Each block is stored in three copies in different DataNodes. To access

HDFS satellite data, it is necessary to firstly obtain its metadata information through the NameNode node, read the relevant parameter information, and finally realize the data management functions.

## 2.2 Platform layer

A platform layer provides service and management similar to that of the operation system. That is to say, developers don't care about which machine is used to run the program they edited. Through geometric correction, image cropping, and other treatments, raw satellite data could be processed to obtain apparent reflectance level-one data products. Using primary satellite data and the most widely used NASA dark pixel algorithm, secondary data products with AOD (Aerosol Optical Depth, AOD) could be retrieved. Generally, AOD is the most widely used parameter of atmospheric remote sensing, which represents the whole-layer extinction capacity of particulate matter from the ground to the air. Secondary data—AOD products—can be operated to obtain the revised level-three remote sensing products—$PM_{2.5}$—through vertical correction, and business statistics of retrieval level-three products; it provides users a quick check of the $PM_{2.5}$ distribution at a certain time throughout the country or at a certain region. Based on the retrieval result of the previous step, level-four $PM_{2.5}$ applications could be acquired through statistical analysis or other users defined before. In this paper, an attempt has been made to apply MapReduce parallel processing methods to the process mentioned above in order improve the efficiency of obtaining products of each level; and conduct the retrieval calculation of the primary and business statistics of retrieval level-three products, it provides users a quick check of the $PM_{2.5}$ distribution at a certain time throughout the country or at a certain region.

## 2.3 Application layer

The application layer lies on top of the infrastructure layer and the platform layer, whose design follows the philosophy of the cloud computing service software, namely, service (Software as a Service, SAAS) philosophy, in which all the resources and functions are available to users in the form of services. With regard to the needs of the entire system and the data after disposal of the MapReduce stage, the application layer is responsible for statistics of which areas are covered by different levels of $PM_{2.5}$ concentrations and producing briefing and other services (as shown in Table 1).

According to the data stored in the HDFS output stage, we could know that which level of different concentrations a certain region belongs to, and then call the template to produce briefings.

Table 1. Air quality standard of $PM_{2.5}$.

| $PM_{2.5}$ Average daily concentrations ($\mu g/m^3$) | Air quality levels | Air quality type |
|---|---|---|
| 0–35 | 1 | Excellent |
| 36–75 | 2 | Good |
| 76–115 | 3 | Slightly polluted |
| 116–150 | 4 | Moderately polluted |
| 151–250 | 5 | Heavily polluted |
| >250 | 6 | Hazardous |

## 3 CORE ALGORITHM

MapReduce parallel processing was respectively used in the following two processing procedures; one is the process of transferring from the primary data product with apparent reflectivity obtained from GF-1 to level-three $PM_{2.5}$ remote sensing products (referred to as the first stage) and the other is the process of acquiring level-four applications of $PM_{2.5}$ by statistically classifying level-three $PM_{2.5}$ remote sensing products according to a certain time interval (referred to as the second stage). In the first stage, the processing time the employment of MapReduce consumed has been reduced apparently, and in the second stage, parallel processing is employed to obtain statistical results on the basis of saving time. This section will introduce the two parts of parallel processing in detail.

## 3.1 $PM_{2.5}$ satellite production process

Given the total amount of primary data product, the primary data can be seen as a data block (approximately 300 MB), all of which becomes the data input in the Map stage. The map stage is mainly responsible for transferring from a primary data product to a $PM_{2.5}$ level-three satellite product through various disposals. This procedure includes the following two satellite retrieval methods (as shown in Figure 2):

1. Dark Pixel Algorithm: based on apparent reflectance observed by a satellite, it can solve problems such as data preprocessing, cell identification, ground reflection contribution, aerosol mode and several other issues to achieve the inversion of aerosol optical depth and conduct pixel operations separately with satellite data. Due to the massive radiation transferring and interpolation, the process is quite time-consuming. Therefore, using the distributed computing model cloud platform can greatly shorten the computation time.

2. Vertical Correction and hygroscopic Correction: since AOD represents the extinction capacity of particulate matter of the whole layer and

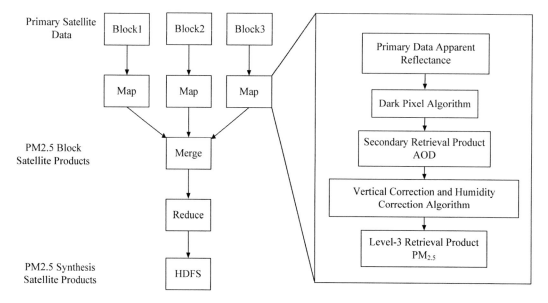

Figure 2.    The product flow of a PM$_{2.5}$ based-satellite using MapReduce.

PM$_{2.5}$ stands for mass concentrations near the dry ground, a two-step vertical correction and moisture absorption correction are required to transfer the satellite retrieval of AOD into PM$_{2.5}$. Through this process, the pixel scales of PM$_{2.5}$ distribution products are obtained.

After a sort-merge operation, the output in the map stage could be regarded as an input in the reduce stage. In the reduce stage, the principal task is to convert and merge blocked PM$_{2.5}$ products and then obtain PM$_{2.5}$ synthetic products throughout the country or in a certain region at a certain time, and finally store them in the HDFS (as shown in Figure 2).

## 3.2   PM$_{2.5}$ statistical production processes

This section focuses on monitoring the PM$_{2.5}$ distributions for the whole country or some regions at a certain time based on Hadoop by using synthetic PM$_{2.5}$ products and taking advantage of MapReduce to achieve statistical and other functions, which mainly includes four stages, namely data input, map stage, reduce stage and data output (as shown in Figure 3).

Specific procedures are listed as follows:

1. Data input stage: taking the total amount of the level-three products and each data block into account, the size of each data block is set to 64 MB as default. Because PM$_{2.5}$ level-3 remote sensing products are in .tiff format and Hadoop is more suitable for file operations, as

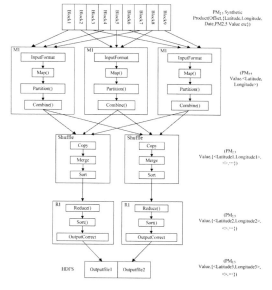

Figure 3.    The statistical product flow of PM$_{2.5}$ by using MapReduce.

a result, it is necessary to transform each data block into text file format when these are sent as input, as shown in Figure 2. In this case, the text files are stored in rows by offset, latitude, longitude, date, and PM$_{2.5}$ values, among which each row is a record. The MapReduce parallel programming model contains a variety of classes and interfaces for the development of parallel processing programs. Here, the text format is selected as the input format, which uses InputFormat default format—TextInput Format, in order to input each row of the data as a text value. Key is the offset calculated from the text's starting position to the row's starting position. Thus, the data format of the input stage is (offset, {latitude, longitude, date, $PM_{2.5}$}). In the data input stage, the transformation from the data block to the key–value key/value pair can be realized.

2. Map stage: according to the requirements of specific applications and customized map function, this model uses the map function for the sake of extracting the $PM_{2.5}$ values, latitude and longitude, and date information, and outputs them in the form of key–value pair ($PM_{2.5}$ values, <latitude, longitude, date>), while filtering out the key–value pair in which value of $PM_{2.5}$ is 0 at the same time. Intermediate results of the map stage will be written to disk. Before writing to disk, the thread will divide the disk into different partitions based on the date in the key–value pair, and partition operation needs to call the getPartition() method in the partitioner class. In each partition, the data will be sorted by the key-$PM_{2.5}$ value. In order to reduce the amount of the network transmission rate and the tasks of each reduce, we would add a combiner onto each map task, which could be regarded as the local reduce. Each map task makes use of the combine() method to merge the data based on the key–value pair, the combined data show the key value list form like ($PM_{2.5}$ values, {<latitude 1, longitude 1, the date>, <2 latitude, longitude 2, the date>, …}). Under the unified control of job tracker, these data could be assigned to different computing nodes to conduct calculations in the reduce stage.

3. Reduce stage: In the reduce stage, intermediate results need to be read from the disk in the map stage, and the results from multiple map tasks are confronted with merging and sorting operations based on the different time intervals of each reducer by using the key value. With the reduce() function designed depending on specific applications, this model uses the hash function to hash the key value of a certain day onto a corresponding reducer to conduct standardization and merge, and then realize statistical analysis about the $PM_{2.5}$ distribution conditions of a certain time interval throughout the whole country or at a certain region. For example, at the same time period, when the value of $PM_{2.5}$ is 80, on the basis of the result obtained by the reduce stage, we could obtain how many cities that the $PM_{2.5}$ value is equivalent to 80 there are in a certain region or the number of cities of

the concentration equal to 80 accounts for the whole country. According to China's air quality standards (as shown in Table 1), on the basis of the results obtained at the reduce stage; the proportion that the $PM_{2.5}$ value of each range has taken in the whole country or a certain region is achieved.

4. Data output stage: since a reduce task corresponds to one output file, we would use the output format class in the MapReduce model to classify many output files and store the merged files in HDFS.

## 4 EXPERIMENT AND ANALYSIS

### 4.1 *Performance analysis*

Based on the GF-1 original satellite data collected from January to March 2014 in the Beijing–Tianjin–Hebei region, this experiment obtained a primary data product whose size was approximately 183 GB; after various middle-management measures, the amount of the $PM_{2.5}$ level-three remote sensing products reached 3.3 GB in total. Four test machines are monitored to process data and all of them are under the same configuration, specifically as follows: processor AMD Athlon (tm) X4 750 k quad-core 3.4 GHz, memory 4.0 GB, and hard drive 1T. By using the VMware virtual machine, install JDK1.7.0_17, Hadoop 1.2.1, and the programming platform is Eclipse 3.7. The virtual machine is used to virtualize one physical machine into several virtual machines, forming 8 nodes or 16 nodes of the Hadoop cluster, respectively. Among these nodes, there is one node whose name is Master, as NameNode and Job tracker, and the remaining nodes as DataNode and Task tracker. Different nodes are used to simultaneously carry out the procedure to transform from level-one data products to level-three remote sensing products, and the calculating time is shown in Figure 4.

As shown in Figure 4, at least 10 hours are needed to deal with the procedure of transformation from primary data products to level-three remote sensing products with a traditional single machine approach; when working nodes are expanded to 16 nodes, it only takes two hours at most. Thus, time is greatly reduced by using parallel processing computation.

The calculating time of transformation from 3.3G $PM_{2.5}$ level-three remote sensing products to level-four application products by using different nodes in parallel processing, as shown in Figure 5.

As shown in Figure 5, with traditional single machine mode, processing the mentioned data needs at least 67 min; when extended to 16 nodes, not more than 10 min is needed.

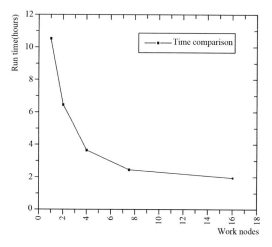

Figure 4. Comparison of computations performed by using different nodes from level-3.

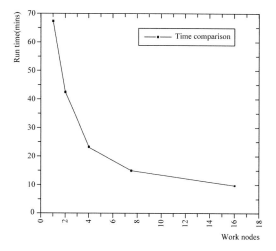

Figure 5. Comparison of computations performed by using different nodes from level 3 to level 4.

With an increase in the number of working nodes, the transmission time between nodes also slightly increased. Though the running time is not strictly reduced by half, the time for processing the data of the two parts is quite obviously declined and the speed-up ratio reaches to 5 and 6.5, respectively. Thus, it is very suitable to use the parallelization approach to process the above-mentioned data, that is, the speed-up ratio will be greatly enhanced with more worker nodes.

### 4.2 *Applications*

Analyzing the distribution statistics of $PM_{2.5}$ in the Beijing–Tianjin–Hebei region from January to March in 2014, by using this system, we could

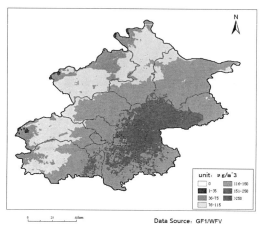

Figure 6. Beijing $PM_{2.5}$ satellite remote sensing figure as on March 18, 2014.

obtain all latitude and longitude pairs which correspond to different $PM_{2.5}$ values and all corresponding latitude and longitude pairs and numbers when the values are the same. On 18 March 2014, we displayed the distribution condition of $PM_{2.5}$ concentrations of Beijing by calling a template, which is illustrated in Figure 6. According to China's air quality standards (Table 1), the values of $PM_{2.5}$ in different ranges correspond to different colors, through which we could broadly see the characteristics of $PM_{2.5}$ pollution in Beijing that it is more consistent with northeast–southwest direction and from west to east, the problem of pollution becomes more and more serious.

### 5 SUMMARY

In this paper, the system's design concept, the overall architecture, and the use of this system to monitor the $PM_{2.5}$ pollution across the country are mainly introduced. Due to the scalability of nodes in the Hadoop platform, with the increase of remote sensing data, under the conditions of not affecting the system design cost, only increasing the number of nodes in the cloud platform could make the processing performance of the whole system remain the same and achieve monitoring of the counting activity of the full-year distribution of $PM_{2.5}$ throughout the country. The experiment can generally provide good monitoring results, which could reveal $PM_{2.5}$'s spatial distribution of different concentrations across the country in macro. Taking different treatment measures in the light of different pollution degrees, we could make a better contribution to the $PM_{2.5}$ pollution issues. In future, we will standardize the satellite data processing based on the MapReduce processing proposed in this paper, extend it to a more remote sensing inversion algorithm, and then accelerate remote sensing products' operation and application.

## ACKNOWLEDGMENTS

This work was supported by the National Key Technology Support Program (grant number: 2014BAC21B00) and National Natural Science Foundation of China (41571417).

## REFERENCES

Dean, J. & Chemawat, S. 2008. MapReduce simplified data processing on large cluster. *Communications of the ACM*. 51(1): 107–113.

Fang, Y., Min, F. & Yunqiang, Z. 2013. Research on vector spatial data distributed computing using Hadoop projects. *Computer Engineering and Application*. 49(16): 25–29.

Graham, E. Edgar, G. & George, B. 2004. Extending the MPI specification for process fault tolerance on high performance computing system. *In Proceedings of the 2004 International Supercomputing Conference (ISC2004)*.

Hongbin, Y. Xudong, Z. & Hongyu, W. 2012. Study progress on $PM_{2.5}$ in atmospheric environment. *Journal of Meteorology and Environment*. 28(3): 77–82.

Kaufman, Y.J. Waid, A.E. & Remer, L.A. 1997. The MODIS 2.1 $\mu$m channel-correlation with visible reflectance for use in remote sensing of aerosol. *IEEE Transactions on Geoscience and Remote Sensing*. 35(5): 1286–1298.

Liangfu, Chen. Shenshen, Li. & Jinhua, Tao. 2011. *Research and Application of Aerosol Quantitative Remote Sensing Inversion*: 92–99. Beijing: Science Press.

Mchenry, J.N. Vukovich, J.M. & Hsu, N.C. 2015. Development and implementation of a remote-sensing and in situ data-assimilating version of CMAQ for operational $PM_{2.5}$ forecasting. Part 1: MODIS Aerosol Optical Depth (AOD) data-assimilation design and testing. *Journal of the Air & Waste Management Association*. 65(12): 1395–1412.

Qulin, T. & Yun, S. 2000. Application of remote sensing technology to environmental pollution monitoring. *Remote sensing technology and application*. 4(15): 246–251.

Rajarama, A. & Ullman, J.D. 2012. *Mining of Massive Datasets*: 16–4. Beijing: Posts and Telecom Press.

Shenshen, L. Liangfu, C. & Xiaozheng, X. 2013. Retrieval of the haze optical thickness in north China plain using MODIS data. *IEEE transaction on Geoscience and Remote Sensing*. 51(5): 2528–2540.

Wei, W. 2008. Study of grid enabled high performance aerosol quantitative retrieval from remotely sensed data. Beijing: The Institute of Remote Sensing Application.

White, T. 2010. *Hadoop: The Definitive Guide 2nd*. Beijing: Tsinghua University Press.

YanLong, Z. Zhuang, L. & Kai, Y. 2013. High Performance massive data computing framework based on Hadoop cluster. *Computer Science*. 40(3): 100–103.

Yue, Z. Siqi, J. & Junkui, Z. 2011. Research of urban traffic carbon emission data mining based on Hadoop. *Application Research of Computers*. 28(11): 4213–4215.

Zifeng, W. 2010. *Research on the algorithm of satellite remote sensing estimation of particle concentration near the ground*. Beijing: The Institute of Remote Sensing and Digital Earth.

*Advances in Energy and Environment Research – Achour & Wu (Eds)*
*© 2017 Taylor & Francis Group, London, ISBN 978-1-138-62682-9*

# A novel design of the MANFIS-PSO PID controller for quadrotors

Chun-Chih Ko, Chung-Neng Huang & David T.W. Lin
*Graduate Institute of Mechatronic System Engineering, National University of Tainan, Tainan, Taiwan*

Jenn-Kun Kuo
*Department of Greenergy Technology, National University of Tainan, Tainan, Taiwan*

ABSTRACT: Nowadays, the exploration technologies mainly rely on the satellite positioning or the video capturing of Unmanned Aerial Vehicles (UAVs). Here, for the UAV is with the natures of high mobility and secrecy in operation, low cost, and without the limits on terrain obstacles, UAVs are numerously employed in military applications to decrease casualties. However, weather condition is the major factor of uncertainty that affects the video capturing of a UAV. In order to obtain high-definition video data, how to level up the balancing technology for the UAV becomes one of the most important topics in the UAV development. For the above-mentioned target, a novel PID tuning technology based on the integration of MIMO-ANFIS and PSO is designed in this paper. The control effectiveness is confirmed through the Matlab/Simulink simulation study on a quadrotor model.

## 1 INTRODUCTION

A quadrotor is small, light-weighted, and disposable. As the development of smart phones, it brings forward the development of electronic gyroscope, GPS, fly-by-wire flight control system, and takes advantage of the electric motor installed with accelerator with fast speed response time as its power system; hence, it has hovering, vertical take-off, and landing capabilities. It is often used in civilian and military fields, which include scouting, monitoring, environmental defecting, rescuing activities, and aerial photography (Trong, 2015; Yang, 2016). A quadrotor's control mechanism is to use the propeller for balancing and steering, to use a gyroscope for detecting the angle of deflection, increasing or decreasing the speed of the propeller to achieve balance or deflection. By using the PID (proportion—integral—differential) control method, it helps a quadrotor improve flight stability. It also applies other control methods, such as adjusting PID parameters to get parameters' most optimal values or wave filtering to reduce noise to increase its stability. In addition, the quadrotor's dynamic equation is to apply the Newton—Euler equation to establish 6 degrees of freedom dynamic equation of motion, so as to facilitate control and simulation and further research (Trong, 2015; Yang, 2016; Salih, 2010).

When the quadrotor flies in the air, its aircraft is easily influenced by external factors, and its aircraft hull is also easily affected by loading, flight time constraint, and dynamic nonlinear and uncertainty of setting parameters; therefore, affects flight stability (Jeong, 2014). In control, the quadrotor is installed with a 6-output and 4-input under-actuated system that has features such as multivariable, nonlinear, strong-coupling, and interference sensitivity; therefore, it makes the flight control system hard to design and makes its accelerator control hard to be precisely accurate and prompt.

During the flight, an unmanned aerial vehicle would encounter many nonlinear factors; therefore, it is very important to choose the type of algorithm that is capable of fast convergence and excellent traceability. PSO (Particle Swarm Optimization) is the most optimal search algorithm built on the concept of birds and can be used as an auto-adaptive tool for adjusting PID parameters. It is also often used as comparison to the GA algorithm method in the academic field; because these two interpret a large search quantity as a biomorphic population. When the particles move, it will move according to individuals' or groups' most optimal experiences. Also, a number of metrics has similar concepts to the genetic algorithm's "mating" operator; one of their biggest differences is that, when the particle moves, it uses the concept of random selection and avoids choosing "best solution in the region". However, flight speed and memory ability that exists in particles in PSO do not exist in GA algorithm method. PSO and GA's comparative advantages are that, it only requires simple calculation, does not need to setup too many parameters, and uses a minimum code to run in computers.

PSO's most optimal solution search, in the late iteration period, has not been stable; hence, in this article, we use both MANFIS and PSO for adjusting the PID and make this system more stable. We also utilize mixed types of control and combine virtues of all control methods and make the result more reliable. Basically, MANFIS is composed by using the multiple artificial neural network and fuzzy control (ANFIS). Because MANFIS has an outstanding learning ability and fuzzy control can simulate human thinking, combining these two has a learning ability that is able to adapt to environmental changes, and in the mean time, has fuzzy inference's thinking, and can be used in situations that require predictions (Andreas, 2012). Hence, this article explores further into quadrotors, the aircraft flight control system is used as the control, a PID controller is added, and fuzzy control is used, and PSO and PSO-MANFIS control methods are used to design an auto-adaptive PID controller. We expect the improvement and enhancement of quadrotors for better stability and balance and lowering response time.

## 2 THE QUADROTOR CONTROL MODEL

Quadrotors external form is X-shaped. On its four edges, there are four rotors at the same height that are used as the flight's direct power source. Four rotors have the same structure and radius. Four motors are symmetric forms that are installed separately on the aircraft's end supports. When the flight takes off, rotor 1 and rotor 3 rotate counterclockwise, rotor 2 and rotor 4 rotate clockwise, and the flight control board and other external equipment are placed in the central space inside end supports (refer Fig. 1).

### 2.1 Quadrotor dynamics model (Mckrrow, 2004)

At first, we setup separately the following two basic coordinate systems: an inertial coordinate system E (OXYZ) and an aircraft coordinate system B (o'xyz). These two define separately the Eulerian

angles' aircraft coordinate system and the inertial coordinate system models (1) (c:cos, s:sin):

$$
R = Rx \cdot Ry \cdot Rz
$$
$$
= \begin{bmatrix} c\psi c\phi & c\psi s\theta s\phi & c\psi s\theta c\phi + s\psi s\phi \\ s\psi c\theta & s\psi s\theta s\phi & s\psi s\theta c\phi - s\phi c\psi \\ -s\theta & c\theta s\phi & c\theta c\phi \end{bmatrix} \tag{1}
$$

where R denotes matrix transformation; $\psi$ denotes Yaw; $\theta$ denotes pitch; $\phi$ is the roll angle. $F_f$, as quadrotors work together, is given by equation (2), which is as follows:

$$
F_f = \begin{bmatrix} c\varphi c\psi s\theta + s\psi s\varphi \\ c\varphi s\psi s\theta - s\varphi c\psi \\ c\varphi c\theta \end{bmatrix} \sum_{i=4}^{4} F_i \tag{2}
$$

$$
F_i = C_L \omega_i^2 \tag{2-1}
$$

i = 1, 2, 3, 4

$C_L$ is the lift coefficient; and $\omega_i$ is the motor angular speed.

The general coordinate transformation is given by equation (1) and the conjunction with the propeller is given by equation (2). Its kinetic model is given by equation (3), which is as follows:

$$
\left.\begin{aligned}
\ddot{x} &= -\left(c\varphi s\theta c\phi + s\varphi s\phi\right)\frac{u_1}{m} \\
\ddot{y} &= -\left(c\phi s\theta s\varphi - s\phi s\varphi\right)\frac{u_1}{m} \\
\ddot{z} &= g - \left(c\phi c\theta\right)\frac{u_1}{m} \\
\ddot{\phi} &= \dot{\theta}\dot{\varphi}\left(\frac{I_y - I_z}{I_x}\right) - \frac{I_r}{I_x}\dot{\theta}g\left(u\right) + \frac{L}{I_x}u_2 \\
\ddot{\theta} &= \dot{\phi}\dot{\varphi}\left(\frac{I_z - I_x}{I_y}\right) - \frac{I_r}{I_y}\dot{\phi}g\left(u\right) + \frac{L}{I_y}u_3 \\
\ddot{\varphi} &= \dot{\phi}\dot{\theta}\left(\frac{I_x - I_y}{I_z}\right) + \frac{1}{I_z}u_4
\end{aligned}\right\} \tag{3}
$$

where $I_x$ denotes the x axis of the moment of inertia; $I_y$ denotes the y axis of the moment of inertia; $I_z$ denotes the z axis of moment of inertia; $I_r$ denotes the motor's moment of inertia; $m$ denotes the quadrotor's quality; and $\omega_i$ denotes the motor's angular speed.

$$
g\left(u\right) = \left(\omega_1 - \omega_2 + \omega_3 - \omega_4\right) \tag{3-1}
$$

where $g(u)$ is the Gyro torque; $u_1$ is the total of four rotors lift; $u_2$ is the pitching moment; $u_3$ is the roll torque; and $u_4$ is the Yaw moment;

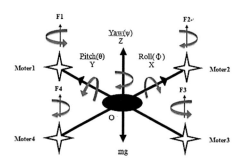

Figure 1.  Schematic of a quadrotor' structure.

And can be used in equation (4), as follows:

$$\begin{bmatrix} u_1 \\ u_2 \\ u_3 \\ u_4 \end{bmatrix} = \begin{bmatrix} C_L & C_L & C_L & C_L \\ C_L & 0 & C_L & 0 \\ 0 & -C_L & 0 & C_L \\ K_d & -K_d & K_d & -K_d \end{bmatrix} \begin{bmatrix} \omega_1^2 \\ \omega_2^2 \\ \omega_3^2 \\ \omega_4^2 \end{bmatrix} \qquad (4)$$

where $K_d$ is the drag coefficient.

## 3 MANFIS-PSO PID CONTROLLER

### 3.1 *PID controller*

A PID controller is the most widely used close-loop controller in the industry. Its PID algorithm can be expressed as equation (5). Suppose Y(t) is the control output (refer Fig. 2),

$$G(S) = K_P + \frac{K_I}{S} + K_D S \qquad (5)$$

where $K_p$ is the proportion; $K_I$ is the integral; and $K_D$ is the differential.

Because these three parameters would mutually affect each other; hence, it is difficult to achieve system stability and obtain most optimal value by setting these three parameters ($K_p$, $K_I$, $K_D$) in PID controllers. Even if the system uses a PID controller, it neither guarantees to achieve the most optimal control nor system stability. If the system requires enhancement in its stability, then it is essential to find the most optimal PID parameters.

### 3.2 *PSO-PID*

The Particle Swarm Optimization algorithm (PSO) is obtained from the birds' habit of foraging and migrating. Particles that move in the space would have an adaptive value obtained from an objective function. Every particle would have an initial speed that determines its moving direction and distance. Hence, when a group of particles rely on their individual success experiences and the current population's best particle's footstep and fly in the air, as

its individual obtains the function's most optimal value, it will record the best search's variables into its individual memory and replaces its previous value. After it completes iteration calculation, based on the particle group's adaptive value, it calculates the most optimal solution to the question. Its calculation formula is expressed as equation (6).

$$v_{id}^{n+1} = w \cdot v_{id}^n + c_1 \cdot rand(\ ) \cdot \left( p_{id}^n - x_{id}^n \right) \\ + c_2 \cdot rand(\ ) \cdot \left( p_{gd}^n - x_{id}^n \right) \qquad (6)$$

$v_{id}^{n+1}$ denotes the i A particle, at $n+1$ flight speed; $v_{id}^n$ denotes the i A particle, at $n$ flight speed; $c_1, c_2$ denote the constant current location with the best location and under the influence of a moment flying speed weight; $rand(\ )$: Between the random function 0~1 decimal; $p_{id}^n$ denotes all particles' best value found in the path; $p_{gd}^n$ denotes all the best values of the particles found in one of the best paths; $x_{id}^n$: denotes the i particle current location; and $w$ is expressed as a weight value.

Therefore, in this article, PSO features are utilized. Let every particle represent PID's three parameters' candidate values. Set the range and base on the system's transfer function to determine three parameters, such as $K_p$, $K_I$, $K_D$.

### 3.3 *MANFIS-PSO PID*

MANFIS is mainly derived from ANFIS. The main structure of these two is mixed type control of fuzzy control and artificial neural network. Fuzzy control is a type of theory derived from imitating humans' decision-making model. The fuzzy theory's main point is to establish a mathematics model, which utilizes a target system's control knowledge and simulates human behaviors, or makes human-like decisions, learns how to handle uncertainty in decision-making like humans, and furthermore designs control with better robustness and adaptability than traditional control (Sona, 2012).

As shown in Fig. 3, in this article, PID is used to perform self-adaptability to the flight status. PID

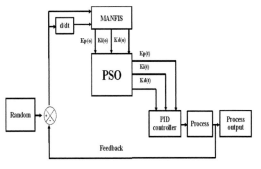

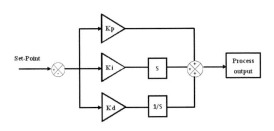

Figure 2. Schematic of the PID inversion algorithm structure.

Figure 3. MANFIS-PSO PID chart.

requires three outputs, and therefore, it is unable to utilize the MANFIS control method derived from ANFIS to perform learning prediction, and combines with PSO to adjust PID parameters and to fix some errors for PSO after the late iteration period; therefore, a PID controller can obtain the most optimal solution.

## 4 SIMULATION STUDIES

This article performs a model establishment for quadrotors and control system. It proves that the MANFIS-PSO control method is an efficient and more optimized reference. As for its postural control balance, it sends input signals from the input end separately to ROLL (as shown in Fig. 4), PITCH (as shown in Fig. 5), and YAW (as shown in Fig. 6), and discovers that MANFIS-PSO is the most optimal control. It also proves effectiveness in

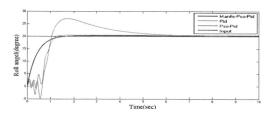

Figure 4. Graph showing the roll angle input and output.

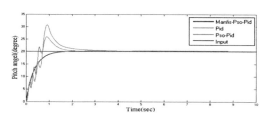

Figure 5. Graph showing the pitch angle input and output.

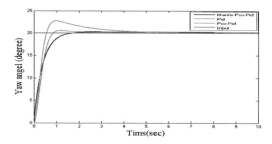

Figure 6. Graph showing the Yaw angle input and output.

self-adaptive PID control, in that it has a shorter response time than the general PID control and can boost stability of quadrotors and understands differences between these two types of control.

## 5 CONCLUSIONS

In this article, quadrotors are used as the research target, Matlab is used to establish a six-degree-of-freedom non-linear model, and a self-adaptive MANFIS-PSO PID controller design is presented. Through analysis result, it shows that MANFIS-PSO-PID and PID control methods prolong the flight system's stable time; therefore, it provides better stability to the flight during sudden change of postures.

## REFERENCES

Andreas Bortfeldt, Sep. 2012, "A hybrid algorithm for the capacitated vehicle routing problem with three-dimensional loading constraints", *Computers & Operations Research*, Vol. 39, no. 9, pp. 2248–2257.
Jeong, S. H., S. Jung, Dec. 2014, "A quad-rotor system for driving and flying missions by tilting mechanism of rotors", *Mechatronics*, Vol. 24, no. 8, pp. 1178–1188.
Le Hoang Sona, Nguyen Duy Linha, Hoang Viet Longb, Mar. 2012, "A lossless DEM compression for fast retrieval method using fuzzy clustering and MANFIS neural network", Engineering Applications of Artificial Intelligence, Vol. 29, pp. 33–42.
Mckrrow, P., Apr. 2004, "Modelling the Draganflyer four-rotor helicopter", *International Conference Robotics and Automation, IEEE,* Vol. 4, pp. 3596–3601.
Salih, A.L., May 2010, "Modelling and pid controller design for a quadrotor unmanned air vehicle", *International Conference Automation Quality and Testing Robotics (AQTR), IEEE,* Cluj-Napoca, Vol. 1, pp. 1–5.
Thang Nguyen Trong; Yang Sheng Xuan, Aug. 2015, "The quadrotor MAV system using PID control", *International Conference Mechatronics and Automation (ICMA), IEEE,* Beijing, pp. 506–510.
Yueneng Yang, Ye Yan, Apr. 2016, "Attitude regulation for unmanned quadrotors using adaptive fuzzy gain-scheduling sliding mode control", *Aerospace Science and Technology,* Vol. 54, pp. 208–217.

*Advances in Energy and Environment Research – Achour & Wu (Eds)*
*© 2017 Taylor & Francis Group, London, ISBN 978-1-138-62682-9*

# A mechanism study of power grid ancillary service cost change by intermittent power supply incorporated into the power grid

Kaijun Wang & Guoying Wu
*State Grid Zhejiang Electric Power Company, Hang Zhou, China*

Zhendong Du
*State Power Zhejiang Economic Research Institute, Hang Zhou, China*

Dunnan Liu & Qi Li
*School of Economics and Management, North China Electric Power University, Beijing, China*

ABSTRACT: In recent years, wind power and photovoltaic power generation have undergone rapid development in the world and the grid ancillary service demands are also getting higher and higher. It includes peak load regulation service cost and reserve capacity service cost. The reserve capacity cost adopts constant reliability of the power supply power method to determine reserve capacity and then to determine the cost change mechanism. Load adjustment cost is mainly analyzed through the change of load adjustment demand and then by using the load adjustment cost change mechanism. In this paper, through analyzing the change mechanism of ancillary service cost caused by the intermittent power supply connected in the system, a reference for determining the operating value of intermittent power supply is provided.

## 1 INTRODUCTION

Intermittent power supply has functions such as energy conservation and emission reduction, which is also the value of operation of intermittent power supply (Zhou, 2006). Due to the uncertainty and intermittent output of wind power and photovoltaic generation, the load peak valley difference increases, resulting in the problem of power grid load adjustment (Zhang, 2010; Zhang, 2011).

Generally speaking, with an increasing proportion of the intermittent power supply, the demand of peaking capacity also increased, by which the impact on conventional power also continues to grow. In our country, peaking power is thermal power and hydropower, in some areas there are gas-fired units. Hydropower peaking by seasonal is relatively large. The most common peaking power is the thermal power unit. When the connection's intermittent power ratio is small, only a small part of the thermal power generating unit makes pressure peak load adjustment, load adjustment cost change is not obvious, and the intermittent power supply operation value is relatively high. When the grid's intermittent power ratio is small, only a small part of the thermal power generating unit make pressure peak load adjustment, and peak shaving cost change is not obvious. The intermittent power supply operation value is relatively high.

Along with an increase in the proportion, the peak valley difference of equivalent load also increases, thermal power units which take part in peak adjustment also increases, and the corresponding coal consumption will gradually increase. When intermittent power increases to a certain extent, the pressure load cannot meet the peak adjustment demand and therefore, small thermal power units will adjust the load by carrying out the start and stop activity. But the cost of the start and stop load adjustment is significantly higher than that of the pressure load adjustment, thus causing the load adjustment cost to increase dramatically. Because the prediction accuracy of PV and wind power is not high, the prediction accuracy of the equivalent load is increased, in order to maintain the reliability of the system, which leads to the increase of spare capacity, so as to increase the cost of the reserve in the system (He, 2013).

## 2 CHANGE MECHANISM OF THE CAPACITY COST AFTER INTERMITTENT POWER SUPPLY INCORPORATED INTO THE POWER GRID

### 2.1 Deviation of equivalent load forecasting after intermittent power supply connection is increased

The reserve capacity can be classified as the load operating reserve and emergency reserve from the function, in which the load operating reserve is the spinning reserve; due to the load forecast error, we set the reserve capacity. The more accurate load forecasting is the load operating reserve capacity is smaller. The emergency reserve is to ensure electricity users without being seriously affected when the accident happened and maintain a normal standby power supply in the system. Generally about 10% of the maximum power load, but not less than the biggest capacity of the generating unit in the system.

In the actual operation, the dispatching department regards the intermittent power supply as a negative load but not as power equipment. Therefore, according to the classification of reserve capacity, intermittent power generation will increase the reserve capacity for load and does not change the accident reserve capacity.

The statistics show that the probability of the random fluctuation around the load exhibits normal distribution and the variance of the load is related to the accuracy of the load forecasting. As shown in Figure l, the load probability density function which fluctuated around the predicted value can calculate the outage probability by using load fluctuation

$$P_{LOLP} = \int_{P_1}^{\infty} \frac{1}{\sqrt{2\Pi}\sigma_{load}} \exp\left(-\frac{P^2}{2\sigma_{load}^2}\right) dp \qquad (1)$$

In this formula, $\sigma_{load}$ represents the standard deviation of the load fluctuation near the predicted value, $P_1$ represents the load reserve capacity in the system.

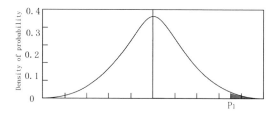

Figure 1. Graph showing the load fluctuation density function.

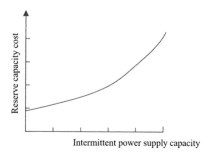

Figure 2. Relationship diagram of intermittent power capacity and spare capacity cost.

After incorporating an intermittent power supply into the power grid, due to its own forecast accuracy also showed normal distribution, so that we can set it as a prediction of standard deviation of the intermittent power generation capacity.

$$\sigma_{new} = \sqrt{\sigma_{load}^2 + \sigma_r^2} \qquad (2)$$

As for a new load-predicted standard error, in order to maintain given outage probability $P_{LOLP}$, we can figure out the corresponding load reserve capacity by using the normal distribution table, which is as follows:

$$P_2 = \sigma_{new}\Phi^{-1}(1 - P_{LOLP}) \qquad (3)$$

In equation (3), $\Phi^{-1}$ represents the inverse function of the standard normal distribution.

Thus, it can be seen that when the system reserve capacity cost is C, the increasing reserve capacity cost during time t is as follows:

$$C = (P_2 - P_1)C_0 t \qquad (4)$$

Thus, it can be seen that with an increase in the proportion of an intermittent power grid, the equivalent load prediction's standard difference is increasing more and more; in order to maintain the reliability of power supply, the system reserve must increase gradually and the reserve cost gradually increases.

The approximate relationship between the standby cost and the intermittent power supply is shown in Figure 2.

## 3 THE LOAD ADJUSTMENT DEMAND VARIATION MECHANISM AFTER INCORPORATING INTERMITTENT POWER INTO THE POWER GRID

### 3.1 Concept of peaking capacity ratio

The peaking capacity is defined as the total maximum capacity that can be an adjustable output and the total minimum output power difference in the peaking power plant in the system, the ratio of the peaking capacity, and the rated installed machine capacity is the peaking capacity ratio. The equation is as follows:

$$R_G = \frac{P_{G\max} - P_{G\min}}{P_N} \times 100\% \qquad (5)$$

In this equation, $R_G$ represents the peak capacity ratio of the generator unit; $P_N$ represents the rated generating capacity of the generator unit; $P_{G\max}$ represents the maximum technical output of the generator unit; generally, the maximum technical output of the general thermal power unit is equal to the rated output; and $P_{G\min}$ represents the minimum technical output of the generator unit.

### 3.2 Relationship between the intermittent power supply capacity and equivalent load

All the power outputs in the region always meet the load levels of the area. Figure 3 is the schematic diagram of the power balance.

The load level, contact line plan, local non-adjustable unit output, and intermittent energy generation output cannot be adjusted or is difficult to adjust; they belong to one class and the local adjusting unit output returns for another class.

In order to meet the power balance, the scope of the equivalent load should be within the scope of the local adjustable unit output, so as to ensure the adequate peaking capacity, as shown in Figure 4 below.

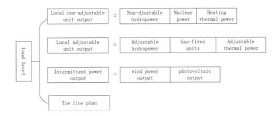

Figure 3.   Schematic diagram of power balance.

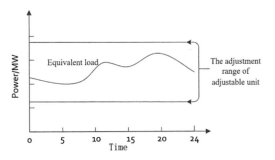

Figure 4.   The relationship between the adjustment range of the adjustable unit and equivalent load.

Due to the instability and intermittent power supply of wind power and photovoltaic power generation output, after being incorporated into the power grid, they may play a positive role in peaking, may also play a role in anti-peaking. Anti-peaking refers to the intraday increase or decrease trend of the intermittent power output is opposite to the system's load curve, the peak and off-peak differences of the equivalent load curve would increase after gaining an intermittent power access to the system; peaking refers to the intraday increase or decrease trend of the intermittent power output which is nearly the same as the system load and the peak and off-peak difference of the output is less than that of the system; but, after access to the system, the peak and off-peak differences of equivalent load curve would decrease.

According to a number of research results, the peak and off-peak differences of equivalent load is normally larger than the original load after the grid connection of an intermittent power supply. And also, with the increase of the proportion of the intermittent power supply, this phenomenon, which has a certain impact on the system peaking capacity, is more and more obvious.

First, the peak and off-peak differences of the system will change after the grid connection of intermittent power supply, $P_{i\max}$ and $P_{i\min}$ are the maximum value and minimum value of system's original load in the ith day, respectively, $P'_{i\max}$ and $P'_{i\min}$ are the maximum value and minimum value of the equivalent load in the ith day, respectively, D is the number of stimulation days, $P_{V_i}$ and $P'_{V_i}$ are the peak and off-peak difference of the original load and equivalent load, respectively.

$$P_{V_i} = P_{i\max} - P_{i\min} \qquad (6)$$

$$P'_{V_i} = P'_{V\max} - P'_{i\min} \qquad (7)$$

In the equation, $i \in [1, D]$. In all the number of days, $P'_{V\max}$ is defined as the maximum peak and off-peak difference of the system under the 99% confidence level. $\sigma = \frac{1}{D}\Sigma_{i=1}^{D} P'_{V_i}$ is the average peak and off-peak difference of the equivalent load, which can reflect the change of the system's peaking capacity after the grid connection of intermittent power supply. Figure 5 is the diagram of equivalent load after the grid connection of low proportion of intermittent power supply. The basic peaking line is the sum of the system gas-fired units' peaking capacity and the thermal power unit's normal peaking capacity. In this case, the peaking cost of the system and the peaking cost before the wind power access to the system is not very different.

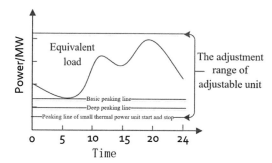

Figure 5. The diagram of equivalent load after the grid connection of intermittent power supply.

## 4 THE COST OF THE THERMAL POWER UNIT WITH DIFFERENT PEAKING WAYS

### 4.1 Analysis of the thermal power unit's start-stop cost

The traditional calculation method of the thermal power unit start-up cost is composed of 3 parts: the fuel oil cost that is consumed during the start-up process, the plant power consumption cost, and the depreciation cost of the unit during the start-up process. The fuel oil cost and auxiliary power cost are relatively easy to calculate, during the cold start and hot start of the turbine, the alternating thermal stress induced in the rotor will directly affect the life of the unit and therefore, the depreciation cost cannot be ignored, but the amount is also difficult to be determined. In short, the unit's start-up cost is related with the unit capacity, start-up efficiency, unit performance, and other factors; it is difficult to regulate them with a uniform standard. The traditional thermal power start-up cost analysis is carried out by using the equation below, it is the sum of the consumption of oil, coal, water, and human resources during the unit's start-up process (without considering the benefits during the start process).

$$F_T = \sum_{i=1}^{n} Q_i \times V_i \tag{8}$$

where, $F_T$ is the cost of the unit's single start-stop cycle; $Q_i$ is the physical consumption during the process of unit's start-stop cycle, including coal, oil, water, auxiliary power, and human resource; and $V_i$ is the price corresponding to the above consumption.

### 4.2 Analysis of thermal power compression load cost

The thermal power unit uses the compression load to achieve peaking in that the thermal power unit

constantly adjusts the output to meet the load's demand according to the load change. Wind power is used as an example in this paper to introduce the cost situation of the thermal power unit's peaking with compression load. The sum of the power of wind power is $g_W$; all these parts of the power are acquired by the thermal power unit 1 to n (n ≥ 1) by pressure for the thermal power unit with compression load, the output reduction amount of the thermal power unit i (I = 1, 2 ... n) is directly proportional to the installed capacity of the unit.

If wind power not connected to the grid, the thermal power unit i can get more load, $g_W \cdot \frac{f_1}{\Sigma f_1}$, where $f_1$ is the installed capacity of the thermal power unit i. The thermal power unit output is as follows:

$$p_{io} = p_{iw} + g_W \cdot \frac{f_1}{\sum f_1} \tag{9}$$

where $p_{io}$ is the thermal power unit's output without wind power access to the system (MW) and $p_{iw}$ is the thermal power unit's output with wind power access to the system (MW).

According to the relevant information, if regardless of the coal consumption change of the unit during the process of load increasing and decreasing, comparing with the wind power integration, the sum of the coal consumption of the thermal power unit that participated in peaking with compression load can be calculated by using the following equation:

$$F_{11} = \sum_{i=1}^{n} p_{iw}(b_{iw} - b_{io})\tau \times 10^{-3} \tag{10}$$

where $F_{11}$ is the coal's loss amount of the thermal power unit with wind power access to the system (t); $b_{iw}$ is the output of the thermal power unit f with wind power access to the system; the corresponding power supply coal consumption rate (g/kWh);

$b_{io}$ is the output of the thermal power unit i without wind power access to the system; the corresponding power supply coal consumption rate (g/kWh);

$\tau$ is the wind power integration operation duration (h).

## 5 THE CHANGE OF THE PEAKING COST AFTER ALLOWING INTERMITTENT POWER ACCESS TO THE SYSTEM

With an increase in the intermittent power's proportion, the peak and off-peak difference of the equivalent load will increase further and the low load is very likely over the basic peaking line. At this time, the thermal power unit will peak deeply with oil, the deep peaking has a bigger influence on the unit, and the cost of fuel is more expensive than coal and therefore, the system's peaking cost will be slightly higher. And when the intermittent power's proportion increases further, in low-end cases, the peak and off-peak difference will be so large that the low load of the equivalent load will rise over the deep peaking line. At this time, the system must carry out the start-stop cycle in the small thermal power unit to ensure the safety, or to abandon the wind power, resulting in the loss. The start-stop cost of the small thermal power is very high and therefore, the peaking cost of the system will increase greatly. The peaking cost curve is shown in Figure 6.

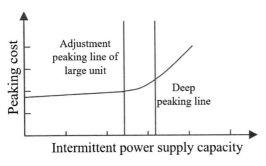

Figure 6. The relationship between the intermittent power capacity and peaking cost.

## 6  CONCLUSIONS

By studying the intermittent and prediction accuracy of wind power and photovoltaic output, it is concluded that the peak and off-peak differences and the spare capacity of the system have increased. This is caused by the intermittent power connect to the grid, so as to determine the relationship between the intermittent power and the peaking cost and the standby cost, obtain the change cost mechanism of ancillary service and then, the intermittent power supply operation value was analyzed based on the above information, thereby providing a reference for regional intermittent power supply capacity development.

## REFERENCES

He Yang, Hu Jun-feng, Yan Zhi-tao, Shang Jin-cheng. Compensation Mechanism for Ancillary Service Cost of Grid-Integration of Large-Scale Wind Farms [J]. Power System Technology, 2013, 12: 3552–3557.

Zhang Hong-yu, Yin Yong-hua, Shen Hong, He Jian, Zhao Shan-shan. Peak-load Regulating Adequacy Evaluation Associated with Large-scale Wind Power Integration [J]. Proceedings of the CSEE, 2011, 22: 26–31.

Zhang Ning, Zhou Tian-rui, Duan Chang-gang, Tang Xiao-jun, Huang Jian-jun, Lu Zhan, Kang Chong-qing. Impact of Large-Scale Wind Farm Connecting with Power Grid on Peak Load Regulation Demand [J]. Power System Technology, 2010, 34(1): 152–158.

Zhou Shuang-xi, Wang Hai-chao, Chen Shou-sun. Analysis on Operation Value of Wind Power Resources [J]. Power System Technology, 2006, 14: 98–102.

*Advances in Energy and Environment Research – Achour & Wu (Eds)*
© *2017 Taylor & Francis Group, London, ISBN 978-1-138-62682-9*

# A probabilistic evaluation method of energy-saving benefit for daily generation based on the Monte Carlo stochastic simulation method and wind power and load uncertainties

Lin Guo, Junmei Wang & Chao Ma
*State Grid Chongqing Electric Power Corporation, Chongqing, China*

Chuncheng Gao
*Nari Group Corporation/State Grid Electric Power Research Institute, Nanjing, China*

Dunnan Liu & Mo Yang
*School of Economics and Management, North China Electric Power University, Beijing, China*

ABSTRACT: Currently, assessment methods for energy efficiency in energy-efficient scheduling cannot meet the needs of wind power output and load power uncertainty for saving management in random environments. To solve this problem, in this paper, day scheduling probabilistic assessment methods are proposed to achieve energy efficiency based on the Monte Carlo stochastic simulation method. First, the method used the Latin hypercube sampling to simulate the random state of wind power output and maximum power load node. And then, a day energy-efficient scheduling optimization model with network security and contractual capacity constraints was established in this paper, to simulate the output state and start and stop activity of thermal power as well as the wind turbine output status. For the large integer variables and constraints of network security that the model features contain, the problem is broken down into sub-problems and main issues with the iterative mechanism based on the principle of Benders' decomposition is dealt with to improve computational efficiency. On the basis of a large number of samples and random state simulation unit, the Daily Dispatch Program of energy efficiency assessment of the probability is finally realized in this paper. Finally, a provincial power grid is used in this paper as an example to verify the effectiveness of the energy efficiency probabilistic assessment of the proposed method.

## 1 INTRODUCTION

Now, due to energy shortage, saving energy and reducing consumption have been incorporated into the national strategy and these have become long-term constraints of the national economic development. Energy-efficient scheduling is an important way to realize saving energy and reducing consumption. The energy-saving benefit evaluation is the tool of the optimization of the power grid operation, mining and energy saving, and provides theoretical basis for energy saving assessment and management (Wen, 2013).

With the construction of the smart grid, large-scale access and load demand diversification of wind power, the impact of wind power output and load power, and other random factors also significantly increased, and thus maintain the need for a balance of the power system, network security constraints, etc. The nature of the thermal power output is that it also possesses random characteristics.

The energy level and the corresponding output of thermal power units have a defined function. An appropriate implementation of its energy efficiency also reflects the stochastic nature. In practice, however, due to the lack of awareness of energy efficiency characteristics probability, this leads, at the end, only to "blackouts" to the save energy assessment of the phenomenon, which requires us to acquire a profound reflection of limitations of the method to evaluate the energy efficiency of current deterministic (Yan, 2010). Energy-efficient scheduling is an important means of the energy saving power system, but it also needs to introduce energy efficiency probabilistic assessment methods (Li, 2012).

In this article, a probabilistic evaluation method of the energy-saving benefit for the daily generation dispatching schedule considering wind power and load uncertainties is put forward. This method firstly adopts the efficient Latin Hypercube Sampling technique (Latin Hypercube from, LHS), sampling simulation wind maximum power output, and the node load random state, and then builds including network security constraints and the contract of the energy-efficient scheduling optimization model of power constraints, to simulate the thermal power unit start-stop cycle and state of output, the wind generator, and the output state (Chen, 2008).

## 2 SIMULATION METHOD OF WIND POWER OUTPUT AND RANDOM SITUATION OF THE LOAD STATE

Wind power capacity and the node load power in the state of randomness both obey the normal distribution that can be thought of in each period. Now, the wind power output is taken as an example to show the above random factors in the state of the random simulation method.

Assume that the current in each period is scheduling cycle wind turbines (for convenience, in this paper, a wind farm is equivalent to a typhoon electricity unit) statistical regularity of the efforts to satisfy normal distribution, and the expected value is the predicted value of the wind turbine output. In each period, as a result, the probability density function of the efforts of wind turbines can be expressed as follows:

$$z(P_W^t) = \frac{1}{\sigma_W^t \sqrt{2\pi}} \exp\left[ \frac{-(P_W^t - P_{W,u}^t)^2}{2(\sigma_W^t)^2} \right] \quad (1)$$

In this equation, $P_{W,u}^t, P_W^t$ represent the output capacity of wind turbines in period t expectations and random value. $\sigma_W^t$ represents the time t output's standard deviation of wind turbines.

The wind power output and the premise of expectations and standard deviation of the node load in each period is known, and can, through the corresponding random state probability model using the Monte Carlo simulation method, obtain a scheduling cycle of the wind power output power curve and the node load. Based on the Monte Carlo stochastic simulation method of probability assessment, a large number of random factors state of random samples are required for repeated sampling to calculate the simulation. Given the LHS technology from the mechanism to ensure that all sampling areas of a random variable can be covered by the sample point, when compared with random sampling Monte Carlo's technique, the sampling efficiency and precision obtained in this method are high. Therefore, by using the LHS technology for scheduling the plan for the state of the wind power output and the node load to simulate specific factors, a reported previously (Yu, 2009).

## 3 SIMULATION METHOD OF ENERGY SAVING SCHEDULING PLAN

### 3.1 Energy-efficient scheduling plan optimization model

#### 3.1.1 Modeling ideas

Under the influence of random factors of wind power output and load power, in order to achieve the probability of the energy-saving benefit evaluation, the day energy-efficient scheduling optimization model is used to simulate the unit's start-stop cycle and output state.

The physical connotation of the energy-efficient scheduling plan optimization model can be described to ensure completion of the contract quantity of the power plant, and the maximum power output and the node load of wind randomness, on the premise of meeting the tide of network security constraints, through simulating the thermal power unit's start-up state and output power, wind turbine output state, mining, and energy saving potential to achieve the energy-saving benefit maximization.

### 3.2 Establishment of the optimization model

#### 3.2.1 Objective function

The objective function of the energy-saving scheduling can be expressed as the total coal consumption which is as low as possible during the scheduling cycle (that is, the energy saving benefits are as large as possible). Thus, the objective function can be expressed as follows:

$$\min \sum_{t=1}^{T} \left[ \sum_{i=1}^{N_G} \left( u_i^t f_i \left( P_{i,G}^t \right) + ST_i^t u_i^t \left( 1 - u_i^{t-1} \right) \right) \right] \quad (2)$$

In this equation, $P_{i,G}^t$ represents the output of the $u_i^t$ thermal power unit I at time t, $f_i(P_{i,G}^t)$ represents the energy consumption of the thermal power unit I's curve function (general conic). $ST_i^t$ represents thermal power unit I's start consumption curve function (exponential function associated with immobilized time), represents the start-stop state of the thermal power unit I in time t ("1" represents operation, "0" said halt), T represents the number of hours of the scheduling cycle, the model T takes 24 hours, and NG represents thermal power unit sets.

#### 3.2.2 Constraints
① System balance constraints

1. Each time, power balance constraints are given as follows:

$$\sum_{i=1}^{N_G} P_{i,G}^t + \sum_{j=1}^{N_W} P_{j,W}^t = \sum_{k=1}^{K} P_{k,D}^t \quad t \in T \quad (3)$$

In this equation, $P_{j,\mathrm{W}}^t$ represents the scheduling power of the wind turbines j at time t, $P_{k,\mathrm{D}}^t$ represents node k in period t of the random load power (the numbers are obtained in section 1 of the simulation), K represents the load node number, and Nw represents the number of wind turbines.

2. In each period, the system's up and down spinning reserve constraint.

In view of the randomness of wind power, two kinds of spinning reserves, up and down, should be considered, assuming that the up and down spinning reserves are supplied by the thermal power unit, and then the system's up and down spinning reserve can be expressed as follows:

$$\sum_{i=1}^{N_G}(P_{i,\mathrm{G,max}} - P_{i,\mathrm{G}}^t) \geq U_{\mathrm{SR}}^t \quad t \in T \tag{4}$$

$$\sum_{i=1}^{N_G}(P_{i,\mathrm{G}} - P_{i,\mathrm{G,min}}^t) \geq D_{\mathrm{SR}}^t \quad t \in T \tag{5}$$

In this equation, $P_{i,\mathrm{G,max}}$ $P_{i,\mathrm{G,min}}$ represent the maximum and minimum of the thermal power unit I respectively, $U_{\mathrm{SR}}^t U_{\mathrm{SR}}^t$ represents the up and down spinning reserve system at time t, respectively.

② Unit operation constraints

1. Upper and lower bounds technology of the thermal power unit output

$$u_i^t P_{i,\mathrm{G,min}} \leq P_{i,\mathrm{G}}^t \leq u_i^t P_{i,\mathrm{G,max}} \quad t \in T; i \in N_G \tag{6}$$

In this equation, if the thermal power unit is downtime, then its output is limited to 0; if it is turned on, the upper and lower reserves bound for its technical output.

2. Wind farm output limit constraints [6, 7]

$$0 \leq P_{j,\mathrm{W}}^t \leq P_{j,\mathrm{W.max}}^t \quad t \in T; j \in N_W \tag{7}$$

In this equation, $P_{j,\mathrm{W}}^t, P_{j,\mathrm{W,max}}^t$ represent the scheduling power of the wind turbines j at time t as well as the maximum power output (maximum power output is obtained by the method of section 1 simulation), respectively.

3. Power supply contract constraint of coal-fired power plants

$$W_{m,\mathrm{G,min}} \leq \sum_{P_{i,\mathrm{G}} \in \varphi_m} \sum_{t=1}^{T} P_{i,\mathrm{G}}^t \leq W_{m,\mathrm{G,max}} \quad m \in M_G \tag{8}$$

In this equation, $P_{i,\mathrm{G}} \in \varphi_m$ represents the thermal power unit I that belong to the coal-fired power plant m, $W_{m,\mathrm{G,max}}$, $W_{m,\mathrm{G,min}}$ represents the maximum and minimum contract quantity of the thermal power plants m, $M_G$ represents the number of coal-fired power plants, $W_{m,\mathrm{G,max}}$, $W_{m,\mathrm{G,min}}$ may represent the decomposition of the power plant's long-term contract quantity value on the basis of an appropriate relaxation (such as +/−5.0%), as a daily scheduling plan of inequality constraints.

4. Thermal power unit grade ability constraints are given by the following equations:

$$P_{i,\mathrm{G}}^t - P_{i,\mathrm{G}}^{t-1} \leq R_{i,\mathrm{u}}\Delta T \quad t \in T; i \in N_G \tag{9}$$

$$P_{i,\mathrm{G}}^{t-1} - P_{i,\mathrm{G}}^t \leq R_{i,\mathrm{d}}\Delta T \quad t \in T; i \in N_G \tag{10}$$

In this equation, $R_{i,\mathrm{u}}, R_{i,\mathrm{d}}$ represent thermal power unit I's maximum rising and falling rates, $\Delta T$ is the thermal power unit I's run in a certain period of time.

5. Thermal power unit's minimum for starting and stopping time constraints are given as follows:

$$(T_{i,\mathrm{on}} - T_{i,\mathrm{min,on}})(u_i^t - u_i^{t+1}) \geq 0 \quad t \in T; i \in N_G \tag{11}$$

$$(T_{i,\mathrm{off}} - T_{i,\mathrm{min,off}})(u_i^{t+1} - u_i^t) \geq 0 \quad t \in T; i \in N_G \tag{12}$$

In this equation, $T_{i,\mathrm{on}}$ $T_{i,\mathrm{min,on}}$, represent the thermal power unit I's continuous boot time values and the corresponding constraints, $T_{i,\mathrm{off}}$ $T_{i,\mathrm{min,off}}$, represent thermal power unit I's continuous down time values and the corresponding constraints.

③ Security constraints of the network trend are given by the following equation:

$$P_{l,\mathrm{min}} \leq \sum_{i=1}^{N_G}G_{l-i}^t P_{i,\mathrm{G}}^t + \sum_{j=1}^{N_W}G_{l-j}^t P_{j,\mathrm{W}}^t - \sum_{k=1}^{K}G_{l-k}^t P_{k,\mathrm{D}}^t$$
$$\leq P_{l,\mathrm{max}} \quad t \in T; l \in L \tag{13}$$

In this equation, $P_{l,\mathrm{max}}, P_{l,\mathrm{min}}$ represent the transfer distribution factors of up and down limits on the trend of transmission of branch I, L represents the total number of grid branches, $G_{l-i}^t$ represents the thermal power unit at time t, the node to the branch I generator output power, $G_{l-j}^t$ represents the transfer distribution factor of the wind generator at time, the node j of branch l generator output power, $G_{l-k}^t$ represents the transfer distribution factor of the in-period t to the branch node k's l generator output power.

### 3.3 Daily scheduling plan optimization based on Benders' decomposition and IMAIV

On the basis of the principle of Benders' decomposition, in order to improve the computing efficiency of the model, a single state simulation, the energy saving optimization, and power grid safety feasibility analysis break up into the main and sub-problems of alternating iterative mechanisms, thus creating prerequisites for the daily scheduling plan of energy-saving benefit evaluation methods of evaluating the efficiency.

### 3.3.1 Solving thought of the main problem of Benders' decomposition

Based on the Benders' decomposition principle, the unit's start-up and output is the main problem optimizing at the same time. As a result, the original model and the main problem of model can be expressed as follows:

$$\max z = c(\mathbf{u}, \mathbf{p}) \tag{14}$$

$$s.t. \ A(x) \geq b \tag{15}$$

$$v + \sum_{l=1}^{L} \left[ \lambda_{t,l} \left( \sum_{i=1}^{N_G} G_{l-i}^t \left( P_{i,G}^t - \tilde{P}_{i,G}^t \right) \right. \right.$$
$$\left. \left. \sum_{j=1}^{N_W} G_{l-j}^t \left( P_{j,W}^t - \tilde{P}_{j,W}^t \right) \right) \right] \leq 0 \quad t \in T \tag{16}$$

In these equations, u represents the thermal power unit's start-stop variable vector in the original model, p represents the thermal power and wind power active output vector of the unit in the original model, v represents subproblems of the optimal value after Benders' decomposition, namely all branch trends are combined; $\tilde{P}_{i,G}^t$ represents the iterative active output of the thermal power unit I in time t the last round of Benders, $\tilde{P}_{j,W}^t$ represents the iterative scheduling active output of wind turbines j at time t the last round of Benders, $\lambda_{t,l}$ represents. Equation (14) represents the original objective function of the optimization model, equation (15) represents the original model to remove network security constraints tide, equation (13) represents other's constraints, and equation (16) represents the Benders' cut set of sub-problems return. Its physical meaning is described as the amount of output needed from the generating unit to make all branches trend to eliminate the limit of the total.

### 3.3.2 Solving thought of the sub-problem of Benders' decomposition

In the sub-problems, the tide trend can be the problem of the main and the sub-problem-associated constraints, to ensure the feasibility of the optimization results in the main problem. Thus, the objective function of the feasibility test sub-problem model can be expressed in all branches; current supply is limited as small as possible for 24 hours, and the constraint can be expressed as

the network trend of 24 hours security constraints, thermal power, and wind turbines upper and lower limits.

$$\min v = \sum_{l=1}^{L} \left( P_{l,t}^+ + P_{l,t}^- \right) \tag{17}$$

$$\sum_{i=1}^{N_G} G_{l-i}^t P_{i,G}^t + \sum_{j=1}^{N_W} G_{l-j}^t P_{j,W}^t - \sum_{k=1}^{K} G_{l-k}^t P_{k,D}^t$$
$$- P_{l,t} + P_{l,t}^+ - P_{l,t}^- = 0 \tag{18}$$

$$P_{l,\min} \leq P_{l,t} \leq P_{l,\max} \quad l \in L \tag{19}$$

$$P_{i,G,\min} \leq P_{i,G}^t \leq P_{i,G,\max} \quad i \in N_G \tag{20}$$

$$0 \leq P_{j,W}^t \leq P_{j,W.\max} \quad j \in N_W \tag{21}$$

$$P_{l,t}^+, P_{l,t}^- \geq 0 \quad l \in L \tag{22}$$

In these equations, $P_{l,t}$ represents the trend of branch I in time t and $P_{l,t}^+, P_{l,t}^-$ represent forward and reverse trend limitations of branch I in time t.

By using the Benders' decomposition principle, if the optimization target is 0, the main problem can pass the feasibility of the inspection; otherwise, the corresponding Benders' set must be cut back to the main problems to optimize the unit's start-up in the output state. Between the main and sub-problems, the assessment has to be optimized so many times, until the objective function of 24 small sub-problems are all zero.

Analysis of the sub-problems, with objective function and constraints of 24 small sub-problems as linear equation, can solve the linear programming.

## 4 ENERGY-SAVING BENEFIT PROBABILITY EVALUATION METHOD OF THE DAILY SCHEDULING PLAN

### 4.1 The energy-saving benefit daily scheduling plan evaluation index

Through the optimization of the model to solve the unit's start-stop activity and output status, evaluating the average coal consumption rate of the thermal power unit in the system and setting the coal consumption rate of thermal power units as the standard, by using the thermal power unit's coal saving rate to assess the daily scheduling plan of energy saving efficiency.

$$J = \frac{J_b - J_s}{J_b} \times 100\% \tag{23}$$

In this equation, $J$ represents the coal saving rate of the thermal power unit coal, $J_s$ represents the average coal consumption rate for thermal power unit, $J_b$ represents the set by using the thermal power unit's benchmark coal consumption rate.

### 4.2 Calculation steps of the daily scheduling plan probability evaluation method of the energy-saving benefit

① Setting the initial number: set the number M of the random variable, Latin hypercube sampling matrix scale $N_{max}$, and the unit state simulation counter n = 0.

② Generate the Latin hypercube random state simulation matrix: according to the random probability model of the wind power output power and the node load state, by using sampling, sorting, and steps in the LHS technology, an $M \times N_{max}$ order Latin hypercube random factors random state simulation matrix was generated.

③ Stochastic simulation of wind power output power and load state: from the Latin hypercube random state simulation matrix to extract a day samples, the wind turbines maximum output curve and each node's daily load power curve are obtained; at the same time, the unit state simulation counter is given by the following equation:

n = n + 1.

④ The main problem dimension reduction of Benders' decomposition: to relax and linearize for the energy-efficient scheduling plan optimization model of the Benders' decomposition's main problems, the relaxation linear programming model is solved based on the results of the main problems on dimension (namely the IMAIV method).

⑤ To solve the main problem of the Benders' decomposition: using the MILP solver can solve the linear mixed integer programming model after dealing with the step (4) dimension reduction of the main problems. To get the output without feasibility of Benders' decomposition sub-problems of test generators, go to step 6.

⑥ Perform the feasibility test to main problem of Benders' decomposition: judge whether the output meets the Benders' decomposition sub-problems' feasibility test of each unit in each period.

⑦ Energy-saving benefit calculation: according to step 6, obtain the state of the thermal power unit's start-stop and output, based on equation (23) to complete a sample energy-saving benefit calculation.

⑧ Convergence judgment: if n acuity $N_{max}$, stop the simulation of the unit state, output results,

count the random sample that corresponds to the energy conservation benefit change interval and different sample between zones, and draw the energy-saving benefit probability distribution curve. Otherwise, return to step (3) to simulate.

## 5 THE EXAMPLE ANALYSIS

### 5.1 Basic data

Based on provincial power grid data, the validity of the proposed energy efficiency probability evaluation method is verified. This power grid includes 9 heat-engine plants, 17 corresponding thermal power units, one coal consumption unit, grid structure and tie line power and other basic data are known. The daily load forecasting curve is shown in Figure 1 (minimum and maximum load of 8000 mW, 6000 mW, respectively), after deducting the outside network plan, the load will be undertaken by the thermal power unit in the grid and wind power plant. Containing 1 wind power plant (which is equivalent to 1 typhoon power plant), the daily output power prediction curve is shown in Figure 2 (wind power penetration is 7.5%, in this

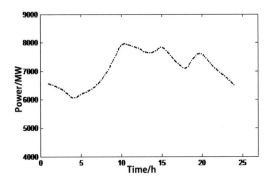

Figure 1.   Daily load curve of the provincial power grid.

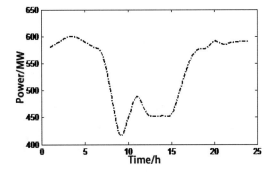

Figure 2.   Daily wind power output curve.

paper, wind power penetration is expressed as the ratio of the biggest wind power prediction to the daily biggest load output). The simulation parameters are as follows: 5.0% of up and down spinning reserve system load in every period; 2.0% of corresponding expectations of the load standard deviations; 50.0% of corresponding expectations of the wind power output standard deviations; LHS scale of 300 times; the benchmark rate of coal consumption of the coal saving rate is 350.0 g/kW·h.

### 5.2 Simulation analysis

#### 5.2.1 Application value of the daily scheduling plan energy-saving benefit of the probability assessment

For ease of comparison with the deterministic method of the energy-saving benefit assessment, in this article, in the model, the random stroke electricity output and node load random power are replaced with their expectations, thus making the model not contain random factors. Thus, by using deterministic scheduling plan energy-saving benefit evaluation methods, the calculated daily scheduling plan energy-saving benefit evaluation index is 1.91%.

Based on the original data of section 4.1, considering the randomness of load power and wind power output and by using the proposed energy-saving benefit daily scheduling plan probability evaluation method and performing an evaluation of energy-saving benefit, the energy-saving benefit daily scheduling plan probability distribution curve is shown in Figure 3. Among them, the horizontal coordinate represents the sample's energy efficiency evaluation index and the vertical coordinate represents the unit energy efficiency interval on the proportion of the total sample size P, after the same.

As shown in Figure 3, the daily dispatch plan energy efficiency probability distribution's curve center value is about 1.90%. It can be observed

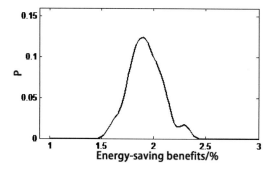

Figure 3. Probability distribution curve of energy-saving benefits.

that, in random environment scheduling plan of energy saving, the distribution essentially is random and the existing deterministic energy-saving benefit evaluation methods cannot reflect the essential characteristics of the energy-saving benefit of the scheduling plan. Based on the proposed date of scheduling the scheme energy saving efficiency evaluation method, the energy saving management department or scheduling mechanism can determine the risk level of the energy saving index of a scheduling plan. The risk level can be described by the probability of the occurrence of risk consequences to describe the equation [6], this article will be defined as follows:

$$R_J = \frac{m_s}{M_s} 100\% \qquad (24)$$

In this equation, $R_J$ represents the daily scheduling energy saving indicator risk level which waits to be set, $m_s$ represents the sample number below the set energy saving index of the daily dispatching plan energy efficiency probability distribution curve, and $M_s$ represents the random state simulation samples of the daily scheduling scheme energy efficiency probability assessment.

As is shown in Fig. 3, if the uncertainty of the energy-saving benefit assessment method is used, the index of energy saving scheduling plan is set to 1.91%. Combined with (24), it can be observed that the level of risk was 53.2%, and an apparently higher level of risk may not meet the needs of the energy-saving management. And the energy efficiency of the probability assessment method, by means of obtaining the probability distribution information of energy efficiency, the energy management department can easily combine the need of the energy-saving management to set energy-saving targets. For example, the calculation of the scheduling scheme of the energy saving index is 1.50% and the risk level is 3.21%.

To sum up, the energy efficiency evaluation method based on the daily scheduling plan is more reasonable, practical, and universal.

## 6 CONCLUSION

1. Under load power and wind power output conditions, in a random environment, scheduling of energy-saving benefit is achieved essentially with random characteristics. The probability of the energy-saving benefit assessment method can provide energy saving benefits probability distribution information, to provide a scientific and theoretical basis for energy efficiency management.

2. The greater the randomness of load power and wind power output, the larger the probability

distribution of energy saving scheduling will be, and the probability of the energy saving efficiency evaluation method is introduced at this time.

3. The strong grid structure of the energy-saving benefit of energy-saving scheduling implementation has an important influence, improving the grid wind power consumptive ability that has an important practical significance to realize energy-saving benefit.

4. To increase the penetration of wind power, which will improve the energy efficiency of the energy saving scheduling, and to make the energy efficiency more random, the energy efficiency evaluation method of probability is particularly required.

## REFERENCES

Chen Zhixiang, Xie Kai, Zhang Jing. Optimal model and algorithm for day-ahead generation scheduling of transmission grid security constrained convention dispatch [J]. Automation of Electric Power Systems, 2008, 32(4): 9–93.

Li Canbing, Lü Su, Cao Yijia, et al. A New Method for day-ahead unit commitment based on energy-saving generation dispatching [J]. Proceedings of the CSEE, 2012, 32(16): 71–77.

Li Wenyuan. Power system risk assessment: models, methods, and Applications [M]. Science Press, Beijing, 2006.

Wen Xu, Yan Wei, Wang Junmei, et al. Regional Energy-Saving Electricity Market and Stochastic Programming Power Purchasing Model Considering Assessment on Energy Consumption in Trans-Provincial Transaction [J]. Power System Technology, 2013, 37(2): 500–506.

Yan Wei, Wen Xu, Yu Juan, et al. Opportunities and challenges faced by electricity market in smart grid [J]. Power System Protection and Control, 2010, 38(24): 224–231.

Yu H, Chung C.Y., Wong K.P., et al. Probabilistic load flow evaluation with hybrid Latin hypercube sampling and Cholesky decomposition [J]. IEEE Transactions on Power Systems, 2009, 24(2): 661–667.

*Advances in Energy and Environment Research – Achour & Wu (Eds)*
© *2017 Taylor & Francis Group, London, ISBN 978-1-138-62682-9*

# Research and design of excavation face remote reproduction and remote control systems

C. Liu, J.G. Jiang, W.H. Qu, Z.Z. Zhou, J.J. Li, H. Liu, S. Ye & S.Y. Yao
*School of Electronic, Information and Electrical Engineering, Shanghai Jiao Tong University, Shanghai, P.R. China*

ABSTRACT: This paper referred to the National Key Basic Research Program of China (grant no. 2014CB046306), *Key Fundamental Research on the Unmanned Mining Equipment in Deep Dangerous Coal Bed*, as the topic background and showed the unmanned fully mechanized excavation face for deep coal seam proposed for higher requirements related to efficiency, automation, and intelligentization of the roadheader. Research and development on fully mechanized excavation face remote reproduction and remote control systems is discussed in this paper. The systems comprise the airborne control system, the industrial supervisory control computer, the Linux operating systems and the CAN bus communication. By the optical fiber and the airborne computer, the supervisory control software platform is used to transmit data and control remotely the openness of signals for the roadheader; thus, it has extensive application prospect, which provides technical support for unmanned excavation equipment, which really achieve development of fully mechanized excavation face machinery aimed at the direction of upsizing, intelligentization, and unmanned property.

## 1 INTRODUCTION

### 1.1 *Domestic resent situation of the research and analysis of automatic control system of the roadheader*

Many domestic scholars and relevant scientific research institutions have made researches on the automatic control systems of the roadheader. Its pose, cutting head position detection, and its motion control are explored. Some schemes about the automatic control system of the roadheader are put forward as well as its relevant patents, for example, the patent of 'full-automatic roadheader" applied by Wu Jialiang and Liang Jianyi from SANY Heavy Equipment Co., Ltd. (WU, 2008.4.30). A new scheme of intelligent control of walking and cutting of boom-type road header (Chen, 2006) has been researched and proposed by Chen Jiasheng from Xi'an University of Science and Technology. In this scheme, the digital hydraulic cylinder is used instead of the traditional hydraulic cylinder. By receiving the digital pulse signal sent by PLC to drive action of the stepping motor, the servo motor, and the hydraulic valve in the digital hydraulic cylinder, the oil cylinder's length vector control is realized; according to the automatic forming theory of the boom-type roadheader and its scheme of control strategy suggested by Li Junli, et al. from Taiyuan University of Technology, by means of trajectory tracking control of the roadheader, the ideal driving force

of hydraulic cylinder calculated by a upper computer is used to work out the control voltage of the electro-hydraulic proportional valve, then the electro-hydraulic proportional valve is driven to control the hydraulic cylinder. In addition, accurate tracking to the route of the drive oil cylinder and the joint corner is realized. In the boom-type roadheader tunnel cross-section forming control system proposed by Xiong Shuanghui, et al. from Xi'an University of Science and Technology (Xiong, 2009), the speed signal of the angling cylinder piston rod is taken as feedback, the electro-hydraulic proportional valve is controlled via the PID so as to further control the hydro cylinder to finish the working of the roadheader. As for the boom-type roadheader automatic cutting forming control system researched by Tian Jie, et al. from China University of Mining and Technology (Tian, 2010), the angular transducer made of the SMR magnetic sensitive component with coding function is adopted, the electromagnetic proportional valve is controlled by the PLC to drive the oil cylinder to finish the excavation. According to the horizontal axis roadheader electro-hydraulic control method proposed by Tong Xiaodong et al. from Anhui University of Science and Technology (Tong, 2007), the linear displacement sensor is selected to detect the route of the hydro cylinder. Then, the mathematical model of the hydraulic control system has been developed. Moreover, the PID controller is set up using the cut-and-trial method. Finally, the cutting head action is completed by controlling the electromagnetic proportional valve.

## 1.2 *Overall introduction of fully mechanized excavation face remote reproduction systems*

Research and development on fully mechanized excavation face remote reproduction and remote control systems is discussed in this paper. The systems comprise the airborne control system, the industrial supervisory control computer, the Linux operating system, and the CAN bus communication. By the optical fiber and the airborne computer, the supervisory control software platform is used to transmit data, control remotely the openness of signals for the roadheader, open and close oil pump, and automatically cut and stop the cutting action (Li, 2005). In addition, the system also shows kinds of the status information related to the roadheader in real time (such as the oil cylinder pressure, the oil temperature, the voltage and the current of the electrical cutting machine, the machine posture, and the cutting head position), evaluates the working and health conditions of the roadheader by comprehensive analyses of the information, and guarantees its safe and effective operation to launch high-frequency signals by using its remote control handle (Wang et al., 2010). After being received by the remote signal receiver of the roadheader, the signals pass through RS232 to the airborne computer control system with the fiber conversion module in order to realize remote control function of the roadheader. The real control distance for the system is dependent on the fiber length and the control distance shall be longer than 500–1000 m. When the working conditions under the mine are relatively complicated, the system can completely realize monitoring for the roadheader and one-button remote control of a keyboard, a mouse, or a remote control handle in the remote control chamber, which is far from excavation face at a certain distance; thus, the unmanned excavation process for fully mechanized excavation face can be achieved, which eliminates hidden dangers of personal safety due to unexpected accidents such as water inrush and outburst of gas and to deal with excavation security problems in fully mechanized excavation face of deep coal seam (Xu et al., 2015). When excavating the roadway or in the coal petrography cutting process, the operators can be far away from the working face and take advantage of the remote control system to monitor and grasp the conditions of the roadheader as well as understand excavation situation in real time. In the working process, the unmanned operation for the working face can be realized. The system layout and the condition of the roadway are shown in Figure 1.

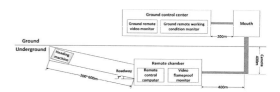

Figure 1. System arrangement and roadway condition graph.

## 2 SYSTEM COMPOSITION AND OPERATING PRINCIPLE

The airborne controller is placed in the body of the roadheader and connected with the monitoring upper computer by the communication modes such as the CAN bus communication and the industrial Ethernet. In addition, the monitoring upper computer is linked to the looped network and, through it, connected to a ground monitoring computer, thus the ground monitoring computer can show the synchronously working situations of the roadheader. The remote control systems comprise the remote signal receiver and emitter, and the remote signal receiver is located near the monitoring upper computer and directly connected to the upper computer through RS-232. By connection of the Ethernet and the video monitoring upper computer, the airborne camera collects on-site videos in real time and sends them to the video monitoring upper computer; all the systems are combined dynamically and mutually cooperated to realize the condition monitoring and the automatic control for the roadheader. The remote control system composition is shown in Figure 2.

## 2.1 *Multi-channel flameproof camera and remote visual monitoring and diagnosis system*

The system includes four wired mining flameproof network cameras in both sides of cutting arm and insertion board and two wireless mining flameproof network cameras in a machine body to observe first- and second-spot operating states. Concrete parameters of camera: Watec-137LH black & white CCD; the horizontal definition: 570TVlines; the minimum illuminance: 0.002Lux F1.4; the wireless visual transmission distance: 150 m. The system can transmit local working images of the roadheader to the remote control spots in real time from multiple views with camera characteristics of good shake-proof performance and short image delay time (Tian and Wang, 2012).

All monitored parameters of the roadheader can be collected and processed by the airborne computer control system and then linked to looped network through the module of Ethernet changed

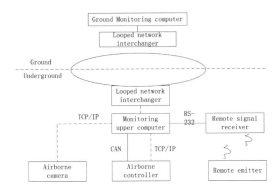

Figure 2. Remote control system composition.

by CAN. The system can monitor and represent the real-time operational status parameters such as the information of the roadheader posture on the ground in any spots with functions of remote fault diagnosis, alarm, and historical data conservation, representation and printing. In addition, the system can realize the visual monitoring and diagnosis for any spots on the ground, observe, and record all status and control the parameters as well as underground videos and images and guarantee safe and effective operation of the roadheader.

## 2.2 Airborne controller

2003 series PCC (Programmable Computer Controller) of BampR is adopted as the airborne controller. PCC is a computer-control system composed of the CPU, the I/O interface, the power supply, and the communication module. Besides, the PCC realizes the design of serialization, modularization, and standardization, thus it is easy to be designed and installed with short debugging cycle and simple maintenance (Qi and Xiao, 2005). All the input and output interfaces of the PPC used the optoelectronic isolation method, which can effectively restrict the influences of an external interference source on the PPC and is suitable for the poor environments under the mine. In addition, due to abundant module variety, perfect communication interference, and simple configurations, it is easy for the PPC to establish networks for communication (Zhou et al., 2010). When the PPC is installed in the anti-explosion chamber of the roadheader body and all I/O interferences are correspondingly connected to input equipment like the sensor and the executive devices like the solenoid valve, it is convenient to change the control method for the roadheader from completely manual and hydraulic control to electrical control.

There are two communication modes between the PPC and the upper computer, namely CAN bus communication and Ethernet communication. Between these two modes, the CAN communication is used as the main communication link, while the Ethernet communication is considered as the redundant one. With strong anti-interference performance and good instantaneity, the CAN communication is suitable for the communications of information control, but the Ethernet communication can be applied to transmission of real-time roadheader status information due to its relatively low reliability but high speed and flexible usage (Hiller, 1996, Das and Dülger, 2005). The roadheader remote control technology and the monitoring system composition framework are shown in Figure 3.

## 2.3 Monitoring upper computer

The hardware platform of the monitoring upper computer is a PC/104 embedded industrial computer with advantages of high reliability, flexible configurations, low power consumption, and small volume. Meanwhile, it is relatively easy to manufacture the anti-explosion machine shell and can be applied to the bad underground conditions. Therefore, the hardware platform of the monitoring upper computer utilizes the high-performance CPU and high-capacity memorizer in order to realize the expansion of the PC/104 platform with interferences of the CAN bus line, the RS-232 and the Ethernet by expanding modules. In addition, the hardware platform can be connected with kinds of equipment in working face and looped networks to achieve reliable data transmission.

The software platform of the monitoring upper computer adopts the Linux operating system for its relatively high performance, high stability, and flexible customizability (Cai et al., 2004). The kernel of the Linux release version is recompiled and shaped to establish a stable and reliable software platform with full functions. The boom-type roadheader remote control software is the kernel of the monitoring upper computer, which is used to represent all kinds of status information of the roadheader as well as to integrate and analyze the information in order to evaluate its working and health conditions and guarantee the safe and

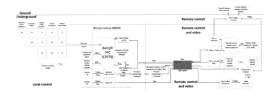

Figure 3. Roadheader remote control technology and monitoring system composition framework.

the effective operation. Meanwhile, the users can take advantage of the upper computer to configure and set various kinds of parameters for the road-header such as information of the sectional form, and to control actions of the roadheader to realize its automatic control.

## 3  SOFTWARE DESIGN

The monitoring software operating in the monitoring upper computer is the kernel of the whole remote control system, which largely determines reliability of the whole system, and thus, reliability and robustness of the software is the foundation for the system.

### 3.1  *Customized Linux operation system*

The monitoring software operates in the embedded Linux platform and the platform's robustness is the foundation for that of the monitoring software. In order to guarantee the robustness of the platform, the Linux operation system shall be customized.

First, the Linux kernel shall be shaped and recompiled, and thus, the kernel can be applied to the hardware platform to the greatest extent, most optimally support equipment outside the system, reduce the occupation of systematic resources as far as possible, and decrease the power consumption. The Linux adopts the monolithic kernel system structure and all the parts of kernel shall be concentrated together. This kind of structures leads to a relatively high communication efficiency and good real-time property among parts of the system. However, its kernel is generally big, and therefore, the unnecessary functional modules are removed by kernel reconfiguration and, finally, an effective and steady operating system kernel can be obtained after compilation.

Second, the file system is customized in order to guarantee its reliability and small volume. There are usually three ways to customize the file system, namely manual establishment, customization by software tools, and shaping for release-version film system (Sun et al., 2010). If the functions are comprehensively realized, the method of shaping release-version film system will be widely adopted.

Third, the system startup shall be reconfigured. Therefore, the services and processes shall be customized elaborately in order to keep low resource occupation and high efficiency of the system.

### 3.2  *Remote control system software structure*

The monitoring software takes advantage of the double-kernel structural design (Matthew and Stones, 2010), and therefore, the multiple processes

are operated at the same time as shown in Figure 4. The software possesses two kernel control processes and they can determine the participation method for control according to operation situations. When in the normal condition, the main control process participates in the control, namely full-line direction in Figure 4 as information flow. However, when the main control process responds too slowly or has abnormal situations due to some causes, the spare control process is involved in the control, namely imaginary-line direction in Figure 4 as information flow. By this double-kernel program design, the robustness of software and control reliability can be guaranteed. Although the abnormities of software appear, they will not affect the real working conditions of the roadheader in order to confirm its safe operation. Meanwhile, the monitoring software uses the GUI and independent framework technology for the control process, and thus, even the user interface is collapsed or has other abnormal phenomena, the control function of system will not be affected to guarantee the system security. The remote control system software structure is shown in Figure 4.

There are many methods for the IPC (Inter-Process Communication) and the common IPC methods include pipeline, named pipe, shared memory, semaphore, message queue, and socket (Li et al., 2001). Data transmission by the IPC method of shared memory existed only in fast-speed internal storage, and data transmission among processes is not related to the kernel in order to reduce system setup time and improve program efficiency. Therefore, this communication mode realizes rapid and efficient IPC. The shared memory allows numerous irrelative processes to visit the same part of the logical memory; thus, its flexibility is relatively high. Therefore, the paper adopts the shared memory as the main IPC method and the semaphore as the auxiliary one.

### 3.3  *Complement*

The PPC and the monitoring software transmit and control information by reliable CAN

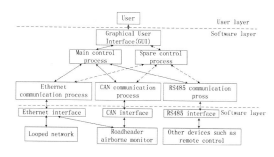

Figure 4.  Remote control system software structure.

communication protocol, and the PPC internal programs take advantage of design by the state machine thought for steady and reliable operation. Before operating, the PCC will examine the security of the executive actions and, meanwhile, feed back to the monitoring software in order to keep its security. By calculation, it takes about 100 ms for the executive security examination before actions and the feedback. However, for mechanical actions, the time can be ignored; thus, this communication mode will not create large time delay, but can improve reliability and security.

## 4 CONCLUSION

The system structure where the airborne controller and the monitoring upper computer are considered as the main parts and all the parts are mutually integrated for control can be applied not only to the remote control for roadheader, but also to remote monitoring and automatic control of various kinds of complicated large-scale machineries. The airborne controller hardware with hardware and software platforms of the monitoring upper computer in the monitoring platform has excellent reusability. The designers recompile the airborne controller program software and the upper computer monitoring software to conveniently apply this platform to the control of other complicated large-scale machineries, especially to coal industry since the platform possesses advantages of high reliability and flexibility.

The constructed remote reproduction and control systems for fully mechanized excavation face in deep coal seam have good performance in terms of reliability, stability, and adaptability in ground and underground tests; thus, it has extensive application prospects, which provide technical support for unmanned excavation equipment.

## ACKNOWLEDGMENTS

This work was supported by the National Key Basic Research Program of China (grant no. 2014CB046306), the National Technology Pillar Program of China (grant no. 2011BAA04B03), and the National High Technology Research and the Development Program of China (grant no. 2014 AA052602). Corresponding author: Jianguo, Jiang, Email: Jiang@sjtu.edu.cn

## REFERENCES

Cai, M., Lai, X., Ren, F. & Li, Y. 2004. Study on fracturing development in overburden stratum of mined-out area by borehole observation under xitian river [A]. *8th CSRME*, 843–846.

Chen, J. 2006. Reseacrh on Intelligentized Control of Driving and Cutting of Boom-type Roadheader [D]. *Journal of Xi'an University of Science and Technology*, 3.

Das, M. T. & Dülger, L. C. 2005. Mathematical modelling, simulation and experimental verification of a scara robot. *Simulation Modelling Practice and Theory*, 13, 257–271.

Hiller, M. 1996. Modelling, simulation and control design for large and heavy manipulators. *Robotics and autonomous systems*, 19, 167–177.

Li, K., Mannan, M., Xu, M. & Xiao, Z. 2001. Electrohydraulic proportional control of twin-cylinder hydraulic elevators. *Control Engineering Practice*, 9, 367–373.

Li, X. 2005. Study of dynamic behaviour of horizontal-axial-roadheader as transverse cutting [J]. *Chinese Journal of Construction Machinery*, 1–18.

Matthew, N. & Stones, R. 2010. *Beginning Linux Programming*, Posts & Telecom Press.

Qi, R. & Xiao, W. 2005. *Programmable computer controller technology*, Publishing House of Electronics Industry.

Sun, X., Wu, X., Yu, Y. & Yang, G. 2010. Design of the Hydraulic Support Monitoring and Control system based on Linux. *Microcomputer Information*, 44–45.

Tian, J. 2010. *Research on boom-type roadheader auto cutting and profiling control system [D]*. China University of Mining and Technology, Beijing.

Tian, Y. & Wang, D. 2012. Research of Performance Testing Method of Section Monitoring System for Roadheader. *Industry and Mine Automation*, 20–23.

Tong, X. 2007. *Study on electro-hydraulic control simulation of the horizontal shaft-type roadheader [D]*. AnHui University of Science and Technology.

Wang, Y., Li, W., Fu, X., Wang, Z. & Liu, Y. 2010. Roadheader Remote Monitoring System Design. *Coal Mine Machinery*, 31, 129–130.

Wu, G. 2008.4.30. *Full automatic roadheader*. China patent application.

Xiong, S. 2009. *Study on Boom-type Roadheader Laneway Section Shaping Control System*. Xi'an University of Science and Technology.

Xu, N., Tong, M., Xue, F., Jia, C. & Deng, S. 2015. Remote Monitoring System of Roadheader Based on King View. *Coal Mine Machinery*.

Zhou, G., Wu, Y., Lin, H. & Li, Y. 2010. Optimization of 6.2 m high mining technology in complex and extra-thick coal seam. *Journal of Xi'an University of Science and Technology*, 30, 397–401.

*Advances in Energy and Environment Research – Achour & Wu (Eds)*
© *2017 Taylor & Francis Group, London, ISBN 978-1-138-62682-9*

# An automatic electricity sales quota allocation algorithm applied in national companies with branches

Siqi He
*North China Electric Power University, Beijing, China*

Shangyuan Wu
*State Grid Information and Telecommunication Group Co. Ltd., Beijing, China*

ABSTRACT: This paper describes an automatic electricity sales quota allocation algorithm, which is suitable to apply in national companies with branches. Owing to the business demand of huge power, an algorithm which can reflect the business developing trend and cover the market factors is needed for company's yearly plan making and quota allocation. This paper is concerned with researches on the business demand of the State Grid Corporation of China and invention of the algorithm which has the ability to predict the sales quota for every branch while taking all factors into consideration. Meanwhile, this algorithm can support the optimization of every branch's sales quota based on its industrial research and electricity consumption prediction. Moreover, it could make the reassignment after the optimization and produce the final electricity sales quota allocation for branches to follow, so that the power company can make sure of achieving better profit in the future.

## 1 PURPOSE OF THE ALGORITHM

### 1.1 Design of this algorithm

The national power companies are often the large ones that possess many branches spread across the country or states. Therefore, naturally, the headquarter needs to make business plan for the whole company, and break it down to sub-goals for its branches to follow, which demand that the business plan needs to be both operable and adjustable. With many factors affecting the business, specifically speaking, for the power company, the policy, the economy, and the weather will all have significant effects on how much electricity a branch can sell in one year, and some of the factors simply could not be changed by marketing activities.

Hence, in order to predict the sale amount and adjust itself in the market, the State Grid Corporation of China prepared to develop an automatic algorithm to allocate the electricity sales quota among its twenty-seven branches, which is meant to fulfill the ideal business plan for the headquarter in a single year and make sure the electricity sales quota is affordable for each branch.

### 1.2 Functions of this algorithm

To achieve the purpose of this electricity sales quota allocation algorithm, it must equip with the following functions:

1. The algorithm needs to have the ability to reflect the business development trend, that is, how much

electricity one area may consume under the specific operation environment, which means the algorithm should be able to predict how much electricity can be used by regular power-users.

2. The algorithm needs to have the ability to distinguish the relationship between the electricity sales and consumption markets, and therefore, it can predict the sales quota according to the electricity consumption and taking the factors that are going to affect the electricity consumption in consideration, by which makes the algorithm adjustable.

3. The algorithm needs to have the ability to support the reassignment of the electricity sales quota among the twenty-seven branches. Under the given target of the headquarter, if there were some natural forces that inevitably change one branch's ability to profit, such as local economic recession or natural disaster, the algorithm should be able to adjust the *Electricity Sales Quota Allocation*, taking both the damaged branch and the surplus profitability of other branches into consideration.

## 2 PRINCIPLE OF THE ALGORITHM

### 2.1 Allocation of the electricity sales quota to each branch

In this part, the algorithm followed the logic which is shown in Figure 1 to allocate the electricity sales quota to each branch.

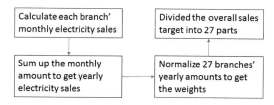

Figure 1. The logic of allocating the electricity sales quota to each branch.

As shown in Figure 1, the main logic of this algorithm is to calculate the yearly electricity sales of 27 branches and normalize them to get the weights for each branch; therefore, when the headquarter sets a total target for the whole company, there will always be a quantitative method to divide the total target into subgoals. Therefore, dividing the total target into 27 parts created the problem of predicting how much electricity each branch can sell in one year.

Some papers have used yearly data to predict a yearly number, but, in this paper, to bring in more sales information and make the prediction more accurate, the algorithm used monthly electricity sales data of each branch to predict the sales amount in the next business cycle.

For example, taking the January electricity sales data of Beijing from 2008 to 2015, and using the following formula:

$$\hat{Y}_{t+1} = \alpha Y_t + \alpha(1-\alpha)Y_{t-1} + \alpha(1-\alpha)^2 Y_{t-2} \\ + \cdots + \alpha(1-\alpha)^{t-1} Y_1 \qquad (1)$$

in theory, the Beijing's January sales amount in 2016 can be predicted.

While applying formula (1), the parameter $\alpha$ has to be decided first. Here, this algorithm used the A/B test to decide what $\alpha$ would be, like using the January electricity sales data of Beijing from 2008 to 2014 to simulate the value of 2015, and comparing it to the real value, if the error is large, then change the value of $\alpha$, until the error is reduced to 5%.

Therefore, after the iterative calculation, this algorithm determines that the parameter $\alpha$ should change among 0.65 to 0.75.

After all the calculation as above, this algorithm predicts the January electricity sales data of Beijing in 2016. Similarly, this algorithm predicts the electricity sales of Beijing in each month of the year 2016. Summing all these numbers up, the yearly number of Beijing's electricity sales in 2016 can be obtained. Following the logic above, all yearly electricity sales of 27 branches can be predicted.

As there must be some errors in the predicted values, this paper puts up with a recursion formula to optimize the predicted result. On the one hand,

this formula can reduce the error, and, on the other hand, it pays more attention to the newest data.

In this section, we still take the Beijing branch as an example. Based on the prediction logic discussed above, not only the value of 2017 can be predicted, but also the value of 2016 can be predicted. Naturally, there will be an error between the predicted value and the real value of 2016, and the recursion formula will use the relative error to optimize the predicted value of 2017, which is to comprehend all the affecting factors in the recent year to adjust the prediction.

The recursion formula is as follows:

*The optimized electricity sales of $Y_{t+1}$*

$= the\ predicted\ value\ of\ Y_{t+1} \cdot \dfrac{the\ real\ value\ of\ Y_t}{the\ predicted\ value\ of\ Y_t}$

By applying this formula after calculating the yearly prediction of all the 27 branches, this algorithm will get the optimized value as $P = (p_1, p_2, \cdots, p_{27})$.

$p_i$ represents the sales ability of each branch, and hence, if normalized $P = (p_1, p_2, \cdots, p_{27})$, then the weights that can describe how much one branch can afford in the whole yearly sales target can be measured.

The normalization formula is as follows:

$$w_i = \frac{w_i}{\sum\limits_{i=1}^{n} w_i} \qquad (2)$$

After the normalization, this algorithm can get the vector as $W = (w_1, w_2, ..., w_{27})$ and, based on that, it can be used to divide the overall target into 27 parts. This result would be the *Electricity Sales Quota Allocation* which all branches need to follow in the following year.

## 2.2 Adjustment of quota for one branch when affected by inevitable factors

Electricity is no more than a consumption commodity, and therefore, the sales amount will be affected by the market. Specifically, how much electricity can be sold by one branch mostly depends on how much electricity will be needed by the users in that area. Therefore, if the consumption can be detected, then the adjustment of the *Electricity Sales Quota Allocation* can be made scientifically.

According to the above-mentioned logic, this algorithm takes in the major industrial electricity consumption data to predict how much electricity consumption will occur in each branch's area of operation. The major industry includes agriculture, industry and manufacture, architecture, transport, software, business, financial and public utilities.

The algorithm uses the same logic and methods to predict and optimize the yearly electricity consumption value for each industry.

Hence, for one branch, this algorithm will calculate how much each industry will consume in the coming year. Summing the consumption of all industries up, there will be a ratio to describe the relationship between the electricity sales and consumption market. Moreover, if some factors occur in the market, it will change the consumption of an industry or some industries, and this affection will reflect on the total consumption, also reflecting on the electricity sales through the ratio.

Therefore, when one branch needs to adjust its Electricity Sales Quota, first it should research the factors like the local policy, economy trends, the development plan and so on, and found out how they affect the industries. By speculating which industry is going to need more electricity and which industry is going to reduce its consumption, one branch can sum up the exact number from all industries to predict whether it could achieve the given Electricity Sales Quota Allocation or not.

If it cannot achieve the given quota, then it needs to report the industrial electricity consumption values as an argument to the Headquarter for reassigning the Electricity Sales Quota Allocation.

If the application is a solid and true one, the Headquarter will change the Electricity Sales Quota Allocation for not only the branch who ask, but also the other branches; therefore, the whole sales target of the company can be guaranteed.

### 2.3 Reassigning the electricity sales quota among the branches

If one branch applies to change its Sales Quota and if it is a reasonable application, then the Electricity Sales Quota Allocation needs to be reassigned by the following logic.

Assumed that the overall target is Q and the original weights of the Electricity Sales Quota Allocation is $W = (w_1, w_2, \cdots, w_{27})$, the assignment of 27 branches is $x_1, x_2, \cdots, x_{27}$.

When the branch $i$ needs to reduce its sales quota by $a_i$, its sales quota becomes:

$$x_i' = x_i - a_i \tag{3}$$

To make sure that the overall target do not change over one branch's adjustment, the other 26 branches should fulfill the new overall target $Q'$ as $Q - x_i + a_i$.

Therefore, the algorithm enters a new quota allocation mission, where the new target is $Q - x_i + a_i$, and there are 26 branches to divide this number.

As we already knew how much electricity each of the 26 branches is going to sell in the coming year, we apply the same logic, and there will be new $W'$ to divide the $Q - x_i + a_i$ into 26 parts to form a reassigned Electricity Sales Quota Allocation.

## 3 EMPIRICAL ANALYSIS

We take the electricity sales data from 2008 to 2016 to apply the algorithm. Using formula (1) to predict every branch's monthly electricity sales value, this algorithm found that when the parameter $\alpha = 0.65$, the predicted value is more accurate.

Based on the prediction result, this algorithm applied formula (2) to normalize the 27 predicted values into the weight which would be used as the basis to divide the whole target into the original Electricity Sales Quota Allocation.

Beijing was taken as an example to read Table 1. If the overall target is 100 million kwh, then the Beijing branch needs to sell 2.44 million kwh to fulfill the demand of the headquarter.

In 2017, the headquarter will set the whole sales target as 30000 million kwh, and therefore, the Beijing branch will have to afford the quota as 30000 * 2.44% = 732 million kwh.

As this algorithm already knew how to divide the whole target, it needed to know how to adjust one branch's quota.

In the calculation given above, this algorithm has already predicted each branch's yearly sales value; still taking Beijing as the example, its sales value in 2017 would be 693.75 million kwh.

Meanwhile, bringing in the industry consumption data, this algorithm knows how much electricity each operational area can consume, and also knows that the ratio between the sales and the consumption is 0.92.

If Beijing found that, owing to their policy, the industry, architecture, software, and finance were

Table 1. Branch's name and their original sales weights.

| Branch name | Weight | Branch name | Weight |
| --- | --- | --- | --- |
| Beijing | 2.44% | Henan | 6.28% |
| Tianjin | 1.84% | Jiangxi | 2.23% |
| HeBei | 4.26% | Sichuan | 4.73% |
| Jibei | 3.79% | Chongqing | 1.64% |
| Shanxi | 4.22% | Liaoning | 4.67% |
| Shandong | 10.15% | Jilin | 1.47% |
| Shanghai | 3.28% | Heilongjiang | 1.96% |
| Jiangsu | 12.06% | Mengdong | 0.83% |
| Zhejiang | 8.25% | Shanxi | 2.45% |
| Anhui | 3.46% | Gansu | 2.68% |
| Fujian | 4.18% | Qinghai | 1.74% |
| Hubei | 3.58% | Ningxia | 2.02% |
| Hunan | 2.83% | Xinjiang | 2.89% |
| | | Tibet | 0.08% |

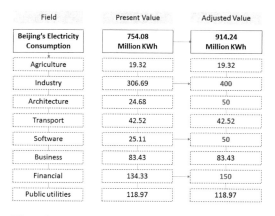

| Field | Present Value | Adjusted Value |
|---|---|---|
| Beijing's Electricity Consumption | 754.08 Million KWh | 914.24 Million KWh |
| Agriculture | 19.32 | 19.32 |
| Industry | 306.69 | 400 |
| Architecture | 24.68 | 50 |
| Transport | 42.52 | 42.52 |
| Software | 25.11 | 50 |
| Business | 83.43 | 83.43 |
| Financial | 134.33 | 150 |
| Public utilities | 118.97 | 118.97 |

Figure 2.

Table 2. Branch's name and adjusted sales weights.

| Branch name | Weight | Branch name | Weight |
|---|---|---|---|
| Tianjin | 1.88% | Jiangxi | 2.29% |
| HeBei | 4.36% | Sichuan | 4.85% |
| Jibei | 3.89% | Chongqing | 1.68% |
| Shanxi | 4.33% | Liaoning | 4.79% |
| Shandong | 10.41% | Jilin | 1.51% |
| Shanghai | 3.36% | Heilongjiang | 2.01% |
| Jiangsu | 12.36% | Mengdong | 0.85% |
| Zhejiang | 8.46% | Shanxi | 2.51% |
| Anhui | 3.55% | Gansu | 2.75% |
| Fujian | 4.28% | Qinghai | 1.78% |
| Hubei | 3.67% | Ningxia | 2.07% |
| Hunan | 2.90% | Xinjiang | 2.96% |
| Henan | 6.44% | Tibet | 0.08% |

going to use more electricity in the coming year, and the summed up consumption value would be adjusted to 914.24 million kwh, as shown in Figure 2.

Beijing's industrial electricity consumption changed from 754.08 to 914.24 million KWh, and therefore, the Beijing branch should change its sales goal to 914.24 * 0.92 = 841.1 million KWh.

Comparing with the assigned quota of 732 million kwh, it increased by 109.1 million, and therefore, the other 26 branches only need to fulfill the new goal as 30000 − 109.1 = 29890.9 with the new weights as Table 2:

Hence, after the adjustment, Beijing branch's sales goal is 841.1 million KWh in 2017, and the Tianjin's sales goal is 29890.9 * 1.88% = 591.9 million KWh in 2017, and so for the rest of the braches.

Therefore, the algorithm had calculated the reassigned *Electricity Sales Quota Allocation* for the company in 2017 after necessary adjustments.

After the empirical analysis, the algorithm proved its accuracy and use, and hence, it has been deployed into the business system, making it an automatic electricity sales quota allocation algorithm supported by the information system.

Now, the calculation process, the adjustment process, and the adjustment reasons are all documented in the system. The *Electricity Sales Quota Allocation* is shown as visualization graphics and the results are saved as formatted report.

## 4 CONCLUSIONS

This algorithm is designed to solve the sales quota allocation problems for large companies with braches. In the research process, this paper found that the traditional way of weight calculation does not have a very good accuracy, and therefore, this paper treats each branch as a separate unit, by predicting its own sales value; it gets how much electricity one branch may sell in the coming year. Meanwhile, with all 27 predicted sales values being calculated, the weights can precisely describe each branch's sales ability. Through the consumption value, the algorithm covers all the factors that may affect the branch, and by predicting the consumption value and building the relationship between the sales prediction and the consumption prediction, this algorithm can make accurate adjustment according to the market factors. It can also reassign the sales quota allocation, when there are forces that change one branch's operating environment. This algorithm has a generic logic and do not need much data preprocessing, which can reduce the data information loss and improve the availability of this algorithm. In general, this algorithm provides a method to help large companies with branches to make the Electricity Sales Quota Allocation, any other company with the similar structure can take this algorithm as a reference.

## REFERENCES

Cui Herui, Wang Di. A Study of China's Energy-Economy-Environment system based on VAR Model. J Journal of Beijing Institute of Technology (Social Sciences Edition) 2010, 2: 23–28.

Han Zhiyong, Wei Yiming, Jiao jianning, Fan Ying, Zhang Jiutian. Co-ingreation analysis and an error correction model of China's energy consumption and economy grownth. J System Engineering 2004; 24: 17–21.

Mayeres. I, Van Regemorter. D. The introduction of the external effect of air pollution in AGE model: towards the endogenous determintation of damage valution and its application to GEM-E3. R Final report of the GEM-E3 elite project of the EU Joule research program 1999.

Mehrzad Zaman. Energy Consumption and Economic Activities in Iran [J]. Energy Economics, 2007, 29(06): 1135–1140.

Paresh Kumar Narayana, Russell Smythb, Arti Prasad. Electricity Consumption in G7 Countries: A Panel Cointegration Analysis of Residential Demand Elasticities [J]. Energy Policy, 2007, 35(09): 4485–4494.

Stern D.I. A Multivariate Cointegration Analysis of the Role of Energy in the US Macroeconomy [J]. Energy Economics, 2000, 22: 267–283.

Wang Haipeng, Tian Peng. The study of the relationship between China's energy consumption and economic growth based on time varying parameter model. J Application of Statistics and Management 2006; 03: 253–258.

Wei Yiming, Wu Gang, Liu Lancui. Progress in modeling for Energy-Economy-Environment complex system and its applications. J Chinese Journal of management 2005; 22: 159–170.

Yu Mengquan, Meng Weidong. The relationship of China's energy consumption and economic growth based on panel data. J System Engineering 2008; 06: 68–72.

Zhao Fang. Economic explanation for the unharmonious development of 3Energy-Economy-Environment. J China Population Resources and Environment 2008: 184: 67–72.

Zhao Fang. Empirical study on the coordinated development of China's Energy-Economy-Environment (3E). J Economist 2009; 2: 35–41.

Zhao Tao, Li Hengyu. On the Coordinating Evaluation Model for Energy-Economy-Environment System. J Journal of Beijing Institute of Technoligy (Social Sciences Edition) 2008; 102: 11–16.

*Advances in Energy and Environment Research – Achour & Wu (Eds)*
*© 2017 Taylor & Francis Group, London, ISBN 978-1-138-62682-9*

# Cooling of oil-filled power equipment

O.S. Dmitrieva
*Department of Theoretical Bases of Heat Engineering, Kazan National Research Technological University, Kazan, Russian Federation*

A.V. Dmitriev
*Department of Theoretical Bases of Heat Engineering, Kazan State Power Engineering University, Kazan, Russian Federation*

ABSTRACT:  In the hot period of the year, oil-filled power equipment exploitation is often accompanied by incidents of overheating as well as cooling system disorders. As a result, there are disruptions in the energy supply; moreover, the replacement of equipment in a short period of time is technically impossible and expensive. A possible solution is to upgrade transformers and small devices development to cool them. The device is mounted in existing and new equipment; it is compact and is reliable. The operational concept is based on the fact that, at night, when the temperatures of ambient air are minimum, the tank accumulating cold that is used in the heating process. We have analyzed the process of ice formation on the thin metal fins in the water tank. The calculations showed that maximum temperature difference between the hot and cold side of the converter is observed at minimum values of heat transfer coefficient.

## 1 INTRODUCTION

At a certain stage of industries and enterprises, expansion and industrial zones development often have problems with connection of new facilities and power consumers to the existing power grid due to limited design capacity of existing substations and load limits achievement of step-down transformers. The lack of transformers power reserve at substations has extreme effects in case of failure of one of the transformers in the period required to replace it on backup, or to return after repair, when operating transformers are not able to redistribute and, at least temporarily take additional load of released down transformer. In addition, transformers operation in the heat is often accompanied by incidents of overheating and cooling system disorders. As a result, there are disruptions in the energy supply and the huge monetary losses due to the short supply of electricity. However, the replacement of equipment in a short period of time is technically impossible and expensive (Peredel'skij, Kolbasov, Sadovnikov, & Jakimov 2009, Larin 2015).

Under these conditions, upgrading of transformers and development of compact devices for their cooling are very important. The transformer's load capacity increasing during modernization can be achieved by improving cooling system efficiency for diffusion growing total losses at no-load operation and short-circuit increases with load capacity. However, existing cooling systems are not effective, complex, and expensive; because of the shortage of places for their installation, the project implementation is difficult (Peredel'skij, Kolbasov, Sadovnikov, & Jakimov 2009, for details, see Miheev, Efremov, & D. Ivanov 2013, Bashirov, Minlibayev, & Hismatullin 2014, Hismatullin, Vahitov, & Feoktistov 2015, Merem'janin, Bushuev, Bushueva, & Volod'kina 2007).

## 2 USE OF A THERMOELECTRIC DEVICE FOR ADDITIONAL COOLING OF THE OIL-FILLED TRANSFORMER

A possible solution is to install, on the surface of the cooling fins, oil-filled power equipment cascade of semiconductor thermoelectric converters. Introduction of a thermoelectric device (Dmitriev et al. 2016) for additional cooling of the oil-filled transformer will solve the problem of heat removal from the windings, as the consequence is to increase the life time and reliability of the equipment. The device is mounted in existing or new equipment, and it is compact and reliable. It consists of a thermally insulated tank filled with water, at inner and outer sides of which are present metallic fins; moreover, thermoelectric transducers are present outside of the tank (see Fig. 1). The device can effectively operate at relatively high ambient temperatures.

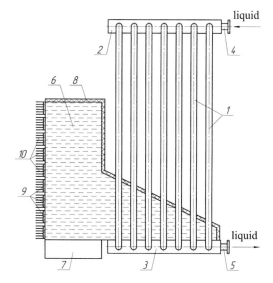

Figure 1. Scheme of device for additionally cooling the oil of the transformer, *1*–the vertical pipe, *2*–upper horizontal manifold, *3*–lower horizontal manifold, *4*, *5*–nipples, *6*–container, *7*–support, *8*–isolation, *9*–thermoelectric module, *10*–heat sink fins.

At night, heat is removed from the water by thermoelectric transducers which increase the heat transfer driving force, and portion of the water becomes ice, which is formed on the fins disposed in the water. During the daytime, the ice melts, taking away the excess heat from the transformer. At the coldest time of the day, the nominal cooling system provides necessary heat dissipation, and hence, at this time, thermoelectric converters accumulate cold. Structurally, the device can be provided with ribs of different lengths and thicknesses.

Studies (Dmitrieva & Dmitriev 2015) showed that, at the peak equipment load modes using thermoelectric converters, cooling is effective for a certain volume of water contained in the tank up to a temperature of 0°C, freezing the remaining capacity, i.e. the use of water-ice phase transition phenomenon.

## 3 CALCULATION PROCEDURE

We analyzed design parameters and operating modes of thermoelectric devices provided by manufacturers (for details, see Thermoelectric Modules Crystal LTD S-127-14-15-L2 specification). Dependency of heat flow and voltage supplied to the converter from temperature gradient are linear. Summarizing the available data, Equations 1 and 2 were obtained.

For the heat flow, the equation will be written as

$$Q = A(I)\Delta t + B(I),\qquad(1)$$

where $A(I) = -0.059\,I - 0.42$ and $B(I) = 9.59\,I + 6.3$ are the coefficients depending on the current strength; $\Delta t = t_h - t_c$ is the temperature difference between values of hot and cold sides of thermoelectric converter, K; and $I$ is the current strength, A.

Voltage dependence applied to the thermoelectric converter will be written as

$$U = C(I)\Delta t + D(I),\qquad(2)$$

where $C(I) = 0.0019\,I + 0.0536$ and $D(I) = 2.23\,I - 0.0077$ are the coefficients depending on the current strength.

The value of the reliability approximation of represented dependencies was less than 5%. An important stage of evaluation of the proposed device at solving issues on the cooling system upgrade is the analysis of the process of ice formation on the thin metal fins in the water tank.

Therefore, it is necessary to obtain the dependence of the ice formation thickness on the technological parameters.

Design scheme is shown in Figure 2. The temperature of hot side of the thermoelectric converter (see Fig. 2) can be determined according to Equation 3

$$t_h = t_A + \frac{Q}{\alpha F_A},\qquad(3)$$

where $F_A$ is the contact area of thermoelectric converters with air, m²; $t_A$ is the air temperature, °C; and $\alpha$ is the coefficient of heat transfer to the air, W/(m²·K).

The temperature of the cold side of the thermoelectric converter (see Fig. 2) is defined as

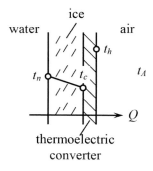

Figure 2. Calculation model.

$$t_c = t_n - \frac{\delta_k Q}{\lambda_k F_k}, \qquad (4)$$

where $F_k$ is the contact area of ice with water, m²; $t_n$ is the temperature of ice surface, °C; $\lambda_k = 2.21 - 0.0046\,(t_n + t_c)$ is thethermal conductivity coefficient of ice, W/(m·K); and $\delta_k$ is thethickness of ice, m.

Then, the current strength can be calculated according to Equation 1:

$$I = \frac{Q - 6.3 + 0.42\Delta t}{9.59 - 0.059\Delta t}. \qquad (5)$$

Further, the voltage according to Equation 2 is determined and compared with the set value. The calculation is complete if the convergence is satisfactory. The error of calculation was ± 0.5°C.

## 4 RESULTS AND DISCUSSION

The calculations showed that the maximum temperature difference between hot and cold sides of the converter is observed at minimum values of heat transfer coefficient. For values of the heat transfer coefficient to the air more than 50 W/(m²·K), its effect on the temperature difference between the hot and cold sides of the converter is significantly reduced. The same effect occurs with increasing thickness of ice (see. Fig. 3). This is due to the increasing thermal resistance.

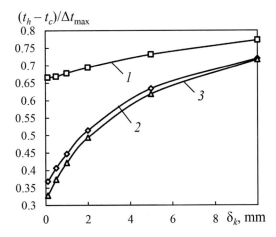

Figure 3. Dependence of difference in ratio between the values of hot- and cold-side temperatures of the thermoelectric converter on the maximum temperature difference of the thickness of ice, as stated by the manufacturer. α, Вт/(м²·K): 1–5; 2–50; 3–100.

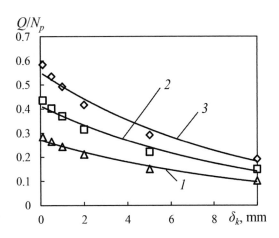

Figure 4. Dependence of ratio of the heat flow to rated power of the thickness of ice. U, W: 1 – 7; 2 –10; 3 – 13.

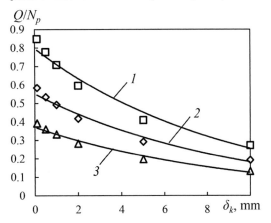

Figure 5. Dependence of the ratio of the heat flow to the rated power of the thickness of ice. $t_A$, °C: 1 – 0; 2 –20; 3 – 35.

To assess the cooling effectiveness of the transformer using thermoelectric converters, we can use the ratio of rejected heat flow to the power consumed. Owing to peculiarities of the thermoelectric converter with the increasing voltage applied to the device, the amount of rejected heat is reduced (see. Fig. 4). Device operation is desirable at the largest values of the applied voltage, since this leads to a significant increase in heat flow.

The thickness of the layer of ice is distributed along the edge in a parabola. The ratio of the maximum thickness and the length of the zone of freezing does not depend on the thermal conditions, and almost exclusively from the thermal conduction ribs. The thickness of rib is increased, which causes a uniform distribution of the ice layer along the length of the ribs in the same interval of time. It is necessary to increase the heat conductivity by the thickness of the ribs, and the material.

The effectiveness of heat removal is significantly increased at a lower ambient temperature (see. Fig. 5). Hence, the maximum load on thermoelectric converters must be submitted in the coldest period of the day.

Studies have demonstrated that the contact area of thermoelectric transducers with air practically has no effect on the efficiency of the cooling process. The thickness of ice is largely dependent on the characteristics of the fin itself—its thermal conductivity and thickness.

## 5 CONCLUSIONS

1. An additional device can be used for cooling oil transformers to increase the capacity of substations and improve the reliability of power supply.
2. Cooling system modernization using thermoelectric converters provides more efficient cooling of the liquid, as a refrigerant is used as an additional fluid in the tank, for example, water.

    In this case, the heat transfer coefficient from the water to the wall at free movement is higher than the coefficient of heat transfer at pipes cross flow of turbulent air in the 3.57–9 times.
3. By a proposed method of cooling transformers, using thermoelectric converters ensures effective heat dissipation from the current-carrying elements at the short-term overload, which greatly increases the reliability of the equipment in heat.

## ACKNOWLEDGMENTS

This reported study was funded by the President of Russian Federation grant, according to the research project no. MK-5215.2016.8.

## REFERENCES

Bashirov, M., M. Minlibayev, & A. Hismatullin (2014). Increase of efficiency of cooling of the power oil transformers. *Neftegazovoe delo*. 2, 358–367. (in Russ.).

Dmitriev, A. et al. (2016). Zajavka na poleznuju model' No 2016112002 ot 30 marta 2016 goda. A thermoelectric device for the additional cooling of oil transformer. (in Russ.).

Dmitrieva, O., & A. Dmitriev (2015). The cooling system of the oil transformer using thermoelectric modules. *Izvestija vysshih uchebnyh zavedenij. Problemy jenergetiki. 11–12*, 56–59. (in Russ.).

Hismatullin, A., A. Vahitov, & A. Feoktistov (2015). The study of heat transfer in industrial high-voltage gas-insulated transformers cooled under the influence of vibration. *Uspehi sovremennogo estestvoznanija, 12–0*, 173–176. (in Russ.).

Larin, V. (2015). World trends in transformer equipment (according to the 45-th Session of CIGRE). *Jelektrichestvo, 8*, 20–26. (in Russ.).

Merem'janin, Ju., V. Bushuev, L. Bushueva, & O. Volod'kina (2007). Patent No. 2292603 Russian Federation, MPK H01F 27/12. Cooling unit oil transformer (published January 27, 2007). 2 p. (in Russ.).

Miheev, G., L. Efremov, & D. Ivanov (2013). Ways to improve the energy efficiency of power transformers. *Vestnik Chuvashskogo universiteta, 3*, 212–217. (in Russ.).

Peredel'skij, V., V. Kolbasov, V. Sadovnikov, & V. Jakimov (2009). Modernization of power transformers to increase their load capacity. *Jelektro. Jelektrotehnika, jelektrojenergetika, jelektrotehnicheskaja promyshlennost', 5*, 33–37. (in Russ.).

Thermoelectric Modules Crystal LTD S-127-14-15-L2 specification. URL: http://www.crystalltherm.com/upload/iblock/9e3/TM%20S-127–14–15-L2%20SPEC.pdf.

*Advances in Energy and Environment Research – Achour & Wu (Eds)*
*© 2017 Taylor & Francis Group, London, ISBN 978-1-138-62682-9*

# Integration of SolidWorks and Simulink models of an open-closed cycle air-fueled Stirling engine

Ali Rehman & Weiqing Xu
*School of Automation Science and Electrical Engineering, Beihang University (BUAA), Beijing, China*

Fatima Zakir
*School of Economics and Management, Beihang University (BUAA), Beijing, China*

Maolin Cai
*School of Automation Science and Electrical Engineering, Beihang University (BUAA), Beijing, China*

ABSTRACT: Liquid air-powered vehicles are playing a vital role in the area of new energy vehicles, due to the reason of its advantages over electric vehicles. Electric vehicles have short battery life, long charging time, heavy metal pollution, and spontaneous combustion in comparison with liquid air-powered engines. In this paper, a comprehensive approach is established to describe the integration of SolidWorks model with SIMULINK environments of LN2-fueled open-closed cycle Stirling engine. This Stirling engine is 'cryogenic heat engine', which is fueled by cryogenic medium, that is, liquid nitrogen, and considered as zero-emission engine. A proposed open closed cycle Stirling engine is constructed on SolidWorks and it is integrated with SIMULINK via SimMechanic link. These integrated models provide us certain simulation results to visualize the body movements and the respective position, velocity, and acceleration as well. Specifically, motion analysis of power piston and axis of proposed Stirling engine is taken into account.

## 1 INTRODUCTION

In current scenario of technology development and advancement, consumers demand for the devices having elevated qualities, efficiencies, and functions. This competitive environment leads the engineers to design the products with a certain level of qualities and having a number of supporting functionalities. It is impossible to develop the prototype of every device for testing and analysis. In this situation, computer-aided software is used, for example, SolidWorks. SolidWorks is not only useful to make geometrical models of the system but can also perform operations like kinematics and dynamics on different parts of the body (Rusiński, 2002; E.berhard P., 1998). To obtain the visualization and control on different body parts of the working machine, one is needed to integrate the SolidWorks model with the Simulink environments. In this way, we can control as well as visualize the motion of machine parts. Matlab renders to different types of toolboxes, which furnish us to elaborate our models and to solve the optimization problems. In this case, the most extensive tool is SimMechanics, which engender the chain of kinematics and support for dynamics simulation and results visualization. Simulink provides the platform to define the control of the integrated models and display the results of simulation in specified time intervals. In the graphical window, one is able to create complete control mechanism having assistance of various blocks, for example, signal sources and measuring blocks. In this paper, authors present the research of simulation that is performed on the proposed liquid nitrogen Stirling engine.

Outline of the paper is as follows: Section I introduces the integration of two modules, Section 2 defines the open-closed cycle of Stirling engine, Section 3 describes the construction of SolidWork model, Section 4 explains the implementation details, Section 5 accounts the details of simulations, and Section 6 describes the experimental results and discussion.

### 1.1 Open-closed cycle Stirling engine

Zero-emission vehicles are much demand due to elevated levels of global warming. In this regard, air-powered engines perform critical functions to reduce air pollution. In addition, these engines have very low cost of operation and less recharging time. These are very close enough to battery engines in comparison to energy density (Ordonez, 2002). In this paper, the geometrical model of liquid nitrogen-fueled Stirling engine is established on SolidWorks. This Stirling engine belonged to the cryogenic heat engine operated on the basis of open-closed Stirling cycle. In the mechanism of this engine, on the time of expansion stroke heat is absorbed by the gas of hot space from the atmosphere. In the compression stroke, gas in the cold space is cooled by the liquid air. This mechanism forms the closed Stirling cycle having liquid nitrogen as the heat sink and atmosphere as the heat source. Other way round, in hot space an exhaust port is established which can be observed in the geometrical model. The vaporized gas is allowed to discharge, while expansion in the hot space and open Stirling cycle is established (Xu, 2000).

## 2 CONSTRUCTION OF THE SOLIDWORKS MODEL

The geometrical body of proposed unconventional open-closed cycle Stirling engine is first constructed on the CAD program SolidWorks. To obtain the open-closed mechanism, exhaust ports are introduced on the cold space body. Expansion in the hot space leads the power piston to move in the direction of +y-axis and it can be observed as given in Figures 1–2. In this way, position of power piston and angular displacement of axis-rod can be visualized and analyzed. After every cycle of axis-rod around its axis, exhausts ports become open to discharge the gas. These ports are designed to open after every cycle with the help of port-sensors. Constructed model is the assembly of

Figure 1. SolidWorks model of the Stirling engine (isometric-view).

Figure 2. SolidWorks model of the Stirling engine (front-view).

different parts of the engine. An assembly of different body elements would be used solely for the purpose of simulation analysis. It is due to that in such cases there is always present an interaction between individual parts, which leads to introduction of the control system and operation of different motion analyses. Thus, every simulation model contains the assembly of individual sub-assemblies.

Construction of every part of geometric model in SolidWorks is similar to that one which is most currently used for that type of applications, and it starts with defining a 2D geometry that constructs the surface or solid after the completion of one of the main operations (pocket or extrusion). Then, relying on modification or expansion of the achieved element constructed by removing or adding material, the creation of parts is continued. Most of the CAD systems present normalized elements, which can be used to construct assemblies. The final assembly of different elements has a permanent link to each single part file. It shows that alteration created in one of the files is by default associated with the linked files, respectively. The products do not contain its own geometry, rather, it contains a set of links to the elements and constraints, which were used to connect these parts of the machine. The obtained geometric models ought to be parametric that allows the changing positions and dimensions of components that are relative to one another to perform analysis of motion. These geometrical models are not only used for the creation of the simulation model, but they can also be used to perform other different analyses, for example, modal or stress analysis.

# 3 IMPLEMENTATION OF THE SOLIDWORKS MODEL IN THE SIMULINK ENVIRONMENT

The next step in this procedure of building the simulation model is the implementation of the SolidWorks assembly model in Matlab/Simulink environment (Fig. 3). Before doing so, SimMechanics Link for SolidWorks is needed to install and to link it with the SolidWorks. In the very step, the SimMechanics Link exporter is utilized to create the Physical Modeling XML files. These XML files contain the information about mass and inertia of every part of the assembled bodies. It also has the STL (stereo-lithographic) files to represent the surface geometries of the bodies of the assembly. This approach can be summarized by the given schematic diagram Fig. 3 (Dawid, 2014).

The second step includes the loading of the obtained files into Simulink environments. The very first of the model will be the block of different elements of the engine. These blocks are adjoined with different types of SimMechanics blocks, for example, revolute joints, weld, and bodies. This simulation model appears as the complex net of connecting SimMechanics blocks. To avoid complexity, different blocks are adjoined to create subsystems. These subsystems are named on main parts of the engine. After that, option of masking is executed on each subsystem to create the image of each subsystem. These images are associated and replica of the respective main parts of the subsystem. It is obvious in the schematic that the input force ($Fy$) is applied on the power piston and as output position power piston and angular displacement of the axis are obtained. This schematic of subsystems is shown in Figure 3.

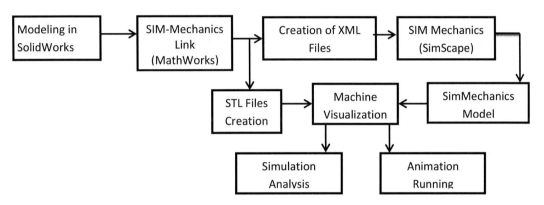

Figure 3.   Implementation process of CAD model in Matlab/Simulink environment.

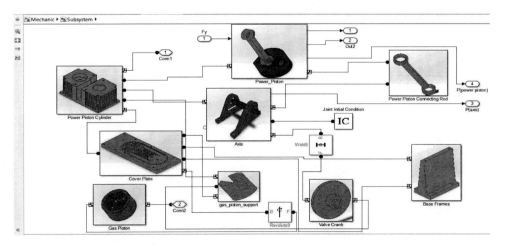

Figure 4.   The modified and simplified SimMechanics Simulation model.

## 4 SIMULATION OF THE MODEL

When the above given integrated model is run in Simulink environments, we can get the animation of the running model. The alignments of different parts of the engine in coordinate axis can be observed in the animation image of Fig. 5. The two main concerned parts of this research are axis and power piston of the proposed Stirling engine. We can observe that the movement of the power piston is the translational motion.

The direction of movement of the power piston is in the +y-axis direction. On the other hand, the movement of axis is rotational and it rotates around the x-axis. The maximum range of axis is, therefore, from −180 degrees to +180 degrees.

## 5 EXPERIMENTAL RESULTS AND DISCUSSION

Simulation tests are executed with the agency of input control signals. In this case, kinematics excitations are done with the help of input force signal. This input force signal is applied on the power piston cup. As it is known that power piston can move along the +y-axis direction only, the input force along the +y-axis direction only is taken into account, that is, $F_y$. This input force is applied on the actuators with the help of joint actuator blocks. The movements of the joints are, then, analyzed by the joint sensors. Bodies are actuated with the body blocks and are analyzed with the help of the body sensor blocks. In this research, the primary targets are axis and power piston of the Stirling engine. Translational displacements as well as velocity of the power piston are analyzed after applying the input force. On the other hand, as known that axis of the engine moves in the rotational motion; therefore, its angular displacement as well as angular velocity is taken into account.

### 5.1 Kinematics analysis of the axis of the engine

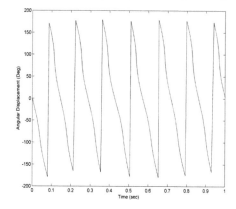

Figure 6.    Angular displacement of the axis.

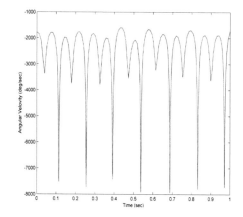

Figure 7.    Angular velocity of the axis.

### 5.2 Kinematics analysis of power piston of the engine

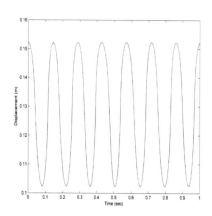

Figure 8.    Displacement of power piston.

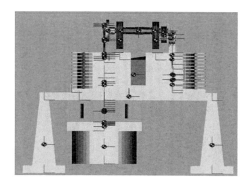

Figure 5.    Animation of the simulation of the integrated model.

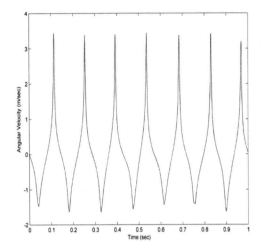

Figure 9.   Velocity of power piston.

## 5.3   *Effect of input force $F_y$*

It can also be observed that there is some effect on simulations when input force $F_y$ is applied and when this force is ignored. Authors tested both cases and the result as the differences between two tests can be scrutinized. The output result is obvious in the Figs. 10–11.

## 6   CONCLUSION

In this paper, we elaborated the integration of SolidWorks and MATLAB/Simulink that is executed through the implementation of the CAD model of the proposed unconventional liquid-fueled Stirling engine. This model was created previously in a SolidWorks program and then is performed into the Matlab/Simulink environment. The model is able to execute different simulation researches that will lead to visualization of

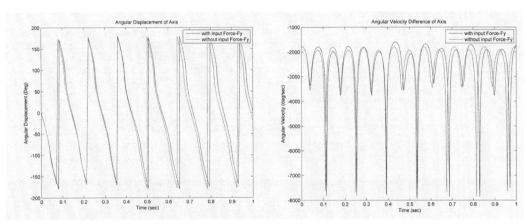

Figure 10.   Effect of input force Fy on the motion of the axis.

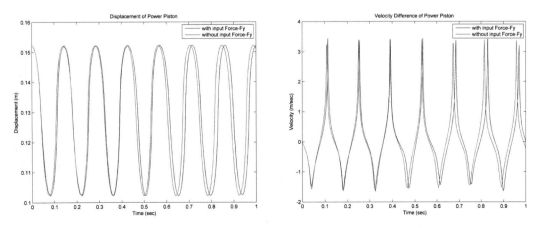

Figure 11.   Effect of input force Fy on the motion of power piston.

movements, tracking of the trajectory, velocity, and acceleration of any point of the system of the proposed Stirling engine (Dawid, 2014). Tracking of any point in the model has become possible establishing the simulation model by putting functions and scripts (Posiadała, 1990; Posiadała, 1990). In this scenario, we can conclude that:

1. Angular displacement and angular velocity of the axis of the engine determine the efficiency of the proposed Stirling engine. Tests and analysis on axis of the engine lead the reader to probe into general behavior of the overall performance of the engine.
2. Translational velocity of the power piston leads the reader to examine the behavior of pressure of the working gas in the system.
3. Effect of any input force on the system can be determined and compared with the help of the comparison presented in this paper.

## REFERENCES

Chaturvedi D.K., Modeling and Simulation of Systems Using Matlab and Simulink. CRC Press, 2010.

Dawid Cekus, Bogdan Posiadała, Paweł Waryś Integration of modeling in solidworks and Matlab/Simulink environments. Archive of Mechanical Engineering, vol. lxi 2014 no.1.

Eberhard P., Schiehlen W., Hierarchical modeling in multibody dynamics. Archives of Applied Mechanics, 1998, 68, pp. 237–246.

Gran R.J., Numerical Computing with Simulink, Volume I, Creating Simulation. Society for Industrial and Applied Mathematics, Philadelphia, 2007.

Mrozek B., Mrozek Z., MATLAB and Simulink. User's guide. Helion, 2004 (in Polish).

Mrozek B., Mrozek Z., MATLAB universal environment for scientific and technical calculations. CCATIE, Kraków, 1995 (in Polish).

Ordonez CA., Liquid nitrogen fueled, closed Brayton cycle cryogenic heat engine. Department of Physics, University of North Texas, Denton, TX, USA.

Posiadała B., Skalmierski B., Tomski L., Theoretical, computational model of the telescopic mechanisms of the cranes. Prace Naukowe CPBP 02.05. Wydawnictwa Politechniki Warszawskiej, Warszawa, 1990 (in Polish).

Posiadała B., Sklamierski B., Tomski L., Motion of the lifted load brought by a kinematics forcing of the crane telescopic boom. Mechanism and Machine Theory, 1990, Vol. 25, Issue 5, pp. 547–556.

Rusiński E., Design principles for supporting structures of self-propelled vehicles. Oficyna Wydawnicza Politechniki Wrocławskiej, Wrocław, 2002 (in Polish).

Weiqing Xu, Jia Wang, Maolin Cai, Liquid air fueled open-closed cycle Stirling engine, Beihang University of Aeronautics and Astronautics, Beijing, China. http://www.mathworks.com/help/physmod/smlink/index.html (access 25.07.2013).

*Advances in Energy and Environment Research – Achour & Wu (Eds)*
© 2017 Taylor & Francis Group, London, ISBN 978-1-138-62682-9

# 3D surface MRS method based on variable geometry inversion

Yue Zhao, Jun Lin & Chuandong Jiang
*College of Instrumentation and Electrical Engineering, Jilin University, Changchun, China*

ABSTRACT:   when used to invert three-dimensional (3D) surface nuclear magnetic resonance (surface NMR) data, the current Fixed Geometry Inversion (FGI) method cannot clearly identify the boundaries of water-bearing structures and can easily cause artifacts. In this study, we propose a horizontal smoothness-constrained Variable Geometry Inversion (VGI) method that allows the occurrence of sharp boundaries in the vertical direction and provides easier identification of the geometric boundaries of 3D water-bearing structures. Based on a 3D synthetic model designed in the present study, we evaluate the ability of VGI and FGI models to identify 3D water-bearing structures as well as the accuracy of the inversion results obtained using these models. Results show that the VGI model has far fewer parameters and higher calculation efficiency than the FGI model.

## 1  INTRODUCTION

The VGI method used for 1D surface NMR was first proposed by Mohnke and Yaramanci (2002). Hertrich (2008) introduced and summarized the FGI and VGI methods in detail. FGI results vary with varying types and orders of smoothness constraints. In comparison, no smoothness constraints are imposed on the VGI model. Thus, VGI results are relatively simple and stable. The VGI model assumption of 1D layered sedimentary strata is reasonable. However, when the VGI model is used to image 2D and 3D water-bearing structures, multiple 1D inversion results are generally pieced together to generate the results for these multi-dimensional structures. The ambient noise in the measurements and model equivalencies often result in a relatively large difference in the inversion results of adjacent models. Generally, a Laterally Constrained Inversion (LCI) method or a Spatially Constrained Inversion (SCI) method is used to add smoothness constraints to a quasi-2D or quasi-3D model. Using an LCI method, Auken et al. (2005) processed a set of segmented 1D electrical resistivity tomography data and improved the capacity of identifying 2D electrical resistivity anomalies. Subsequently, they applied the LCI method in the inversion of time domain and frequency transient electromagnetic data (TEM and FEM), time-domain and frequency-domain Induced Polarization (IP) data and surface NMR data, forming a set of mature, stable and general quasi-2D inversion algorithms (Auken et al., 2014). Viezzoli et al. (2008, 2009) expanded the lateral constraints to spatial constraints that could be imposed on inversions and used an SCI method to invert airborne

transient electromagnetic (AEM) data to form a quasi-3D model. Similarly, SCI methods are suitable for inverting various types of electromagnetic measurement data. However, quasi-2D and quasi-3D models are all based on 1D forward results, and 1D forward results are different from 2D and 3D forward results. Auken and Christiansen (2004) proposed a real 2D laterally constrained electrical resistivity inversion algorithm and noted that the "real" 2D VGI results were superior to both the quasi-2D VGI results and the 2D FGI results.

In this study, based on the scope of LCI and SCI methods, we introduce horizontal smoothness constraints on 3D surface NMR survey data and create a 3D VGI method. We discus the advantages of VGI using a synthetic model and evaluate different mesh size used for the kernel calculation. Finally, through synthetic example we verified the advantages of the VGI model.

## 2  SURFACE TECHNOLOGY

Surface NMR takes advantage of the NMR phenomenon (Levitt, 1997) in the presence of the Earth's magnetic field $B_0$. In the subsurface, the nuclear spin of hydrogen protons precess around $B_0$ at the Larmor frequency $\omega_L = |\gamma B_0|$, where $\gamma_0$ is the gyromagnetic ratio of the proton. On the surface, an alternating current $I(t) = I_0 e^{-j\omega_L t}$ is passed through a transmitter loop ($T_x$) for a period of $\tau_p$, which generates a highly inhomogeneous magnetic field $B^T$ throughout the subsurface. After switching off the current, the precession of M generates a small but perceptible alternating magnetic field $B^R$ that can be measured as an induced voltage, refers to Free Induction Decay (FID), in a receiver loop ($R_x$) deployed at the surface.

For convenient representation, we define a kernel function G, so that the voltage response at time t equal zero is written as

$$V(q,t) = \int G(q,r) \cdot w(r) d^3 r \qquad (1)$$

The integral kernel G, therefore, defines the sensitivity to the water content w(r) in the subsurface and is a function of the pulse moment q, the resistivity distribution in the subsurface, the loop sizes, the shapes and orientation of $T_x$ and $R_x$ and the properties of the Earth's magnetic field.

## 3 VARIABLE GEOMETRY INVERSION METHOD

### 3.1 Horizontal smoothness-constraint model

The smoothness constraints imposed by an FGI method act between adjacent elements. In relation to the entire inversion space, the constrained area is relatively small. As a result, the horizontal continuity of a large-scale water body cannot be ensured even if an anisotropic coefficient is added. To solve this problem, we propose a horizontal smoothness-constrained VGI model based on the LCI and SCI methods. Figure 1 shows the proposed model.

In the horizontal direction of the proposed VGI model, the plane is divided into several homogeneous elements. Relatively few layers with unfixed boundaries are set in each element in the vertical direction. Thus, the unknown parameters of the inversion process include the water content and variable boundaries of each layer in the horizontal fixed elements. As shown in Figure 1, the model consists of three layers overall. The water content and boundaries in the horizontal direction vary among these three layers. To ensure the continuity of a large scale water body, constraints are imposed on the inversion parameters of adjacent elements in the horizontal direction. In each layer, the boundaries are constrained by smoothness conditions, and the inversion results approximate to a layered structure.

In the horizontal direction of the proposed VGI model, the plane is divided into several homogeneous elements. Relatively few layers with unfixed boundaries are set in each element in the vertical direction. Thus, the unknown parameters of the inversion process include the water content and variable boundaries of each layer in the horizontal fixed elements. As shown in Figure 1, the model consists of three layers overall. The water content and boundaries in the horizontal direction vary among these three layers. To ensure the continuity of a large scale water body, constraints are imposed on the inversion parameters of adjacent elements in the horizontal direction (as shown in Figure 2). In each layer, the boundaries are constrained by smoothness conditions, and the inversion results approximate to a layered structure. In addition, the water content also exhibits continuous change.

### 3.2 Calculation of the model increment

For the VGI forward models, expression (1) is rewritten in the form of the accumulation of each hexahedral element:

$$V_0(q) = \sum_{i=1}^{I} \sum_{j=1}^{J} V_{i,j}(q) \qquad (2)$$

$$V_{i,j}(q) = \sum_{n=1}^{N} \int_{z_{i,j,(n-1)}}^{z_{i,j,n}} G(q,z) \cdot w_{i,j,n} dz \qquad (3)$$

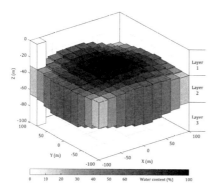

Figure 1. A horizontal smoothness-constrained variable geometry inversion (VGI) model in 3D.

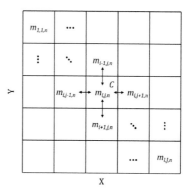

Figure 2. The horizontal smoothness constraints used in the VGI model.

where $V_{i,j}$ represents the initial amplitude of the surface NMR signal generated by the (i, j)th element on the horizontal plane. The unknown parameters of the nth layer in the VGI model include the water content ($w_n = [w_{1,1,n}, w_{1,2,n}, \ldots, w_{1,J,n}, w_{2,1,n}, \ldots, w_{I,J,n}]^T$ (n = 1,2,...,N)) and the boundary ($z_n = [z_{1,1,n}, z_{1,2,n}, \ldots, z_{1,J,n}, z_{2,1,n}, \ldots, z_{I,J,n}]^T$ (n = 1,2,...,N)). The corresponding Jacobian matrix is calculated from the following expressions:

$$J_{i,j,n}(q) = [J^w_{i,j,n}(q), J^z_{i,j,n}(q)] \quad (4)$$

$$J^w_{i,j,n}(q) = \frac{\partial V_{i,j}(q)}{\partial w_{i,j,n}} = \int_{z_{i,j,(n-1)}}^{z_{i,j,n}} G(q,z)dz \quad (5)$$

$$J^z_{i,j,n}(q) = \frac{\partial V_{i,j}(q)}{\partial z_{i,j,n}} = [w_{i,j,n+1} - w_{i,j,n}] \cdot G(q, z_{i,j,n}) \quad (6)$$

From expression (6), we know that the Jacobian matrix used for solving the boundaries ($z_n$) includes the water content ($w_n$) of each layer. Therefore, the VGI model is nonlinear and needs to be solved iteratively. In the kth iteration, a conjugate gradient least square solver using constraints, data weighting and parametric mapping (CGLSCDP) is employed to calculate the increment of the model ($\Delta m_k$) (Gunther et al., 2006):

$$\begin{aligned} (J_k^T D^T D J_k + \lambda C_m^T C_m)\Delta m_k &= J_k^T D^T D \Delta V_k \\ &- \lambda C_m^T C_m \cdot m_k \end{aligned} \quad (7)$$

## 4 INVERSION RESULTS

To verify that the 3D VGI method can improve the identification of water-bearing structures, we designed a 3D water-bearing structure model for synthetic analysis (Figure 3). The space of the synthetic model was set to 200 × 200 × 50 m³. Five groups of separated loops were laid on the ground (Figure 3(a)). Each group of separated loops included one transmitter loop (solid blue lines) and five receiver loops (dotted black lines). The 3D water-bearing structure was an ellipsoid structure (Figure 3(b)), and its center was located underground at (0 m, 0 m, −15 m). The corresponding semi-axis lengths of the ellipsoid structure in the Cartesian coordinate system were 75 m, 50 m and 10 m, respectively. The ellipsoid structure had a water content of 40%. The remaining areas were dry environments with water contents of 1%.

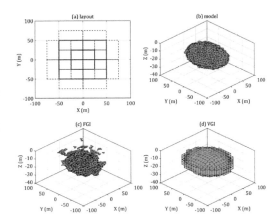

Figure 3. A 3D water-bearing structure model.

For the forward calculation of surface NMR signals, an FGI model is more suitable for water-bearing structures with an irregular shape. Therefore, 20 pulse moments were set in the range of 0.1–10 As based on the logarithmic distribution and expression (1). 30 nV Gaussian white noise was added to the data to simulate the field environment where the case study was carried out. Figure 3(c) and (d) show the inversion results for the FGI and VGI methods, respectively. The cyan cells in Figure 2 are elements with a water content greater than 10%. The elements with water contents less than 10% were considered to be dry areas. Furthermore, the corresponding transparency was set based on the water content. The FGI model has 168,736 parameters. The water content was limited within the range of 0–1.1. The element size of the kernel function for the VGI model was set at a moderate level (4 × 4 × 2 m³). The underground space was assumed to consist of three layers with a total of 12,500 model parameters (water content and boundary). The water content was limited within the range of 0–1.1. No limitations were placed on the boundaries. The first layer of the underground space used to select the initial value model was located between 0 and −14 m and had a water content of 1%. The second layer was located between −14 m and −16 m and had a water content of 10%. The third layer was located between −16 m and −50 m and had a water content of 1%. The initial model was established this way because it is easy to thicken a thin aquifer but very difficult to reduce a thick aquifer in the VGI model. The initial model is also used as the preconditioned matrix, thereby ensuring the stability of the algorithm. A weight of 1 was assigned to the horizontal smoothing matrix of the water contents, and a weight of 0.5 was assigned to the horizontal smoothing matrix of the boundaries.

347

Figure 3(c) and (d) show the results obtained using the two inversion methods, both of which can invert the general location of the water body. However, in the FGI results, multiple depressions and protrusions appear at the boundaries of the water-bearing structure. In comparison, the shape of the water-bearing structure can be more accurately identified from the VGI results, indicating that the VGI results are more stable and reliable than the FGI results at this noise level.

## 5   DISCUSSION

In this study, through a simulation study, we prove that horizontal smoothness-constrained VGI results are stable and reliable when using surface NMR survey data to image a 3D object body. Compared to traditional FGI methods, the VGI method allows the occurrence of sharp boundaries in the vertical direction and provides geometric shapes of water-bearing bodies that can be more easily identified. The horizontal smoothness constraints imposed on the VGI model improve the imaging quality in low-resolution areas and reduce high or low water content artifacts that appear in the FGI results. The water content is more evenly distributed within the water-bearing body in the VGI results. The synthetic results prove that the optimum VGI results can be obtained only by selecting a suitable element size. An excessively large element size leads to relatively coarse imaging results, and an excessively small element size leads to fluctuations in the imaging results; the real model cannot be reflected in either of these situations. In addition, the VGI model with the kernel function in a moderate element size has far fewer parameters than the FGI model as well as a low memory footprint when calculating the kernel functions and a high inversion calculation speed.

Similar to other geophysical inversion methods, a suitable element size, a suitable initial model, a suitable preconditioned matrix and a suitable horizontal smoothness coefficient weight must be selected to obtain VGI results that are consistent with the real model. Therefore, it is necessary to further study the criteria and empirical pattern of selecting these parameters or improving the stability of the VGI method by improving the inversion algorithm. Finally, the VGI method proposed in the present study only realizes the variable geometry model in the vertical direction and neglects the horizontal direction.

## REFERENCES

Auken, E., and A. V. Christiansen, 2004, Layered and laterally constrained 2d inversion of resistivity data: Geophysics, 69, 752–761.

Auken, E., A. V. Christiansen, B. H. J. N. Foged, and K. I. Srensen, 2005, Piecewise 1d laterally constrained inversion of resistivity data: Geophysical Prospecting, 53, 497–506.

Auken, E., A. V. Christiansen, C. Kirkegaard, G. Fiandaca, C. Schamper, A. Behroozmand, AndrewBinley, E. Nielsen, F. Effers, N. B. Christensen, K. Srensen, and Nikolaj Foged1 Giulio Vignoli1, 2014, An overview of a highly versatile forward and stable inverse algorithm for airborne, ground-based and borehole electromagnetic and electric data: Exploration Geophysics.

Chevalier, A., A. Legchenko, J. F. Girard, and M. Descloitres, 2014, Monte carlo inversion of 3-d magnetic resonance measurements: Geophysical Journal International, 198, 216–228.

Coscia, I., S. Greenhalg, N. Linde, J. Doetsch, L. Marescot, T. Günther, and A. Green, 2011, 3D crosshole apparent resistivity static inversion and monitoring of a coupled river-aquifer system: Geophysics, 76, G49-G59.

Dlugosch, R., T. Günther, M. Müller-Petke, and U. Yaramanci, 2014, Two-dimensional distribution of relaxation time and water content from surface nuclear magnetic resonance: Near Surface Geophysics, 12, 231–241.

Günther, T., C. Rücker, and K. Spitzer, 2006, Three-dimensional modeling and inversion of dc resistivity data incorporating topography - Part II: Inversion: Geophys. J. Int., 166, 506–517.

Hertrich, M., 2008, Imaging of groundwater with nuclear magnetic resonance: Progress in Nuclear Magnetic Resonance Spectroscopy, 53, 227–248.

Hertrich, M., M. Braun, T. Günther, A. G. Green, and U. Yaramanci, 2007, Surface Nuclear Magnetic Resonance Tomography: IEEE Transactions On Geoscience And Remote Sensing, 45, 3752–3759.

Hertrich, M., A. G. Green, M. Braun, and U. Yaramanci, 2009, High-resolution surface-NMR tomography of shallow aquifers based on multi-offset measurements: Geophysics, 74, G47-G59.

Jiang, C., M. Müller-Petke, J. Lin, and U. Yaramanci, 2015a, Imaging shallow three dimensional water-bearing structures using magnetic resonance tomography: Journal of Applied Geophysics, 116, 17–27.

Legchenko, A., 2013, Magnetic Resonance Imaging for Groundwater: John Wiley & Sons.

Legchenko, A., M. Descloitres, C. Vincent, H. Guyard, S. Garambois, K. Chalikakis, and Dimensional Magnetic Field and NMR Sensitivity Computations Incorporating Conductivity Anomalies and Variable-Surface Topography: IEEE Transactions on Geoscience and Remote Sensing, 49, 3878–3891.

Levitt, M. H., 1997, The signs of frequencies and phases in NMR: Journal of Magnetic Resonance, 126, 164–182.

Mohnke, O., and U. Yaramanci, 2002, Smooth and block inversion of surface NMR amplitudes and decay times using simulated annealing: Journal of Applied Geophysics, 50, 163–177.

Müeller-Petke, M., and U. Yaramanci, 2010, QT inversion-Comprehensive use of the complete surface NMR data set: Geophysics, 75, WA199-WA209.

Müller-Petke, M., R. Dlugosch, and U. Yaramanci, 2011, Evaluation of surface nuclear magnetic resonance-estimated subsurface water content: New Journal of Physics, 13, 095002.

Patrick, K., 1986, Moderate-degree tetrahedral quadrature formulas: Computer Methods in Applied Mechanics and Engineering, 55, 339–348.

Viezzoli, A., E. Auken, and T. Munday, 2009, Spatially constrained inversion for quasi 3d modelling of airborne electromagnetic data an application for environmental assessment in the lower murray region of south australia: Exploration Geophysics, 40, 173–183.

Walsh, D. O., 2008, Multi-channel surface NMR instrumentation and software for 1D/2D groundwater investigations: Journal of Applied Geophysics, 66, 140–150.

Yaramanci, U., M. Hertrich, and G. Lange, 1999, Examination of Surface-NMR within an integrated geophysical study in Nauen, Berlin: Presented at the Proceedings of the 5th EEGS conference.

Yu, J., 1984, Symmetric gaussian quadrature formulae for tetrahedronal regions: Computer Methods in Applied Mechanics and Engineering, 43, 349–353.

*Advances in Energy and Environment Research – Achour & Wu (Eds)*
*© 2017 Taylor & Francis Group, London, ISBN 978-1-138-62682-9*

# Hierarchy and development framework of military requirement system

Ji Ren
*The Equipment Demonstration Center, Beijing, P.R. China*

Xiaolei Zheng
*The Logistics College, Beijing, P.R. China*

ABSTRACT: A Military Requirement System (MRS) is not a disorderly, nonlogical, and simple stacking of all requirements but a diversified and multihierarchical system. This paper presents the objectives of establishing an MRS hierarchy and its source. In addition, in this study, an analysis is conducted considering different perspectives, and the concept of a hierarchical tree is introduced to define the MRS hierarchy. Moreover, the ontology of military requirements, the approach in the development of military requirements, and the descriptions of military requirements in the MRS development framework are elaborated.

## 1 INTRODUCTION

The Military Requirement System (MRS) hierarchy divides the elements contained in the MRS and defines the common characteristics of the elements in each group. Moreover, it analyzes the interelement relation within a component and establishes component structure. Finally, it describes the intercomponent relations and identifies the diagram of core relationship among components.

In this study, an MRS development framework is established based on the MRS hierarchy. Various concepts involving military requirements are classified and defined, describing the relationship among the core elements. In addition, the development of military requirements and technologies may be studied according to the contents formulated in the development framework, thus yielding satisfactory results on developmental requirements..

## 2 HIERARCHY OF THE MRS

### 2.1 *Objectives of establishing the MRS hierarchy*

The following objectives may be fulfilled by establishing the MRS hierarchy:

- Promote the accuracy and reusability of the hierarchy for the development of military requirements. Standardize the relations among various elements in the hierarchy by defining related concepts, thus accurately standardizing the description of military requirements.

- Facilitate the cooperation among requirement development teams. Consolidate the understanding of requirements proposed by development teams by defining concepts of an MRS hierarchy, thereby providing a platform for requirement development and management.
- Be conducive to a modularized development approach. The MRS hierarchy may enable MRS to be independent of specific development approaches. This is conducive to the introduction of new technologies for requirement development.
- Facilitate the cooperation between military personnel and developers. The MRS hierarchy is versatile and universal, thus enabling communication and cooperation between military personnel and developers.

### 2.2 *Source of MRS hierarchy and its correlated concepts*

By considering the related information on software requirements and in connection with the concept of military requirements, MRS hierarchy is divided into three aspects: combat operational, user, and system requirements.

Combat operational requirement refers to the combat functions required of the system to fulfil military objectives or accomplish military missions, including combat missions and actions as well as necessary combat actions and functions required to accomplish these missions.

User requirement indicates the missions or actions, which must be accomplished by the user by using the military system under the guidance of combat operational requirements.

System requirement includes the military system functions that must be implemented to support the users to fulfill missions requisitioned through combat operational requirements.

Moreover, system requirements may be further subdivided into functional and nonfunctional requirements and constraint conditions.

Functional requirements mainly interpret the nature of interaction between various subsystems and the different functional components of the military information system and environment, that is, the functions expected to be delivered by the system.

Nonfunctional requirements reflect the additional requirements regarding quality and characteristics of the military information system and describe the system behaviors exhibited to users and executed manipulations.

Constraint conditions refer to the system performance and structural constraints resulting from environmental, technical, or human factors during the design and development of a military information system.

Thus, MRS depicts the overall user objectives based on these three aspects, as illustrated in Fig. 1.

In Fig. 1, prior to the establishment of the military LIIS system by the user, all the requirements corresponding to a target point in the 3D space are proposed, respectively generating projections over dimensions of combat operational, user, and system requirements. Thus, requirements at the three objective levels are developed, depicting the overall user requirements at different levels.

## 2.3 Hierarchy of the MRS

In this study, five hierarchy trees are introduced to describe the following five aspects of the MRS: hierarchy trees of combat operational requirements, user requirements, functional requirements, nonfunctional requirements, and constraint conditions. A hierarchy tree includes two parts: the concept set and tree structure. The concept set is a set of all concepts employed in the hierarchy tree, and the tree structure reflects the relations among the tree's core elements in a centralized manner, as shown in Fig. 2.

The concept sets for the five hierarchy trees are defined as follows:

The concept set of hierarchy tree of combat operational requirements: {combat operational requirements, combat entity, combat mission, and combat capability}.

Concept set of hierarchy tree of user requirements: {user requirements, activity, behavior, and behavioral sequence}.

Concept set of hierarchy tree of functional requirements: {functional requirements, scenario, agency, objectives, and manipulation}.

Concept set of hierarchy tree of non-functional requirements: {performance, constraint aging, spatial constrains, safety, reliability and availability, and maintainability}.

Concept set of hierarchy tree of constraint conditions: {history, background, roles, and functions}.

MRS hierarchy depicts the relationship among the five hierarchy trees, as shown in Fig. 3.

In Fig. 3, the undirected line segments denote the parent–child relation, and the directed line segments represent the guidance relations of starting and ending points. Moreover, the thickness of line segments reflects the closeness of relations between two elements.

The figure shows that a dependency relationship exists among the five hierarchy trees of which

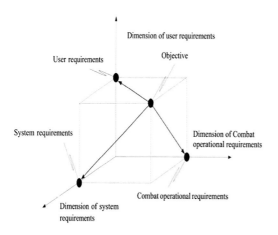

Figure 1.   3D schematic view of MRS hierarchy.

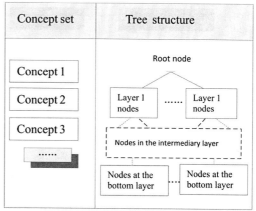

Figure 2.   Descriptive format of the hierarchy tree.

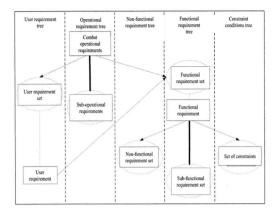

Figure 3. Schematic hierarchy of MRS.

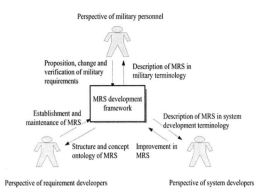

Figure 4. Functions of MRS development framework from various perspectives.

combat operational requirement is the most crucial element in the relationship diagram that guides the establishment of user and system requirements. In the hierarchy tree of combat operational requirements, each node denotes one operational requirement. Moreover, a particular combat operational requirement identifies separate groups for user and functional requirements.

The user and functional requirements are dependent on particular combat operational requirements, the tree structures of which are generally similar to those of combat operational requirements.

Functional requirements remain the core part of system requirements, whereas nodes in the hierarchy tree of nonfunctional requirements are completely dependent on nodes of functional requirements. Further, the internode relation within the hierarchy tree is relatively loose. The set of constraints has characteristics identical to that of nonfunctional requirements.

Fig. 3 presents a visualized description of all correlated elements of the MRS hierarchy, and substantially improves the level of accuracy when describing it.

## 3  THE DEVELOPMENT FRAMEWORK OF MRS AND ITS ESSENTIAL ELEMENTS

### 3.1  *Definition and functions of MRS development framework*

Zachman considered the MRS framework to be a logical structure for organizing and classifying complex information. A robust framework generally includes the following: unified design contents and forms of expression of objects; rules and guidelines on using the framework; and other contents required to be observed uniformly when describing objects.

Based on the earlier framework concept, the MRS development framework is defined as follows:

The MRS development framework functions as a standard to denote and describe the ontology of military requirements, MRS hierarchy, development approach of military requirements, and statement of military requirements, ensuring a unified standard in terms of development, maintenance, and understanding of MRS.

The framework serves as a standard for the development, description, and acquisition of military requirements. Moreover, it provides guidelines on the establishment of MRS, reflecting its contents in various aspects and their interrelations, the function of which is depicted through the following three perspectives, as illustrated in Fig. 4.

- Perspective of requirement developers: The framework provides a standardized MRS hierarchy, clearly identifying its content, and defining the military requirement ontology, which delivers a platform for developers to uniformly appreciate requirements. Moreover, it enables an interface for the developers to establish corresponding MRSs to address specific issues and perform system maintenance.
- Perspective of military personnel: The MRS development framework should provide standard wording conventions familiar to military personnel; describe requirements in established military terminologies for their easy reading; and allow military personnel to make changes in requirements, which is also conducive to the verification of requirements.
- Perspective of system developers: The MRS development framework should describe requirements in terminologies specific to system development, propose concrete requirements on the system design, and reserve a certain unspecified part to be described by system developers regarding knowledge in the field of system development.

## 3.2 MRS development framework

To satisfy the functions of the abovementioned three perspectives based on the MRS hierarchy, the MRS development framework should include four components: MRS hierarchy and ontology, development approach, and description of military requirements, as shown in Fig. 5.

- MRS hierarchy: This defines the contents contained in the MRS and the relations among various contents and elements. Moreover, it forms the main part of MRS, on which the establishment of the development framework and other constituent elements are based.
- Ontology of military requirements: This defines all the concepts involved in the MRS. The definition of ontology ensures the concept reflects the nature of what is described with better structural stability.
- Development approach of military requirements: This defines the MRS development process. The modularized structure enables requirement developers to select different development approaches, engage in specific developments to address particular issues, and ensure standardized development results, thereby guaranteeing the validity of contents in the requirement system.
- Description of military requirements: This defines the MRS specifications. Further, it provides different personnel types with relevant contents in the respective terminologies, thus enabling them to understand MRS and maintain its consistency intrinsically.

## 4 ESSENTIAL ELEMENTS OF MRS DEVELOPMENT FRAMEWORK

### 4.1 Ontology of military requirements

In the past one or two decades, ontology has been adopted in the field of computer science for knowledge expression, sharing, and reuse.

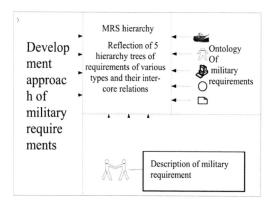

Figure 5.   MRS development framework.

Ontology of military requirements refers to a conceptualized content system of various elements (e.g., requirement, target, and function) in the problem domain of military requirements. Moreover, it was established to deliver standardized definitions of MRS concepts and terminologies and achieve consistent understanding. Furthermore, it provides unified views for MRS and renders support for its integration, communication, adaptability, and supportability.

- Integration: The ontology integrates different viewpoints and information acquired through MRS
- Communication: It ensures the sharing of MRS descriptions within an organization and various missions
- Adaptability: The ontology allows the adjustment of MRS to cater to changes in the environment and requirements
- Supportability: It assists requirement developers to understand MRS information, thus avoiding errors.

The ontology of military requirements is an important aspect of the MRS development framework. However, this study does not present an in-depth discussion on this topic. Relevant literature may be referred to for further study.

### 4.2 Development approach of military requirements

As founders and maintainers of the MRS, the functions of the developers of military requirements include studying the needs of armed forces, obtaining the requirements according to the correct method, organizing the requirement contents, and accomplishing the establishment of the requirement system. Therefore, the development of requirements should resolve the mapping relation between the requirement system and external environment as well as that within the requirement system, as illustrated in Fig. 6.

A complete set of the developmental approach must explicitly identify technologies to implement various directed line segments, as shown in Fig. 6, thereby satisfying the following "five essentials" in the development of military requirements.

- Sufficiency in the development of military requirements: This refers to the constraints imposed on the process according to the needs of the armed forces and the target system. The requirements are categorized into the following: combat operational, user, and system requirements. These requirements are all acquired according to the descriptions of military forces. There also exists a mapping relation between the contents acquired and target system.

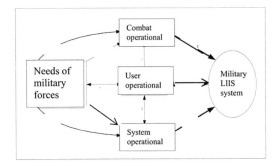

Figure 6. Model for the development approach of military requirements.

- Consistency for the development of military requirements: This refers to the requirements on the descriptive documentation regarding the needs of armed forces in various forms. Further, the needs of armed forces provided to requirement developers must be complete.
- Relative consistency for the development of military requirements: This refers to the constraints imposed on the process from the three types of requirements to the target system.
- Military sufficiency of the MRS: This refers to the constraints imposed on the process considering the needs of armed forces to combat operational requirements.
- System sufficiency of the MRS: This refers to the common constraints on the process of system requirements from all three aspects, that is, needs of the armed forces, combat operational requirements, and target system.

It should be noted that the model for the developmental approach (Fig. 6) depicts issues to be addressed by requirement developers; however, technologies employed to address issues are not specified.

### 4.3 Description of military requirements

Description of military requirements refers to various standards to be addressed while formulating and depicting the MRS, and is specifically represented by the formats of military requirement documents. The significance of formulating the description of military requirements is presented as follows:

- Guiding requirements developer to author requirement documents
- Assisting military personnel and system developers to read requirement documentation
- Provision of a friendly "user interface" for the MRS

To fulfill these objectives, the standardized description of requirements should include the following:

- Stipulation of categories of requirement documents
- Formulation of wording conventions for documentation
- Provision of uniform document formats
- Diverse forms of expression and statements

The description of military requirements is an important aspect of the MRS development framework. However, a more in-depth research on this subject was not conducted in this study. Thus, relevant literatures may be referred to for related research contents.

## 5 CONCLUSION

The study on the hierarchy and development of the MRS framework is of great significance in the acquisition of military requirements and development of the military information system. In addition, the hierarchical structure may be implemented as a solution to the requirement development of each section. Moreover, the development approach of military requirements and technologies may be studied according to the contents of the development framework, thereby satisfying the requirements of military clients. Future research will focus on the requirement design, services to implement the development of military requirements, and the development of requirements through man–machine interaction.

## REFERENCES

Griendling K., Mavris D.N., Development of a DoDAF-based executable architecting approach to analyze system-of-systems alternatives[C]. Aerospace Conference, 2011 IEEE, vol., no., pp. 1–15, 5–12 March 2011.

Lizotte M., Bernier F., Mokhtari M., et al, CapDEM-Toward a Capability Engineering Process: A Discussion Paper[R]. Report No. (s): ADA440003, Dec. 2004.

Ramon E. Moore, R. Baker Kearfott, Michael J. Cloud, Introduction to interval analysis [M]. Published 2009 by Society for Industrial and Applied Mathematics in Philadelphia.

Sutcliffe A. The domain theory for requirement engineering. IEEE Trans on Soft Eng, 2008, 24 (3): 174–195.

Svoboda C.P. Structured analysis. In: Thayer R.H., Dorfman eds. Tutorial: System and Software Requirements Engineering. CA: IEEE Computer Society Press, 2009. 218–237.

Zhang Chengbin, Luan Liqiu, Zhang Hongzhou, Multi-agent technology based design of hierarchy model of groups/teams in Air Defense System[C]. Audio Language and Image Processing (ICALIP), 2010 International Conference on, vol., no., pp. 1403–1406, 23–25 Nov. 2010.

Advances in Energy and Environment Research – Achour & Wu (Eds)
© 2017 Taylor & Francis Group, London, ISBN 978-1-138-62682-9

# Research on the WRSPM reference model for the development of system requirements

Xiaolei Zheng
*The Logistics College, Beijing, P.R. China*

Ji Ren
*The Equipment Demonstration Center, Beijing, P.R. China*

ABSTRACT: This paper proposes a WRSPM reference model that defines the software environment (World), hardware requirements, software requirements, non-vendor software requirements, i.e. an IBM system requires a Microsoft product, software specifications, software programming environment and the hardware (Machine) environment that crosses system boundaries defining those machines on which the system will be deployed. This development model is based on related systems that will be integrated into a single system. This model defines the relations between the nodes of the reference model, and provides a formal description of the reference model. Finally, it compares the WRSPM reference model with the Functional Documentation Model (FDM), and analyzes the characteristics of the reference model to provide insight regarding when the model is most appropriate.

## 1 INTRODUCTION

Fred Broohs observed that, in building a software/hardware system, the most difficult decision is deciding precisely what to build and the constraints that limit the construction of the system. The first and most important set of decisions constitute the identification and documentation of the system requirements. These decisions must be "cast in concrete" prior to the start of the development phase. The costs to change the system requirements increases exponentially as the process moves forward as any change in the requirements identified at later phases invalidates all previous development work. The participants of the system development include Project Managers, System Designers, System Programmers (developers), System Testers, experts in System performance measurement, System integrators and producers (writers) of System documentation.

With the expansion of system scale, requirements development already play an increasingly important role in the system development and maintenance process. It permeates the whole life cycle of system development, and its results directly affect the success of system development. In addition, requirements development covers a wider range of fields, and the knowledge expertise of participants is more complex. These factors create an environment characterized by many diverse system roles and associated points of view within the development process. This has led to communication difficulties amongst the diverse set of participants, resulting in sub-optimal system design. To address these problems, we must study the common attributes of end-user requirements, component software requirements, software implementation requirements such as programming language ('C', Java, 'C++', 'C#'), system performance requirements, system testing requirements and customer system acceptance requirements in order to describe the relations between this diverse set. This is necessary in order to identify a unified development platform.

Based on the problem above, this paper proposes a reference model for requirements development, which contains several core elements of requirements development, attributes of the elements, and the relations between the elements. This paper will also provide a formal description of the model. The reference model provides a unified platform for the communication amongst personnel in the diverse fields that are participating in the System Build process.

## 2 CONTENTS OF THE WRSPM REFERENCE MODEL

The WRSPM reference model of requirements development consists of the following five components:

- Domain knowledge (World, W): provides assumed set of environmental facts;

- Requirement description (Requirement, R): specifies user requirements from the system, and describes these requirements based on the impact of the system on the environment;
- Specification description (Specification, S): provides sufficient information for programmers to build the system, and is the core of this paper;
- Programming (Program, P): implements the specifications through the programming environment;
- System entity (Machine, M): provides a description of all heterogeneous hardware requirements and specifications.

## 3 FRAMEWORK OF THE WRSPM REFERENCE MODEL

The WRSPM mode l can be regarded as descriptions of the system in different languages, and each description is based on the terms defined within their own domain. Some terms can be used in two or more domains. To understand the relations between these descriptions, we first need to understand the divisions between the environment and system.

The distinction between environment and system is a classical engineering topic, which has a profound influence on problem analysis. The terms used in WRSPM must first be defined. This process is very important, and can be referred to as the sixth element of the reference model: term designation. It describes the application domain (environment), system entity (system), and the interface between them.

The model framework classifies various phenomena (states, events and individuals) by environment and system, and uses terms to identify them. Some phenomena belong to the environment and

are controlled by the environment, and they are denoted by set $e$; some phenomena belong to the system and are controlled by the system, and they are denoted by set $s$.

At the interface between the environment and the system, some phenomena in set E are visible to the system, and they are denoted by a subset of $e$, $e_v$. Its complement is not visible to the system, and is denoted by $e_h$. Similarly, $s$ can also be decomposed into $s_v$ and $s_h$.

The phenomena in sets $e_v$, $e_h$ and $s_v$ are described by terms in W and R, and they are visible to W and R. The phenomena in sets $s_v$, $s_h$ and $e_v$ are described by terms in P and M, and they are visible to P and M. The sets $s_v$ and $e_v$ that are visible to both environment and system form S. Therefore, S is restricted only to terms describing $s_v$ and $e_v$. In a patient monitoring system, for example, requirement R is a warning system, which notifies nurses that a patient's heartbeat has stopped. System entities M include the sensor that detects heartbeats and the buzzer that sends the alarm signals, which use programming technology P and domain knowledge W. These terms are divided into the following four groups:

$e_h$: Nurses and patients with heart disease
$e_v$: Sounds of heartbeat
$s_v$: Buzzer signal
$s_h$: System internal data
S (Specification): when the sensor detects that the heartbeat has stopped, the buzzer should send a warning signal.

## 4 RELATION BETWEEN ENVIRONMENT AND SYSTEM

Both system and environment have the ability to implement events. W provides constraints for the environmental set $e$ or the relational behaviors of $e$ and $s_v$. R (Requirement) provides more constraints to describe all possible expected actions. If R includes an event in the environment, then the program (M and P) needs to implement this event. The description in logic is as follows:

$$\forall es. W \wedge M \wedge P \Rightarrow R \qquad (1)$$

That is, requirement specification includes environmental events ($e_h$, $e_v$) and corresponding system events ($S_h$, $S_v$), which can occur simultaneously.

This characteristic is known as adequacy. If the environmental assumption does not include all possible events, then **adequacy** is trivial, whereas the opposite case implies that it is non-trivial. Therefore, non-trivial environmental assumption is required, that is, the **consistency** of domain knowledge is required:

$$\exists es. W \qquad (2)$$

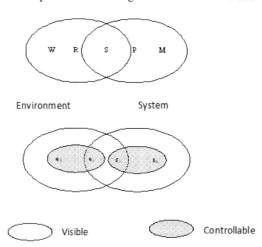

Figure 1.   Framework of WRSPM reference model.

(Detailed description: $\exists e_h e_v s_v$. $W$, and variable $S_h$ is not included by $W$). Furthermore, it also requires a property: if any possible value of the environmental variable visible to the system is consistent with the environmental assumption, it must also be consistent with $M \wedge P$. This property is known as **relative consistency**:

$$\forall e_v.(\exists e_h s.W) \Rightarrow (\exists e_h s.W \wedge M \wedge P) \qquad (3)$$

Consistency is a requirement that needs to be satisfied during the requirements development phase, but it should be noted that an overly strong or overly weak definition of consistency will lead to errors.

For example, assume that $M' = M \wedge P$ and simplify the relative consistency to $\exists es.W \wedge M'$. If the system and the environment are considered as consistent as long as the system realizes some events, then the consistency requirement is too low. If both domain knowledge consistency (Rule 2) and relative consistency (Rule 3) are considered to obtain the relative consistency $\forall es.W \Rightarrow M'$ (that is, if $W$ accepts any potential system behavior, the system to be built must also accept it), the consistency requirement is too high, and it is difficult for the system to implement. Another weakening: $\forall e \exists s.W \Rightarrow M'$ as long as the system realizes some attributes of environmental events, the system and environment are considered consistent. Such a consistency requirement is too low and cannot meet the needs of users.

Similar to the consistency of Rule 3: $\forall e.(\exists s.W) \Rightarrow (\exists s.W \wedge M \wedge P)$, that is, the system must be consistent with all environmental events. This consistency requirement is too high. Take the patient monitoring system as an example. The patient's heartbeat may stop ($e_v$) in the environment, but the nurses ($e_h$) are not notified of this event. The requirement specification that meets this consistency requirement also requires the system to allow the occurrence of this event, which obviously does not meet with expectations.

## 5 REQUIREMENT SPECIFICATION

The implementation process of the requirement is divided into two stages: requirement development and program implementation, which are performed by users and programmers, respectively. Generally, the knowledge indicating user requirements must be filtered from R and W to form a specification, which is implemented as follows: if S takes W into consideration when describing R, and P is an implementation of S in M, then P implements R. These two stages separate the responsibilities of users and suppliers in the process of cooperative development. Users and suppliers

build services based on S, which is also the basis of their communication. This requires a clear description of S.

First, a stipulation is made: S must exist in the common dictionary of the environment and system, that is, elements of S must exist in $e_v$ and $S_v$, but cannot exist in $e_h$ or $S_h$.

Firstly, the adequacy of S is defined:

$$\forall es.W \wedge S \Rightarrow R \qquad (4)$$

Secondly, the relative consistency of S is decomposed:

$$\forall e_v.(\exists e_h s.W) \Rightarrow (\exists s.S) \wedge (\forall s.S \Rightarrow \exists e_h.W) \qquad (5)$$

$$\forall e.(\exists s.S) \Rightarrow (\exists s.M \wedge P) \wedge (\forall s.(M \wedge P) \Rightarrow S) \ (6)$$

Formulas 4 and 5 combined with formula 2 form the user-side responsibilities, and formula 6 forms the supplier-side responsibilities.

Formulas 5 and 6 constrain the relative consistency $\forall s.S \Rightarrow \exists e_h.W$ (the event that occurs in S will definitely occur in W). If these constraints are removed and formula 3 is decomposed directly, the obtained formula 5 is as follows:

$$\forall e_v.(\exists e_h s.W) \Rightarrow (\exists e_h s.W \wedge S) \qquad (7)$$

It seems that the weaker formula 7 may be used to replace formula 5, but this is actually not possible. Consider a "good" specification $S_1$, which can satisfy formula 5 and meet the requirements. At the same time, assume there is a "bad" specification $S_2$ which is completely inconsistent with W. If $S = S_1 \vee S_2$, the formula can satisfy formula 4 and the weaker formula 7, but cannot satisfy formula 5. An actuator can implement $S_2$ to satisfy its responsibilities formula 7, but the implementation of $S_2$ is unpredictable when deployed, as it is not allowed by the extra strength of formulas 5 and 6.

## 6 CHARACTERISTICS OF THE WRSPM REFERENCE MODEL

The WRSPM model has a description method that is similar to the Functional Documentation Model (FDM). In FDM, there are four independent sets of variables: $m$ denotes the environment monitored values, $c$ denotes the system-controlled values, $i$ denotes the register values of program input, and $o$ denotes the register values of program output. Similarly, there are five predicates to denote the necessary documentation: NAT $(m,c)$ describes the nature without any assumption about the system; REQ $(m,c)$ represents the expected system behavior; IN $(m,i)$ denotes the monitored

359

values in the real world and their corresponding internal representations; OUT ($o,c$) denotes the corresponding system-controlled value that the software generates and outputs to the external world; SOF ($i,o$) denotes the relation between program input and program output.

A complete FDM must have the following three attributes:

- Feasibility: requires $\forall m.(\exists c.NAT(m,c)) \Rightarrow (\exists c. NAT(m,c)) \wedge (REQ(m,c))$;
- Adequacy: requires $\forall m.(\exists c.NAT(m,c)) \Rightarrow (\exists i. IN(m,i))$;
- Acceptability: requires $\forall mioc.NAT(m,c) \wedge IN(m,i) \wedge SOF(i,o) \wedge OUT(o,c) \Rightarrow REQ(m,c))$.

The following figure shows the correspondence between WRSPM and FDM:

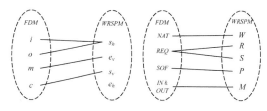

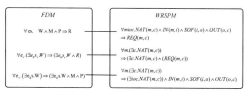

As shown in the figure, the FDM model does not have an element corresponding to $e_h$. NAT and REQ have smaller ranges than W and R, and they can only make judgments about the common phenomena of environment and system. IN and OUT correspond to M only when they represent specific input and output data. In an FDM example, assume that all variables of the FDM are functions with time as variables, and the following five predicates are defined:

$$NAT : (\forall t.c(t) > 0) \wedge (\forall t.m(t) < 0)$$

$$REQ : \forall t.c(t+3) = -m(t)$$

$$IN : \forall t.i(t+1) = m(t)$$

$$SOF : \forall t.o(t+1) = i(t)$$

$$OUT : \forall t.c(t+1) = o(t)$$

These predicates satisfy all attributes of the FDM, but the FDM determines that these predicates cannot be realized because ($\forall mioc. NAT(m,c) \wedge IN(m,i) \wedge SOF(i,o) \wedge OUT(o,c)$). In the reference model, this relation is expressed as:

$$\forall m.(\exists c.NAT(m,c)) \Rightarrow (\exists ioc.NAT(m,c)) \wedge IN(m,i) \wedge SOF(i,o) \wedge OUT(o,c)$$

It satisfies relative consistency. As a result, the WRSPM reference model can prevent the errors in the FDM and determine that these predicates can be realized.

The FDM and WRSPM reference models require that system terms must be strictly separated from environment terms. Other systems, such as Unity and TLA, allow the system and the environment to use the same terms, requiring users to distinguish the system and the environment, and to distinguish requirements and domain knowledge. For example, in the Unity language, E and M can control a Boolean variable B, but E must be set to true, and M must be set to false. The Boolean variable B can be divided into two variables B1 and B2 in WRSPM, and the distinction of terms can facilitate the description of requirements.

## 7 CONCLUSION

The WPSRM reference model proposed in this paper can describe the five elements of requirement development: domain knowledge, requirement description, specification description, programming, and system entity. In addition, it defines the relations between elements and their internal events, and provides a higher-order logical expression of the reference model as a theoretical basis and formal method for requirement analysis. Finally, the paper demonstrated the characteristics of the reference model by comparing it with FDM.

This discussion of the reference model is only limited to concepts and structure, and research on the application of the WPSRM reference model will be conducted in the future.

## REFERENCES

Aluretal, R., "Mocha: Modularity in Model Checking," Proc. 10th Int'l Conf. Computer Aided Verification, A.J. Hu and M.Y. Vardi, eds., Lecture Notes in Computer Science, Springer-Verlag, New York, 1998, pp. 521–525.

Brooks Jr., F.P. The mythical man-month: essay on software engineering (second ed) [M]. Addison Wesley Longman, 2005, 179–203.

Gunteretal, C.A., *A Reference Model for Requirements and Specifications*, tech. report, Univ. of Pennsylvania, Dept. of Computer and Information Science, Jan. 2012.

Maguire, S.A., *Debugging the Development Process: Practical Strategies for Staying Focused, Hitting Ship Dates, and Building Solid Teams*, Microsoft Press, Redmond, Wash., 2011.

Parnas, D.L., and J. Madey, "Functional Documentation for Computer Systems," *Science of Computer Programming*, Vol. 25, No. 1, Oct. 2013, pp. 41–61.

*Advances in Energy and Environment Research – Achour & Wu (Eds)*
© 2017 Taylor & Francis Group, London, ISBN 978-1-138-62682-9

# Study and simulation of titanium pipe deformation in the multi-pass backward spinning process

Z.Y. Xue, Y.J. Ren & Y. Ren
*School of Energy, Power and Mechanical Engineering, North China Electric Power University, Beijing, P.R. China*

W.B. Luo
*School of Materials Science and Engineering, University of Science and Technology Beijing, Beijing, P.R. China*

ABSTRACT: In this study, the tensile test of TA1 titanium was carried out at Room Temperature (RT) and Elevated Temperature (ET). The results show that the strength of TA1 at ET is less than that at RT, whereas the elongation of TA1 at ET is higher than that at RT. The backward spinning process of the titanium pipe was simulated by SuperForm software. Three schemes of spinning process were designed, and the deformations at different locations of the pipe were analyzed. The simulation shows that scheme-3 exhibits good deformation in the three projects, and the pipe thickness is decreased to 1–2 mm with the increase of the number of passes. Furthermore, metal accumulation does not occur after 21 passes of the spinning process. The thickness of the connection between the pipe and the deformation part increases to 13 mm at the end of the process. The deformation of the mouth is higher than that of the other zones, and the deformation of different positions in each pass increases with the increase of the number of passes.

## 1 INTRODUCTION

Titanium possesses remarkable mechanical and processing properties, high strength-to-weight ratio, and very high corrosion resistance and heat resistance. However, its plasticity is poor at room temperature, and the rebound of deflection is high. Therefore, titanium can easily crack and wrinkle in the plastic deformation process (Yang, 2008; Li, 2012; Zhao, 2013). Besides, the pipe deformation of the workpiece of titanium is larger close to the mouth than the other zones, because the ratio of yield strength ($\sigma_{0.2}$) to tensile strength ($\sigma_b$) is high ($\sigma_{0.2}/\sigma_b > 0.8$), Young's modulus is low (only half that of steel), and the plastic deformation temperature range is narrow. Therefore, the deformation of titanium is difficult.

Spinning is widely used in the manufacture of missiles and rockets as a metal plastic forming process (Zhao, 2013; Xu, 2001). In the spinning process, the appropriate process parameters are very important, such as spinning pass and decrement of wall in each spinning pass. Multi-pass spinning facilitates plastic deformation. Manufacturing defects such as metal accumulation, wrinkle, and larger gap between the mould and the workpiece are significantly decreased with the increase in the number of spinning passes (Li, 2012; Zhao, 2013; Xu, 2001). Two types of spinning deformation exist: forward spinning and backward spin-

ning. Stress concentration zones are more and the deformation of the last zone is larger in the backward spinning process than the forward spinning process (Shan, 2009; Tian, 2015). However, the displacement distribution is not uniform in the cross-section of workpiece in the backward spinning process. It is difficult to measure the stress and strain of the workpiece in the practical spinning process. However, it is easy and convenient to simulate the backward spinning process and observe the stress and strain of the workpiece using software, although the process conditions are changed (Liu, 2003; Zhang, 2013; Song, 2000; Zhan, 2008).

In this paper, the mechanical properties of TA1 were tested at different temperatures, and software SuperForm is used to simulate the backward spinning process of a TA1 titanium metal pipe. The pipe deformation of the spinning process was studied systematically in different passes.

## 2 TESTS OF MECHANICAL PROPERTIES AND SIMULATION

### 2.1 *Tests of mechanical properties*

Table 1 shows the mechanical properties of TA1 titanium at Room Temperature (RT) and Elevated Temperature (ET). Although the Tensile Strength (TS) and Yield Strength (YS) of TA1 at RT are

Table 1. Mechanical properties of TA1 at RT and ET conditions.

| No. | Tempera-ture/°C | TS/MPa | YS/MPa | Elonga-tion/% |
|---|---|---|---|---|
| 1 | 20 | 507 | 426 | 25.5 |
| 2 | | 492 | 417 | 27.0 |
| 3 | 500 | 107 | 106 | 23.5 |
| 4 | | 108 | 107 | 28.0 |
| 5 | 600 | 85 | 81 | 74.0 |
| 6 | | 86 | 84 | 80.5 |
| 7 | 700 | 33 | 31 | 89.5 |
| 8 | | 32 | 30 | 92.5 |

Table 2. Simulating scheme of the TA1 pipe backward spinning process (mm).

| Scheme\Pass | 1–15 | 16 | 17 | 18 | 19 | 20 | Total |
|---|---|---|---|---|---|---|---|
| 1 | | 6 | 6 | 0 | 0 | 0 | |
| 2 | 6 | 3 | 3 | 3 | 3 | 0 | 102 |
| 3 | | 3 | 3 | 2 | 2 | 1 | |
| 2 | | 3 | 3 | 3 | 3 | 0 | |

high, the elongation is low (i.e. 25.5%). With the increase of temperature, the strengths of TA1 reduced dramatically. It is only 107 MPa when the temperature reaches 500°C; however, the elongation of TA1 slightly changes. When the temperature is increased to 700°C, the elongation of TA1 dramatically increased to 91%. When the temperature is higher than 700°C, i.e. 700–800°C, the strength of TA1 titanium becomes very small, but the elongation becomes very high, which is easy to deform under such circumstances.

## 2.2 Simulating scheme

The outer radius of the pipe is 250 mm, and there is a step of height 2.7 mm in the inner surface of the pipe. The thickness of the front pipe wall is 10 mm, the end part is 7.3 mm, and the total deformation length is 110 mm. There are three simulating schemes of the pipe spinning, as shown in Table 2. Scheme-1 has 17 passes to form the mouth of the pipe, whose radius is 23 mm. However, in scheme-2, the process increased from two passes to 19 passes, but the thickness decreases with each pass, and reaches 3 mm in the last two passes. Scheme-3 has two passes higher than scheme 2, and the decrement of thickness of each pass is only 1 mm in the last two passes. The shell cover is elliptical, and the long and short axes are 125 and 88 mm, respectively. The decrement of thickness in each pass is 6 mm (Y direction).

The following are the basic parameters of the spinning process: spinning temperature,

700–800°C; rotating speed of the principal axis, 200 rpm; and dimension of the spinning roller $\phi$400 mm × 100 mm, whose moving speed is 500 mm/min, and the angle of attack is 0°.

## 3 BUILDING MODULE

### 3.1 Finite element modeling

According to the cross-sectional dimension, the titanium pipe was simulated using the software mentioned above. The three-dimensional grid of the pipe could be obtained after extending the mesh on the circumference section around the pipe axis (extending once each 3°); the total number of units are 17,520 (146 × 120), and there are 23,280 nodes. Spinning roller was indicated as a rigid body, which rotates the generatrix around its axis (as shown in Figure 1).

Saving the pipe mesh units after each pass of spinning, the new model was built by the next spinning pass. However, it needs to divide new units under the circumstances that mesh distorts

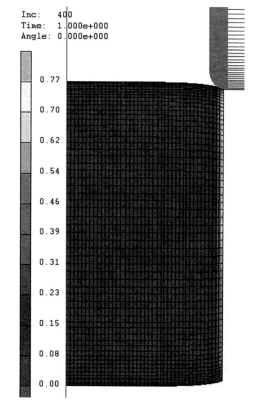

Figure 1. The first pass of backward spinning (the first second).

severely in the spinning process; at this moment, the units are divided along the outline of distortion. The decrement of thickness of each pass is same in 1–15 spinning passes.

## 4 RESULTS AND DISCUSSION

### 4.1 Deformation analysis of the workpiece in the spinning process

The decrement of thickness is same for the first 15 passes in the three schemes, and deformation of the pipe occurs. However, in scheme-1, the decrement of thickness of the 16th and 17th passes is 6 mm. It can be found that metal accumulation takes place in the front part of the pipe (Figure 2(a)), and the spinning process is hard to finish. In the whole spinning process, the pipe radius reduces from 250 to 70 mm, the maximum thickness is 12.37 mm, and the resistance of pipe deformation is very large. Because the deformation zone of the metal is pushed to move forward along the axis, metal accumulation occurs in the front part of the pipe. Because of the distorted grid, the front part of the pipe forms folded defects.

In scheme-2, the decrement of the thickness of each pass is 3 mm in the 16th to 19th pass processes. Metal accumulation occurs in the front part of the pipe, but the resistance of pipe deformation is less than that in scheme-1 (as shown in Figure 2(b)). This shows that stress distribution is not uniform.

In the process of scheme-3, the decrement of thickness of each pass is only 1 mm in the 20th to 21st pass. The result shows that metal accumulation does not occur in the front part of the pipe. The size of the TA1 workpiece in scheme-3 would be better than that in scheme-1 and scheme-2. Figure 2(c) shows the pipe deformation after 21 spinning passes in scheme-3.

According to the above statements, the resistance of pipe deformation of TA1 will focus on the last spinning passes. Therefore, each decrement of thickness should be reduced. The results show that the pipe mouth could be completely formed in scheme-3 spinning process.

### 4.2 Change of thickness in the spinning process

The deformations of the pipe are same in 1–15 passes in the three schemes. The mouth is completely formed only in scheme-3, and hence the change of thickness in scheme-3 is discussed. In the spinning process, the wall thickness of the mouth increases in 1–10 passes with the decrease of the outer radius, and the thickness reaches 12.3 mm after the 10th pass, as shown in Figure 3, while the length of the pipe mouth increases

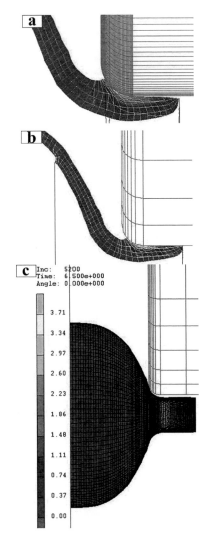

Figure 2. Pipe deformation in different schemes: (a) the 16th pass in scheme-1 when the decrement of thickness is 6 mm; (b) the 18th pass in scheme-2 when the decrement of thickness is 3 mm; (c) the 21st pass in scheme-3 when the decrement thickness is 3 mm.

significantly in the 11th pass. The wall thickness of the pipe mouth gradually decreases to 10.26 mm at the end of the 21st pass. Meanwhile, the length of the pipe mouth is about 166.9 mm (as shown in Figure 4). The thickness of the transition zone between the mouth and the undeformed zone is maximum along the axial direction of the pipe, i.e. 13 mm. The inner part of the mouth zone is trumpet-shaped, and the length of the straight zone is 46 mm. When the spinning process is finished, the radius of the step is increased from 3.7 to 4.1 mm.

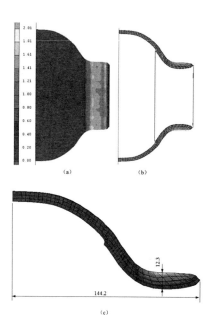

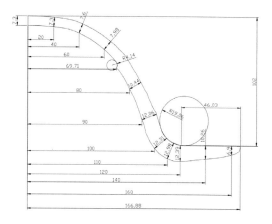

Figure 3. Deformation of the pipe after the 10th pass: (a) the external view; (b) the cross-section along radial direction; (c) the upper part of the cross-section along the radial direction.

Figure 4. Schematic of the deformation zone after the 21st spinning pass.

## 5 CONCLUSIONS

1. TA1 titanium exhibits remarkable mechanical properties at room temperature. The strength of TA1 decreases gradually with the increasing of temperature. However, the increase of elongation of TA1 is higher in ET than that at RT. When the temperature is 700°C, the UTS reduces to 33 MPa, and the elongation significantly increases to 91%. The resistance of pipe deformation is very small when the temperature is in the range of 700–800°C, which facilitates plastic deformation.

2. The resistance of pipe deformation is very high when the spinning process is not finished in scheme-1 and scheme-2 processes. Metal accumulation is absent even after the spinning process is finished in scheme-3.

3. The pipe deformation is not uniform in the spinning process. It is larger closed to the mouth than the other zones.

## ACKNOWLEDGMENTS

This study was financed by the Program for New Century Excellent Talents in University (NCET-12-0849) and the Fundamental Research Funds for the Central Universities (13ZD12, 2014ZZD03).

## REFERENCES

Li, Q. J., Fan, K. C., Wang, Q., Huang S. Y. & Zhao, S. Q. 2012. Factors influencing spin forming of large-diameter, thin-walled TC4 alloy tube. *Aerospace Materials and Technology* (1): 86–88. (in Chinese).

Liu, J. H. 2003. Research on forming mechanism of multi-pass conventional spinning process and influence of roller-trace on the process. Shanxi: Northwestern Polytechnical University, 2003. (in Chinese).

Shan, D., Yang, G. & Xu, W. 2009. Deformation history and the resultant microstructure and texture in backward tube spinning of Ti-6Al-2Zr-1Mo-1V. *Journal of Materials Processing Technology* 209(17): 5713–5719.

Song, Y. Q. 2000. The prospect of successive partial plastic forming. *China Mechanical Engineering* 11(1): 65–66. (in Chinese).

Tian, Y., Zhang, H. F. & Tian, W. X. 2015. Numerical simulation analysis of power spinning for metallic. *Foundry Technology* 36(1): 191–194. (in Chinese).

Xu, Y., Zhang, S. H., Li, P., Yang, K., Shan, D. B. & Lu, Y. 2001. 3D rigid-plastic FEM numerical simulation on tube spinning. *Journal of Materials Processing Technology* 113(1–3): 710–713.

Yang, Y. L., Guo, D. Z., Zhao, Y. Q., Zhao, H.Z. & Su, H. B. 2008. Progress on the spin-forming technology of titanium in china. *Rare Metal Materials and Engineering* 37(4): 625–629. (in Chinese).

Zhan, M., Li, H., Yang, H. & Chen, G. 2008. Wall thickness variation during multi-pass spinning of large complicated shell. *Journal of Plasticity Engineering* 15(2): 115–121. (in Chinese).

Zhang, C., Yang, H. C., Han D, Wang, X. J., Mo, R., Lu, X. R. & Gong, J. S. 2013. Applications and development of titanium alloys spinning technology in domestic aerospace field. *Journal of Solid Rocket Technology* 36(1): 127–132. (in Chinese).

Zhao, B. 2013. Research on the load and drum of thin walled tube during the spinning process. Hunan: Central South University. (in Chinese).

*Advances in Energy and Environment Research – Achour & Wu (Eds)*
*© 2017 Taylor & Francis Group, London, ISBN 978-1-138-62682-9*

# Nonlinear mechanical analysis for contact between the main shaft and the supporting wheel in a large mine hoist

Jia-Chun Li, Ji-Chao Cao, Jie Hu & Xue He
*College of Mechanical Engineering, Gui Zhou University, Guizhou, China*

ABSTRACT: Finite Element Method (FEM) is the most effective numerical means for structural analysis. The stress distribution and contact force are finite elements simulated using the ABAQUS software in the large mine hoist assembly process. The simulation results are used to evaluate the interference fit and pressing force, which also provide some reference for the joint strength analysis and optimum design in the interference fit and effectively reduce testing time and cost.

## 1 INTRODUCTION

2JK-5 type large-scale mine hoist spindle is the main drive part in the main shaft device, and is also the main bearing part of the hoist. Because the spindle has borne all hoist torque, and its strength, trans-mission, stability, and reliability have directly affected the safety of the whole hoist system (Zhang, 2010), it is very important to strictly grasp the main shaft strength and study connected assembly methods.

The traditional connection mode of the main shaft in the mine hoist is that one side of the wheel hub is connected by the tangential key and the other side is the clearance fit. The tangential key, which has easily loosed in the driving process, will cause axial movement of the entire drum. Thus, the stability and safety of the entire drive system will be weakened (Qi, 2012).

In view of the above problems, the 2JK-5 large hoist spindle and supporting wheels are connected by means of interference connection. The interference fit has been used on both sides of the fixed drum. On the basis of considering the convenient installation, the main shaft and the left branch of the portal vein used a round $\varphi700H7/U6$ the $\varphi680H7/U6$ employed by the spindle and right supporting wheel, and the connection is shown in Figure 1. Handbook of mechanical design defined the interference range of 0.66–0.79 mm. The supporting wheels and the reel are fixed by the high-strength bolt connection to ensure the reliability of the transmission and the safety of the whole lift system.

Interference connection is realized by interference fit of the parts. The connection has the following advantages: simple structure; high centering precision; resistance to torque, axial force, or the composite load; and reliable working under shock and vibration load (Zhang, 2010). The types, characteristics, and application of the interference fit are shown in Table 1.

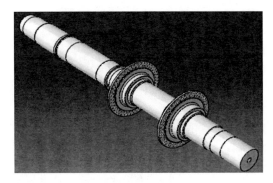

Figure 1. The interference connection structure between the main shaft and the left (right) supporting wheel.

Table 1. The types and characteristics of interference fit.

| Interference coupling type | Characteristics | Application |
|---|---|---|
| Cylinder interference fit | The structure is simple and the processing is convenient | Widely used in shaft and hub connection, etc. |
| Conical surface interference fit | The assembly of conical surface by relative axial displacement | Applicable to large parts such as steel rolling machinery and propeller shaft |

Table 2. Assembling method, characteristics, and application of interference fit.

| Method of interference fit | Characteristics | Application |
|---|---|---|
| Pressing in method | Press in method is simple | Generally used in a small amount of surplus |
| Temperature difference method | The process is complicated, and the surface is not easy to scratch | Suitable for heat-treated or coated surfaces |
| Hydraulic method | The process is complicated, and the surface is not easy to scratch | The fitting is mainly used for the interference of the conical surface |

## 2 THEORETICAL BASIS

### 2.1 Model simplification

The interference fit between the spindle and the supporting wheel can be simplified to a thick-walled cylinder with inner radius a and outer radius c. As shown in Figure 2, it is composed of two layers in the cylindrical shell, the inner radius of the inner cylinder is a, the outer radius is $(b + \delta_1)$, the inner radius of the outer cylinder is $(b − \delta_2)$, and the outer radius is c.

An uniform compressive stress is generated on the surface of the two-layer cylindrical shell, which is the set pressure. The value of the set pressure is related to the amount of interference, and the interference of the two cylinders is their radial displacement. Figure 2 shows a combined thick-walled cylinder with an inner cylinder and set pressure. According to the elastic mechanics thick-walled cylinder theory formula, for the plane stress problem, the outer radius of the radial displacement is:

$$u|_{t=b}\delta_1 = -\frac{pb}{E_2}\left(\frac{b^2 + a^2}{b^2 - a^2} - \varepsilon_2\right) \tag{1}$$

where
$\varepsilon_2$: Poisson's ratio for supporting materials;
$E_2$: plastic model for a supporting wheel in formula (1).

The set pressure is the internal pressure based on the theory of elastic mechanics thick-walled cylinder. For the plane stress problem, the radial displacement of the inner radius is:

$$u|_{t=b}\delta_2 = \frac{pb}{E_1}\left(\frac{c^2 + a^2}{c^2 - b^2} + \varepsilon_1\right) \tag{2}$$

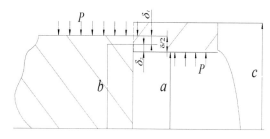

Figure 2. The spindle and the supporting wheel.

where
$\varepsilon_1$: Poisson's ratio for spindle material;
$E_1$: elastic modulus of the principal axis in formula (2).

As $\delta_1$ is the reduced radius, its direction is opposite to that of the radius. The geometric condition of the set is written as:

$$\delta = 2\left(-\delta_1 + \delta_2\right) \tag{3}$$

The pressure of the set can be obtained by substituting the formulas of $\delta_1$ and $\delta_2$ into formula (3):

$$p = \frac{\delta E_1 E_2}{2b\left(E_1\left(\frac{b^2 + a^2}{b^2 - a^2} - \varepsilon_2\right) + E_2\left(\frac{c^2 + a^2}{c^2 - a^2} + \varepsilon_1\right)\right)} \tag{4}$$

Formula (4) shows that the pressure after assembly in the cylinder with inner radius a and outer radius c can be determined from the amount of interference and layered radius b. According to the support wheel set pressure and the coefficient of static friction f, the torque transferred by the supporting wheel is:

$$T = \int_0^L 2\pi b^2 p_i f dl \tag{5}$$

where
b: radius of the contact surface between main shaft and supporting wheel;
L: the length of the interference fit between the supporting wheel and the main shaft;
$p_i$: interference fit between the spindle and the support of a small section of the set pressure within the wheel;
dl: the main shaft and supporting wheel with a small section of the axial displacement;
f: coefficient of static friction between the spindle and supporting wheel.

### 2.2 Establishment of finite element model

The software ABAQUS has been widely used in many countries in machinery, civil engineering, water conservancy, aerospace, shipbuilding, electrical, automotive, and other engineering fields. In recent years, ABAQUS has attraction much attention in China, because of its simplicity and ability to be finished in the interface. The software provides a good application platform for beginners. In addition, ABAQUS easily establishes models of complex problems. For most numerical simulations, the user needs to provide only simple geometric conditions, boundary conditions, material properties, loads, and other engineering data. For the analysis of nonlinear problems, ABAQUS can automatically select the appropriate load increment and convergence criterion and adjust the parameters in the analysis to ensure the accuracy of the results (Chen, 2012).

The contact model is established to accurately reflect the mechanical characteristics of the contact area when the wheel and axle contact analysis is established. ABAQUS is a globally accepted and advanced large-scale general finite element software with strong nonlinear analysis ability. In this paper, the contact algorithm is used to simulate the interference fit between the main shaft and the supporting wheel of the large mine hoist.

### 2.2.1 *Determine material properties*

The parameters of the 2JK-5 single rope winding type of mine hoist spindle $45MnMo$ are $b_1 = 1030$ $MPa$, $\sigma_{s1} = 835$ MPa, elastic modulus $E_1 = 2.06 \times 10^5$ $Mpa$, and Poisson's ratio $\varepsilon_1 = 0.3$, whereas those of the supporting wheel material ZG310-570 are $b_2 = 570$ $MPa$, $\sigma_{s2} = 310$ $MPa$, $E_2 = 1.75 \times 10^5$ $MPa$ elastic modulus, and Poisson's ratio $\varepsilon_2 = 0.3$.

### 2.2.2 *Grid division*

Because of the symmetry of the structure and the 1/4 model analysis, four-node quadrilateral bilinear nonconforming axisymmetric element is chosen. The size of parts is larger, the global size of 10 mm is selected, the number of units is 36437 units, and the grid is shown in Figures 3 and 4.

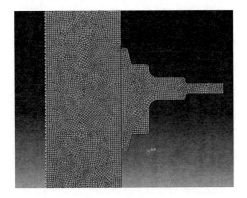

Figure 3.   The spindle and the left wheel meshing.

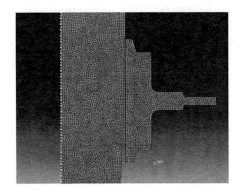

Figure 4.   Meshing of spindle and right supporting wheel.

### 2.2.3 *Defining boundary conditions*

The principal axis and the supporting wheel are symmetrically flexible bodies, and the only displacement of the rigid body is U2. The boundary conditions of U2 = 0 exist on the main shaft bottom. The axial displacement of the supporting wheel is eliminated by friction force.

The press head is set as an axisymmetric rigid body. Motion is determined by the reference point, and the displacement and rotation of the rigid body in each direction can be eliminated by the boundary conditions. The boundary conditions imposed on the head 1 are as follows:

Step 1:   the pressure head 1 to move up 0.1 mm (U1 = 0, U2 = 0.1, UR3 = 0);

Step 2:   analysis of the pressure head 1 to fit length L1 = 500 mm mobile left wheel and spindle (U1 = 0, U2 = 500, UR3 = 0);

Step 3:   the pressure head 2 to move up the 0.1 mm (U1 = 0, U2 = 0.1, UR3 = 0);

Step 4:   analysis of the pressure head 2 to fit length L2 = 760 mm mobile left wheel and spindle (U1 = 0, U2 = 760, UR3 = 0).

### 3   FINITE ELEMENT SIMULATION RESULTS OF THE ASSEMBLY PROCESS

The spindle and the supporting wheel assembly are three-dimensional transformed and simulated.

1. The stress nephogram for main shaft and the left wheel interference fit is shown in Fig. 5, The stress nephogram of the main shaft and the right supporting wheel interference fit are shown in Fig. 6.

2. The contact pressure nephogram: when the supporting wheel is pressed into the spindle, the spindle and the supporting wheel are contacted with the node CPRESS > 0, and no contact with the node CRRESS = 0. In the whole process of

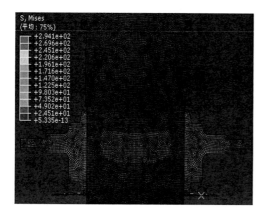

Figure 5.   The spindle and the left wheel assembly simulation stress.

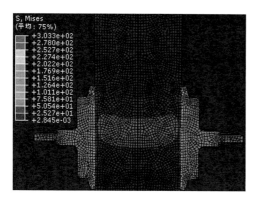

Figure 6.   Principal axis and right supporting wheel assembly stress nephogram.

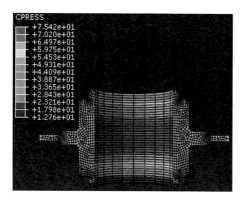

Figure 7.   The spindle and the left wheel contact pressure nephogram.

assembly, the area of CPRESS > 0 is expanding continuously. Principal axis and left and right supporting wheel contact pressure cloud hidden in the main spindle are shown in Figures 7 and 8.

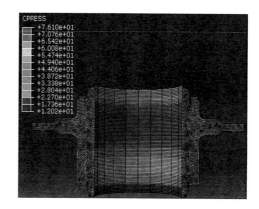

Figure 8.   The contact pressure between the main shaft and the right supporting wheel.

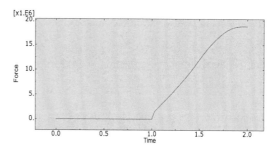

Figure 9.   Head 1 axial contact force between the wheel and the left supporting wheel.

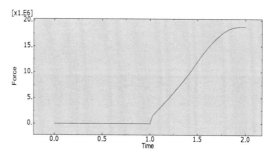

Figure 10.   The axial contact force between the pressure head 2 and the right supporting wheel.

3. Contact force: the pressure curve of the contact force between the head and the inner ring. The axial force between the pressure head 1 and the left wheel is shown in Figure 9, and the axial force between the pressure head 2 and the right supporting wheel is shown in Figure 10. According to the figures, the contact area between the shaft and the supporting wheel is increased with pressure head press-in, and the axial force is larger but nonlinear.

## 4 SUMMARY

1. For the axisymmetric plane stress or plane strain problem, the three-dimensional problem can be simplified as a two-dimensional problem, thereby significantly reducing the size of the model and computation time.
2. By means of finite element analysis, the main force between the spindle and the left wheel can be calculated as $\sigma_{max1}$ = 294.1 MPa < $\sigma_{s2}$ = 310 MPa, and the main stress between the spindle and the right supporting wheel is $\sigma_{max2}$ = 303.3 MPa < $\sigma_{S2}$ = 310 MPa. These results show that, in the wheel, no plastic deformation occurs between large mine lift shaft and the left and/or right supporting wheel.
3. Finite element solutions can more truly reflect the distribution of contact pressure, which is not uniformly distributed. The central distribution is uniform and smooth, but both ends are smaller. This is caused by the edge effect of the structure.

## ACKNOWLEDGMENTS

This work was supported by the Major science and technology projects of Guizhou Province of China via the Key technology research and industrialization of a new type of mine hoist (Grant No. [2013]6018).

## REFERENCES

Da-peng Zhang, Rob Ciugene, Xing-ping Liu, et al, Vehicle wheel shaft interference fit assembly stress A nonlinear contact finite element analysis, J, Tibet's Science & Technology, 7(2010) 78–80.

Ming-JieJing, Zi-Gui Li, Mine hoist machinery, M, Machinery Industry Press, Beijing, 2011.

Xiao-jie Chen. Interference fit based on ABAQUS finite element analysis of the contact sleeve, J, Mechanical Engineering, 2(2012) 79–80.

Xiao-jie Qi. Cylindrical interference fit coupling in Mine Hoist spindle apparatus, J, Liaoning University of Technology, 32(2012) 169–172.

Wei Qi. ABAQUS6.14 super learning Guide, M. Beijing: Posts & Telecom press, 2016.

*Advances in Energy and Environment Research – Achour & Wu (Eds)*
*© 2017 Taylor & Francis Group, London, ISBN 978-1-138-62682-9*

# Dynamic analysis and optimization of the extracting cartridge case system of Breechblock

Yongmei Zhu & Luyao Ni
*School of Mechanical Engineering, Jiangsu University of Science and Technology, Jiangsu, China*

ABSTRACT: The effect of shape parameters of rocker arm on the cartridge case extraction is studied in this paper. The aim is to provide theory basis for the optimization design of naval gun. This paper studies the influence of the thickness of rocker arm on the kinetic and internal energy of cartridge case extraction and stress distribution of the extractor, and optimizes the thickness of the rocker arm, achieves the goal of reducing the maximum stress of the extractor dangerous point. The research finds that when rocker arm worn by 1 mm, the average speed of extraction reduces by 21.46%, the curve of speed is delayed about 0.5 ms, the maximum of internal energy reduces by 19.1% and the maximum of kinetic energy reduces by around 3%. The overall tendency is consistent with before. When the thickness of rocker arm is increased by 10 mm, the maximum stress of the extractor danger points is minimal.

## 1 INTRODUCTION

The breech mechanism mainly consists of breech block, breech, extractor, rocker arm and extracting template (Tan, 2014). When the breech crashed into the extracting template at the speed of 3 m/s, the cartridge-case will be pulled out in a very short time (about 6 ms~10 ms). However, it has been found that the breech mechanism cannot work normally because of the fracture of extractor (Zhang, 2012). One way to solve this problem is to analyze and optimize the cartridge case extraction.

For example, Han (2012) carried out innovative design of a typical artillery gun amount, based on CAE technology and optimization design. By Patran and Nastran software, Yao (2014) analyzed the vibration characteristics of artillery, optimized the installation position and structure parameters of the damper, and reduced the vibration amplitude of the muzzle. Gao et al. (2016) investigated the effects of aluminum shell thickness and density of aluminum foam on energy absorption properties of the filling structure.

The thickness of rocker arm is calculated by empirical formula and the simple mechanical model in traditional design method. This paper analyzed the influence of rocker arm's wear on dynamic behavior of cartridge case extraction, and optimizes the thickness of the rocker arm based on the finite element technology.

## 2 NON-LINEAR DYNAMIC ANALYSIS

Based on the symmetry of structure and load, 1/2 of the whole structure is analyzed and calculated (Tang, 2011). The model of cartridge case extraction contains 45793 elements and 57055 nodes. The minimum elements size is 2.5 mm. The model is shown in Figure 1. The extractor is a key part in the cartridge case extraction. During the prototype test of initial design, it appeared plastic deformation in many times, sometimes even fractured. The result of this finite element analysis showed that maximum stress of the dangerous area is 1115.41 MPa, and it beyond the yield limit about 980 MPa, plastic deformation is easier to occur. Therefore, when we optimize the cartridge case extraction reducing the stress should be first considered.

Figure 1. Finite element model.

## 3 DYNAMIC CHARACTERISTICS ANALYSIS OF CARTRIDGE CASE EXTRACTION BEFORE AND AFTER THE WEAR OF ROCKER ARM

In order to solve the problem of wear of rocker arm in the prototype test, we research on the changes of cartridge case extraction's dynamic parameters before and after the wear. The parameters include internal energy, kinetic energy, dynamic stress of dangerous area of extractor and speed of extraction, etc. The element type of model is C3D8I (Xu, 2011). The mesh deformation technology is used in analysis. According to the prototype test, rocker arm is worn by 1 mm in average. The wearing section is shown in Fig. 2. Fig. 2(a) is the sketch of the wear areas in rocker arm. Fig. 2(b) is the 3D model of the wear areas in rocker arm. Then the finite element model of cartridge case extraction after wear is built.

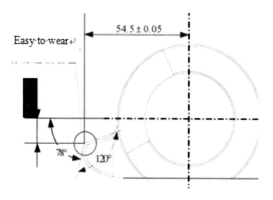

(a) The sketch of rocker arm

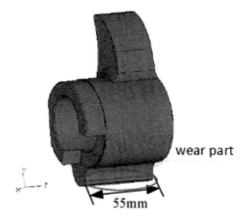

(b) The 3D model of rocker arm

Figure 2. Wearing section of rocker arm.

Through the dynamic analysis of cartridge case extraction, it can be concluded that the tendency of the internal energy and the kinetic energy after wear is consistent with before wear. The maximum value of the internal energy is 658.952 J before wear, and it is 533.066 J after wear, which is decreased by 19.1% in average. At first, the kinetic energy of cartridge case extraction remained almost unchanged, and it decreased by 3% after 3.7 ms. The result is shown in Fig. 3.

The extraction speed curve is shown in Figure 4, and it shows the effect of wear on rocker arm. It can be concluded that the average speed of extraction is reduced by 21.46% after wear, and the curve is delayed by about 0.5 ms. When rocker arm worn by 1 mm, the energy of the system has been reduced and the speed of extraction is obviously reduced. This will greatly decrease the efficiency of the launch and fighting power for naval gun.

## 4 OPTIMIZATION OF ROCKER ARM THICKNESS

### 4.1 *A comparative analysis of thickening and thinning*

Through the analysis of the wear of the rocker arm in the working process, it is known that the shape of rocker arm can influence on the cartridge case extraction. The effects of rocker arm thickness on the internal energy, the kinetic energy and the speed of extraction are studied in this paper.

The analysis is carried in ABAQUS. Increasing and reducing the thickness of rocker arm by 6 mm, the finite element models are built. The energy change curve of cartridge case extraction is shown in Figure 5. The tendency of kinetic energy and internal energy are similar, whenever increasing and reducing the thickness of rocker arm. The maximum value of the kinetic energy is same, and the change of the internal energy of the system is almost same. In addition, the average speed of the cartridge is consistent with the average speed of the original model. But, the time of the kinetic energy fluctuation is small different during the analysis.

The maximum stress for dangerous areas is shown in Table 1. After reducing the rocker arm thickness, the maximum stress is 1142.746 MPa, increased by 2.48%. After increasing the rocker arm thickness, the maximum stress is 1042.611 MPa, reduced by 6.5%. The result showed that the maximum stress for dangerous areas can be reduced after increase the rocker arm thickness, and reduce the thickness of rocker arm can't improve the strength of cartridge case extraction.

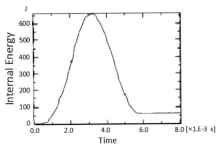

(a) Curves of internal energy before wear

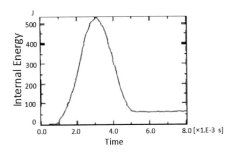

(b) Curves of internal energy after wear

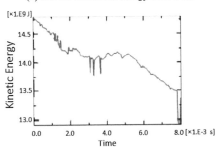

(c) Curves of kinetic energy before wear

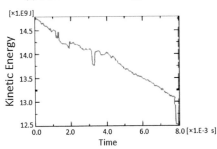

(d) Curves of kinetic energy after wear

Figure 3.    The energy curve before and after wear.

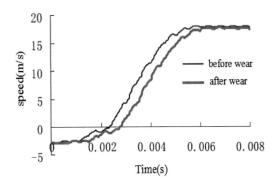

Figure 4.    The speed curve of extraction.

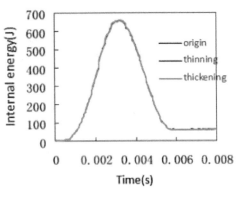

(a) Curves of internal energy

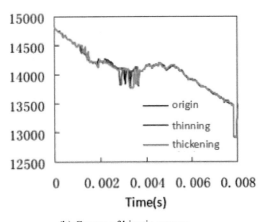

(b) Curves of kinetic energy

Figure 5.    Curves of energy.

## 4.2  *Optimization of rocker arm thickness*

From the above research, it is concluded that system energy and pumping shell speed basically aren't change with the thickness of rocker arm, but the stress in dangerous area of extractor can reduce.

In this paper, the thickness of the rocker arm is increased by 2 mm each time, and analysis the maximum stress in the dangerous area of extractor. The result of six kinds of thickness is shown

Table 1. Maximum stress of dangerous point.

|  | Origin | After thinning | After thickening |
|---|---|---|---|
| Maximum stress (MPa) | 1115.092 | 1142.746 | 1042.611 |

Table 2. The maximum stress of the extractor.

| The thickness of the rocker arm | Initial stress (MPa) | The maximum stress (MPa) |
|---|---|---|
| Thicken 2 mm | 1115.092 | 1103.981 |
| Thicken 4 mm | – | 1098.425 |
| Thicken 6 mm | – | 1031.759 |
| Thicken 8 mm | – | 993.647 |
| Thicken 10 mm | – | 972.981 |
| Thicken 12 mm | – | 998.055 |

in Table 2. The rocker arm is thickening by 2 mm, 4 mm, 6 mm, 8 mm, 10 mm, 12 mm, the maximum stress is gradually lower than the initial stress. When the thickness is increased by 10 mm, the maximum stress is 972.981 MPa, which is less than the material yield limit. It is the minimum stress of dangerous area in Table 2. When the thickness is increased by 12 mm, the maximum stress is 998.055 MPa, which is slightly above the yield limit.

In general, when the thickness is increased the maximum stress in the dangerous point of the extractor will be decreased. After thickening 10 mm, the maximum stress of extractor is less than the yield strength. And plastic deformation will not happen on the extractor. In the prototype test, the problem is the plastic deformation in extractor. Thickening 10 mm can solve this problem. Therefore, thickening 10 mm is the best choice for the thickness of the rocker arm.

## 5 CONCLUSIONS

1. Through the analysis of nonlinear dynamics of cartridge case extraction, the most dangerous areas of the extractor was found. The maximum stress of dangerous areas exceeded the yield limit of material, and it can cause plastic deformation.
2. In this paper, we have analyzed the influence of the wear in the rocker arm on the dynamic characteristics of cartridge case extraction. We found that the kinetic energy and internal energy of cartridge case extraction were decreased, and the speed of extraction was also decreased after the rocker arm wore 1 mm. Meanwhile, it had little effect on the stress in the dangerous areas of the extractor.
3. The thickness of the rocker arm has been optimized. When the thickness of the rocker arm was changed, system energy and the speed of extraction were almost unchanged, but increase the thickness of rocker arm can reduce the maximum stress in the dangerous areas of extractor. When thickening the rocker arm's thickness by 10 mm, the plastic deformation won't appear in the dangerous areas of extractor. This paper studies the effect of shape parameter of rocker arm on the cartridge case extraction, which can provide a theoretical basis for the optimization design of naval gun.

## ACKNOWLEDGMENTS

A part of this study is supported by the ship research foundation of China (No. 2014JX007G).

## REFERENCES

Gao Meng & Xu Peng. Pressing-crack energy absorption for aluminum foam filled aluminum shell under high value impact [J]. Ordnance Material Science & Engineering, 2016, 01: 45–49.

HanYapeng. Static Stiffness Analysis and Optimization Design of Typical Artillery Gun Mount [D]. Dalian University of Technology, 2012.

Tan Bo & Hou jian. 2014. Simulation Analysis of Naval Gun Extractor Mechanism in Process of Cartridge Case Extraction. 35(04): 49–52.

Tang Wenxian & Zhang Jian. Cartridge-case extracting analysis of some breech mechanism based on flexible body dynamics method [J]. Applied Mechanics and Materials, 2011, 43(10): 338–341.

Xu Yaochun, Nonlinear Structural Dynamics Analysis of Naval Gun Extractor Systems [D]. Jiangsu University of Science and Technology, 2011, 03.

Yao Haoze. Design Optimization of Artillery Carriage and Vibration Absorber for Firing Accuracy Improvement [D]. Dalian University of Technology, 2014.

Zhang Jian & Tang Wenxian. Structural Optimization design for Breechblock Extractor [J]. Acta Armamentarii, 2012, 06: 647–651.

Advances in Energy and Environment Research – Achour & Wu (Eds)
© 2017 Taylor & Francis Group, London, ISBN 978-1-138-62682-9

# Anvil configuration optimization of an underwater pile driving hammer

Liquan Wang & Donghua Chen
*College of Mechanical and Electrical Engineering, Harbin Engineering University, Harbin, China*

ABSTRACT: The anvil configuration model was established for impact system of 1200 kJ underwater pile driving hammer. By the numerical simulation of impact system with different configuration anvil, the rule between dynamic performance of impact system and the configuration parameters of anvil was discovered. On the basis of radial basis function, the approximate functions between dynamic performance and configuration parameters were obtained, and then the optimization model of anvil configuration was established. The optimization model was solved by the nonlinear optimal tool fmincon in MATLAB, whose results were analyzed by numerical simulation. The maximum relative error between optimized results and numerical results is 1.69%, which stated that the approximate functions established were valid. The optimized results meeting the dynamic strength were obtained for the impact system of 1200 kJ underwater pile driving hammer.

## 1 INTRODUCTION

The pile driving hammer is an energy transformation device. It converts other forms of energy into kinetic energy of ram and then converts it into kinetic energy of pile by means of the collision of ram and anvil. The configuration of anvil influences the energy transmitting efficiency of the hammer, penetration of pile, and durability of the impact system. Well-known pile driving companies own their anvil patent (Wang 2003). Figure 1 shows the impact system model of underwater pile driving hammer. This paper aims at revealing the effect of anvil configuration on the performance of the impact system, and then optimizing the configuration of anvil to conform to the strength of the component of the impact system and realize maximization of the energy transmitting efficiency.

In order to realize configuration optimization, the functional relations between the performances of impact system and the configuration parameters are required. For the anvil of determined configuration parameters, the energy transmitting efficiency and the maximum stress of the impact system component can be acquired by numerical simulation. By the simulation of anvil with different configuration parameters, large amounts of performance data of impact system in the corresponding configuration parameters are acquired, and the approximate functional relationships are obtained by the method of function fitting with these data. Fitting scattered data is the core problem of anvil configuration optimization. Radial

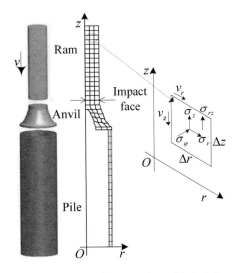

Figure 1. Axisymmetric geometric model of the impact system.

basis function method is effective in resolving such problem. The method has been used in many research fields, such as physical design (Zhang et al. (2007), Xu et al. (2010), Chen et al. (2015)), image reconstruction (He et al. (2010), Zhang et al. (2014)), and geological research (Yamamoto (1998), Cai et al. (2010), Wang et al. (2014)). Optimization model is built by the fitting function. The optimization model is solved to realize the optimal design of anvil configuration.

## 2 PROBLEM DESCRIPTION

The first step in realizing the optimal design of anvil configuration is to build the parameterized model of anvil, and then determine the configuration parameters. Figure 2 shows the anvil structure diagram of the underwater pile driving hammer. The impact energy of the underwater pile driving hammer is 1200 kJ. It is evident from the figure that the height of the anvil is 1700 mm, diameter is 1300 mm at the end of anvil collision with the ram, inner diameter is 970 mm, outer diameter is 2510 mm at the end of anvil contact with pile, thickness of the up-transition layer of anvil is 300 mm, and the thickness of the down-transition layer of anvil is 150 mm. The main section of anvil lies between the two transition layers, which is determined by the curves $l_n$ and $l_w$.

The anvil is an axisymmetric component, which can be described by its cross-section. The anvil has rectangular coordinates $rOz$, the $z$ axis of $rOz$ is the symmetry axis of the anvil, as shown in Figure 3.

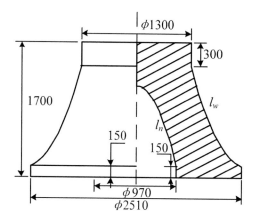

Figure 2. Structure diagram of anvil.

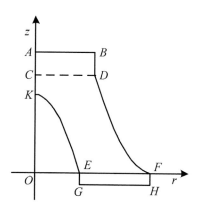

Figure 3. Axisymmetric geometrical model of anvil.

The rectangles $ABCD$ and $EFGH$ denote the up-transition layer and down-transition layer of the anvil, respectively. The configuration of the anvil is obtained by determining the curves between points $K$ and $E$ and points $D$ and $F$.

The coordinates of $A$, $B$, $C$, $D$, $E$, $F$, $G$, $H$ can be obtained from Figure 2. Suppose both the inner curve $KE$ and outer curve $DF$ are part of the curve of quadric function. Suppose the $z$ axis is the symmetry axis of curve $KE$ and point $K$ is the vertex of curve $KE$. The solid degree of anvil can be changed by changing the $z$ coordinate of point $K$. Expressions (1) and (2) represent curves $KE$ and $DF$, respectively:

$$z = a_n r^2 + c_n \qquad (1)$$

$$z = a_w r^2 + b_w r + c_w \qquad (2)$$

The coordinates of points K and E are $(0, r_k)$ and $(485, 0)$, respectively, and the coefficients $a_n$ and $c_n$ in (1) can be found from (3). According to the $r$ coordinates of points $K$ and $E$, the definitional domain of (1) is [0,485]:

$$\begin{cases} a_n = \dfrac{-r_k}{485^2} \\ c_n = r_k \end{cases} \qquad (3)$$

The coordinates of points $D$ and $F$ are (650, 1250) and (1255, 0), respectively, and the coefficients $b_w$ and $c_w$ in (2) can be found from (4). According to the $r$ coordinates of points $D$ and $F$, the definitional domain of (2) is [650, 1255]:

$$\begin{cases} b_w = \dfrac{1250 - a_w(650^2 - 1255^2)}{650 - 1255} \\ c_w = \dfrac{1250}{605} - \dfrac{650^2 + 604 \times 1255^2}{605} a_w \end{cases} \qquad (4)$$

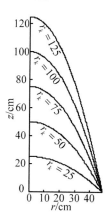

Figure 4. Family curves of the anvil inner contour.

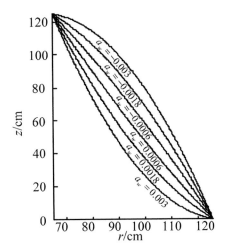

Figure 5. Family curves of the anvil outer contour.

Table 1. Different parameters of anvil vs. energy transmitting efficiency ($\eta$) (%)

| | $r_k$ | | | | | |
|---|---|---|---|---|---|---|
| $a_w$ | 0 | 250 | 500 | 750 | 1000 | 1250 |
| −0.003 | 86.07 | 87.06 | 87.56 | 87.88 | 87.98 | 88.17 |
| −0.0018 | 86.72 | 87.67 | 87.99 | 88.37 | 88.52 | 88.67 |
| −0.0006 | 87.30 | 88.21 | 88.48 | 88.68 | 88.96 | 89.13 |
| 0.0006 | 87.67 | 88.52 | 88.82 | 89.06 | 89.16 | 89.37 |
| 0.0018 | 87.93 | 88.56 | 88.83 | 88.91 | 89.05 | 89.21 |
| 0.003 | 87.83 | 88.10 | 88.14 | 88.13 | 88.33 | 88.61 |

Table 2. Different parameters of anvil vs the maximum stress ($\sigma_a$) in anvil (unit: MPa).

| | $r_k$ | | | | | |
|---|---|---|---|---|---|---|
| $a_w$ | 0 | 250 | 500 | 750 | 1000 | 1250 |
| −0.003 | 428.5 | 422.15 | 429.45 | 431.7 | 432.02 | 430.45 |
| −0.0018 | 434.8 | 427.62 | 439.99 | 437.13 | 444.51 | 451.47 |
| −0.0006 | 370.71 | 365.78 | 369.69 | 366.72 | 379.22 | 385.05 |
| 0.0006 | 351.72 | 351.64 | 331.08 | 326.36 | 327.19 | 327.22 |
| 0.0018 | 362.16 | 358.94 | 338.94 | 331.29 | 290.95 | 285.94 |
| 0.003 | 368.63 | 369.05 | 342.85 | 339.4 | 295.44 | 298.58 |

By the process described above, the parameterized model of anvil is established; $r_k$ and $a_w$ are the configuration parameters of anvil. The value range of $r_k$ is [0, 1250], whereas that of $a_w$ is [−0.003, 0.003]. The anvil is full solid when $r_k = 0$.

Figures 4 and 5 show the family curves of KE and DF, respectively. The cavity volume of anvil can be changed by the configuration parameter $r_k$, as shown in Figure 4. The concavity of curve DF can be changed by $a_w$, as shown in Figure 5.

## 3 ORTHOGONAL EXPERIMENT STUDY OF ANVIL

The impact system of underwater pile driving hammer needs to improve the energy transmitting efficiency, thereby ensuring the safety of the components of the impact system. Therefore, the performance parameters of the impact system are the energy transmitting efficiency and the maximum stress in anvil and pile during the collision. von Mises stress is associated with the yield of experiments conducted in materials such as steel, copper, and aluminum [12], and hence it is adopted in the dynamic stress analysis of the components.

Six samples from $r_k$ and $a_w$ are uniformly selected independently. The 36 sets of configuration parameters are acquired by orthogonal combination of the 12 selected samples. Each set of the configuration parameters correspond to a configuration of anvil. Each anvil is simulated and the impact energy of each simulation is 1225 kJ.

Through solving the dynamic model of impact system, the energy transmitting efficiency ($\eta$) of 36 anvils are obtained, as shown in Table 1.

According to the data in Table 1, the energy transmitting efficiency is affected by the configuration of anvil. The energy transmitting efficiency increases with the increase of cavity of anvil. Its value reaches 89.37% when $a_w = 0.0006$ and $r_k = 1250$. The size of the cavity of anvil will have an impact on the weight of anvil. So from the perspective of decreasing the weight of pile driving hammer, the anvil with larger cavity should be selected.

During the collision of hammer and anvil, the anvil will bear tremendous alternate loading. The maximum stress in the anvil $\sigma_a$ is the main factor affecting the life of anvil. Under impact energy of 1225 kJ, the maximum stress of anvil $\sigma_a$ with different configuration parameters is as shown in Table 2. The maximum stress is 451.47 MPa, the minimum stress is 285.94 MPa, the median is 367.675 MPa, and the average value is 373.45 MPa, as shown in Table 2. The range and the standard deviations presented in Table 2 are 165.53 MPa and 49.59, respectively. The data are highly fluctuated by the variation of configuration parameters. Above all, the maximum stress $\sigma_a$ is significantly affected by the variation of anvil configuration.

Table 3 shows the maximum stress in pile $\sigma_p$ during the simulation of collision. As shown in the table, the maximum stress is 352.50 MPa, the minimum stress is 261.73 MPa, the median is 295.55 MPa, and the average value is 301.56 MPa. The range of data in Table 3 is 90.77 MPa. The standard deviation

Table 3. Different parameters of anvil vs the maximum stress ($\sigma_p$) in pile.

| $a_w$ | $r_k$ | | | | | |
|---|---|---|---|---|---|---|
| | 0 | 250 | 500 | 750 | 1000 | 1250 |
| −0.003 | 281.06 | 313.06 | 314.74 | 335.92 | 340.45 | 267.18 |
| −0.0018 | 297.94 | 317.77 | 332.84 | 275.29 | 261.73 | 272.91 |
| −0.0006 | 315.49 | 275.48 | 273.75 | 271.39 | 274.05 | 282.49 |
| 0.0006 | 280.66 | 284.78 | 276.37 | 287.65 | 288.76 | 301.5 |
| 0.0018 | 285.06 | 290.5 | 293.16 | 311.36 | 310.99 | 327.53 |
| 0.003 | 329.92 | 323.92 | 315.33 | 343.96 | 352.5 | 348.68 |

of data in Table 3 is 26.06. The maximum stress $\sigma_p$ changes with the change of the configuration parameters $r_k$ and $a_w$. According to the above data analysis, the maximum stress in pile is largely affected by the configuration parameters of the anvil.

## 4 OPTIMIZATION OF ANVIL CONFIGURATION

The range of configuration parameters $r_k$ and $a_w$ differ by orders of magnitude, and hence the radial basis function hardly fits the approximate function of impact system performances. The function fitting is realized by establishing the transformation of the anvil configuration parameters as (5):

$$\begin{cases} r_k' = r_k / 1000 \\ a_w' = 1000 a_w \end{cases} \tag{5}$$

Approximate response surfaces of the performance parameters related to the configuration parameters are built by the Gaussian radial basis functions. Figure 6 shows the energy transmitting efficiency ($\eta$) of the approximate response surface with the configuration parameters of the anvil. The fitting function of the energy transmitting efficiency ($\eta$) and the configuration parameters $r_k'$ and $a_w'$ is $\eta(r_k', a_w')$. Figure 7 is the maximum stress in anvil $\sigma_a$ approximate response surface with the configuration parameters of the anvil. The fitting function of the maximum stress in anvil $\sigma_a$ with the configuration parameters $r_k'$ and $a_w'$ is $\sigma_a(r_k', a_w')$. Figure 8 shows the maximum stress in pile $\sigma_p$ approximate response surface with the configuration parameters of the anvil. The fitting function of the maximum stress in pile $\sigma_p$ with the configuration parameters $r_k'$ and $a_w'$ is $\sigma_p(r_k', a_w')$. According to Figures 6, 7, and 8, these approximate response surface have appreciable smoothness. The performance change rule of the impact system with the configuration parameters can be reasonable described by the fitting functions. These fitting functions are obtained by Gaussian radial basis functions.

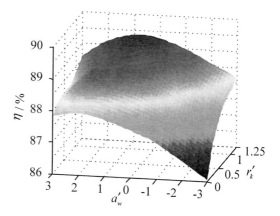

Figure 6. Fitting function surface of $\eta$.

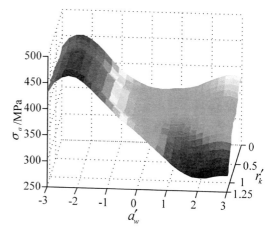

Figure 7. Fitting function surface of $\sigma_a$.

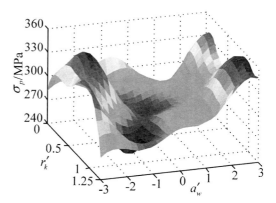

Figure 8. Fitting function surface of $\sigma_p$.

The optimization model of anvil is described in (6). The fitting function obtained above is used in the optimization model. The yield stress of the anvil material is 490 MPa, the yield stress of pile

378

Table 4. Results of optimization.

| $r_k$ | $a_w$ | $\eta/\%$ | $\sigma_d$/MPa | $\sigma_p$/MPa |
|---|---|---|---|---|
| 615.3 | 0.0009281 | 90.40 | 324.33 | 283.92 |

Table 5. Comparison of optimization and simulation results.

| $\eta$ FDM/% | RE $\eta$/% | $\sigma_a$ FDM/ MPa | RE $\sigma_d$/% | $\sigma_p$ FDM/ MPa | RE $\sigma_p$/% |
|---|---|---|---|---|---|
| 88.90 | 1.69 | 327.53 | −0.98 | 287.39 | −1.21 |

material is 355 MPa, and the safety factor is 1.25. The allowable stress of anvil material $\sigma_{as}$ is 392 MPa and the allowable stress of pile material $\sigma_{ps}$ is 284 MPa. The optimization model is solved by the nonlinear optimal tool fmincon in MATLAB optimal toolbox. The initial configuration parameters of the anvil are $r_k = 1250$ and $a_w = 0$. On solving the dynamic model of the impact system consisting the anvil, the transmitting energy transmitting efficiency is 89.32%, the maximum stress in anvil is 348.11 MPa, and the maximum stress in pile is 293.73 MPa. The maximum stress in pile is higher than the allowable stress of the pile material:

$$\begin{cases} \max \eta(r_k', a_w') \\ s.t.\ \sigma_a(r_k', a_w') < \sigma_{as} \\ \sigma_p(r_k', a_w') < \sigma_{ps} \\ 0 \leq r_k' \leq 1.25 \\ -3 \leq a_w' \leq 3 \end{cases} \quad (6)$$

Table 4 shows the performance parameters of the impact system after anvil optimization. Both the maximum stress of anvil and pile are reduced. Especially, the maximum stress of pile is less than the allowable stress of the pile material.

The impact system with optimized anvil is simulated. Table 5 compares the optimization and simulation results. The FDM indicates the simulation result and RE indicates the relative error between optimization and simulation results. The maximum relative error is 1.69%, so the approximate function between performances of impact system and the configuration parameters are well fitted by the radial basis function.

## 5 SUMMARY

In this paper, the configuration model of anvil is established. The anvil is the most important component of underwater pile driving hammer with 1200 kJ impact energy. By the numerical simulation of different configuration anvil, the performances of the impact system are largely affected by the configuration parameters of anvil: (1) the configuration parameters of anvil have some effect on transfer efficiency of impact energy; (2) those parameters have significant effect on the maximum stress of anvil and pile during collision. The approximate functions of performance parameters of the impact system with the configuration parameters are obtained by the radial basis function, and then the optimal model of anvil is built. The optimization model is solved by the nonlinear optimal tool fmincon of MATLAB, and the optimized anvil meets the requirements of the impact system.

## ACKNOWLEDGMENT

The research work presented in this paper was supported by the National High-tech R&D Program (863 Program) under Grant No. 2013AA09A217, which is gratefully acknowledged.

## REFERENCES

Cai, Z.C., Zheng, C.M., Tang, Z.S., & Qi D. 2010. Lunar digital elevation model and elevation distribution model based on Chang'E-1 LAM data. *Sci China Tech Sci* 53: 2558–2568.

Chen, Z., Cao, F., & Li, M. 2015. Scattered data quasi-interpolation on spheres. *Mathematical Methods in the Applied Sciences* 38(12): 2527–2536.

He, Q., Sun, F.R., Geng, J.Q., Yao, G.H., & Zhang, Y. 2010. Simulation Study on Volume Data Reconstruction in 3D US System Using Radical Basis Function. *Journal of System Simulation* 22(08): 1992–1996.

Wang, C.H., Zhu, H.H., & Qian, Q.H. 2014. Application of Kriging methods and multi-fractal theory to estimate of geotechnical parameters spatial distribution. Rock and Soil Mechanics 35: 386–392.

Wang, Q.M., & Wang, J.T. 2003. Application of Hydraulic Pile Driver in Piling Works for Sea Ports. *China Harbour Engineering* 23(6): 9–11.

Xu, J.T., & Liu, W.J. 2010. Surface Reconstruction and Capacity Measurement of Rail Tankers Based on Scattered Points. *Journal of Mechanical Engineering* 46(7): 154–159.

Yamamoto, J.K. 1998. A Review of Numerical Methods for the Interpolation of Geological Data. *Anais da Academia Brasileira de Ciencias* 70(1): 90–116.

Zhang, F., Lü, Z.Z., & Zhao, X.P. 2010. Novel Method Based on Sequence Shepard Interpolation for Structual Reliability Analysis. *Journal of Mechanical Engineering* 46(10): 176–181.

Zhang, Q., Zheng, G., & Zhou, D. 2014. Comparison study of Gauss, MQ and TPS for interpolation application. *International Journal of Industrial and Systems Engineering* 18(2): 185–198.

Advances in Energy and Environment Research – Achour & Wu (Eds)
© 2017 Taylor & Francis Group, London, ISBN 978-1-138-62682-9

# Study on thermo-mechanical coupling characteristics for linear rolling guideway

Yaoman Zhang, Shuxian Jia & Xianzhan Zhu
*School of Mechanical Engineering and Automation, Northeastern University, Shenyang, Liaoning, P.R. China*

ABSTRACT: A kind of linear rolling guideway was taken as the research object of the paper, and the key technology of the thermo-mechanical coupling characteristic modeling method and its simulation analysis were studied. Based on Hertz contact theory and the kinematics model of the linear rolling guideway, the friction mechanism was studied, the theoretical model of friction heat for the linear rolling guideway was derived, and the temperature of the contact region was obtained. Then the contact parameters and the thermal boundary conditions of the linear rolling guideway were calculated based on the friction heat model and the convective heat transfer formula. The finite element analysis model of the linear rolling guideway was established, and the thermo-mechanical coupling characteristics of the linear rolling guideway was carried out. By comparing the thermal analysis result with the structural analysis result, the influence of the thermal characteristics on the mechanical properties was studied. The temperature calculated by the theoretical model were found to be in close agreement with the simulation results and the validity of the theoretical model was verified. The research work lays a foundation for improving the structure design and performance optimization of the linear rolling guideway.

## 1 INTRODUCTION

Linear rolling guideway is one of the widely used functional components of CNC machine tool, its performance will affect the final performance of NC machine tools directly. Hence, it's necessary to carry out the further research on the characteristics of the rolling guideway. Lots of study on the performance of the rolling guide rail have been carried out by the scholars at home and abroad, and a lot of meaningful results have been achieved. Some studies on thermal characteristics of the linear rolling guide rail are listed in the paper. The FEM was established by Won Soo Yun et al. (1998) to estimate the thermal behavior of the guideway. It was found that thermal deformation of the guideway introduces the angular, straightness as well as straightness positioning error. Later, friction force decreases of over 30% caused by the thermal distortion of the slide table was proposed by Sun-Kyu Lee et al. (2003) based on the investigation of the typical behavior of the friction coefficient in response to both the external load and the linear speed. The thermal deformation of the machine tool feed system was analyzed through the method of multiple regression model, in which the key points of the thermal source were selected as independent variables in the work of Shyh Chour Huang (1994). The thermal contact resistance between balls and rings of bearings was investigated by Katsuhiko Nakajima (1995) based

on the assumption of heat transfer through the smooth elastic contact region of bearings under the axial, radial and the combined loads and the steady temperature. The evolution of temperature with time in the ball bearing under the lubrication conditions was studied both analytically and experimentally by Jafar Takabi & M.M. Khonsari (2012). His work indicated that large temperature gradient and thermally-induced preload can be caused by the high speed, oil viscosity and housing cooling rate. In the work of Haolin Li et al. (2011), uncertainties of time-variant thermal errors for the guideway were studied by the calculation method of combining the approximate model with numerical simulation technology.

From the above, most of the research on thermal characteristics of the linear rolling guideway was carried out without consideration of its interaction with mechanical properties. However, the thermo-mechanical coupling characteristics of the linear rolling guideway were mainly studied in this paper.

## 2 STRUCTURE SPECIFICATIONS

The linear rolling guideway of HSR series produced by a factory was the research object. It includes three parts, namely, the carriage, rail and ball. The main structural parameters of the HSR20A guideway are shown in Figure 1.

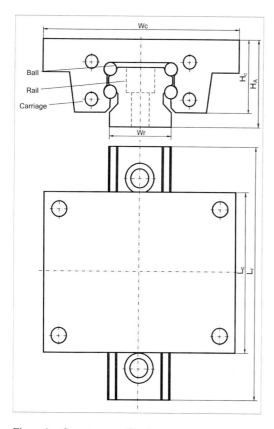

Figure 1. Structure specifications.

$$H_s = 2 \sum_{i=1,4} \sum_{j=1}^{Z} \left( F_{scj} v_{scj} + F_{srj} v_{srj} \right) \tag{1}$$

$$H_a = 2 \sum_{i=1,4} \sum_{j=1}^{Z} (w_{ac} M_{pcj} + w_{ar} M_{prj}) \tag{2}$$

$$H_{xz} = 2 \sum_{i=1,4} \sum_{j=1}^{Z} (w_{xzc} M_{\varepsilon cj} + w_{xzr} M_{\varepsilon rj}) \tag{3}$$

where $v_{scj}$ and $v_{srj}$ are speeds of balls with respect to the carriage and rail, respectively, $w_{ac}$ and $w_{ar}$ are the spin angular velocities of balls in the contact region between the ball and carriage and guideway respectively. $w_{xyc}$ and $w_{xyr}$ are the normal angular velocities of balls in the contact region between balls and the carriage and rail, respectively. The suffix $j$ refers to the ball number at each groove. $F_{scj}$ and $F_{srj}$, $M_{pcj}$ and $M_{prj}$ and $M_{\varepsilon cj}$ and $M_{\varepsilon rj}$ are the differential sliding friction forces, spin motion friction torques and pure rolling friction torques between the ball and the carriage and the rail, respectively. See Equations 4–6.

$$F_{sj} = fQC \tag{4}$$

$$M_{pj} = \frac{3QaE(e)}{8} f \tag{5}$$

$$M_{\varepsilon j} = \frac{3Qb}{16} \tag{6}$$

where $Q$ is the equivalent contact forces between the contact surfaces, $f$ is the dry friction coefficient, $a$ is the major axis of the contact ellipse, $E(e)$ is the second type of elliptic integration, and the value of $C$ is $1.8558 \times 10^{-3}$ determined by the geometrical shape.

The temperature of the contact surfaces between balls and raceways will increase and the temperature gradient will be generated in the reciprocating motion process of the linear rolling guideway, thus the contact surfaces between the balls and raceways are regarded as the heat source. The temperature of the contact zone is related to material properties of the guideway as well as the geometrical shape of the groove, in other words, the method of calculating the radius of an equivalent circle with the same area as the contact ellipse is feasible in case of the certain material properties. It's assumed that the equivalent circle area with the radius $m$ is heated by the moving heat source with constant heating rate, the temperature rise of ball surfaces in the contact area can be obtained by Equation 7.

$$\Delta T = \frac{2Hm^{1/2}\pi^{1/2}\gamma}{2\pi m^2 \pi^{1/2} (k\rho c V)^{1/2}} = \frac{0.318H\gamma}{m(mk\rho c V)^{1/2}} \tag{7}$$

Where the assembly height $H_A$ is 25.34 mm, the length $L_c$, width $W_c$ and height $H_c$ of the carriage are 50.8 mm, 63 mm and 22.18 mm, respectively, the rail length $L_r$, width $W_r$ and height $H_r$ are 600 mm, 19.96 mm and 18 mm, respectively, and the ball radius $r_0$, the groove radius $r$, the number of grooves $i$, the number of the balls at each groove $Z$ and the norminal contact angle $\alpha_0$, are 2.003 mm, 2.12 mm, 4, 11 and 20°, respectively.

## 3 FRICTION HEAT GENERATION AND CONTACT REGION TEMPERATURE

The complexity of heat generation mechanisms is caused by the complexity of friction force generation mechanisms. However, the friction heat generated by the differential sliding, spin motion and pure rolling from the elastic hysteresis were mainly considered in the paper. The total heat generation of the system can be obtained by calculating the heat generation of each contact zone. The friction heat generation of the differential sliding, spin motion and pure rolling unit time were expressed as Equations 1–3.

where $H$ is total heat per unit time, $k$ is the thermal conductivity, $\rho$ is the density, $c$ is the specific heat, $V$ is the ball speed with respect to the raceway and $\gamma$ denote the heat partition coefficient referring to the distribution ratio of the total heat for the contact surface. $\gamma$ is determined by the Peclet number Pe, expressed as Equations 8–9.

$$\gamma = k_2/(k_1 + k_2)(Pe \leq 1) \qquad (8)$$

$$\gamma = 1 - \frac{1/2(1-1/\sqrt{2}) + k_1/k_2\sqrt{\pi/2Pe}}{1 + k_1/k_2\sqrt{\pi/2Pe}}(Pe > 1) \quad (9)$$

where $Pe = Vm/2\alpha$, $\alpha$ denotes the thermal diffusion coefficient.

The temperature evolution of the contact region with the carriage running speed based on the theory above are depicted in Figure 2 and Figure 3. Where, A and C refer to the ball surface temperature in the ball and carriage contact zone of the 1st and 4th groove, respectively while E and G are the carriage surface temperature, respectively. Similarly B and D refer to the ball surface temperature in the ball and guide rail contact zone of the 1st and 4th groove, respectively while F and H represent the rail surface temperature, respectively.

Shown in Figure 2, the ball surface temperature in the 1st raceway groove is higher than that in the 4th raceway groove and the temperature of ball surfaces contacting with the rail is higher than that contacting with the carriage. The ball surface temperature varies between 26°C and 27°C with the carriage speed of 0.2 m/s. Shown in Figure 3, the temperature of the 1st raceway is higher than that of the 4th raceway, and the temperature of the rail surface is higher than that of the carriage surface within the same raceway. The temperature of rail and carriage surfaces varies between 25.5°C and

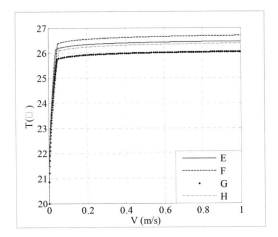

Figure 3.   Carriage and rail surface temperatures.

26.5°C under the carriage speed of 0.2 m/s. It's indicated that temperatures of the ball, guideway and carriage surface rise greatly with the carriage speed of less than 0.05 m/s, and by contrast, the temperature rise of the guideway and carriage surface slows down and tends to the steady state with the carriage speed increasing until larger than 0.05 m/s by comparing with the two figures. However, the ball temperature rise is significantly larger than that of the rail and carriage. Moreover, the surface temperature of the guideway and carriage is higher than the ball surface temperature when the carriage speed is relatively small, and conversely, when the speed beyond a certain value.

## 4   FINITE ELEMENT SIMULATION ANALYSIS

### 4.1   Operating conditions and material properties

These assumptions of operating conditions for the linear rolling guideway are made due to the complexity of the model as well as the limitation of the finite element analysis software. The carriage does the uniform linear motion at the speed of 0.2 m/s in the reciprocating process and the stop and start at both ends of the rail are ignored. All the degrees of freedom for the bottom surface of the guideway are constrained. The heat radiation and the effect of oil lubricate are both ignored. The initial temperature of the linear guideway system is 20°C while the temperature of the ambient air is 22°C.

The rail and carriage are from the same material and the property values of the rail, carriage and balls are expressed as following, respectively. The thermal conductivity, specific heat, density, elastic modulus, Poisson's ratio, linear expansion coefficient and the thermal diffusion coefficient for the

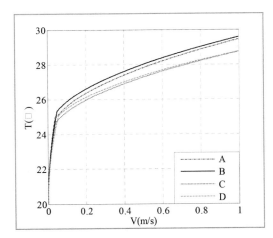

Figure 2.   Ball surface temperatures.

rail and carriage are 41.84 W/(m · °C), 481 J/(kg · °C), 8149 kg/m$^3$, 2.06 × 10$^{11}$ Pa, 0.3, 1.18 × 10$^{-5}$°C$^{-1}$ and 1.6074 × 10$^{-5}$ m$^2$/s while those for balls are 50 W/(m · °C), 460 J/(kg · °C), 7845 kg/m$^3$, 2.04 × 10$^{11}$ Pa, 0.28, 1.06 × 10$^{-5}$°C$^{-1}$ and 1.3855 × 10$^{-5}$ m$^2$/s.

### 4.2 Thermal boundary conditions

Parameters for the contact area of the 1st and 4th raceway grooves and balls are calculated based on the previous theoretical model and material properties, respectively. The major axes of contact ellipse $a_1$ and $a_4$ are 0.4135 mm and 0.3535 mm, minor axes of the contact ellipse $b_1$ and $b_4$ are 0.06445 mm and 0.0551 mm, the radius of the equivalent circle $m_1$ and $m_4$ is 0.1632 and 0.1396 mm, the differential sliding friction forces $F_{s1}$ and $F_{s4}$ are 2.856 × 10$^{-3}$ and 1.784 × 10$^{-3}$ N, the pure rolling friction torques $M_{\varepsilon1}$ and $M_{\varepsilon4}$ are 1.24 × 10$^{-3}$ and 6.621 × 10$^{-4}$ N·m, spin friction torques $M_{p1}$ and $M_{p4}$ are 2.468 × 10$^{-4}$ and 1.318 × 10$^{-4}$ N·m. After calculation, it's found that the heat flux of the same ball surfaces contacting with the rail and carriage, is closely, thus the average heat flux is obtained 266.33 W/m$^2$ and 188.66 W/m$^2$ for the 1st and 4th raceway grooves, respectively.

There are three ways of heat transfer including heat conduction, heat convection and heat radiation according to the basic law of the heat transfer. Only the heat conduction and convection are considered here. The convection coefficient can be expressed as Equation 10.

$$h = \frac{Nu \cdot k}{L} \quad (10)$$

where $Nu$ is the Nusselt number, $k$ denotes the thermal conductivity of the lubricant and $L$ is the characteristic length.

The formulas available to this research object are shown as Equations 11–12.

$$Nu = 0.54(Gr.Pr)^{1/4} \quad (11)$$

$$Nu = 0.664\,Re^{0.5}\,Pr^{1/3} \quad (12)$$

where, $Gr$ is the Grash number, $Pr$ is the Prandtl number and $Re$ is the Reynolds number.

A part of the guide is applied to the finite element analysis to avoid the large amount of calculation and the heat convection coefficient is shown in Table 1.

### 4.3 Finite element result analysis

The three-dimensional model of the linear rolling guideway was established by the SolidWorks with the secondary features ignored, such as screw holes and chamfer angles. Then the model was imported into the finite element analysis and generated mesh by the Hex Dominant method with overall size of 2 mm, the raceway surface size of 1 mm and the ball surface size of 0.5 mm. The total number of nodes and elements were 112880 and 30681, respectively. The contact pairs of balls and raceway grooves were established with the ball surfaces as contact surfaces the raceway surfaces as the target surfaces according to the rules. Moreover, the friction coefficient was set as 0.015, the control formulation was set as the Augmented Lagrange and the normal stiffness factor was set as 0.1 to simulate the bending deformation. In the temperature field analysis, the fixed support was applied to the bottom of the guide rail and the simulation time was set up to 6 hours with three load steps including 1 × 10$^{-6}$ s, 4260 s and 21600 s. And Time Integration was on for the transient analysis, the heat flux was applied to ball surfaces and the convention coefficient was applied to guideway and carriage surfaces (shown in Table 1). Finally, results of the temperature analysis as the body load were applied to the guide rail and the vertical downward force 600 N was applied to the upper surface of the carriage, shown as Figure 4.

The structural and thermal Von-Mise stress are shown in Figure 5 and Figure 6, respectively. It can be found that the maximum stress transfers from the contact area to the bottom of the guide rail under the temperature load with the value of the maximum Von-Mise stress rising from 6.88776 to 58.187 MPa. The maximum Von-Mise stress appears where there is the lower temperature based on the temperature analysis results. This phenomenon is mainly caused by smaller expansion and

Table 1. The convection coefficient of the linear rolling guideway.

| Name | L mm | Gr – | Re – | Gr/Re$^2$ × 10$^{-5}$ | Nu – | h W/(m$^2$ · °C) |
|---|---|---|---|---|---|---|
| Upper surface of guideway | 80 | 63.7 | 1058.7 | 5.768 | 19.209 | 6.238 |
| Side surface of guideway | 80 | 965.1 | 1058.7 | 8.610 | 19.209 | 6.238 |
| Upper surface of carriage | 50.8 | 743.7 | 672.3 | 115.5 | 15.307 | 7.828 |
| Side surface of carriage | 50.8 | 476.0 | 672.3 | 105.3 | 15.307 | 7.828 |

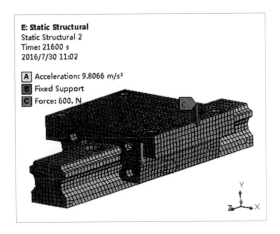

Figure 4. The FEM of the linear rolling guideway.

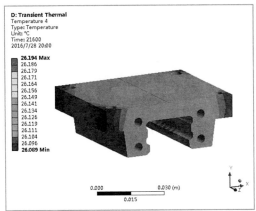

Figure 7. Carriage temperature distribution.

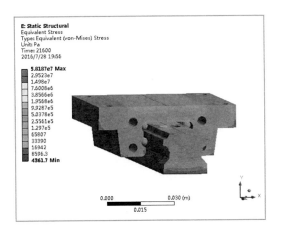

Figure 5. Structual Von-Mise stress.

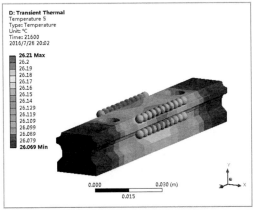

Figure 8. Rail temperature distribution.

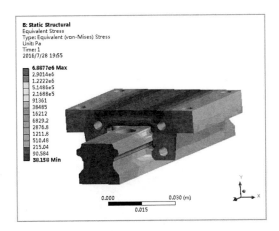

Figure 6. Thermal Von-Mise stress.

surface area with the less effect of the temperature. The stress distribution of the carriage is affected by temperature loads little and greater stress mainly appears on the contact area as the result of the interaction between balls and the rail and carriage. By contrast, the Von-Mise stress distribution of the rail is affected by the temperature load significantly, especially the contact area, the differences between the contact area and its surroundings become smaller obviously.

The steady state of the transient temperature field distribution after running 6 hours for the carriage of the rolling guideway is demonstrated in Figure 7 and Figure 8. Obviously, the highest temperature occurs on contact areas of balls and rail, about 26.23°C. It's also found that the temperature value is higher in the middle of the raceway groove and lower at both ends of the raceway groove along the axial direction of the carriage and guide rail which is similar to the normal distribu-

Table 2. Theoretical and simulation temperature.

| Temperature | $T_1$ °C | $T_2$ °C | $T_3$ °C | $T_4$ °C | $T_5$ °C | $T_6$ °C | $T_7$ °C | $T_8$ °C |
|---|---|---|---|---|---|---|---|---|
| $T_T$ | 26.353 | 26.338 | 26.602 | 26.551 | 25.963 | 25.931 | 26.127 | 26.241 |
| $T_S$ | 26.188 | 26.194 | 26.210 | 26.204 | 26.180 | 26.171 | 26.199 | 26.194 |
| $E_A$ | 0.165 | 0.144 | 0.392 | 0.347 | 0.217 | 0.240 | 0.072 | 0.047 |

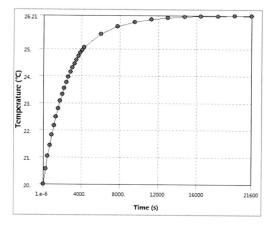

Figure 9. Temperature-time of the linear rolling guideway.

tion. This phenomenon is mainly caused by the friction heat generation in the contact areas.

The simulation temperature is between 26.069°C and 26.23°C. Based on equations (7), (8) and (9), surface temperature values of the contact area can be obtained by the Matlab simulation with the carriage speed of 0.2 m/s. Comparative results between the theoretical calculation and finite element simulation are shown in Table 2. Where $T_1$, $T_2$, $T_5$ and $T_6$ are the temperature of the balls and the carriage in the 1st and 4th raceway grooves of the carriage and balls, respectively. And $T_3$, $T_4$, $T_7$ and $T_8$ are the temperature of balls and the rail in the 1st and 4th raceway grooves, respectively. $T_T$, $T_S$ and $E_A$ denote the theoretical temperature, finite element analysis temperature and the absolute errors.

The transient temperature evolution of the linear rolling guideway with time are shown in Figure 9. It's indicated that temperature rises significantly up to about 25°C within the initial 1 hour. Although the temperature shows a rising trend 1 hour later, the rate slows down gradually. And the temperature trends to be steady and stay constant about 26°C 3.5 hours later finally.

## 5 CONCLUSIONS

In this paper, the friction heat model of the linear rolling guideway was established and the relevant parameters of the contact region were obtained by the theoretical calculation and the finite element simulation of thermo-mechanical coupling properties was carried out. The following conclusions are drawn from theoretical and simulative results and comparisons:

1. Temperature evolutions of the contact region with the carriage running speed were studied based on the friction heat generation model of the linear rolling guideway, which has great reference values for further study on thermal properties of the rolling guideway.
2. The thermo-mechanical characteristics of the linear rolling guideway was simulated by the finite element analysis. It indicated that the temperature distribution along the carriage and rail axial direction is closely to the normal distribution. Moreover, the mechanical properties of the guideway are affected by the temperature distribution differently. Hence, thermal errors in the operation of the linear rolling guideway shouldn't be negligible.
3. The temperature calculated by the friction heat generation model was consistent with the simulation results, which validated the theoretical model of the linear rolling guideway. This work has the certain guide significance to the establishment and solution of thermo-mechanical characteristic models for the linear rolling guideway.

ACKNOWLEDGMENT

This work was financially supported by the National Natural Science Foundation of China (51575092) and the Fundamental Research Funds for the Central Universities (N140104001) and Project of Education Department of Liaoning Province (L20150184).

REFERENCES

Haoli Li, Xingjuan Ying & Yulun Chi. (2011). Optimal Calculation for Thermal Errors of Machine Guideway Way with Approximation Model Method. *J. Journal of Mechanical Engineering*, 22(4): 423–427.
Harris T.A. (2001). Rolling bearing analysis. New York: Taylor & Francis Press.

Hiroyuki Ohta & Keisuke Tanaka. (2010). Vertical stiffnesses of preloaded linear guideway type ball bearings incorporating the flexibility of the carriage and rail. *J. Journal of Tribology*, 132(1): 547–548.

Jafar Takabi & M.M. Khonsari. (2012). Experimental testing and thermal analysis of ball bearings. *J. Tribology International*, 60(7): 93–103.

Jiwei Luo & Tianyu Luo. (2009). Analysis and application of rolling bearing. Beijing: China Machine Press.

Katsuhiko Nakajima. (1995). Thermal contact resistance between balls and rings of a bearing under axial, radial, and combined loads. *J. Journal of Thermophysics & Heat Transfer*, 9(1): 88–95.

Liming Xie, Bin Zhang & Lan Jin. (2014). Analysis of thermal characteristics of slide guideway of the compound machining center considering the junction surface thermal resistance. *J. Modular Machine Tool & Automatic Manufacturing Technique*, 15(02): 28–31.

Shyh Chour Huang. (1994). Analysis of a model to forecast thermal deformation of ball screw feed drive systems. *J. International Journal of Machine Tools and Manufacture*, 35(8): 1099–1104.

Sun Kyu Lee, Jae Heung Yoo & Moon Su Yang (2003). Effect of thermal deformation on machine tool slide guideway motion. *J. Tribology International,* 36(1): 41–47.

Won Soo Yun & Soo Kwang Kim (1998). Thermal error analysis for a CNC lathe feed drive system. *J. International Journal of Machine tools and manufacture*, 39(7): 1087–1101.

*Advances in Energy and Environment Research – Achour & Wu (Eds)*
*© 2017 Taylor & Francis Group, London, ISBN 978-1-138-62682-9*

# Effect of superfine grinding on the physicochemical properties of Moringa leaf powder

Yupo Cao, Neng Liu, Wei Zhou, Jihua Li & Lijing Lin
*Zhanjiang, Guangdong, China*

ABSTRACT: In this paper, the physicochemical properties of vibration mill-comminuted superfine Moringa leaf powder were investigated. Results showed that with increasing grinding time, the Moringa leaf powder gains smaller particle diameter and lower reducing sugar content, water holding capacity, and oil holding capacity. Protein, vitamin C, and total flavonoid content first increased and then decreased. Different grinding times of Moringa leaf power lead to differences in the physicochemical properties. Moringa leaf power showed the optimal index at the ultrafine grinding time of 20 min.

## 1 INTRODUCTION

Moringa, native to India, widely grows in many tropical and subtropical countries of the word. It is commonly known as "drumstick tree" or "miracle tree" (Rockwood & Anderson 2013). Moringa is considered as a high nutrient source because of its significant amounts of amino acids, fatty acids, vitamins, and nutrients, and its constituents such as leaf, flower, fruit, and bark have been anecdotally used as herbal medicines in treatments for inflammation, paralysis, and hypertension (Nesamani 1999, Chuang et al. 2007, Pari & Kumar 2002). With the high nutritional value and potential healthcare function, the "miracle tree" was explored for its commercial purpose by many researches. Extracts from all parts of the plant show pharmacological properties, recognized by popular use and corroborated by the scientific community (Rerreira & Farias 2008). Leaf extracts show hypotensive, bradycardic (Gilani et al, 1994), hypocholesterolemic (Mehta et al, 2003) and antiulcerative activity (Pal et al, 1995). In China, the Moringa has been mainly introduced in Yunnan, Fujian, and Guangdong provinces, and the long-distance transportation of untreated Moringa leaf increases the cost significantly. However, only a few studies focus on such wastage. Recognizing this limitation, the superfine grinding technology, a new type of food processing, is used to produce powders with outstanding characteristics, such as surface effect, mini-size effect, and chemical and catalytic properties.

Recently, many researchers have used the superfine grinding technology in functional food processing, such as red grape pomace powder, Qingke (hull-less barley), *Lycium barbarum*

polysaccharides. Zhang et al. (2014) reported that superfine grinding treatment could improve the antioxidant activities of *L. barbarum* polysaccharides. Zhao et al. (2014) found that superfine grinding treatment could decrease bulk density, tapped density, and fluidity of grape pomace powders, but improve the solubility. The extract of grape pomace powder with a particle size of less than 18.83 µm showed that the highest total polyphenolic content and flavanol content accompanied with the best antioxidant activity through all antioxidant assays. Despite the increase in the application of superfine graining technology in the processing of agricultural products, no research explored the effect of superfine grinding on the physicochemical properties of Moringa leaf powder. In this work, a vibration mill was used to obtain superfine Moringa leaf powder. The physicochemical properties were studied to investigate the effect of superfine graining on relevant properties. The main objective of this paper was to establish the theoretical foundation for deep processing, storage, and transportation of Moringa leaf powder.

## 2 MATERIALS AND METHODS

### 2.1 Materials

Moringa leaves were obtained from Zhanjiang, China. Catechin, rutin, and ascorbic acid were obtained from Sigma. Analytical grade reagents such as glucosum, 3,5-dinitrosalicylic acid, alcohol, NaOH, NaNO$_2$, AlCl$_3$, and HCl were obtained from Sinopharm Chemical Reagent Company, China.

## 2.2 Methods

### 2.2.1 Preparation of superfine Moringa leaf powder

Fresh Moringa leaves were dried using an electric constant-temperature drying oven (DHG-9426A, Jinghong, Shanghai, China) at 50°C. The moisture content of the dried leafs was measured by a moisture analyzer (MRS120-3, KERN Company, Germany). When the moisture content reached 8%, the triple preliminary comminuting of dried leaves was performed using an universal high-speed smashing machines (FLB-500A, Filibo, Shanghai), with comminuting time of 15 s and interval time of 1 min. Six preliminary power samples (20 ± 0.01 g) were then respectively comminuted for further powdering using Vibro Mill (KCW-10, Kunjie Yucheng, China) at six different processing times: 0, 10, 20, 30, 40, and 50 min.

### 2.2.2 Physical measurement of superfine Moringa leaf powder

The moisture content of final superfine Moringa leaf powder was measured by a moisture analyzer (MRS120-3, KERN Company, Germany).

The particle size was measured using a laser particle analyzer (Nicomp 380ZLS, PSS, USA) according to the research of Zhang & Li (2009).

The water holding capacity was determined based on the study of Zhao et al (2010). Approximately 1.0 g ($A_0$) of superfine Moringa powder was dissolved in 20 ml of distilled water. After equilibration at room temperature for 30 min, the samples were centrifuged using a centrifuge (TDL-5-A, Anting Scientific Instrument Factory, Shanghai, China) at 4,000 rpm for 25 min to remove the supernatant. The weight of the sediment and centrifuge tube was recorded ($A_2$). The water holding capacity was calculated as follows:

$$Q = \frac{A_2 - A_1 - A_0}{A_0} \qquad (1)$$

where Q is the suspended particles' water holding capacity (g/g), $A_0$ is the weight of superfine Moringa powder (g), A1 is the weight of the centrifuge tube (g), and $A_2$ is the total weight of the sediment and centrifuge tube (g).

The oil holding capacity was assessed as follows. Approximately 1.0 g of superfine Moringa powder was placed in a centrifuge tube, and then 20 ml ($V_1$) of food-grade oil was added. After equilibration at room temperature for 30 min, the samples were centrifuged at 4,000 rpm for 25 min and the volume of the supernatant ($V_2$) was recorded. The oil holding capacity was calculated as follows:

$$O = V_1 - V_2 \qquad (2)$$

where O is the oil holding capacity (mL/g), $V_1$ is the volume of added oil, and $V_2$ is the volume of the supernatant.

The following description shows the nutritive content of superfine Moringa leaf powder.

Protein. The protein content was determined using a Kjeldahl apparatus (KDY-08C, Ruizheng, Shanghai).

Vitamin C. The content of vitamin C was determined according to Zhou (2014). Briefly, 1 mL of the treated sample was placed in a 10 mL colorimetric tube followed by addition of 0.3 mL of EDTA (0.25 M), 0.5 mL of acetic acid (0.5 M), and 1.25 mL of fast blue B salt (2 g = L) in sequence. Then, the mixture was diluted to 10 mL with deionized water. After 20 min, the above mixture was observed spectrophotometrically at 420 nm using a UV–Vis spectrophotometer (T6, Beijing Purkinje General Instrument Co., China).

Reducing sugar. The reducing sugar was quantified by the DNS method. Superfine Moringa powder (1 g) was dissolving in 30 ml of distilled water at 50°C for 20 min to extract reducing sugar. The extract liquid was centrifuged for 25 min at 4,000 rpm and the supernatant was diluted with water to 100 ml. The diluent (0.2 ml) was transformed to a volumetric flask (10 ml), and 1.8 ml of distilled water and 1.5 ml of DNS were added sequentially. The mixed liquid was immediately incubated in a boiling water bath for 5 min, and the sample was cooled down to room temperature and 10 ml of water was added. Then, the solution was observed spectrophotometrically at 420 nm used a UV–Vis spectrophotometer.

Total flavonoid. The total flavonoid was determined spectrophotometrically by a method based on the formation of a complex flavonoid, with absorbtivity maximum at 430 nm. Rutin was used to obtain the calibration curve. Diluted sample (1 ml) was separately mixed with 1 ml of 2% aluminum chloride methanolic solution. After incubation at room temperature for 15 min, the absorbance of the reaction mixture was measured at 430 nm with a UV–Vis spectrophotometer and the flavonoid content was expressed in milligrams per gram of rutin equivalent.

## 3 RESULTS AND DISCUSSIONS

### 3.1 Effect of superfine graining time on particle size and moisture of Moringa leaf powder

The moisture content of the superfine graining powder showed a linear decrease with the increase of comminuting time, as shown in Table 1. Such

Table 1. Effect of superfine grinding time on particle size and moisture content of Moringa leaf powder.

| Time/min | Moisture content/% | Particle size/μm |
|---|---|---|
| 0 | 6.72 ± 0.13 | 105 ± 9.12 |
| 10 | 6.23 ± 0.15 | 28.91 ± 3.28 |
| 20 | 5.78 ± 0.12 | 24.04 ± 2.54 |
| 30 | 5.57 ± 0.14 | 22.23 ± 2.25 |
| 40 | 5.23 ± 0.13 | 20.55 ± 2.86 |
| 50 | 4.87 ± 0.12 | 16.76 ± 2.07 |

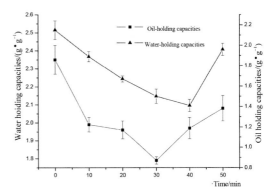

Figure 1. Water and oil holding capacities of Moringa leaf with different grinding times.

decrease becomes stable when the comminuting time reached 20 min. The reason might be cell breakage and the structure destruction of fibrin and other macromolecular materials during the comminuting process. The consequent release of free water and partly bound water coursed the sharp decrease of moisture content. However, such decrease became stable because of the limitation of the water release. Mean particle size also appeared to decrease sharply with the increase of comminuting time. It was reported that the particle size of superfine powder food was 10–25 μm for suitable solubility, dispersibility, and fluidity (Hu et al, 2012). Thus, the results show that the Moringa superfine powder could reach prescriptive level at comminuting times longer than 20 min.

### 3.2 Effect of superfine graining time on water holding capacity and oil holding capacity of Moringa leaf powder

Figure 1 shows the plot of the water and oil holding capacities against processing time. It is evident from the figure that these two capacities show a similar trend, that is, they first decrease to minimum and then increase. The water holding capacity reached minimum at the comminuting time of 40 min, while the oil holding capacity reached minimum at 30 min.

Further analysis shows that the decrease of water and oil holding capacities was coursed by the structure destructing of porous network structure during the strong mechanical processing. Although the two opposite forces can increase the surface area of the powder particles, the water and oil holding capacities will still decrease with the destruction of fiber structure (Zhu et al, 2015).

However, with the increase of comminuting time, the water holding capacity becomes higher. It might be due to the change of surface properties, that is, the increase of surface area and surface energy after superfine grinding. Moreover, the exposed hydrophilic groups in the cellulose and hemicelluloses of the powder resulted in an easy integration with water, and finally the value of water holding capacity increased (Zhao 2009).

The oil holding capacity was mainly based on the combination between oil and protein. The protein structure was broken due to the violent mechanical processing. Thus, the exposed hydrophobic groups of the protein enhanced the combination of oil.

### 3.3 Effect of superfine graining time on protein and reducing sugar contents of Moringa leaf powder

Protein and reducing sugar contents showed a generally decreasing trend when the comminuting time was increased (as shown in Figure 2). It was interesting to see a significant increase of protein content at a shorter processing time. It was probably due to the nutrition released from the broken cell during the short-time superfine graining process. However, with the increase of processing time, the Maillard reaction was promoted at a high internal temperature and resulted in a reduced protein content.

### 3.4 Effect of superfine graining time on vitamin C and flavonoid contents of Moringa leaf powder

The contents of vitamin C and total flavonoid had a transient reducing and subsequent increasing trend. Interestingly, the contents became stable at processing times longer than 30 min. Such phenomenon maybe highly related to the decreased particle size and increased specific surface area after comminuting. On the contrary, the violent mechanical effect led to cell disruption. However, with time prolonged, the collisions and frictions between particles were intensified by the violent rotation in the mill. The generated high temperature resulted in the degradation of vitamin C and flavonoids. The long exposed time aggravated the oxidation of vitamin C and flavonoids, which directly reduced the content.

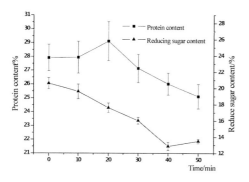

Figure 2. Protein and reducing sugar content of Moringa leaf with different grinding times.

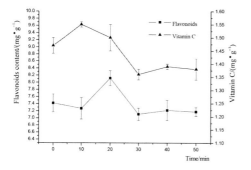

Figure 3. Vitamin C and total flavonoid contents of Moringa leaf with different grinding times.

## 4 CONCLUSION

The physical properties and nutrition contents of Moringa leaf powder were significantly improved after superfine comminuting processing. This work showed that Moringa leaf power had the optimal index when the ultrafine grinding time was 20 min. In this condition, the particle size of Moringa powder decreased from 105.03 to 24.08 μm, which conformed the prescriptive level of superfine graining powder. Meanwhile, the expanded surface area and the highly increased contents of total flavonoids, protein, and vitamin C also suggested 20 min was the optimal comminuting time. Furthermore, the decreased moisture content benefits storage. Our research provided a new approach for further processing of Moringa leaf powder.

## ACKNOWLEDGMENTS

This research was supported by the Fundamental Scientific Research Funds for Chinese Academy of Tropical Agricultural Sciences (Project No. 1630062015015) and the Project of Guangxi Science Research and Technology Development (Project No. 15248003-18).

## REFERENCES

Chuang P.H., Lee C.W., Chou J. M., Chou J.Y., Murugan M., Shieh B.J. & Chen H.M. 2007. Anti-fungal activity of crude extracts and essential oil of Moringa oleifera Lam. Bioresour. Technol. 98: 232–236.

Gilani A.H., Aftab K. Suria A, Siddiqui S., Salem R., Siddiqui B.S. & Faizi S. 1994. Pharmacological studies on hypotensive and spasmolytic activities of pure compounds from Moringa oleifera. Phytother. Res. 8(2): 87–91.

Hu J., Chen Y., Ni D. 2012. Effect of superfine grinding on quality and antioxidant property of fine green tea powders. LWT-Food Sci & Technol, 45(1): 8–12.

Mehta L.K., Balaraman R., Amin A.H., Bafna P.A. & Gulati O.D. 2003. Effect of fruits of Moringa oleifera on the lipid profile of normal and hypercholesterolaemic rabbits. J Ethnopharmacol. 86(23): 191–195.

Nesamani S. 1999. Medicinal Plants (vol. I). State Institute of Languages, Thiruvananthapuram, Kerala, India. p. 425.

Pal S.K., Mukherjee P.K. & Saha P. 1995. Studies on the antiulcer activity of Moringa oleifera leaf extract on gastric ulcer model ulcer models in rats. Phytother Res. 9(6): 463–465.

Pari L. & Kumar, N.A., 2002. Hepatoprotective activity of Moringa oleifera on antitubercular drug-induced liver damage in rats. J. Med. Food. 5(3): 171–177.

Rerreira P.M.P. & Farias D.F. 2008. Moringa oleifera: bioactive compounds and nutritional potential. Revista de Nutrição. 21(4): 431–437.

Rockwood J.K. & Anderson B.G. 2013. Potential uses of Moringa oleifera and an examination of antibiotic efficacy conferred by M. oleifera seed and leaf extracts using crude extraction techniques available to under-served indigenous populations. J. Phytothearpy Res. 3: 61–71.

Zhang L., Li S. Effects of micronization on properties of Chaenomeles sinensis (Thouin) Koehne fruit powder. Innov Food Sci Emerg, 10(4): 633–637.

Zhang M., Wang F., Liu R., Tang X., Zhang Q. & Zhang Z. 2014. Effects of superfine grinding on physicochemical and antioxidant properties of Lycium barbarum polysaccharides, LWT-Food Sci. & Technol. 58: 594–601.

Zhao X., Du F., Zhu Q., Qiu D, Yin W. & Ao Q. 2010. Effect of superfine pulverization on properties of Astragalus membranaceus powder. Powder Technol, 203(3): 620–625.

Zhao X., Yang Z., Gai G., Yang Y. 2009. Effect of superfine grinding on properties of ginger powder. J Food Eng, 91: 217–222.

Zhao X., Zhu H., Zhang G. & Tang W. 2015. Effect of superfine graining on the physicochemical properties and antioxidant activity of red grape pomace powders. Powder Technol, 286: 838–844.

Zhou W., Liu W., Zou L., Liu W., Liu C., Liang R., Chen J. 2014. Storage stability and skin permeation of vitamin C liposomes improved by pectin coating. Colloid Surfaces B, 117: 330–337.

Zhu F., Du B., Xu B. 2015. Superfine grinding improves functional properties and antioxidant capacities of bran dietary fibre from Qingke (hull-less barley) grown in Qinghai-Tibet Plateau, China. J Cereal Sci. 65: 43–47.

*Advances in Energy and Environment Research – Achour & Wu (Eds)*
© *2017 Taylor & Francis Group, London, ISBN 978-1-138-62682-9*

# Analysis and discussion of a household heat pump dryer

Zhen Yan & Lanxiang Hou
*Zao Zhuang University, Zaozhuang City, Shandong Province, China*

ABSTRACT: Heat pump is a high-efficiency heat transfer device. The heat energy is transferred from the evaporator to the condenser by consuming a small quantity of electric energy. This paper combines the vapor compression heat pump and air circulation. The circulating air flows through the evaporator, and after being cooled and dehumidificated, it flows through the condenser. Then, it turns into high-temperature and low-relative humidity air. Finally, the air enters the oven for drying clothes. The heat pump makes full use of the heat and the refrigerating output, thereby reducing the energy consumption with high drying effect. In this paper, we performed theoretical analysis and calculation of heat pump dryer and compared the power consumption and drying rate of heat pump dryer and electrical heating dryer. Furthermore, we tested the heat pump dryer, verified its effectiveness, and designed the major components. The SEC can be improved by analyzing the drying effect at different stages and enhancing the drying at initial and later stages. Finally, some methods to optimize the heat pump cloth dryer are introduced.

## 1 INTRODUCTION

Heat pump dryer operates by the combination of heat pump system and air circulation system, which has several advantages, such as simple structure, convenient operation, low energy consumption, and less pollution to the environment. As a result, the heat pump dryer conforms to the policy of energy conservation and emissions reduction in China. Since the 1980s, China has been developing the heat pump drying technology. At present, the heat pump drying technology has been widely used in various industrial applications (Braun, 2002). However, the mainly used device is large drying device, and the research on small household cloth dryer and heat pump dryers is still ongoing.

## 2 THE AMOUNT OF MOISTURE IN THE MATERIAL

Heat pump drying device operates by two cycles: the heat pump cycle and air circulation. Heat pump, as a type of high-efficiency heating device, consumes a small amount of energy and transfers heat energy from low temperature to high temperature. The basic structure principle is shown in Figure 1.

The moisture content in materials can be described in two ways: (1) by using dry radical moisture content represented by $x_m$ and (2) by using wet radical moisture content represented by $\omega_m$. In any material, the ratio of the weight of water $m_{wm}$ to the weight of dry matter $m_{dm}$ (pure substance) is the called dry radical moisture content $x_m$, which is calculated as:

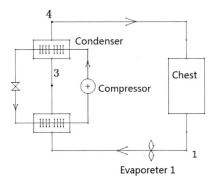

Figure 1. The basic structure diagram of a heat pump dryer.

$$x_m = \frac{m_{wm}}{m_{dm}} \tag{1}$$

The ratio of the weight of water $m_{wm}$ to the total weight of the material (the sum of water $m_{wm}$ and dry matter $m_{dm}$) is called the wet radical moisture content, which is represented by $\omega_m$. The wet radical moisture content is also known as humidity, which is expressed as:

$$w_m = \frac{m_{wm}}{m_{xm} + m_{dm}} \tag{2}$$

The capacity of the heat pump dryer is designed so as to match the cloth weight after being dried by the household washing machine. After drying, the water weight in the wet clothes is set to 1.3 kg.

Because the drying time is different for clothing materials, it is usually set to 1 h post drying.

To select the main components of the heat pump system, we need to determine the operation condition of the heat pump system and air circulation. The compressor is selected according to the condensation pressure, evaporation pressure, and the refrigerating capacity. Then, the structure of the condenser and evaporator is designed according to the volume of circulating air, wind speed, and the temperature difference in and out of the heat exchanger. Set calculation of the standard operating mode is shown in Figure 2, that is, area 1-3-4.

In Figure 2, the temperature of point 3, which represents the condition of air coming out from the evaporator, is 10°C, and the relative humidity is 100%. On point 3, the enthalpy and moisture content are 28.7 kJ/kg and 6.9 g/kg, respectively. After entering the condenser, the air condition is up to point 4, and the moisture content remains unchanged. On point 4, the temperature is up to 50°C. At this point, the relative humidity and enthalpy are found to be 9.7% and 68.6 kJ/kg, respectively. When dry hot air enters the drying cabinet, it exchanges heat-wet with the dry material. When the air escapes the drying cabinet, its enthalpy drops to 61.74 kJ/kg, about 90% of the original value, and the temperature drops to 30°C.

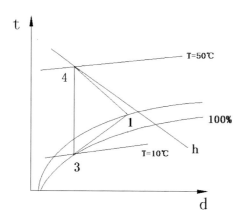

Figure 2. Enthalpy of wet air circulated in standard operation.

Table 1. Each state point parameters of air circulation.

| State points | Temperature (°C) | Moisture content (g/kg) | Relative humidity (%) | Enthalpy (kJ/kg) |
|---|---|---|---|---|
| Out of evaporator 3 | 10 | 6.9 | 100 | 28.7 |
| Out of condenser 4 | 50 | 6.9 | 9.7 | 68.6 |
| Out of the oven 1 | 30 | 11.4 | 46 | 61.74 |

At this time, the moisture content is 11.4 g/kg and the relative humidity is 46%. The air circulation parameters of each state point are summarized in Table 1.

# 3 SELECTION OF THE COMPRESSOR

Because of the effect of the moisture enthalpy's descent on drying, we can assume that wet air after drying wardrobe maintains a constant enthalpy.

In this cycle, cooling quantity $U_d$ required for drying every gram of water is given by:

$$U_d = \frac{\Delta U}{\Delta d} = \frac{h_1 - h_3}{d_1 - d_4} = 7.15 \, kj/g \tag{3}$$

The total cooling quantity U required for drying 1.3 kg of water is given by:

$$U = 1300 \times 7.15 = 9295 \, kJ \tag{4}$$

The drying finished in 1 h, and the refrigerating quantity Q of the heat pump system is given by:

$$Q = \frac{U}{3600} = 2.59 \, kW \tag{5}$$

In the process of drying, the heat of air equals the latent heat of evaporation required. Therefore, even when the condenser's releasing heat is higher than its evaporation and heat sent out constantly, the latent heat discharge is not needed when it does not enter the wardrobe. As a result, when the heat pump system is designed, people may install auxiliary heat exchanger at the back of the wardrobe or introduce fresh air from the front of the evaporator and the back of the air duct chest. This is done to ensure full use of the heat and solve the problem of the increasing circulation heat simultaneously (Chua, 2005).

# 4 THE EVALUATION INDEX OF THE HEAT PUMP DRYER

The two important indicators of drying equipment are the energy consumption of dehumidification and drying rate. Of them, the energy consumption of dehumidification is the power consumed for drying every unit mass of water, usually abbreviated to SEC (Yadav, 2008) expressed as:

$$SEC = \frac{W_{tot} \cdot \tau}{M_{de}} \tag{6}$$

where $M_{de}$ indicates the amount of water removed from the material (kg), $W_{tot}$ represents the total power consumption (kW), and $\tau$ is drying time (h).

Another parameter of concern is the drying rate. The instantaneous drying rate is calculated as:

$$W_{de} = Q_{air} \times (d_1 - d_4) \qquad (7)$$

In addition to the energy consumption of dehumidification and drying rate, several indices are used to measure the heat pump drying effect.
1. Heating coefficient of heat pump ($COP_H$)

Heat pump heating coefficient is defined as the ratio of heat produced to drive energy consumed, denoted by $COP_H$ (Klocker, 2001):

$$COP_H = \frac{Q_c}{W_c} \qquad (8)$$

where $Q_c$ indicates the heat produced by the heat pump (kW) and $W_c$ denotes the power consumed by heap pump (kW).
2. Thermal efficiency of the dryer ($\eta_t$)

The thermal efficiency is defined as the ratio of the heat of water vaporization to the total heat provided by the material (Bojic, 1997). It is denoted by $\eta_t$ and calculated as:

$$\eta_t = \frac{Q_{de}}{Q_t} \qquad (9)$$

where $Q_{de}$ represents moisture evaporation needed in the material (kW) and $Q_t$ denotes the heat source to provide heat (kW).

# 5 THE MAIN PARAMETERS FOR THE COMPARISON OF HEAT PUMP DRYER AND ELECTRIC HEATING DRYER

Different drying characteristics exist at different stages of the drying process. In order to contrast commodiously, we select stable drying stage for comparison in this paper. Let us assume that the inlet air temperatures of the two dryers are the same (50°C). Furthermore, their chest structure and exhaust temperature (set at 30°C) are similar. If the environment temperature is 20°C and relative humidity is 60%, the heat loss through the wardrobe is 10%, and both of them can remove 1 kg of water.

For general electric heating by expulsion-type dryer, if we assume environment temperature of 20°C, then the air relative humidity is 60%. Furthermore, the enthalpy is 42 kJ/kg, and the moisture content is 8.7 g/kg. The temperature of the hot air is up to 50°C using PTC element, and the moisture content remains unchanged at 8.7 g/kg. The enthalpy and relative humidity reach 73 kJ/kg

Table 2. State point parameters of circulation electric heating dryer.

| State points | Temperature (°C) | Moisture content (g/kg) | Relative humidity (%) | Enthalpy (kJ/kg) |
|---|---|---|---|---|
| Environment temperature | 20 | 8.7 | 60 | 42 |
| In the oven | 50 | 8.7 | 11 | 73 |
| Out of the oven | 30 | 13.9 | 52 | 65 |

and 11%, respectively. Through clothing surface, the temperature dropped to 30°C. Considering 10% heat loss, the enthalpy dropped to 65 kJ/kg, about 90% of the original value, the moisture quantity becomes 13.9 g/kg, and the relative humidity is 52%. Finally, the air is discharged through the drying box and enters the indoor environment. The air circulation state parameters of electric heating-type dryer are shown in Table 2.

## 5.1 Comparison of power consumption

In the heat pump system, because the heat released by the condenser is much higher than the heat transfer of the evaporator, the difference is equal to the input power of the compressor. There are two types of design schemes of the heat pump drying device:

Scheme 1: Meeting the demand of cold energy. If we wish to meet the demand of cold energy, heat would exceed the designed value. Therefore, excess heat discharges directly, which meets the demand of cold energy 7780 kJ. Let us assume that the refrigeration coefficient of vapor compression refrigeration system COP is 2.5, the input power of compressor is 3113 kJ, the power of draught fan is 0.1 kW, and the power consumption is 360 kJ per hour. Then, the ratio of the energy consumption of dehumidification SEC is 0.97 kWh/kg, about 53% of the power consumed by the electric heating dryer.

Scheme 2: Meeting the demand of heat. If we wish to meet the demand of heat, cold quantity is insufficient, which needs further cooling environment in the air. Let us assume that the heat coefficient of the heat pump is 3.5, the compressor input power of this heat pump system is 2225 kJ, the draught fan power is 0.1 kW, and the power consumption is 360 kJ per hour. Then, the ratio of the energy consumption of dehumidification SEC is 0.7 kWh/kg, about 39% of the power consumed by the electric heating dryer.

The above calculation provides the theoretical value. In fact, when the heat pump dryer operates, it does not emit heat. If the heat which can be used to dry clothes emits directly, more power will be wasted. Furthermore, when the environment temperature is higher, insufficient cooling results. Therefore, it would not supply cooling to the environment cooling system according to the second scheme. In practical application, the above two design schemes should be integrated.

### 5.2 *Comparison of environmental load*

In the electrothermal dryer, the heat emitted to the indoor environment is 6635 kJ for drying 1 kg of water. Moisture evaporation to the indoor environment is 1 kg. For the fully enclosed heat pump dryer taking away 11 kg of water, the heat emitted to the indoor environment is 3475 kJ. If all the moisture is condensed, the wet pollution indoor environment is zero. For semi-closed heat pump dryer taking away 1 kg of water, the heat emitted to the indoor environment is still 3475 kJ. Wet pollution to indoor environment depends on the introduction of new air volume, but it must be less than the electric heating-type dryer.

### 5.3 *Comparison of drying effect*

In the heat pump dryer, because the air before entering the condenser has been cooled and dehumidificated through the evaporator, the humidity of air entering the chest is low. And the corresponding equilibrium moisture content of the clothes will be very low, which indicates better drying effect.

All of the above discussion is under the assumption that the environment temperature is 20°C and the material relative humidity is 60%. In wet weather, for electrical heating in expulsion-type dryer, moisture content of air is higher. Therefore, under the same inlet air temperature, clothing equilibrium moisture content will increase, which can reduce the drying effect.

## 6 CONCLUSION

The energy consumption of the heat pump dryer is lower than that of the electric heating-type dryer. The following questions would be the key to promote the use of the heat pump dryer: how to make the heat pump dryer run stably under different conditions; how to control the environment temperature and humidity throughout the year, which can ensure efficient operation of the heat pump dryer; how to stay dry within different phases of the system of excess heat emissions accurately; and how to determine whether the clothes completely dried? In this paper, optimized solutions are summarized and put forward.

1. Using special heat pump compressor. The operating modes of the heat pump dryer and air conditioning are different. In order to improve drying effect, the heat pump dryer requires larger difference between the high and low temperatures. A compressor with higher compression ratio, exhaust pressure, and exhaust temperature is always favorable.
2. Automation control. Different operation strategies can be adopted for the different stages of drying. Heating is the main purpose of the drying process. Air only through the condenser rather than evaporator achieves rapid heating. In the late drying, the air discharged from the chest still has a high temperature and low relative humidity. Here, a ventilation pipe can be set beside the wardrobe, which introduces some air to the wardrobe and makes the hot dry air reach the clothes continuously.
3. Using regenerative optimization way. In order to obtain a higher air temperature difference, the regenerative optimization way can be used.
4. Using Lorentz circle. In practice, the refrigeration coefficient of Carnot cycle is the highest, but the refrigeration coefficient of Lorentz cycle is the highest theoretically. For heat pump dryer, when the air flows through the evaporator or condenser, it has large temperature change. The use of Lorentz circle makes the refrigerant phase change temperature and warm source match, which will achieve better refrigeration coefficient.
5. On the surface, because the temperature of the compressor is higher, up to 100°C, the structure should be changed to facilitate airflow through the compressor surface into the wardrobe cabinets, which cools the compressor and increases the temperature of air entering the wardrobe.

## REFERENCES

Braun, J. E., P. K. Bansal, E. A. Groll. Energy efficiency analysis of air cycle heat pump dryers. Int J Refrigeration. 2002, 25: 954–965.

Bojic M., Savanovic G., Trifanovic N., et al. Themoelectric cooling of train carriage by using a coldness-recovery device. Energy, 1997, 22(5): 493–500.

Chua K. J., Chou S. K. A modular approach to study the performance of a two-stage heat pump system for drying. Applied Thermal Engineering, 2005, 25(8/9): 1363–1379.

GB/T 23118–2008 Household and similar use drum-type washing dryer technical requirements [S].

Klocker, K., E. L. Schmidt, F. Steimle. Carbon dioxide as a working fluid in drying heat pumps. International Journal of Refrigeration. 2001, 24: 100–107.

Yadav V., Moon C. G., Fabric-drying process in domestic dryers. Applied Energy, 2008, 85(2/3): 143–158.

*Advances in Energy and Environment Research – Achour & Wu (Eds)*
*© 2017 Taylor & Francis Group, London, ISBN 978-1-138-62682-9*

# Fano resonance based on the coupling effect in a metal–insulator–metal nanostructure

X.G. Zhang & M.Z. Shao
*Department of Physics, South University of Science and Technology of China, Shenzhen, China*

ABSTRACT: In this paper, Fano resonance based on metal–insulator–metal structure is proposed and investigated by cascading double cavities, which exhibit high transmission and sharp Fano resonance peaks via strengthening the mutual coupling of the cavities. The simulation results show that with the symmetry breaking introduced in the structure, the number of Fano resonances increases accordingly. This work can find considerable applications in bio/chemical sensors with excellent performance and other nano-photonic integrated circuit devices.

## 1 INTRODUCTION

Fano resonances have attracted increasing attention in recent years. Different from Lorentz resonances (symmetric spectra), one of the main features of the Fano resonance is its asymmetric line profile. The asymmetry originates from a close coexistence of resonant transmission and resonant reflection, and can be reduced to the interaction of a discrete (localized) state with a continuum of propagation modes.

Fano resonance based on plasmonic structure is a classical case (Chen et al. 2014, Zhang et al. 2012, Hayashi et al. 2015). As we know, there have been remarkable progresses in developing plasmonic nanostructures due to their special capability to confine light in sub-wavelength dimensions and overcome the traditional optical diffraction limit (Shen et al. 2014, Huidobro et al. 2014, Thomas et al. 2011, Mark et al. 2010, Kashif et al. 2014, Park et al. 2008). Fano resonances sustained by plasmonic structures depend strongly on the shape and size of the geometry and the surrounding materials. Such properties have potential applications in designing plasmonic devices, especially for various nanosensors. It has been demonstrated that a type of sensor is designed in Metal–Insulator–Metal (MIM) waveguide and an asymmetric rectangular cavity, in which independently double-tunable Fano resonances can be achieved by changing the different parameters of the geometry (Qi et al. 2014). And it has been reported that Fano-type sensors with high sensitivity can be realized in nano-shell clusters deposited on a substrate of $\beta$-SiC/SiO$_2$/Si multilayers. It shows that multilayer substrate plays a fundamental role in the confinement of optical power in the nano-shell

layer, and results in the formation of pronounced Fano dips (Saeed et al. 2015).

As we know, optimizing the plasmonic structures and enhancing the transmission efficiency are a challenge for Fano resonances. In this paper, we propose a type of compact plasmonic structure realized by cascading double asymmetric cavities at the center of an MIM waveguide. By modulating the geometrical parameters appropriately, multiple Fano resonances with high transmission and sharp lineshape can be obtained by breaking the symmetry in the cascaded cavities. Such properties considered here can find many important applications in various bio/chemical sensors, and other nano-photonic integrated circuit devices.

## 2 STRUCTURE DESCRIPTION

In general, there are two types of plasmonic structures used for designing devices: Insulator–Metal–Insulator (IMI) structure and MIM structure. As we know, because of the strong capability to confine light, MIM structure has more applications in optical devices (Zia et al. 2014). Here, we use MIM to design plasmonic structure as shown in Figure 1. It is an MIM nanostructure composed of a main waveguide and two cascaded cavities at the center. The width of the waveguide is set as $w = 50$ nm. The two square cavities have the same scale size. We chose the size $a = 400$ nm. In each cavity, it includes a metal core with a radius of $r = 90$ nm. Figure 1(a) shows the case of the symmetric structure. In Figure 1(b), the left and right metal cores shift along $+y$ and $-y$ axis directions, respectively, with the same deviation $d/2$. As a result, the designed structure changes to symmetric case and

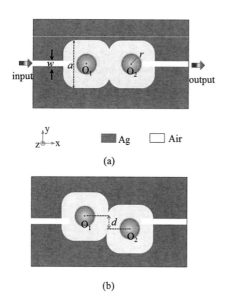

(a)

(b)

Figure 1. (a) A Metal–Insulator–Metal (MIM) nano-structure composed of a main waveguide and two cascaded cavities at the center (symmetric structure) and (b) the left and right metal cores shift along +y and −y axis directions with a distance of $d/2$ (asymmetric structure).

the distance between the two centers is $d$. Therefore, the parameter $d$ can be used to describe the symmetry breaking in the structure. The insulators in the cavities and waveguide are chosen to be air ($n = 1.0$). The metal is silver, whose frequency-dependent complex relative permittivity is characterized by the Drude model (Qu et al. 2016):

$$\varepsilon_m(\omega) = \varepsilon_\infty - \frac{\omega_p^2}{\omega(\omega+i\gamma)} \quad (1)$$

where $\varepsilon_\infty$ is the dielectric constant at the infinite frequency, $\gamma$ is the electron collision frequency, $\omega_p$ is bulk plasma frequency, and $\omega$ is the angular frequency of incident light. The parameters are $\varepsilon_\infty = 3.7$, $\omega_p = 1.38 \times 10^{16}$ Hz, and $\gamma = 2.73 \times 10^{13}$ Hz. In order to excite the surface plasmon polaritons, the input light is set to be transverse magnetic plane wave.

## 3 RESULTS AND DISCUSSIONS

In order to investigate the effect of asymmetry structure ($d \neq 0$) on the transmission properties, for a comparison, the case of the symmetric structure ($d = 0$) was first considered, whose transmission is calculated in Figure 2(a). The corresponding structure is shown in the inset. The figure shows three transmission peaks in the spectrum, denoted

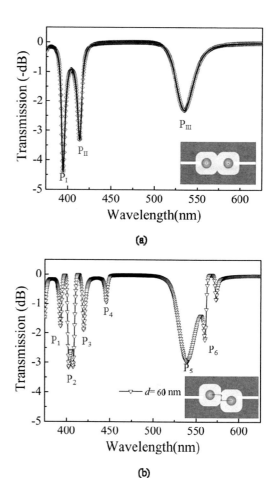

(a)

(b)

Figure 2. Transmissions for the cases of (a) the symmetric structure with the left and right metal core located at the center of its corresponding cavity and (b) the asymmetric structure with the left and the right metal cores, respectively, deviated along the +y and –y axis directions with the distance of d = 60 nm.

by $P_I$, $P_{II}$, and $P_{III}$. It can be considered that they are formed by the resonances and interactions between the two cascaded cavities.

When the asymmetry breaking was introduced in the plasmonic structure, i.e. the left and right metal cores, respectively, shift along +y and −y axis directions with the distance of $d = 60$ nm, the transmission spectra were calculated and the results are shown in Figure 2(b). Direct comparison of these two figures shows that the resonance peaks $P_I$ and $P_{II}$ split into four new resonance peaks, denoted by $P_1$, $P_2$, $P_3$, and $P_4$ and the resonance peak $P_{III}$ splits into two new resonance peaks, denoted by $P_5$ and $P_6$. On the contrary, we can also find that the new resonance peaks $P_5$ and $P_6$ have higher

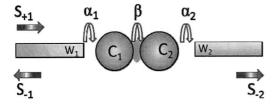

Figure 3. A simplified model of two cascaded cavities coupled with a main waveguide.

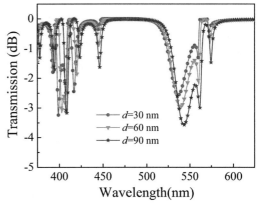

Figure 4. Transmissions for the cases of the asymmetric structure with the distance of d = 30, 60, and 90 nm.

transmission and sharper lineshape than those of the resonance peak $P_{III}$. Meanwhile, the new resonance peaks $P_1$–$P_4$ have higher transmissions and sharper lineshapes than those of the resonance peaks $P_I$ and $P_{II}$.

In order to understand the cascaded structure in theory, the temporal coupled mode theory (Haus 1984, Haus et al. 1991, Piao et al. 2011) is utilized to analyze the Fano resonance. As shown in Figure 3, the two cascaded cavities are denoted by $C_i$ ($i$ = 1, 2), and the mode amplitudes ($A_i$) in cavity $C_i$ can be described by:

$$\frac{dA_1}{dt} = (j\omega_1 - \alpha_1)A_1 + j\sqrt{2\alpha_1}s_{+1} - j\beta A_2 \qquad (2)$$

$$\frac{dA_2}{dt} = (j\omega_2 - \alpha_2)A_2 + j\sqrt{2\alpha_2}s_{+2} - j\beta A_1 \qquad (3)$$

$$S_{-i} = -S_{+i} + j\sqrt{2\alpha_i}A_i (i=1,2) \qquad (4)$$

where $\omega_i$ ($i$ = 1, 2) is the resonator frequency of the cavity $C_i$, $\alpha_i$ ($i$ = 1, 2) is the coupling coefficient between the cavity $C_i$ and its connecting waveguide arm $W_i$, and $\beta$ is the mutual coupling coefficient between the two cavities. The transmission $T$ from left to right port can be calculated from Eqs. (2)–(4) as:

$$T = \left|\frac{S_{-2}}{S_{+1}}\right|^2 = \left|\frac{2j\beta\sqrt{r_1 r_2}}{(j\omega - j\omega_1 + r_1)(j\omega - j\omega_2 + r_2) + \beta^2}\right| \qquad (5)$$

From Eq. (5), it can be seen that the transmission from left to right port is significantly affected by the coupling between the two cascaded cavities. When $\beta$ = 0, $T$ = 0. Fano phenomena can occur by the complex resonances and interactions in the coupling system. The introduction of asymmetry to the structure can effectively modulate the cavity coupling, which results in the corresponding changes in Fano resonances. Then, Finite Element Method (FEM) is used to investigate numerically the transmission properties of the designed structures. In addition, Perfectly Matched Layers (PMLs) are added at the outside of the calculated domain to absorb the electromagnetic wave.

For the asymmetric structure, in order to clearly study the transmission changes for different d, the transmissions of the cases of $d$ = 30, 60, and 90 nm were simulated and the results are shown in Figure 4. It is evident from the figure that multiple Fano resonance peaks with asymmetric lineshape are formed with the increase of d from 30 to 90 nm. It can be considered that the deviation of the metal cores will increase the connectivity between the two cascaded cavities, which can further strengthen the resonances and interactions of the coupling system. Thus, some new Fano resonance modes are introduced due to the complex coupling, and the transmissions can be enhanced.

## 4 SUMMARY

In summary, we propose a new type of plasmonic structure composed of a main waveguide and two cascaded asymmetric cavities at the center. The research results show that the cascaded asymmetric structure is advantageous for obtaining sharp Fano resonance peaks with high transmission through strengthening the mutual coupling of the cavities. The plasmonic nanostructure presented here can find many applications in biological/chemical sensors, optical filters, modulators, switches, and other photonic integrated devices.

## ACKNOWLEDGMENT

This work was financially supported by the Shenzhen Innovation Foundation of Fundamental Research (JCYJ20140901153335318).

# REFERENCES

Chen, Z. & L. Yu (2014). Multiple Fano resonances based on different waveguide modes in a symmetry breaking plasmonic system. *IEEE Photonics J.* 6: 1–8.

Huidobro, P.A., X. Shen, J. Cuerda, E. Moreno, L. Martin-Moreno, F. J. Garcia-Vidal & J. B. Pendry (2014). Magnetic localized surface plasmons. *Phys. Rev. X* 4: 340–342.

Haus, H. A (1984). Waves and Fields in Optoelectronics; Prentice-Hall: Upper Saddle River, NJ, USA.

Haus, H. A. & W. P. Huang (1991). Coupled-mode theory. *IEEE Proc.* 79: 1505–1518.

Hayashi, S., D. V. Nesterenko & Z. Sekkat (2015). Waveguide-coupled surface plasmon resonance sensor structures: Fano lineshape engineering for ultrahigh-resolution sensing. *J. Phys. D Appl. Phys.* 48: 325303.

Kashif, M., A. A. Bakar, N. Arsad & S. Shaari (2014). Development of phase detection schemes based on surface plasmon resonance using interferometry. *Sensors* 14: 15914–15938.

Mark, S. A. (2010). Nonplasmonic surface enhanced raman spectroscopy using silica microspheres. *Appl. Phys. Lett.* 97: 131116.

Park, J. H. Kim, B. Lee (2008). High order plasmonic Bragg reflection in the metal-insulator-metal waveguide Bragg grating. *Opt. Express* 16: 413–425.

Piao, X., S. Yu, S. Koo, K. Lee & N. Park (2011). Fano-type spectral asymmetry and its control for plasmonic metal-insulator-metal stub structures. *Opt. Express* 19: 10907–10912.

Qi, J., Z. Chen, J. Chen, Y. Li, W. Qiang, J. Xu & Q. Sun (2014). Independently tunable double Fano resonances in asymmetric MIM waveguide structure. *Opt. Express* 22: 14688–14695.

Qu, S. N., C. Song, X. S. Xia, X. Y. Liang, B. J. Tang, Z.-D. Hu & J. C. Wang (2016). Detuned plasmonic bragg grating sensor based on a defect metal-insulator-metal waveguide. *Sensors* 16: 784-1–9.

Saeed, G., A. Arash & P. Nezih (2015). Fano resonances in nanoshell clusters deposited on a multilayer substrate of β-SiC/SiO2/Si to design high-quality plasmonic sensors. *J. Lightw. Technol.* 13: 2817–2823.

Shen, X. & T. J. Cui (2014). Ultrathin plasmonic metamaterial for spoof localized surface plasmons. *Laser Photonics Rev.* 8: 137–145.

Thomas, B. A., Z. H. Han & I. B. Sergey (2011). Compact on-chip temperature sensors based on dielectric-loaded plasmonic waveguide-ring resonators. *Sensors* 11: 1992–2000.

Zhang, Z. D., H. Y. Wang & Z. Y. Zhang (2012). Fano resonance in a gear shaped nanocavity of the metal-insulator-metal waveguide. *Plasmonics* 8: 797–801.

Zia, R., D. M. Selker, P. B. Catrysse & M. L. Brongrsma (2004). Geometries and materials for subwavelength surface plasmon modes. *J. Opt. Soc. Am.* 21: 2442–2446.

# Design of a smart grid dispatching operation analysis system based on data mining

Jie Li, Kunpeng Wang, Yaling Yu, Handan Tan & Yi Tang
*Electric Power Research Institute, China*

ABSTRACT: In this paper, development trend of smart grid is considered to put forward the solution of constructing smart grid dispatching operation analysis system by data mining technology according to actual operation condition of power dispatching and business knowledge demand. Data mining model and business functions of dispatching operation analysis system are discussed. System architecture and implementation technology in cases of dispatching operation business information analysis are introduced.

## 1 INTRODUCTION

As automation and information solution of power grid in next generation, smart grid is central for power grid operation and management, and it is regarded as major scientific and technological innovation and development trend of power system. Smart grid aims to achieve more environment-friendly, efficient, and interactive modern power system through updating and transforming power generation, power transmission, power distribution, and power distribution links of original power grid. Various dispatching automatic data information can be handled comprehensively. Various intelligent analysis and auxiliary decision-making tools are provided for dispatchers. Smart grid is characterized by high openness, feasibility, and information safety, which can adapt to various dispatching automation systems of standard information integration frames (Cepeda, 2011).

In this paper, actual situation and business demand of power grid dispatching are combined to propose a solution of constructing application system by data mining technology aiming at construction background of smart power grid in China. It is difficult to extract valuable information quickly from data resources to support decision-making effectively due to information islands formed by several discrete data and technology bottlenecks. The constructed power grid dispatching operation analysis system can eliminate information islands according to the concept of "information integration-model construction-knowledge extraction". Information is mined, and knowledge is extracted from several business data, thereby supporting smart dispatching operation analysis and decision-making.

## 2 DATA MINING MODEL

Data mining refers to a technology of exploring valuable information from vast amounts of data. A series of technologies are applied for extracting hidden and usable information and knowledge from several incomplete, noisy, fuzzy, and random data. The working process includes data integration, model establishment, mining, and knowledge analysis. The data can be structured, such as data in a relational database; the data can also be semi-structured, such as text, graphics, and image data. The event can be heterogeneous data distribution among all businesses of power dispatching. Extracted knowledge can be represented as a concept, rule, model, and other forms (Li, 2011).

Data mining technology has prominent advantages in the aspects of handling massive data of power dispatching and mining in-depth information. Prediction model, multidimensional analysis model, associated analysis model, and other model for discovering knowledge in data mining can be adopted for constructing mining model and realizing knowledge extraction.

### 2.1 *Prediction model*

Time series data and sequence data should be mined in order to grasp the development law of analysis object. Existing data sequence is used to predict future trend. Power load prediction is a common technology in dispatching, where system operation features, capacity increase decision-making, natural environment, and social impact are considered. A set of mathematical methods capable of handling previous and future load systematically is studied and utilized under the condition. Power consumption demand in the future can be predicted according to temperature, humidity, and other factors. The load value of some specific time can be determined under the precondition of meeting certain precision requirement, thereby guiding decision-making in power dispatching (Zhou, 2013; Li, 2013).

The following common methods are adopted in load prediction: time series method, gray prediction method, fuzzy clustering recognition prediction method, neural network prediction method, optimization combination prediction method, etc. Artificial neural network can establish arbitrary nonlinear model, which is suitable for solving the problem of time series prediction. Time characteristics are emphasized and considered in time series model; in particular, levels of time cycle such as day, week, and month are considered. Influence of calendar, such as holidays is also considered, which is suitable for load prediction of power system (Liu, 2013).

## 2.2 Analysis model

Dimension is a level concept corresponding to information. Multi-dimensional analysis is based on dimensions. Data are classified for abstract statistical analysis. Business information classification model is established according to property and features of analysis objects. Multidimensional analysis is based on statistics theory. Data are analyzed after correlation of different dimensions. Data are always analyzed according to time (year, quarter, month, week, and day) and areas (region, province, city, and district) in power dispatching analysis. Common methods include regression analysis and variance analysis.

Electric power dispatching business data can also be statistically analyzed by classification model. Data are divided into three categories for summarization. Information representing common features is extracted: (1) equipment account category: it refers to data for describing inherent attributes of power dispatching objects, including primary equipment, secondary equipment, automation equipment, and communication equipment, and the data are mainly obtained from equipment database of business system, including various parameters and fixed value; (2) action record category: work of power dispatching actually belongs to operation on power grid equipment and actions of handling power grid equipment. The data result from log (scheduling, protection, automation and communication), operation ticket, working ticket, switch displacement, protection fault information system, etc., which can reflect discontinuous process of power system operation; (3) time sampling category: it refers to a series of data for reflecting continuous change process of

power system. It has the most prominent feature of timeliness. Different granularities are set aiming at different application purposes, for example, raw data from SCADA/EMS can reach second-level time interval, including minute-level or hour-level collection power, and statement data are summarized and processed according to day, meadow, month, season, and year (Fan, 2012).

When electric power load characteristics are analyzed, factors affecting load characteristic changes can be divided into two categories: the first category includes factors with long-term influence effect on load; the influence on load is manifested as long-term trend of load change, such as economic development and industry structure change. The other category includes influence factors with short-term effect on load, such as temperature, rainfall, and other climate factors. Study of the influence law of various related factors on power load change in the industries or regions is beneficial for improving prediction precision of power load and ensuring safe and economic operation of power system.

ID3 algorithm is improved in practical application by the system. Information, again due to decision-making attribute, is considered, while the information gain of subsequently selected attribute after selection of the attribute is also considered. Namely, nodes of two levels on the tree are considered at the same time, and high-quality decision-making tree is constructed. The algorithm is utilized for reducing node number, namely tree depth can be reduced for improving mining speed and efficiency. Specific algorithm is shown as follows: A is set as candidate attribute, A has n attribute values, and the corresponding probabilities are $p_1$, $p_2$, ..., $p_n$. Attribute A is expanded according to the principle of minimum information entropy, $\{B_1, B_2, ..., B_n\}$ refers to attribute selected for n child nodes, and the corresponding information entropies are $H(B_1)$, $H(B_2)$, ..., $H(B_n)$, then:

$$H^{\cdot}(A) = \sum_{i}^{n} p_i * H(B_i) \quad (1)$$

Standard of algorithm selection attribute $A^{\cdot}$ is that $H^{\cdot}(A)$ is minimized by $A^{\cdot}$. Detailed steps of the algorithm are shown as follows:

a. It is assumed that A has n attribute values aiming at any unselected attribute A, and the corresponding probabilities are $p_1, p_2, ..., p_n$. Attribute A is extended for generating n child nodes $\{B_1, B_2, ..., B_n\}$. $B_i$ is the consequent attribute of A selected according to the minimum information entropy principle when the ith value is selected for attribute A. The corresponding information entropies are $H(B_1)$, $H(B_2)$, ..., $H(B_n)$;

b. $H^*(A)$ is calculated according to formula (1)
c. $A^*$ is selected for minimizing $H^*(A)$, is regarded as newly selected attribute
d. The calculation result of step (a) is utilized for establishing subsequent node $\{B_1, B_2, ..., B_n\}$ of node A*
e. If all $B_i$ belong to leaf nodes, the node should not be expanded, otherwise, (a)–(e) processes are implemented recursively. Power load is random, and it is impossible to consider all factors in practical prediction, because it is difficult to collect and observe historical data simultaneously. In addition, it is difficult to establish model due to excessive factors, and the following problems can occur: complicated computation, instable value, etc. Therefore, week factors, temperature, humidity, and other weather information with larger influence on load prediction can be selected as main factors for consideration, thereby improving decision-making tree prediction model and overcoming the defects of algorithm calculation, which depends on attributes with high values; the correlation strength among attributes is not enough and the attributes are more sensitive to noise. Excellent prediction results are available. The importance of all influence factors affecting daily characteristic loads is revealed to a certain degree.

# 3 FUNCTION OF DISPATCHING OPERATION ANALYSIS BUSINESS

All collected data must be analyzed carefully, which can be converted into intuitive and simple dispatching experience knowledge in order to achieve safe and economic power dispatching. Therefore, analysis results are observed by dispatching operation knowledge acquisition model based on data mining, and information with high visibility level can be provided, which facilitates dispatching operation personnel to obtain valuable information. Actual dispatching business analysis is supported quickly and effectively.

## 3.1 *Statistical analysis on power grid operation condition*

Macro-information of power grid operation is analyzed and mined, thereby supporting the discovery of power grid operation dispatching knowledge. Future power distribution plan can be arranged rationally according to analysis information, thereby fully exerting its potential in power transmission in different regions, and maximally meeting current power demand. Therefore, 3D power grid operation data perspective and analysis model composed of power consumption area,

voltage grade, and power consumption load can be established for analyzing the route load condition of different routes and different power consumption areas. Preliminary power dispatching plans can be formulated according to analysis results (Dutta, 2012).

A variety of complex retrievals can be utilized for calculating overall mean value, variance, standard deviation, central moment, overall skewness, and overall kurtosis of all characteristics in the data sources in business analysis. Power grid operation information can be extracted. Correlation analysis is implemented according to fault type, fault time, severity, system operation mode, and personnel quality, thereby exploring more dispatching operation experience and guiding actual dispatching operation business.

## 3.2 *Analysis of load characteristics*

Power load has the characteristics of frequent change, that is, hour, day, week, and year. Meanwhile, the load is also changed constantly according to day as unit with larger cyclical feature. Load change is a continuous process. Larger jump cannot be produced generally. However, power load is sensitive to season, temperature, weather, etc. Climates in different seasons and different regions, and temperature change can have prominent impact on load.

It is also necessary to forecast load and electricity in order to select appropriate power grid supply unit type and reasonable power supply structure. In addition, load and electric quantity should be predicted. Load prediction can be divided into ultrashort term, short term, medium term, and long term according to different purposes. Power system load prediction includes prediction of maximum load power, load electric quantity, and load curve. Maximum load power prediction is very important for determining the capacity of power system power generation equipment as well as power transmission and transformation equipment. Therefore, division load trend comparative analysis is established, and change trend laws of power grid load can be mined for guiding power supply load dispatching of power grid (Pang, 2015).

Power optimization scheduling should reduce the power demand of users during power grid peak load period, enabling them to utilize energy more effectively. Power consumption can be reduced when energy services are met at the same time. Peak reduction, valley filling, and other measures, which are mainly implemented currently, change the shape of load curve to a certain extent. Beneficial reference is provided for selecting users in power dispatching, adopting various price measures (such as electricity price at peak and valley, interruptible power price, and peak-avoiding electricity price) to affect user power consumption behavior, and improving system load curve shape through comparing user curve and system curve, thereby promoting the production of electricity system and improving operation efficiency.

User classification based on load is the foundation for analyzing power consumption features. Peak adjustment keys lines in intuitive indication of commercial power consumption and domestic power consumption in some power supply area. Management of the power demand side can be strengthened; various measures should be used for reducing consumption of the two types of electricity in the time section, thereby reducing the maximum load of the power grid.

## 4  SYSTEM ARCHITECTURE AND IMPLEMENTATION TECHNOLOGIES

### 4.1  System architecture

Electric power dispatching operation analysis system integrates data report, Data mining (DM), multidimensional analysis, and other technologies. Information is extracted from historical data for analyzing knowledge valuable for dispatching decision-making[10]. The system architecture is divided into three layers: data layer—structured and unstructured data of various business systems are integrated to establish comprehensive information platform; application service layer—business service objects are integrated according to particle size through ESB integrated business module, thereby realizing knowledge discovered based on data mining model; and exhibition layer—integration of the system access layer can be mainly realized, and a uniform interaction mode is provided for users. Business is linked through information communication and workflow technique. Business flows are integrated across many applications. System architecture structure is shown in Figure 1.

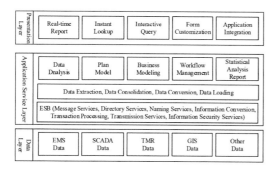

Figure 1.  System architecture.

Electric power enterprises have accumulated several data about scheduling operation and management. The data are accumulated in SCADA, EMS, TMR, GIS, and other system. It is difficult to obtain data of the systems directly due to scattered and heterogeneous features of the data, security zoning, and horizontal isolation requirements. Information or knowledge hidden behind massive data is discovered based on the data. Data should be integrated first in order to discover the information or knowledge. The scattered data are extracted, transferred, loaded, and cleaned. Then, data of related themes are mined for obtaining required knowledge based on the data information platform.

Business integration refers that original "business flow breakpoint" scattered inside or outside the enterprise can be linked, thereby completing automation of business activities. Workflow, information, coordination, and other technologies are utilized for realizing integration across systems. Therefore, coordination treatment of the same affair can be realized in the network environment aiming at business of different departments. Reliance of services on system integration can be reduced. Services can be communicated through simply and accurately defined interface, thereby realizing business flow integration and management among systems.

Integration of information system user interaction layers is realized in the exhibition layer. Stored information from various sources is distributed in the access enterprises. A uniform retrieval and content access control are provided. Internal users can realize access integration of various business applications via the Internet according to different permissions given by the system. Integrated services are provided to realize information inquiry and statement exhibition.

### 4.2  Distributed storage and computing architecture design

Huge pressure is brought to database performance in order to meet system analysis and application expansion. Distributed data storage system is adopted for smart grid dispatching operation analysis system as mass data analysis tool. Data are stored on many independent devices in a scattered mode for computing, thereby reducing massive database I/O operations. In distributed data storage and parallel computing framework, Hadoop, HBase, MapReduce, Hive, and components thereof are utilized for building the service layer with high reliability, high extensibility, high efficiency, and high fault tolerance. Dynamic expansion functions are provided for intelligent power grid dispatching operation analysis system

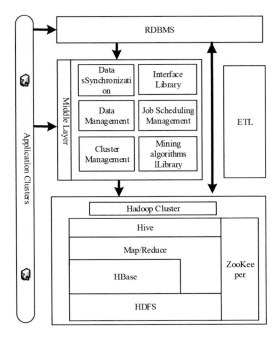

Figure 2. Distributed storage and computing architecture diagram.

based on mass data statistical analysis. Layered processing mechanism is adopted in distributed architecture, including data storage layer and calculation analysis layer. Analysis results are stored in database through analyzing and mining stored data. Software architecture is shown in Figure 2.

### 4.3 Dynamic statement and graphical analysis realization

In the system, dynamic business statements are combined with graphics analysis technology in order to enhance application effect, including bar chart, pie chart, line chart, and other graphics. Data statistical analysis information is displayed in an attractive and intuitive mode, thereby improving the level of system application. Dynamic statement provided by the system can meet the demand of users on complex Web statements. Scattered information and data of the enterprises can be easily integrated; therefore, the data are exhibited in personalized Web statement, thereby facilitating users and business personnel for inquiry, statistics, and analysis. In the process of business analysis, power load, prediction load, and historical average load are displayed in many curves in the same menu. Deviation trend can be analyzed intuitively, thereby supporting decision-making of scheduling.

## 5  CONCLUSION

Practical operation experience shows that the dispatching business analysis system discussed and constructed in this paper can meet the demand of electric power enterprise operation mode and scheduling business function. It is suitable for business analysis of dispatching personnel, and scheduling operation knowledge information can be extracted. However, data mining process involves human–computer interaction, which is repeated for many times. There is a challenge of ultra-large data in application. Methods for correct handling of redundant information and data noise and increasing the effectiveness of the mining results and the availability of information should be constantly improved in the system.

## REFERENCES

[1] Cepeda J C, Colomé D G, Castrillón N J. Dynamic vulnerability assessment due to transient instability based on data mining analysis for Smart Grid applications// Innovative Smart Grid Technologies. 2011:1–7.

[2] Sheng-Li L I, Ren P M, Zhao S R, et al. Construction of Business Analytics System for Smart Grid Dispatching by Employment of Data Mining and Integrated Information Platform. Guangdong Electric Power, 2011.

[3] Zhou H, Liu D, Li D, et al. Operating Analysis and Data Mining System for Power Grid Dispatching// Advanced Materials Research. 2013:611–617.

[4] Li L, Huang Z X. Research on the Smart Grid Dispatching System Based on Cloud Computing. Applied Mechanics & Materials, 2013, 385–386:1730–1733.

[5] Liu Q J, Huang Z G, Wei L I. Best Dispatching Smart Information Analysis and Construction of Integrated Decision-making System Based on Data Mining. East China Electric Power, 2013.

[6] Fan Z, Chen Q, Kalogridis G, et al. The power of data: Data analytics for M2M and smart grid. 2012, 5(11):1–8.

[7] Mukherjee A, Vallakati R, Lachenaud V, et al. Using phasor data for visualization and data mining in smart-grid applications// IEEE First International Conference on Dc Microgrids. IEEE, 2015.

[8] Dutta S. Data mining and graph theory focused solutions to Smart Grid challenges. Dissertations & Theses—Gradworks, 2012.

[9] Pang S, Co I G. Resource Scheduling for Cloud Data Center Based on Data Mining in Smart Grid. Telecommunications Science, 2015.

[10] Yao J G, Yang S, Gao Z, et al. Development Trend Prospects of Power Dispatching Automation System. Automation of Electric Power Systems, 2007, 13: 001.

*Advances in Energy and Environment Research – Achour & Wu (Eds)*
© 2017 Taylor & Francis Group, London, ISBN 978-1-138-62682-9

# Author index

Al-Khalil, S.A.-A. 109

Balin, A. 23
Bao, J.Y. 53
Besirli, I. 23
Bu, L.M. 71

Cai, M.L. 339
Cai, T.T. 175
Cao, J.-C. 365
Cao, Y.P. 389
Chen, C.C. 85
Chen, D.H. 375
Chen, F.S. 71
Chen, H. 71
Chen, H.Y. 71
Chen, J. 159
Chen, J. 285
Chen, J.H. 253
Chen, K. 219, 227
Chen, Q.-F. 165
Chen, Y. 3
Chen, Y. 43, 189
Chen, Y.M. 37
Chen, Z.B. 67
Cheng, D.-X. 159
Cheng, S.P. 269
Cui, X. 269

Dmitriev, A.V. 335
Dmitrieva, O.S. 335
Dong, J. 81
Dong, J. 243
Dong, L. 33
Dou, R.S. 257
Du, Z.D. 309
Duan, M.H. 183

Feng, C.-H. 165
Feng, Z.C. 175
Fu, D. 49, 195, 199

Gaevskaya, Z.A. 129
Gao, C.C. 315
Ge, Q. 297

Geng, Y.D. 183
Guo, H.L. 141
Guo, J. 123
Guo, L. 315
Guo, Z.C. 43

Hamody, A.Z. 109
Hao, J.H. 275
He, S.D. 263
He, S.Q. 329
He, X. 71
He, X. 365
He, X.H. 141
Hou, L.X. 393
Hu, G. 3
Hu, J. 365
Hu, J.F. 279
Hu, L.F. 147
Huang, C.-N. 305
Huang, L.C. 33
Huang, R. 71
Huang, T.Z. 147, 151
Huang, Y.-N. 159

Ji, F.F. 147, 151
Ji, H.D. 203
Jia, S.X. 381
Jiang, C.D. 345
Jiang, J.G. 323
Jin, J.Y. 103
Jing, Q. 279

Kebaitse, K. 27
Ko, C.-C. 305
Kong, F.B. 97
Kong, H. 91
Kong, L.W. 269
Kuo, J.-K. 305

Lai, J.J. 115
Lee, Y.-J. 207
Li, C.H. 33
Li, H. 49
Li, H. 275
Li, J. 401

Li, J.-C. 365
Li, J.H. 389
Li, J.J. 323
Li, J.S. 43
Li, K. 71
Li, Q. 309
Li, T.L. 119
Li, Z. 151
Li, Z.X. 85
Liang, F.X. 33
Lin, D.T.W. 305
Lin, J. 345
Lin, L.J. 389
Lin, S.-C. 207
Lin, Z. 49
Liu, B.G. 119
Liu, C. 323
Liu, D.N. 309, 315
Liu, H. 323
Liu, N. 389
Liu, S. 195
Liu, X.P. 123
Liu, Z.R. 103
Lu, X.W. 183
Luo, M.X. 203
Luo, P.D. 183
Luo, T.A. 103
Luo, W.B. 361
Lv, Y.S. 67

Ma, C. 315
Ma, F.X. 7
Ma, G.S. 285
Mei, R.W. 269
Mityagin, S.D. 129
Motsamai, O. 27

Ni, L.Y. 371
Niu, W.-J. 165

Pei, F.S. 123
Peng, J.J. 135
Peng, J.Q. 43, 189
Peng, R.C. 235, 243
Peng, X. 263

Qing, D.W. 279
Qu, W.H. 323

Rehman, A. 339
Ren, H. 147, 151
Ren, J. 351, 357
Ren, Y. 361
Ren, Y.J. 361

Sert, M. 23
Shao, M.Z. 75, 397
Shen, L.M. 71
Shi, Y. 3
Shubenkov, M.V. 129
Si, Y.D. 297
Song, L. 263
Song, Z.G. 169
Su, L.W. 263
Sun, Q.Q. 203

Tan, H.D. 401
Tang, Y. 401
Tian, Z.H. 135
Tong, Y.J. 119
Tung, C.-M. 207

Wan, L. 115
Wang, G.G. 67
Wang, J.M. 315
Wang, K.J. 309
Wang, K.P. 401
Wang, L. 269, 297
Wang, L.J. 43, 189
Wang, L.P. 155
Wang, L.Q. 375
Wang, W.J. 235
Wang, W.N. 71

Wang, X. 257
Wang, X.L. 135
Wang, Y. 183
Wang, Z.H. 147, 151
Weerasinghe, R. 61
Wei, Y.Y. 141
Wu, C.J. 123
Wu, G.Y. 309
Wu, M. 43, 189
Wu, S.Y. 329
Wu, X.S. 169

Xia, G.R. 123
Xiao, F.M. 53
Xie, J.Q. 291
Xie, M. 279
Xiong, X.Y. 115
Xu, D. 269
Xu, H. 275
Xu, L.J. 279
Xu, S. 297
Xu, W.Q. 339
Xu, X.H. 81
Xu, X.Y. 71
Xue, Z.Y. 361

Yan, G.Q. 43, 189
Yan, Z. 393
Yang, M. 315
Yang, M.J. 263
Yang, W. 91
Yang, Y.F. 291
Yao, S.Y. 323
Yao, Y.Y. 257
Ye, S. 323
Yin, S.H. 67
Yu, G.Y. 169

Yu, Y.L. 401
Yu, Z. 165
Yuan, J.S. 243
Yuan, Y.J. 97

Zakir, F. 339
Zha, Q.Z. 263
Zhang, F.Y. 67
Zhang, J.D. 43, 189
Zhang, J.L. 67
Zhang, L.H. 235, 243
Zhang, S.Y. 81
Zhang, W. 199
Zhang, X.G. 75, 397
Zhang, Y. 269
Zhang, Y.F. 147, 151
Zhang, Y.M. 381
Zhang, Z.W. 53
Zhao, B. 183
Zhao, D. 175
Zhao, J.J. 275
Zhao, N. 37
Zhao, Y. 345
Zhao, Y.Q. 85
Zheng, F.B. 297
Zheng, X.L. 351, 357
Zhong, Q.-Y. 165
Zhou, B.L. 85
Zhou, L.M. 103
Zhou, W. 235
Zhou, W. 389
Zhou, Z.Z. 323
Zhu, X.Z. 381
Zhu, Y.M. 371
Zou, H.B. 103
Zou, M. 17